Statistical Quality Design and Control

Contemporary Concepts and Methods

Statistical Quality Design and Control

Contemporary Concepts and Methods

Richard E. DeVor
Department of Mechanical and Industrial Engineering
University of Illinois at Urbana–Champaign

Tsong-how Chang
Department of Industrial and Systems Engineering
University of Wisconsin–Milwaukee

John W. Sutherland
Department of Mechanical Engineering and Engineering Mechanics
Michigan Technological University

Macmillan Publishing Company
New York
Maxwell Macmillan Canada
Toronto
Maxwell Macmillan International
New York   Oxford   Singapore   Sydney

Editor: David Johnstone
Production Supervisor: Elaine W. Wetterau
Production Manager: Pamela Kennedy Oborski
Text Designer: Jane Edelstein
Cover Designer: Jane Edelstein
Illustrations: Monotype Composition Company

This book was set in Palatino by Ruttle, Shaw & Wetherill, Inc., printed and bound by
R. R. Donnelley & Sons Company. The cover was printed by Phoenix Color Corp.

Macmillan Publishing Company
866 Third Avenue, New York, New York 10022

Macmillan Publishing Company is
part of the Maxwell Communication
Group of Companies.

Maxwell Macmillan Canada
1200 Eglinton Avenue, E.
Suite 200
Don Mills, Ontario M3C 3N1

Library of Congress Cataloging in Publication Data

DeVor, Richard E.
    Statistical quality design and control : contemporary concepts and
  methods / R. E. DeVor, T. H. Chang, J. W. Sutherland.
        p.   cm.
    Includes index.
    ISBN 0-02-329180-X
    1. Quality control—Statistical methods.   2. Process control—
  Statistical methods.   I. Chang, T. H.   II. Sutherland, J. W.
  III. Title.
  TS156.D53   1992   91-3045
  658.5'62'015195—dc20   CIP

    Printing: 4 5 6 7 8   Year: 3 4 5 6 7 8 9 0 1

To:

Jearnice A., Betty, and the memory of Bob,
   Herb, and Jearnice M.

Linda, Alec, Anthony, and Amy

Brenda, Jenny, Beth, Polly, and the memory of **Bill and Jim**

P R E F A C E

Philosophy and Application of the Book

The decade of the 1980s saw a tremendous change take place in the ways in which we view the quality of goods and services. Those forces that have altered America's position in the global economy have caused us to take a hard look at the way we and others have done things in the past and how we all are now dealing with the issue of quality. Central to this quality revolution are two issues that continue to receive increasing attention. One of these has been a growing awareness and understanding of the roles and responsibilities of management in dealing with quality. The other is our increased understanding of both the need for and the concepts and methods required to move the quality issue upstream into product planning and the engineering design process. Although traditional quality, represented by low variation about the nominal specification, has been the price of entry into the world marketplace, the determination of the nominal specification that provides increased customer satisfaction during use will increasingly become the key to capturing and maintaining market share. That quality should be articulated in terms of criteria for product and process design as well as the economic operation of processes rather than criteria for shipping manufactured product seems so clear today but has not been easily understood or embraced in the past. It is the hope of these authors that the presentation of this book will help to provide a basis for engineers and managers to understand and deal with the issue of quality in today's world and in the world of tomorrow.

A combined experience of more than fifty years in teaching and research in an academic environment has been helpful in delineating what we believe are some of the important statistically based concepts and methods for quality design and improvement. But it has been our experience with quality education and practice in the industrial environment over the years, particularly the 1980s, that has reshaped our thinking and provided us with what we hope is a sound basis to project a more practical and unified philosophical and methodological approach to quality design and improvement. Our overarching goal has been to inject that experience in the practice of quality design and improvement into this book.

The need for engineers and managers to embrace the issues of quality design and improvement is acute and broad-based. Curricula in all engineering disciplines and in schools of management are gradually responding to this need. This book has been written in a way that we hope will enable its use across this broad base, both in the academic world and in the industrial setting. Although the methods of statistical process control and design of experiments dominate the content of the book, no prior background in probability and statistics has been assumed. Rather, the necessary background has been included here and is introduced at the needed points of application throughout the book (a "just-in-time" approach!) in contrast to providing it in two or three dedicated chapters. We have developed a presentation of the material that includes the necessary probability and statistics in a way that will provide readers with a basic understanding of the genesis and foundations of the concepts and methods presented while not distracting from the central thrust of the book.

It has been our intention to prepare a textbook that could be used for a course or sequences of courses on quality design and improvement in both schools of engineering and management. Over the years, the book has evolved in support of a two-semester sequence of courses at the advanced undergraduate/first-year-graduate levels in the Department of Mechanical and Industrial Engineering, University of Illinois at Urbana–Champaign. Some additional material on the design of experiments typically covered in the second of these two courses has not been included here. Over the last several years, the material contained in this book has also been used extensively for a variety of short courses and seminars in industrial settings.

Emphasis on Quality by Design and the Use of Case Studies

For many years, quality was viewed as an attribute of a product impacted primarily through the process of manufacture, observed and "controlled" either during or after this stage of the product development life-cycle. In response to this perspective, the field evolved (as did texts on the subject) emphasizing statistical process control and to an even greater degree product control through sampling inspection. Today, it is well recognized that quality, as a design criterion rather than a shipping criterion, can be significantly impacted through the engineering design process. It is with this in mind that this book includes a significant component on the use of designed experiments for quality design and control both at the process and upstream in design. The contributions of Taguchi to the enhancement of the process of engineering design, implemented with the help of rigorous statistical design of experiments concepts and methods, are an important focus of this book. The book provides the reader with an understanding of the importance of the relationship between the establishment of design intent upstream and its realization through the manufacturing process downstream, and therefore, hopefully an understanding of the joint responsibility for quality on the part of all along the product development life-cycle.

In taking a broad-based approach to the study of quality design and improvement, we have chosen to develop and reinforce the major themes of the book through the use of case studies. Three chapters are devoted entirely to presentation of detailed examples of recent implementation experiences. In almost all of the other chapters, concepts as well as important tools and methods are introduced through case studies and examples. It is our intention in taking this approach to make a more lasting impression on the reader of the power and importance of the concepts and methods presented here and the manner in which a range of practitioners have invoked them to improve the quality of products, processes, and services.

Use of Computer-Based Workshop Exercises

Several years ago we began to make use of the computer to generate exercises that would allow the student to experience the complete problem-solving process in a laboratory setting. Today, these computer workshops have become an integral part of the overall presentation format of this book. These PC-based workshops allow students to simulate the operation of an actual process, collect data, develop statistical charts, read the statistical signals, diagnose the cause(s) of process faults, formulate improvement actions, and then witness the impact that those actions have on the performance of the process. We have found that giving students this opportunity to "close the loop," so to speak, provides for a much deeper and lasting understanding of the value and workings of the concepts and methods under study.

Those who adopt the book may obtain the software for these workshops at no charge by sending a request to College Sales Department, Macmillan Publishing Company, 866 Third Avenue, New York, N.Y. 10022. The software may either be distributed to the students or placed on any number of systems in PC-based laboratories.

The computer workshops and the use of the software from the student perspective are described in detail in the text. The solution manual that accompanies this book contains further information regarding the various uses of the software and workshops in both a real-time laboratory setting and as individual or group homework assignments.

Acknowledgments

There are few involved in the teaching and practice of quality design and improvement who have not been influenced by and stand in continued awe of the work of Dr. W. Edwards Deming. He is a constant source of knowledge and inspiration and represents a level of ethics and conscience that is to be strived for continually but probably not reached by others.

Our associations and experiences with a large number of companies over the past ten years have all been invaluable. But it is both fair and necessary to say that the opportunities we have had to interact with many within the Ford Motor Company and Caterpillar, Inc., have had a great influence on our thinking. For that we are forever grateful. At Ford, we are particularly grateful for our associations with many people in the Plastic and Trim Products Division (formerly, the Plastic Products Division, and before that, the Plastics, Paint, and

Vinyl Division) over a period of more than eight years. It is impossible to recognize all of those there who contributed in one way or another to this effort, but certainly the names of Jerry French, Robert Deacon, Peter Belaire, and Roger Sadowski come quickly to mind. Many others have contributed to our understanding of the concepts and methods herein and have participated in many of our efforts during this period. Three of those that should be mentioned are Professor Donald Ermer of the University of Wisconsin–Madison, Dr. Mark Lindsay of Mark Resources, Inc., and Professor Kevin Dooley of the University of Minnesota. We would also like to acknowledge the encouragement and input we have received from several of our colleagues during the preparation of various drafts of this book. In addition to those already mentioned, we should include Professor Herbert Moskowitz of Purdue University and Professor Emanual Sachs of the Massachusetts Institute of Technology.

Dr. Thomas Babin of Motorola, Inc., participated with these authors in a variety of activities and in stimulating discussions that had a tremendous impact on this book. He is a valued colleague and friend, and we most gratefully acknowledge his many contributions to this effort.

The preparation of this manuscript has been a long and humbling experience. The valuable assistance of Theresa Jordan, Mark Tornberg, and Daniel Waldorf in preparing many of the homework problems and solutions is greatly appreciated. The students in our classes over the past several years have provided invaluable comments and suggestions. The contributions of Amanda Horner at the University of Illinois at Urbana–Champaign in helping us to prepare various drafts and work our way through the many steps required in seeing this project through to completion will always be remembered and greatly appreciated. Becky Mullins of Process Design and Control, Inc. (formerly R. E. DeVor and Associates) has over the years provided valued support to many of the activities that have affected much of what is now contained herein.

Finally, the authors are most grateful for the support of their institutions, namely, the University of Illinois at Urbana–Champaign, the University of Wisconsin–Milwaukee, and the Michigan Technological University during this project.

R. E. DeVor
Champaign–Urbana, Illinois

T. H. Chang
Milwaukee, Wisconsin

J. W. Sutherland
Houghton, Michigan

September, 1991

C O N T E N T S

PART

II

Process Control and Improvement

PART

III

Product/Process Design and Improvement

Statistical Quality Design and Control

Contemporary Concepts and Methods

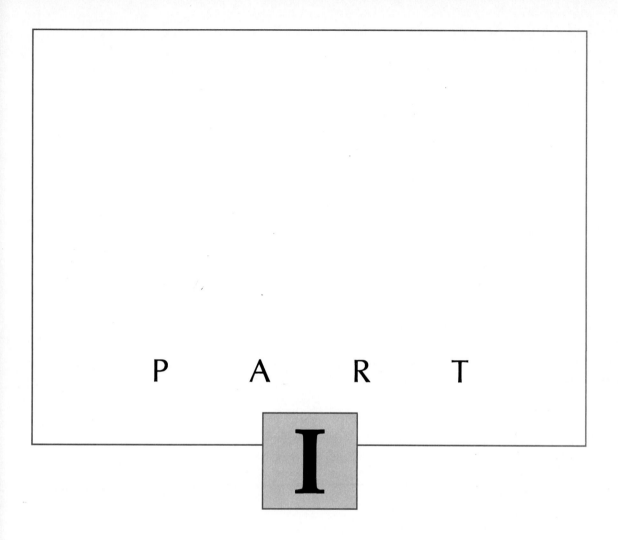

P A R T

I

Fundamental Concepts and Methods

1

Evolution of Quality Design and Control

1.1

Introduction

The field of quality control has undergone a major change over the decade of the 1980s. Since it was first born as a specific discipline in the 1920s, quality control has taken its place as a central activity in the industrial system. Most industrial organizations, at least through the 1970s, had a "quality control" manager of a "quality control" department which had a function to administer "quality control" as outlined in the "quality control" manual. The American Society for Quality Control has been in existence as a major national organization since 1945. Quality control has been taught as a distinct and almost universally present curriculum component in industrial engineering programs at the university level for decades. But the face of quality control has experienced a dramatic change in the last several years. This change has not occurred as the result of a major breakthrough in the technology and science of quality control. Rather, it has occurred out of absolute necessity as America's position in the world marketplace slowly eroded to alarming levels in the late 1970s.

The dramatic change in the meaning and application of quality control as a discipline has both a philosophical and an analytical side. It is the goal of this book to cover both of these aspects in a comprehensive and balanced way. It is in this spirit that we present a historical perspective of quality control in this chapter with the hope that an understanding of its evolution will point to the need for emphasis on the philosophy of quality control as well as the analytical techniques for quality control. That these two aspects of the problem are presented and understood together is essential.

The first two chapters of this book are directed toward the philosophical side of quality control. In Chapter 1 we review the history of the field, introduce some of the pioneers in the field and their basic philosophies, and begin to formulate the conceptual framework for quality design and improvement. In Chapter 2 we review the contributions of Deming and Taguchi, discuss the meaning of quality, and comment on the cost-of-quality issue that has in the past and will continue into the future to get more attention than it probably deserves.

1.2

Quality Revolution of the 1980s

During the 1980s a major revolution took place in the industrial sector in America as manufacturers strove to regain the competitive position once held in the world marketplace. One element of this revolution centered around a renewed emphasis on quality, with an approach aimed at preventing defective materials from being manufactured through improved process monitoring and diagnosis and at designing quality into the product from the very beginning. The concepts and methods of Deming and others have had a profound impact on the way we view quality from the manufacturing/process perspective. The simple but powerful statistical methods for process control developed by Shewhart more than 60 years ago have been successfully revived and applied on a very broad base. In the engineering design arena we have looked to the Japanese and found that the methods of Taguchi, referred to in some quarters as *off-line quality control,* have been used successfully for more than 40 years to provide a sound basis for improved product/process design. From the total system point of view the concept of *company-wide quality control* (CWQC) that has been practiced in Japan for some time is now receiving considerable attention. In particular, recent emphasis has been placed on *quality function deployment* (QFD) as a means to transmit customer needs through the organization both vertically and horizontally.

Major Change in Our View of Quality Control

Central to the quality movement of the 1980s was the tremendous emphasis on the role of statistical thinking and methods. Across the industrial and service sectors in America statistical education and training for employees from the shop floor to the boardroom has been unprecedented in terms of both breadth of exposure and depth of conviction. No other issue in management or science and technology has received such attention over the last several decades. To many, the concepts and methods are totally new. But what's new about the use of statistical methods for quality and productivity design and improvement? It is not the methods themselves, for they have been with us for quite some time. What is new is the mind-set that drives the use of these and other analytical methods toward the never-ending pursuit of quality and productivity design and improvement.

Beginning with the early 1980s, our thinking has undergone considerable transformation. The impetus for this change in thinking lies largely with the teachings of men like W. Edwards Deming, who played a central role in shaping the Japanese approach to manufacturing excellence over the past 40 years, and who during the 1980s had a similar impact on the American manufacturing community. The renaissance lies not with the influx of advanced technology in manufacturing, although the impact of technological change in manufacturing cannot be dismissed as irrelevant here. The renaissance lies, rather, with the management of manufacturing systems within a structure based on those principles that guide how we should view the design and improvement of products and their associated manufacturing systems.

The work of Shewhart in the 1920s led to a sound approach to the scrutiny of process variation and the diagnosis and removal of process faults. However, the statistical approach to the sampling of process output prior to shipping to determine the extent to which it conformed to specifications dominated the quality field from the 1930s through the 1970s. Today, it is recognized that this product control approach to quality assurance contributes little to the enhancement of competitive position. Our recognition that quality and productivity can move together in the right direction only when we attack the process, finding the root causes of process faults and taking action to remove them, is today reshaping the meaning and intent of quality control.

Quality Effort: Comparison of the Approaches of Japanese and U.S. Companies

In recent years many U.S. companies have begun to examine carefully the "quality control element" within their organizations in an effort to identify the necessary new directions that must be cultivated to improve competitive position significantly in the long term. Companies have sought to develop an overarching quality philosophy that can be implemented through policy and operating procedures to ensure that customers needs can be met competitively. One thing that has become clearly evident is the need to push the quality issue farther and farther upstream so that it becomes an integral part of every aspect of the product life cycle (e.g., marketing, product design and engineering, process engineering, production, etc.). The concepts and methods of *quality function deployment* are receiving considerable attention as we try to understand more clearly how to translate the needs of the customer into products that will meet those needs and into management systems that will enable us to deliver those products in a cost-effective manner.

Since about 1980 U.S. companies have gradually but steadily been building a new quality base, moving to accomplish as quickly as possible what has evolved in Japan over a period of about 40 years. It is, therefore, not surprising to find that our current level of maturity in the quality area is still lagging that of our Japanese counterparts. Given the crisis situation in the late 1970s and the way in which the U.S. approach to quality evolved through the 1950s and 1960s, we find that U.S. efforts in quality have for some time been strongly centered in the problem-solving/firefighting arena, with less attention to the other end of the spectrum—product development. Figure 1.1 shows a comparison of the U.S. and Japanese quality effort by activity, as it might be viewed today. It is not nearly as important that the United States is still significantly behind from a maturity standpoint as it is that the United States has a plan in place that will allow its "curve" to move to the same position as the Japanese curve over the next decade. To make this happen, the quality issue must be pushed farther and farther upstream. The recognition of contributions of Genichi Taguchi in this regard will be significant. The United States must do more than innovate; in particular, it must learn to execute and learn patience and persistence, virtues that have not necessarily been prominent in the character of American management.

The general acceptance of the concepts and methods of Taguchi within American industry over the past several years has been impressive, to say the least. Although in some quarters the willingness of industry to embrace the use of statistical methods for process control and design of experiments has been guarded, those advocating the methods of Taguchi have been met with considerably more enthusiasm. The "Taguchi movement" appears to have strong and organized support and has gathered momentum as reports of success have been well documented and disseminated. In particular, the engineering community has been quite receptive to the overall approach while being somewhat slower to embrace the more rigorous statistical approaches to process control and design of experiments. Whereas the latter has found its roots in, and has tended to be promoted by, the statistical community, the methods of Taguchi

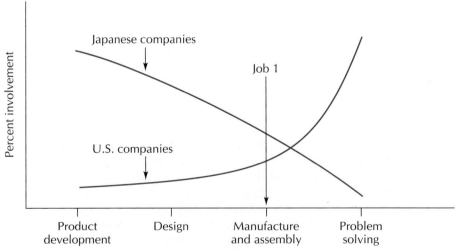

Figure 1.1 Quality Effort by Activity (Adapted from L. P. Sullivan, "The Seven Stages of Company-Wide Quality Control," Quality Progress, May 1986.)

are coming from the direction of the engineering community and therefore appear to be more easily accepted by those in design and manufacturing. Although some view the situation as one of strictly an either/or choice, it is likely that a thoughtful marriage of the two perspectives will provide for a powerful approach to quality design and manufacturing.

1.3

Historical Perspective of the Management of Quality

Many of the systems firmly entrenched in the fiber of present management structure must be changed. They simply are not consistent with the new philosophy of doing business that we must now adopt. They may be, in fact, the very things that kept us from properly viewing this philosophy years ago. We will not attempt to examine management style and structure comprehensively in this book, but we will take a look at a few of the key issues that dictate to a great extent our ability to be successful. The history of quality control goes as far back as history itself, but for our purposes here it is useful to pick up the time line sometime prior to the industrial revolution that began after the Civil War in the United States. Figure 1.2 traces this history forward to today.

Impact of F. W. Taylor

Much of what we now find at the foundation of management of industrial systems was born of necessity as far back as 100 years ago. As the industrial sector began the transformation from the individual craftsmen to cadres of workers organized toward the mass production of products, the principles of scientific management and the methods of shop management began to emerge. Great men such as Frederick Winslow Taylor rose to the task and pioneered the field of industrial management.

The most fundamental change came with the need to divide labor into component tasks. The worker no longer had the opportunity to fabricate from "beginning to end," personally monitoring and controlling the quality and productivity of his or her efforts. Now workers became responsible only for a small element of the complete manufacturing cycle, unable to see clearly and have any impact on what happened upstream or downstream from their new world. They saw only their assigned incremental tasks and often had no conception of the final product and its required function.

As a result of the division of labor and the subsequent narrow scope of involvement of a given worker in the total product life cycle, two basic problems arose. One relates to the quality issue. Since the worker seldom saw the final product or had any contact with its field use and since the tasks were quite repetitive and hence boring, it was more difficult for workers to generate

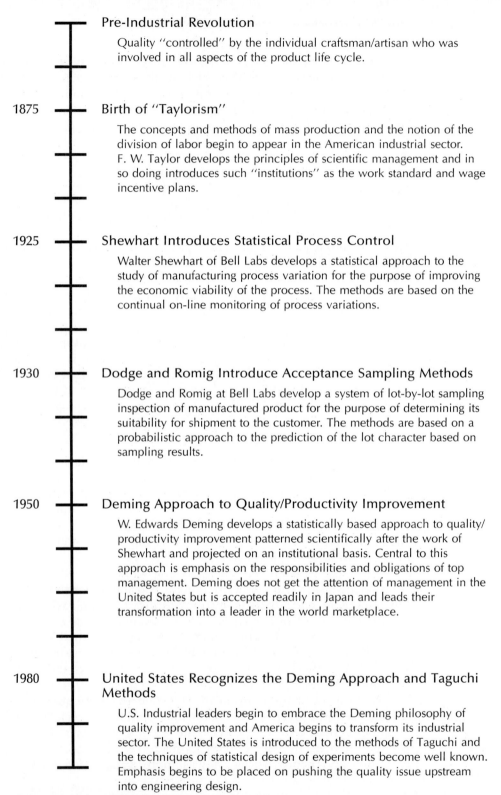

Pre-Industrial Revolution

Quality "controlled" by the individual craftsman/artisan who was involved in all aspects of the product life cycle.

1875 — **Birth of "Taylorism"**

The concepts and methods of mass production and the notion of the division of labor begin to appear in the American industrial sector. F. W. Taylor develops the principles of scientific management and in so doing introduces such "institutions" as the work standard and wage incentive plans.

1925 — **Shewhart Introduces Statistical Process Control**

Walter Shewhart of Bell Labs develops a statistical approach to the study of manufacturing process variation for the purpose of improving the economic viability of the process. The methods are based on the continual on-line monitoring of process variations.

1930 — **Dodge and Romig Introduce Acceptance Sampling Methods**

Dodge and Romig at Bell Labs develop a system of lot-by-lot sampling inspection of manufactured product for the purpose of determining its suitability for shipment to the customer. The methods are based on a probabilistic approach to the prediction of the lot character based on sampling results.

1950 — **Deming Approach to Quality/Productivity Improvement**

W. Edwards Deming develops a statistically based approach to quality/productivity improvement patterned scientifically after the work of Shewhart and projected on an institutional basis. Central to this approach is emphasis on the responsibilities and obligations of top management. Deming does not get the attention of management in the United States but is accepted readily in Japan and leads their transformation into a leader in the world marketplace.

1980 — **United States Recognizes the Deming Approach and Taguchi Methods**

U.S. Industrial leaders begin to embrace the Deming philosophy of quality improvement and America begins to transform its industrial sector. The United States is introduced to the methods of Taguchi and the techniques of statistical design of experiments become well known. Emphasis begins to be placed on pushing the quality issue upstream into engineering design.

Figure 1.2 Historical Evolution of Quality Control

ownership and sustain pride of workmanship in what they were doing. As a result the quality of the product suffered. The other basic problem relates to the productivity issue. For some of the same reasons stated above, workers did not always perform the tasks with the speed required to sustain a profitable situation.

Recognizing the problems associated with this new division of labor, Taylor and others realized that the management of workers must be put on a scientific basis. Workers must know what management expects of them in terms of personal performance. Management must be able to control the balance of flow in the system to achieve a steady and productive output. The simplest measurement of performance was time—how long should a given task take. All workers should be directed in some way to work steadily at a relatively fast pace. So the work standard was born and the methods of time study were developed to establish such standards scientifically. Amazing as it may seem, a look at Taylor's 1911 treatise[1] will reveal the foundations of time study, work standards, and wage incentives, although massaged and embellished over the next 70 years, are playing the same role philosophically in today's world as they did in the infancy of America's industrialization. So entenched did these methods become that even today they are taken for granted in many quarters as essential elements of the management of manufacturing.

Taylor's efforts in the science of management tended to deal with only one of the two problems created by industrialization, the productivity problem. With the introduction of the work standard, a fundamental change took place which may, in part, be viewed as a driving force in bringing us to the point we are at today. Human beings were no longer responsible for the quality of what they were doing, but rather only the speed at which they were able to do it. Time study specialists observed people at work on an established task using predetermined methods and developed a standard time for that task. Certainly, these studies revealed faults with the division of tasks and the assignment of methods and led to improvements, but the preoccupation was on time—time study, time standards. The fundamental purpose for all of this was to invoke incentive pay systems: get people to work harder and pay them more for their efforts.

The establishment of the concept of work standards—a standard time or number of units to be produced per unit time—had a profound influence on the evolution of production management.

We must remember that the concept of the work standard and the methods of the time study were originally created by Taylor to deal with the fundamental issue of wage incentives—the need to make people work harder to drive down the labor cost of manufacture. These principles were, of course, picked up by others and used for other purposes. Gilbreth and others extended time study to the study of motions as the field of work design became formalized. The motivation again was sound: to improve the efficiency of tasks performed by

---

[1] F. W. Taylor, *Principles of Scientific Management*, Harper & Brothers, New York, 1911.

labor and provide a better marriage of worker and machine in the workplace. But work standards were also used as a quantitative basis to negotiate rates of labor and reasonable expectations of workers. Work standards were used by management for internal budgeting purposes and as a basis for many decisions, such as the cost of internal manufacture versus outsourcing. In such instances the work standard is viewed as the accepted benchmark for what we can do at our best and hence will serve to camouflage the presence of improvement opportunity.

Today we recognize the impact of work standards as being completely opposite to that initially intended. As a target or objective, the work standard, in a sense, actually places an upper bound on productivity and stifles the pursuit of improvement beyond the standard. If the standard is being achieved, it would appear that there is no reason or motivation for continually looking for improvement opportunities. The methods introduced later by Shewhart, for example, could not be interpreted properly or adopted broadly, since they were viewed as embracing the issue of quality, not productivity, viewed as a tool to control the quality of the product not the efficiency of the process.

Shewhart's View of Quality Control

In the 1920s at Bell Labs, Walter Shewhart, a physicist, began to formulate a statistically based approach to quality control/improvement. It was Shewhart who first recognized the strong relationship between consistency of performance of products and their associated quality in terms of field performance. Furthermore, it was Shewhart who was able to link this to the manufacturing process in terms of its variation in performance. In his 1931 treatise,[2] Shewhart speaks of the economic control of manufacturing operations and of the use of statistical methods for detecting a lack of control in processes through the study of the variation pattern of product and process quality characteristics over time. That the process was driven solely by a constant system of forces of variation was deemed necessary by Shewhart to guarantee the economic success of the process.

Shewhart's concept was to detect and eliminate the sources of variation in the process that could not be attributed to the routine operation of the process. Unfortunately, again the presence of the work standard probably made it very hard to adopt this new philosophy. If the work standard was being met, the process must have been operating acceptably; we must look, then, for other ways to deal with the quality issue. The work standard had now become a misinterpreted and abused instrument. It was not being used for the purpose that Taylor had intended; instead, it had become an overall measure of performance of the combination of man and machine—a measure of process performance—perhaps *the* measure of process performance.

---

[2] W. A. Shewhart, *Economic Control of Quality of Manufactured Product*, D. Van Nostrand, New York, New York, 1931. Reprinted by the American Society for Quality Control, Milwaukee, Wis.

Inspection and the Quality Control Function

The second profound influence that Taylorism had on manufacturing management was the need to establish an external (to the process) function for the
control of product quality. Although Shewhart, and later Deming, promoted
the establishment of definitions of quality that embraced the process, quality
control was viewed as a screening process separate from manufacturing. It was
the job of production to get the numbers out—meet the quotas established
through the work standards. Since workers had only to achieve a productivity
standard, someone else had to look out for quality.

The quality control function was developed to take care of the business of
detecting and containing bad product—that which was outside of tolerances—
so that it would not reach the customers' hands. The quality control department
became the arch-adversary of the production department, one driving for the
numbers, the other driving for product quality despite the numbers through
inspection of product. And as the numbers grew, it became difficult to inspect
all the parts. There were those who felt further that inspecting all the parts was
unnecessary and ineffective. Probabilistic thinking was adopted to develop
sampling procedures. A field of scientific research and development emerged,
which was richly cultivated for more than 50 years. The underlying philosophy
had not changed, only the methods used to invoke that philosophy.

During the time period that Shewhart was developing and refining his
methods for the economic control of processes, H. F. Dodge and H. G. Romig
also at Bell Labs were developing a system for the lot-by-lot inspection of in-
process and finished goods.[3] Their "acceptance sampling" systems were
founded on a solid probabilistic basis and were designed as a means to pass
judgment on large quantities of goods—lots—using relatively small samples.
Although these methods were probabilistically sound within their framework
of product control, they unfortunately placed overwhelming emphasis and
hence the overwhelming allocation of resources on the "detection and containment of defective material."

One of the by-products of the Dodge–Romig approach to quality control
was the acceptance of the concept of an acceptable quality level (AQL) a target
level of defective material, supposedly based on economic grounds. The most
devastating effect of the AQL concept is that it promotes quality improvement
up to an acceptable plateau beyond which further improvements appear to be
unjustified economically. Scrap became an accepted part of the business, un-
fortunate but inevitable. Certainly, scrap reduction wars were waged from time
to time, but by and large, scrap became an integral part of the production and
business planning activity. Deming has clearly pointed out the fallacies in this
way of thinking, but for many years (the 1930s through the 1970s) this basic
approach to quality control prevailed.

World War II further fueled the fires of sampling inspection as the govern-

---

[3] H. F. Dodge and H. G. Romig, *Sampling Inspection Tables: Single and Double Sampling*, 2nd ed.,
Wiley, New York, 1959.

ment, working in the best interests of the war effort, developed formal standards for sampling inspection and the strict adherence to the use of these standards became part of the contractual agreement signed by all suppliers of military goods. In the short term such methods were undoubtedly necessary and effective in guaranteeing that only high-quality products reached field use. One can even argue that the philosophy on which the Military Standard 105D[4] was founded tended to force the supplier to try to do better when material was repeatedly rejected. But "reject and return" is in most cases not a feasible alternative. A rejected lot is merely a signal that someone has to do better. But how to do better?

It would be unfair to suggest that the quality efforts during World War II were solely of a product control—acceptance sampling—nature. During the famous War Production Board 8-day short courses on statistical methods for quality control, extensive teaching of Shewhart's control chart principles took place. But for numerous reasons that will come out as this book evolves, these methods did not gain widespread popularity in the industrial sector. The most fundamental reason for this may be, as Deming points out, that top management was not aware of the overarching concepts that drive Shewhart's approach and hence did not understand the broad-based benefits of the process control approach.

World War II ended but sampling inspection had only just begun to flourish. Many companies adopted its use both internally and in dealing with suppliers. Military Standard 105D became heavily used and unfortunately abused on both philosophical and technical grounds. So strong was the impact of the military standard for sampling inspection that by the early 1960s an international standard was adopted. If such attention would then have been given to the use of statistical charting for process improvement, one can only speculate about its impact today.

We should not dwell on the use of sampling inspection or inspection by any other means, for that matter, since these are but tools devised to carry out a philosophy that grew out of the establishment of a separate and adversarial function to production—the quality control department. Work standards drove production and the worker, so someone had to take the responsibility for quality. Some quality control departments made use of statistical charting—sometimes to a fetish—diving headlong into the business of making charts. Occasionally, an action might be taken to solve a production problem based on charting information. But in general such work was not effective. How could it be, given the relative roles of production and quality control as defined and supported by management?

Deming: The Need for Change

Yet by and large, through all of this historical evolution, American industry prospered. Since everyone was basically "playing the game" by the same gen-

---

[4] Military Standard 105D, *Sampling Procedures and Tables for Inspection by Attributes,* U.S. Government Printing Office, Washington, D.C., 1963.

eral rules, relative differences in performance tended not to be extreme. But then something happened to change all of this for certain types of industries— a strong competitor in the world marketplace began to emerge. And the competition was not necessarily playing by the same rules—not driven by the same internal forces that had been guiding the evolution of industrial development in America over a period of more than eight decades.

To respond to this new external force it will be necessary to adopt a new philosophy of production management. To this end, W. Edwards Deming developed his "14 points" to show management how to put quality control on an institutional basis rather than a departmental basis.

This phenomenon of "change of necessity" is not a new one. A bit of study of American industrial history shows that it is but a repetition of the past. At the turn of the century, Taylor proposed an important thesis that appears worthy of consideration today. In commenting on the practice of the management and labor in some of the important industries of the day, he said: "In these industries, however, although they are keenly competitive, the poor type of shop management does not interfere with dividends, since they are in this respect all equally bad. It would appear, therefore, that as an index to the quality of shop management the earnings of dividends is but a poor guide." If the United States is to compete effectively in the world marketplace, management must slaughter some of the sacred cows of traditional production management and find new ways of doing business.

One of these authors attended a meeting at a large plant in one of the U.S. automotive companies at which hourly inspectors were presenting some of their results and experiences in using statistical charting for problem solving on the production floor. A few months prior to this meeting the plant management had eliminated the quality control department as such, folding all hourly inspectors and quality control staff personnel into the production function itself. As one hourly inspector explained what he was doing, he became somewhat emotional and began to expound on the change in attitude that had taken place within himself. "I've worked in this plant for 18 years and I can tell you this: I've done more good for this company in the last two months than in all the time combined prior to that!" He went on to explain that not only had management for the first time taught him some tools and methods that enabled him to do an effective job in finding problems and working with others to solve them but that his whole attitude about his work had changed. He said that he used to watch the clock continually, take long breaks and lunch hours, and look for places to "hide." Recently, he caught himself on several occasions forgetting about his break altogether, and on two occasions he had suddenly realized that his shift had been over for about 45 minutes!

Need we look for monetary incentive plans to motivate our work force? Just give the worker some tools to do a better job. Restore pride of workmanship by taking off the handcuffs and removing the barriers that keep people from doing what they really like to do—work together as a team to resolve problems and search for improvement opportunities. In the next chapter, a definitive road map for accomplishing this is described. This road map includes the

Deming philosophy for the enhancement of competitive position and Taguchi's approach to quality engineering.

1.4

Quality and the Engineering Design Process

Traditional View of the Engineering Design Process

There are at least two fundamental issues concerning the engineering design process as we have known it over the years that may need more thought. One has to do with the way we do design itself, the other with the relationship between design and manufacturing. It is conceivable that by addressing the first issue in a fundamental way, with quality design as the focal point, we may really be addressing both issues at the same time.

Figure 1.3 shows the more traditional engineering design process. In this model system design embraces the application of the technology of the day to arrive at a product design configuration and to select the design parameter nominal values. Prototype construction and testing generally follow. Testing in this context is not to be confused with experimentation. Although experimentation generally involves the comparison of design configurations with purposely varying parameter values, usually in some structured design of experiments matrix, testing generally means the placement of numerous prototype design replicates on the life test to determine the functionality and reliability of a specific design configuration, that is, one with fixed design parameter values.

One criticism of the design process model of Fig. 1.3 is that it tends to lead to iteratively looping through the system design and testing phase, often in a somewhat unstructured manner. Furthermore, little concern in this model is directed toward the impact and importance of the manufacturing processes that will be employed to make the product. Once function appears to be served

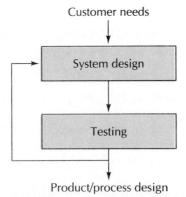

Figure 1.3 Traditional Design Process Model

through the design process, others will need to worry about making the product economically.

The relationship of the design process to manufacturing and the design/manufacturing interface, in particular, has received considerable attention over the past two decades. CAD/CAM has dealt primarily with the computerization of the design and manufacturing processes and in particular with the translation of design specifications into manufacturing procedures and activities. Design modeling and analysis has been strengthened and extended. Standardization and rationalization of both design characteristics and manufacturing process characteristics have been advanced and the efficiency of the interactive processes of design and manufacturing has been improved. The computer has greatly facilitated this translation, and to some extent lead times have been reduced.

The concepts of design for manufacturability and design for assembly have been the subject of considerable research and development. Numerous specific models have been proposed and refined and will not be discussed here. All of this work is aimed at overcoming the difficulties precipitated by the traditional over-the-wall design philosophy depicted graphically in Fig. 1.4.

The magnitude of the problem with the over-the-wall approach to design

"Over the Wall" Design and Manufacturing

Figure 1.4 Traditional (Over-the-Wall) Approach to Design and Manufacturing

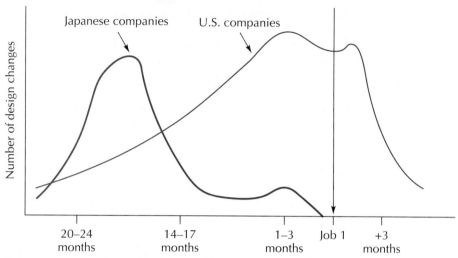

Figure 1.5 Comparison of Japanese and U.S. Product Design Life Cycles (Adapted from L. P. Sullivan, "Quality Function Deployment," Quality Progress, June 1986.)

and manufacturing can perhaps be measured in terms of the number of times the part drawing is thrown back and forth over the wall! As we proceed from the initial design concept through prototype testing and development to final design detailing and ultimately to the initiation of production and beyond, the number of design changes has, in the past, been too high. Figure 1.5 contrasts the typical Japanese and U.S. companies in this regard. This figure suggests that the fundamental way in which we are dealing with the quality issue may be at the heart of the problem. It is certainly clear that what we see in Fig. 1.5 is totally consistent with what we saw in Fig. 1.1 regarding the contrasting emphasis on quality along the product development life cycle.

Alternative Model for the Engineering Design Process

The Taguchi approach[5] to quality engineering through design has a number of significant strengths. Taguchi has placed a great deal of emphasis on the importance of minimizing variation as a means to improve quality and of bringing the mean of the process to the design-mandated target. The idea of designing a product the performance of which is insensitive to environmental conditions, and making this happen at the design stage through the use of design of experiments, has been a cornerstone of the Taguchi methodology for quality engineering.

[5] G. Taguchi and Y. Wu, *Introduction to Off-Line Quality Control*, Central Japan Quality Control Association, Meieki Nakamura-Ku Magaya, Japan, 1979.

Some of the strengths of Taguchi's approach are

1. The center of gravity of the overall approach: the engineering design process.
2. Definition of the relative roles of the factors that influence product/process performance.
3. Robust design: the parameter design concept.
4. The use of the loss function as a means of assessing the economic impact of variation.

We elaborate on each of these points in considerably more detail as this book unfolds.

Taguchi methods focus attention on the engineering design process: in particular, the projection of a three-stage design process model of system design, parameter design, and tolerance design. Figure 1.6 provides a diagrammatic representation of Taguchi's design process model. In the initial stage, system design, the available science, technology, and experience bases, are used to develop/select the basic design alternative to meet customer needs. A variety of techniques may be useful in specifically mapping the relationship between customer needs and the selection of design configuration and parameters that will meet those needs effectively.

At the parameter design stage, interest focuses on the selection of the specific nominal values for the important design parameters. The overarching selection criterion that is used is to "identify those nominal values that minimize the transmission of variation to the output performance as a result of the presence of noise factors operating in the environment in which the product and/or process is functioning." That is, we select the nominal values that are

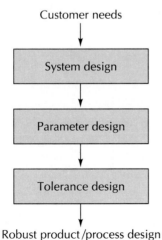

Figure 1.6 Taguchi's Three-Stage Design Process Model

the most robust/least sensitive to noise variation. It is at the parameter design stage that Taguchi strongly advocates the use of design of experiments methods. The tolerance design stage of the design process concentrates on the selective reduction of tolerances to reduce quality loss at the expense of increasing manufacturing cost. The loss function concept of quality is used to provide a basis for striking the economic trade-off.

Parameter Design Concept

The concept of robust design is an intuitively pleasing one and is perhaps one of the most significant contributions of Taguchi in terms of advancing the state of the art of the engineering design process. Figure 1.7 is useful in clearly understanding the notion of robust/parameter design through the transfer function model for the product. In Fig. 1.7(a) we see a representation of a product functioning in the field and we note that it is subject to certain noise factors, the variation of which is transmitted through the product design to the output measure of performance. To improve the quality of the product, we need to reduce the functional variation, that is, improve the consistency of the output response. To do this, the parameter design concept is invoked wherein the nominal values of the control factors—those factors that can be manipulated by the designer—are selected that reduce/minimize the transmitted variability due to the noise factors. This result is depicted in Fig. 1.7(b).

An extremely relevant measure of product quality is the ability of the

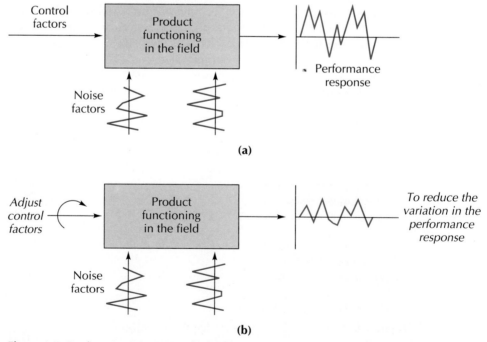

(a)

(b)

Figure 1.7 Product Functioning in the Field

product to function in a sound fashion—according to design intent—over a broad range of conditions. In particular, good quality means that the product function maintains a consistent level of performance in the face of noise factors, sometimes referred to as a *robust* product. In attempting to minimize the effects that noise factors have on product quality, certain countermeasures may be taken. The most important of these countermeasures is that of design. In the past, engineering design in the United States has emphasized system design and tolerance design, particularly tolerance design, to achieve quality at the expense of manufacturing cost. Robust design reduces variation—improves quality—without increasing manufacturing costs.

The design model discussed above and the robust design idea should not circumvent or ignore manufacturing but rather should embrace it in an integral way. On the one hand, the influence of the manufacturing process on our ability to be on target with smallest variation can and should be considered at all three stages of the design process, as portrayed by Taguchi. It is also true that the idea of robustness should be applied to the manufacturing process environment and the noises inherent in this environment that will be transmitted to the resulting product. In fact, it would be useful to apply this concept to both product design parameters and manufacturing process parameters at the same time, in a common setting, during design. In the future, one person will do this, not two, but this person will need both the knowledge and the tools and concepts to make this happen.

1.5

Strategic View of Quality Design and Improvement

We will now begin to lay out a conceptual framework for quality and productivity design and improvement, using the concepts of robust design. In this regard this section begins to provide insight into the following two important issues, which will stand as recurring themes throughout this book:

1. Sources of variation and the consideration of variation as a fundamental measure of product or process performance.
2. The concept of signal-to-noise ratio as a way of thinking about how we measure product or process performance, and how in a general sense we may choose strategically to enhance performance (i.e., increase the signal-to-noise ratio).

On Target with Smallest Variation

Quality is a matter of product function. In Chapter 2 we talk of both traditional and more contemporary definitions of quality. In general, the failure of a product to meet the intended function, as communicated by the customer, can arise from either one or both of two basic problems:

1. Failure to achieve the nominal performance required by design.
2. Excessive variation about the intended nominal performance level.

These two issues may be graphically depicted as shown in Fig. 1.8.

What we should really be interested in is to be on target with the smallest variation. The key is whether we can become sufficiently motivated to pursue the smallest variation part of the question. The importance of this issue—smallest variation—lies at the very center of the concept of never-ending quality and productivity improvement. Two important relationships that today are being better understood hold the answer to the question of motivation. They are

1. The productivity/variability relationship.
2. The quality/variability relationship.

The concept "on target with smallest variation" has been promoted in recent years through the methods of statistical process control. W. Edwards Deming and others have clearly demonstrated the fact that sources of variation are sources of *waste* and *inefficiency* and that for every source of variation identified and removed we will experience increases in both quality and productivity. The concept of quality as measured by loss due to functional variation will further reinforce our newfound desire to strive to be on target with smallest

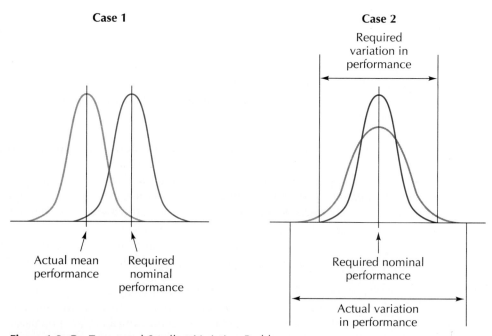

Figure 1.8 On Target and Smallest Variation Problems

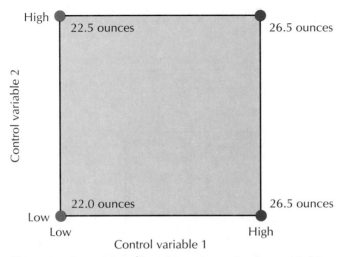

Figure 1.9 Factorial Design of Experiment for the Molded Part Example

variation. The role of experimental design in achieving this goal has been clearly demonstrated by Taguchi and others. But to a great extent, experimental design methods have until recently emphasized the revelation of improvement opportunities on the average, and have therefore been more useful in the "being on target" part of the equation. The following illustration demonstrates this idea and a way we might improve on it in a very simple way.

Suppose that we are attempting to control the part weight of a molded product and we study by a simple experiment how two molding machine parameters influence part weight. We run an experiment and obtain the results shown in Fig. 1.9, where the points on the square correspond to the four test conditions examined. Each weight response is the average of 25 parts molded consecutively, under the given conditions. The results of the experiment seem to indicate that by manipulating control variable 1 we can control part weight. Control variable 2 seems to have little to do with part weight, on average. We can set it (or vary it) however we wish and nothing will happen to change part weight.

Now suppose that someone came in and made a time plot of the individual part weights for the 25 consecutive parts made under each condition. The results could look like those shown in Fig. 1.10. The individual charts in the figure reveal something important that was not known previously—that although control variable 2 does not affect part weight on the average, it does appear to affect *variation* in part weight from part to part. Such could certainly be measured by the standard deviation as a response of the experiment. The results of the experiment seem to be telling us:

1. To control the average part weight, to the target, manipulate control variable 1.

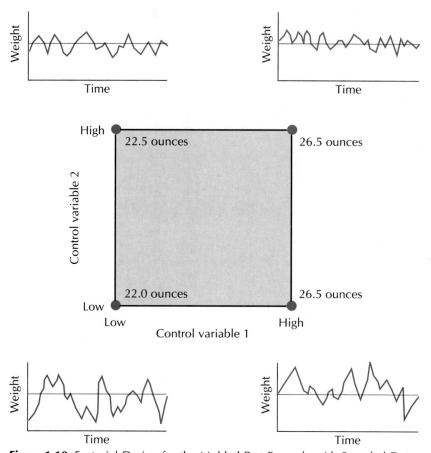

Figure 1.10 Factorial Design for the Molded Part Example with Sampled Data

2. To do item 1 with the smallest variation, set control variable 2 at its high level.

In the past it has been less common for us to use variation as an experimental measure of performance and to seek ways to reduce variation in product/process performance through designed experimentation. Later in the book we will see how this can be achieved.

Concept of Signal-to-Noise Ratio As a Measure of Performance

In the preceding example, which dealt with the control of part weight for an injection molding process, improved quality/product performance is obtained by shifting the process mean to the design nominal via control variable 1. In the terminology of the communications field we say that we are altering the signal. Similarly, in this example, improved quality/product performance is obtained by reducing the variation in performance about the design nominal via control variable 2. In the terminology of the communications field we say that we are reducing the noise. Sometimes it is convenient to put these two aspects of the problem together through use of the signal-to-noise ratio.

In the communications field, the signal-to-noise ratio is commonly expressed as a power ratio in decibels or as a voltage or current ratio, also in decibels. For our purposes at this time, a perhaps more basic and relevant way of examining the signal-to-noise ratio is to use the following definition:

$$S/N = \frac{\text{average}}{\text{standard deviation}}.$$

Figure 1.11 graphically depicts the signal-to-noise ratio concept. In the top part of the figure we have a signal that represents the input to a certain process. For the process to be successful, this input must have a value strictly greater than zero. In Fig. 1.11 it is clear that such is not always the case. There are

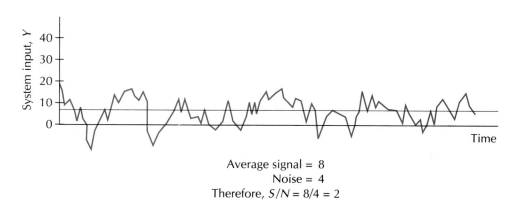

Average signal = 8
Noise = 4
Therefore, $S/N = 8/4 = 2$

Solutions to the problem

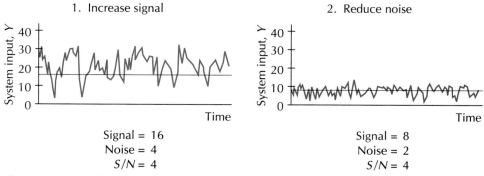

1. Increase signal

Signal = 16
Noise = 4
$S/N = 4$

2. Reduce noise

Signal = 8
Noise = 2
$S/N = 4$

Figure 1.11 Signal-to-Noise Ratio Concept

basically two ways to solve this problem, both depicted in the lower part of Fig. 1.11. To the left in Fig. 1.11 we see that the average signal has been increased, and therefore even with the same amount of variation about the average, the signal is now always above zero. To the right, the problem has been solved by reducing the variation in the input without the need to increase the average value of the signal. Contrast these two strategies for increasing the signal-to-noise ratio in your own mind, thinking of all the problems you may have solved over the years, either personally or professionally. Have you tended to do so by increasing the signal or by reducing the noise?

Signal-to-Noise Ratio Enhancement: A Conceptual Argument

To help you think about the question just posed above, let us suppose that we are attempting to communicate certain information to another party.

150	years ago this may have been two Indians sending smoke signals back and forth across two mountains.
50	years ago this may have been two ship captains flashing light signals from vessel to vessel.
5	years ago this may have been two amateur radio operators talking to each other across the Atlantic Ocean.

All probably experienced difficulties communicating due to the presence of many forces that tended to corrupt or disturb the quality of the signal being sent or received. Sources of noise may include wind, cloud cover, temperature, rolling of the sea, time of day, atmospheric conditions, and so on. In each case it is likely that improvements in communication—improving the quality of the message—would be sought in the following ways:

Build a bigger fire.

Increase the candlepower of the lamp.

Increase the wattage of the transmitter.

Increase the size of the antenna.

All of these countermeasures to improve the signal-to-noise ratios of the systems involved have one thing in common: They do so by increasing the *signal*! This has been the typical approach taken in the Western world.

Figure 1.12 illustrates the signal-to-noise ratio idea in one of the above contexts. Here we consider a person under some pressure who is frantically trying to call for help. Unfortunately, his friend is unable to understand his message. A solution to the problem is found that involves increasing the signal. The signal-to-noise ratio is increased and the message is received. But is the price paid to increase the signal too great?

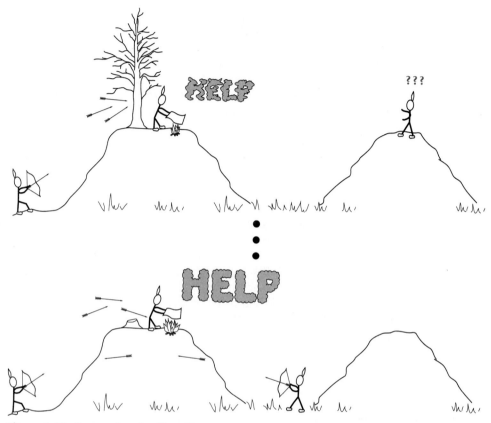

Figure 1.12   "Increasing the Signal"

1.6

Contrasting Approaches to Quality Design and Improvement

It is probably safe to say that from a signal-to-noise ratio point of view the Western world approach to quality and productivity improvement has been much different from that of the Japanese. After World War II the state of affairs in Japan—poor and sparse raw materials, antiquated equipment, manpower and technology limitations—suggested that they must focus on improving quality, improving the signal-to-noise ratio by attacking the denominator in the equation (i.e., reducing the noise). The Japanese effectively used statistical process control as a means of identifying sources of noise or variation and then took action to remove the noise or reduce the amount of variation. They also made effective use of design of experiments in both product and process design via robust/parameter design (to be discussed later) to reduce the transmission from the input (say, raw material) to the output performance of the product, allowing them to use less expensive raw materials.

In contrast, it may be fair to say that the Western world took a much different approach by attacking the numerator in the equation (i.e., increasing the signal). Some of the cases of interest that we will now examine to illustrate this fact are (1) using additional processing resources, (2) parallel redundancy in design, (3) screening inspection/product control, and (4) using more expensive raw materials.

Using Additional Resources

Signal-to-noise ratio could be a measure of the acceptable number of parts produced. Too often, the rate of production of defective parts is large—the process efficiency or yield is low. One way to increase the yield is to make more parts—increase the signal—either by increasing the production rate of the equipment in use or by adding additional equipment/machines (i.e., more machine time) or by reducing the cycle time. This is illustrated in Fig. 1.13. The signal is increased and therefore the usable process output is increased. Another approach would be to use the same equipment/machines but to improve their efficiency—reduce the noise. Statistical thinking and methods, utilizing techniques such as SPC and design of experiments, will provide considerable insight into how such improvements can be made.

Parallel Redundancy in Design

The *mean time between failures* (MTBF) of a given product is an indication of quality similar to the signal-to-noise ratio. Let us suppose that the MTBF,

$$\text{MTBF} = \frac{1.0}{\phi}$$

is unacceptably low, where ϕ is the failure rate (failures per unit time). We can increase the signal-to-noise ratio (MTBF) by including parallel redundancy in the product design (i.e., the system now will be successful if *either* component 1 *or* 2 is functioning), as shown in Fig. 1.14. For parallel redundancy and n components with identical failure rates, ϕ:

$$\text{MTBF} = \frac{1 + 1/2 + \cdots + 1/n}{\phi}.$$

Clearly, parallel redundancy will increase the signal-to-noise ratio by increasing the signal with no regard for the level of noise (ϕ).

An alternative strategy which is receiving considerably more attention today is that of attacking the noise—in this case the component failure rate—to determine what design and/or manufacturing improvements could lead to a reduction in this rate. The techniques of design of experiments and Taguchi's concepts such as robust design are likely to be quite useful in this regard.

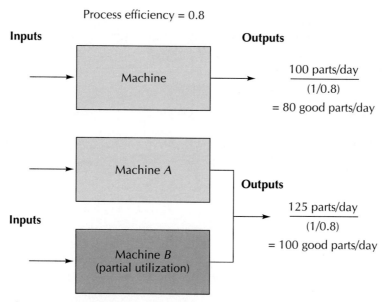

Figure 1.13 Using Additional Machines

Screening Inspection/Product Control

In the field of electronics, signal-to-noise ratio enhancement may be effectively accomplished by filtering out the noise. For example, many signals of interest have relatively low frequencies, with bandwidths extending only over a narrow range, perhaps a few hundred hertz. In these cases a simple low-pass filter can effectively limit the measurement system bandpass to that necessary to pass the desired signal frequencies. Such is shown in Fig. 1.15. In quality control, the analogous situation may be the 100% screening inspection of outgoing (producer) or incoming (consumer) product. In Fig. 1.15, as with the *RC* circuit,

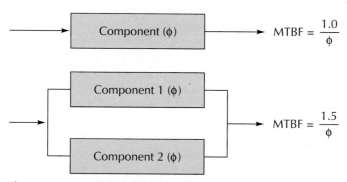

Figure 1.14 Parallel Redundancy

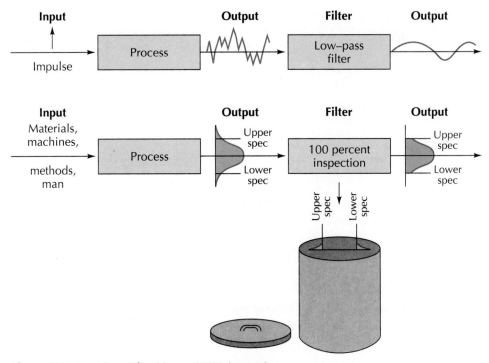

Figure 1.15 Low-Pass Filter Versus 100% Inspection

100% inspection does not remove the source of the noise; it merely filters out the effect of the noise on the output the customer sees.

Earlier we saw that one approach to increasing the signal-to-noise ratio of a manufacturing process would be to increase production by adding resources (more machines, overtime, etc.) that attempt to increase the signal. In this last example, increasing the signal-to-noise ratio is accomplished by reducing the noise by filtering or inspection, which is a postprocessing activity. Actually, these two countermeasures are generally used together; if the defect rate is high, we make more parts and screen the bad ones out. The simultaneous use of these two countermeasures to increase the signal-to-noise ratio is very costly and clearly does not provide opportunities for improvement.

Statistical process control (SPC) is a countermeasure that attempts to attack the noise in the manufacturing process, eliminating sources of waste and inefficiency that produce variation in the function of the product. Statistical process control increases the signal-to-noise ratio by maintaining the target and reducing the noise component. Figures 1.15 and 1.16 contrast inspection (product control) and SPC (process control) as approaches to improving the signal-to-noise ratio. Through process observation, evaluation using such techniques as statistical charting and fault diagnosis to identify the root cause of the problem, sources of variation may be eliminated altogether.

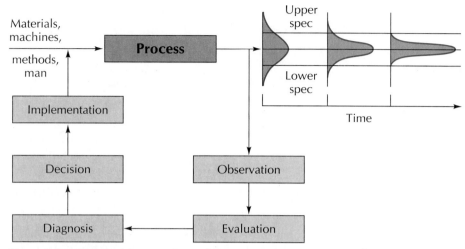

Figure 1.16 Statistical Process Control as a Means to Increase Signal-to-Noise Ratio

Using More Expensive Raw Materials

There are certain situations where increasing the signal-to-noise ratio can be accomplished at the product design stage, again using the idea of filtering. Here, at the design process, through the use of more consistent raw materials at a higher cost we may reduce the effect of the material noise on product performance. We are, in effect, tightening the performance specifications on the material properties to reduce the amount of noise to be transmitted through the product. It is important to note that the use of more expensive raw materials or purchased parts is a sort of filtering process, initiated at the design stage. Better, more expensive materials are themselves the filter that diminishes the effect of the noise on the signal.

We hope to show that such filtering at the design stage is an important strategy but needs to be invoked in a different fashion. Figure 1.17 illustrates two contrasting approaches to the problem. Approach 2 solves the problem of noise transmission by narrowing the material specifications. This may greatly increase the cost of the material. Approach 1 takes a much different tack; it tries to locate a different nominal or target value for the cheaper material that for the same tolerance/input variation reduces the amount of noise transmitted through the product/material to the output performance. In approach 1 we are exploiting the nonlinearity in the system. By applying the same raw material, but in a different way, we reduce the transmitted variation and produce a more robust design.

There are many situations we can point toward that illustrate the Western approach to increasing the signal-to-noise ratio. We have examined several common examples of increasing the signal-to-noise ratio by either increasing the signal or filtering out the noise through postprocessing. The notion of

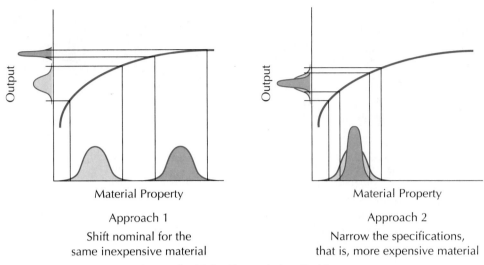

Figure 1.17 Two Approaches to the Noise Transmission Problem

postprocessing here is meant to imply that although the noise still "reaches" the signal and may significantly corrupt it, it may be "removed" after the fact by some form of filtering. As we have just seen, some of these situations relate to product design and some to the manufacturing process, but the overriding theme is always the same—a failure to attack the problem at the root cause and eliminate the forces, through improvement actions, that are producing variation and waste. In the foregoing scenarios the approach to solving the problem was oriented more toward treating the symptom than treating the cause. We hope that the concepts and methods that follow in this book will put us in a better position to treat the causes.

Summary

It has now been over 10 years since the beginning of the renewed emphasis on quality in the United States. What many cynics thought would be a year-or-two phenomenon has sustained its momentum for more than a decade. The fact that several large companies (Ford, Motorola, Xerox, GM, and others) have demonstrated continued drive and leadership in this effort has probably been instrumental in sustaining the movement over the decade of the 1980s. We now have a national quality award, have seen millions of workers, salaried and hourly, trained in the methods of Deming, Taguchi, and others, and have witnessed major advancement in competitive position on the part of many companies. Yet technological growth in the use of the concepts and methods discussed in this book has been hampered by organizational and cultural factors that continue to hold the American industrial enterprise in a straightjacket.

| Exercises |

1.1. Explain the change in the view of quality control in the United States between the decades of the 1950s, 1960s, 1970s, and today.

1.2. What changes must take place for the United States to become competitive again in the world marketplace?

1.3. What are the four major milestone events/circumstances in the history of quality control, each of which has led to a major step forward in the concepts, methods, and implementation of quality control?

1.4. What was the effect of division of labor on the standard of quality of goods manufactured in the United States? Why did this occur?

1.5. Are the use of work standards and/or wage incentives consistent with our view of quality today? Explain.

1.6. What is the danger associated with the use of acceptance sampling for quality control? Why does the type of attitude it promotes hinder the enhancement of competitive position?

1.7. What are some possible effects realized by the general acceptance of the idea of AQL?

1.8. What was Shewhart's basic approach to quality control?

1.9. How did Shewhart's approach to quality control differ from that of Dodge and Romig?

1.10. What is wrong with the phenomenon of "change of necessity"?

1.11. Why has the United States over the last decade become concerned with improving quality? What forces may be at work now that did not exist 20, 30, 40 years ago—or did they?

1.12. Who has responsibility for initiating quality control? What things need to be done?

1.13. Explain why it is important to consider variation about the nominal as a measure of product/process performance; that is, why is it not sufficient or advisable to consider only performance on average?

1.14. How/why do some of the newer concepts/ways of viewing quality encourage manufacturers to reduce the variability about the nominal in their products/processes; that is, what motivates the continual pursuit of variation reduction?

1.15. A major automobile manufacturer advertises that it performs 100% inspection on all processes that produce parts for the brake systems in its cars. It claims that the consumers can feel good about being safe in their cars. What should the consumers feel bad about? Is there any situation in which measuring every part might be useful?

1.16. The percent defective from a process that makes injection-molded plastic keychains has gradually been rising to the point where it can no longer meet production demands in a single 8-hour shift per day. The shop floor manager sends you a memo recommending that a second shift be added on this machine in order to get the product out the door. As his or her boss, how would you respond to this memo?

1.17. After experimenting with a certain product's performance in the field, it is found that better quality raw materials significantly improve the product's performance in all areas. Based on these results, would you recommend that the more expensive raw material be used for this process? Of what is this an example? How might we better deal with this issue?

1.18. Provide an example in your personal or professional life in which you solved a problem by (a) increasing the "signal," and (b) decreasing the "noise" in the signal-to-noise ratio you might use to characterize the performance of the process. Carefully explain your answer.

1.19. Your boss likes the color scheme you specified for the decoration on a new motorcycle helmet and begins production immediately in order to have the first ones off the line in time for his son's birthday. Although you have had no chance to consider design changes that may prove necessary, he assures you that later modifications will cause minimal downtime on the line. Is your boss considering all the costs involved? How would you prove to him that his decision was a poor one?

1.20. Explain the difference between testing and experimentation. Which do you think is more useful in the engineering design process, and why?

1.21. What is meant by the term "robust" in the context of product design?

1.22. What criterion does Taguchi recommend using when selecting nominal values for the controllable variables in a product/process design?

1.23. Relate the differences between the U.S. and Japanese design life cycles (Fig. 1.5) to the differences between the two countries' quality efforts by activity (Fig. 1.1). Which approach would seem to be more costly? Why?

1.24. What do you think is Taguchi's most important contribution in terms of his approach to quality design and improvement relative to the more traditional, Western approach to quality control?

2

Conceptual Framework for Quality: Design and Control

Introduction

In this chapter we deal primarily with the work of two persons whose contributions to quality design and improvement have had a lasting impact on the world we live in today. Although many have contributed in a significant way to the field of quality, the work of Deming and Taguchi provides not only the tools and methods necessary to make a difference but more important, provides the conceptual underpinning necessary to embrace and institutionalize their approaches.

W. Edwards Deming is the person largely responsible for bringing statistical thinking and methods for quality improvement to Japan after World War II. An eminent statistician in America during the 1930s and 1940s, Deming was in the forefront of the statistical quality control scene during World War II. After the war, Deming's philosophies and teachings fell on the deaf ears of American management, and the quality effort fires that burned so brightly during the war years slowly went out. America was in a period of industrial boom. Every product that could possibly be manufactured could be sold. In Japan, the situation was very different. A devastated industrial base was struggling to rebuild, natural resources were scarce, and a reputation for poor, shoddily manufactured products was rapidly solidifying. The Japanese became aware of Deming's work and invited him to Japan, where he met with and caught the attention of top management. A technically oriented management listened patiently to Deming—and the rest is history.

Shortly after World War II, Genichi Taguchi began to cultivate an understanding of how the methods of design of experiments could be used to improve the productivity of research and development work in Japan. Motivated by the need to improve telephone communications, the Japanese government established the Electrical Communications Laboratory (ECL) in 1949 and Taguchi was asked to join the lab and head up an effort to improve the efficiency of R&D activities. He quickly recognized that most of the work of the lab involved costly, time-consuming experimentation, and he set out to see how the already present field of design of experiments might be brought to bear on the problem. As Taguchi began to cultivate the application of design of experiments in the R&D work of the ECL, a unique and profound approach to quality design and improvement slowly began to evolve.

2.2

Deming Philosophy of Never-Ending Improvement

When Deming traveled to Japan in 1950 at the invitation of the Japanese Union of Scientists and Engineers, he found a climate that was quite conducive to the promotion of his concepts and methods. On the one hand, Japan appeared to have a solid base of statistical expertise, although its energies had been directed primarily toward mathematical theory and the application of that theory in nonmanufacturing environments such as agriculture. On the other hand, Deming found an industrial leadership base in Japan very eager to listen to what he had to say. With the lessons learned from his experience in attempting to promote quality improvement in America, Deming's first order of business was to conduct a series of top management training seminars in which he laid out what needed to be done to place quality improvement on an institutional basis within any organization. The obligations and responsibilities of management that he spelled out in these seminars came to be known as his *14 points* for management.

It is perhaps the combination of upper management perspective and a firmly rooted background in mathematical statistics that has enabled Deming to sustain his efforts in a leadership position over more than five decades. Deming has been able to tell management what they ought to do and then provide them with the rigorous analytical tools and methods to carry out his directions. The teachings of Deming are heavily oriented toward the use of statistical thinking and methods to identify opportunities for quality and productivity improvement. Deming has developed a plan for management to follow to enhance competitive position over the long run. This plan, referred to as his 14 points—management's obligations—is discussed briefly below. It should be noted that in the very spirit of his 14 points, Deming continues to this day to refine and improve these tenets. As a result, they themselves are continually changing. An excellent discussion of Deming's 14 points is provided by William Scherkenbach.[1] Some of his interpretations are included below.

Deming's 14 Points

1. Create constancy of purpose for the improvement of product or service.

Over the years Deming has been particularly critical of the shortsighted thinking of American management. Driven by the quarterly report, management has had a tendency to develop strategies that may show profit in the short term but are counterproductive to the long-term well-being of the organization. Central to creation of constancy of purpose is the commitment on the part of management to the long run, making significant investment in research and education so that innovation can be the hallmark of the business plan. Management must believe that the firm will be in business for a long time and therefore develop a business strategy that is based on long-term thinking. This business strategy must be strongly a function of the customer, and this requires that management understand clearly what the needs and wants of the customer are and be able to translate these into the products, processes, and systems that can address those needs competitively.

2. Adopt the new philosophy.

It is clear that we are in a new economic age created, in part, by the leadership of the Japanese in the world marketplace. Management must be willing to discard the old philosophy of accepting defective products. The thought that quality need only be improved up to an acceptable plateau, after which no further improvement is economically justified, fails to recognize the many benefits of variation reduction and the inseparable relationship between variation reduction and productivity improvement. To adopt the new philosophy, management must recognize its roles and responsibilities and assume a leadership posture. To adopt the new philosophy, management must be willing to understand the meaning and importance of the other 13 points and be willing to do whatever is necessary to embrace them fully.

For many years the management of quality was based on the premise that from a cost-of-quality point of view, there existed some optimal level of quality that should be strived for. Beyond this level, the cost of further improvement outweighed the return in terms of better quality. Of course, what "optimal level of quality" really means is an optimal level of defective material! The old philosophy is that of product control, and product control as a way of life is directed toward balancing the costs of quality to define the appropriate percent defective as a corporate objective and then maintaining that objective through

[1] W. W. Scherkenbach, *The Deming Route to Quality and Productivity: Road Maps and Roadblocks,* Mercury Press/Fairchild Publications, New York, 1987.

a quality system built on "detection and containment." Unfortunately, such systems have failed miserably over the last several decades and now must be totally abandoned. The new philosophy is one built on the process control way of thinking and is oriented toward never-ending improvement at the process, on the one hand, and pushing the quality issue farther and farther upstream into engineering design, on the other.

3. Cease dependence on mass inspection for quality control.

In adopting the new philosophy, it is essential that management recognize the need to abandon defect detection and containment as a means of controlling quality and emphasize the importance of defect prevention. Figure 2.1 is a graphical depiction of the product control model for quality control. In the product control model, output from the process is subjected to inspection procedures, usually carried out by the quality control department. Defective product is detected and contained, either to be reworked or scrapped. Although such an approach may keep the outgoing stream of product pretty much defect-free, the cost of such a system is one of the factors that erodes competitive position. More important, such a system never addresses the problem at the origin, and hence the root cause of the defective material remains active in the system.

The product control approach to quality control promotes the notion that quality and productivity are conflicting rather than jointly achievable goals. Under the mass inspection approach the belief is that if quality is to improve, productivity must suffer, and vice versa. The goal of production is to get the numbers out, while the goal of the quality control department is to keep the flow of shipped material free of defectives.

We can cease our dependence on mass inspection when we adopt the process control model for quality control. Figure 2.2 depicts this model graphically. The process control model is directed toward the identification and ulti-

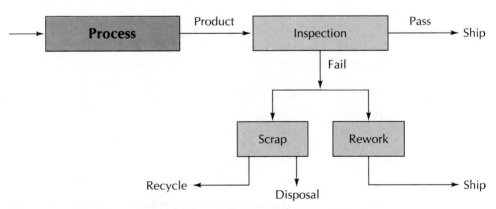

Figure 2.1 Product Control Model for Quality Control

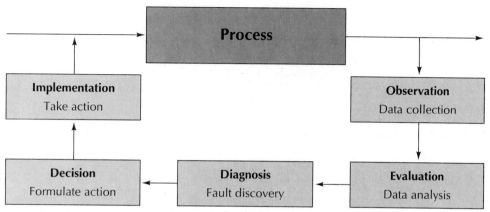

Figure 2.2 Process Control Model for Quality Control

mate removal of the underlying causes of the problem. Hence the central focus of the process control approach is the prevention of faults rather than containment of defective material. As a result, both quality and productivity will be enhanced under this approach.

Competitive position can only be enhanced when the causes of problems can be identified and appropriate remedial actions are taken at the manufacturing process or upstream during the design process. Therefore, it is clear that the process control way of thinking and of managing manufacturing operations must be adopted.

4. End the practice of awarding business on the basis of price tag.

For many years the common practice in American industry was to maintain several suppliers for any given commodity. There were probably two basic reasons for taking this approach. First, it provided multiple sources of goods that could alleviate short-term problems when one or more suppliers could not meet customer schedules because of quality problems. In essence, the multiple suppliers became buffers against their own inability to deliver quality goods in a timely fashion. The second main reason for this approach was related to price. By maintaining multiple relationships on an ongoing basis, the customer could continually apply pressure to reduce the unit costs. By awarding contracts on this basis there was a feeling that the best deal was always being made and the suppliers were continually being kept in line, price-wise.

Multiple sourcing might be likened to a common design strategy—parallel redundancy—as a means to ensure a continually successful system. This analogy is shown in Fig. 2.3, where the mean time between failures (MTBF) of the assembly process due to supplier failure to deliver is

$$\text{MTBF} = \frac{1 + 1/2 + 1/3}{\phi}$$

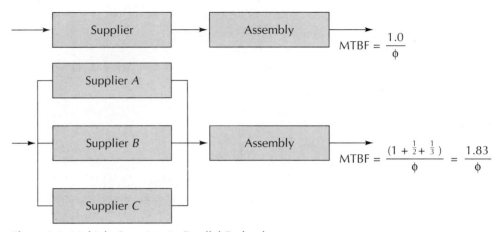

Figure 2.3 Multiple Sourcing As Parallel Redundancy

for three parallel "components" with equal failure rates ϕ. The assembly will be successful if any of the three suppliers is successful (delivers quality goods on time).

Under the way of doing business described above, the relationship between the customer and the supplier was often somewhat adversarial in nature, with a certain amount of mistrust present. The job of the quality assurance representative was to visit the supplier on a frequent basis, spot-checking for bad parts in shipments about to go out and subsequently "beating up" the supplier when it appeared that quality was poor. As a result, suppliers would counter with practices for quality control which were centered around mass inspection to ensure that no bad parts were found. There was often a general feeling that suppliers would try to put one over on customers if they could and so, of course, customers would develop equally devious tactics.

Over the last several years many companies have begun to forge long-term partnerships with their suppliers. The automotive companies, for example, have provided training and implementation assistance in the use of statistical methods to their suppliers. They have developed certification programs such as the Ford Motor Company Q1 program and have rewarded suppliers who have developed and maintained process control-oriented quality programs with sole sourcing and multiyear contracts.

5. Improve constantly and forever the system of production and service, to improve quality and productivity, and thus constantly decrease costs.

The process of continuous improvement is built on the Deming cycle, the following iterative four-step procedure:

1. Recognize the opportunity.
2. Test the theory to achieve the opportunity.
3. Observe the test results.
4. Act on the opportunity.

This cycle is clearly implementable through the process control model and its representation as a feedback control loop as depicted in Fig. 2.2. Figure 2.4 depicts the Deming cycle graphically.

To recognize the opportunity, it is essential that customer needs and expectations are clearly understood. The opportunity is basically defined by the gap between the customer's needs and one's ability to meet those needs. Hence both must be clearly defined in an operational way. Furthermore, data are needed to estimate this gap and must be conceived of and analyzed based on sound statistical principles. To seize the opportunity, one must formulate a plan, a theory for improvement, and the theory must be tested and proven. The use of statistical principles to design an experiment/develop a data collection plan to test the theory is essential. Of course, to realize the gain, one must act on the opportunity. To do this, it is important to institute change in a manner that will allow one to hold the gain. Too often the action on the opportunity is simply an extension of the experiment and may provide further proof of the value of the change but is unable to sustain its presence.

The act of precipitating continual improvement through the cyclic feedback system described above requires that this cycle become an integral part of the system, that is, be put on an institutional basis throughout the organization. This can happen only when two conditions are present. First, the motivation for seeking continual improvement must be clear to the people working within the system. As individuals, each person must come to realize the importance of this approach and must have the tools in hand to achieve this goal. This is

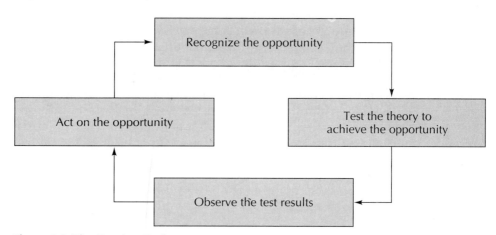

Figure 2.4 The Deming Cycle

a responsibility that management must accept. Second, the management styles and systems, and the operating precedents and technologies of the organization, must be in complete harmony with the concept of never-ending improvement. All such systems must be critically examined by management to ensure that this is the case. If systems are found that appear to inhibit the institutionalization of the new philosophy, they must be changed or eliminated.

6. Institute more thorough, better job-related training.

Training and education are essential to continuous improvement. However, training must be followed by application if it is to be the basis for improvement. Many inhibitors stand in the way of the effective use of training. Management must change the system so that people are encouraged to use the results of training. Some of the major inhibitors to training are: "We probably need to get some of our people trained but we're very busy with new product introductions and haven't got the time to spare"; "It's for my people, not for me"; "SPC training won't help us, our problems are different"; "We rely on experience, our people learn on the job." All of these comments have been used again and again. It is regrettable but true that in many organizations when hard times hit one of the first things to be cut is the training budget—it should be the last. If the Deming cycle is to be implemented on a continuous basis, people must be given the tools to make it happen. Statistical thinking and methods will be essential.

William Conway, former CEO of the Nashua Corporation, tells a story[2] that exemplifies the need for giving people the proper tools so that they can do their jobs effectively. He tells of an employer who decided to reward one of two employees, let's call them Bill and Jack, with a lavish vacation at company expense. The two were to compete for the prize. The employer first called Bill into his office and described the trip, detailing the many extravagant features of this perk. Needless to say, Bill became very enthusiastic, with great motivation to compete to his best ability to win the trip. The employer then called Jack in and proceeded to get him quite excited and motivated about winning the trip. The day came for the competition and both Bill and Jack were eager to perform. The employer told both that the task they were to complete was to secure six screws into a piece of wood. The employer gave Bill the screws, the piece of wood, and a nice screwdriver to aid in completing the task. He turned to Jack, gave him the screws and the wood but told him he would not have the use of a similar tool. On the count of three, both men enthusiastically began to complete the task. In a matter of a couple of minutes, Bill was done, with little effort. In the meantime, Jack broke three of his fingernails, cut open the palm of his hand, and was unable to get even one screw into the board.

It is clear that both Bill and Jack were highly and about equally motivated to do the work assigned to them. Both wanted to do a good job. However, only

[2] Adapted from a videotape presentation by Conway to a group of Ford executives, Ford Motor Company Radio, TV, and Film Dept., File No. 81V44-1, Dearborn, Mich., 1981.

one of them was given the tools by his manager to do the job properly. Motivation and desire are clearly not enough; people must be given the proper tools to do their work effectively. In Chapter 10 we will look at a case similar to the one described above only in an actual manufacturing process setting. In that case we will find that the best efforts of an otherwise skilled operator fell short of the mark simply because she was not given the knowledge and tools to do her job properly. It is clear from this account that while attitude and psychology are important to the never-ending pursuit of improvement, they must be clearly separated from the tools required to do the job properly and efficiently. Training and education are essential to providing these tools.

7. Institute leadership.

For many years Deming referred to this point as: "Improve supervision." But he came to realize that you cannot improve something that does not exist. First you must institute supervision; only then can you improve on it. In this new economic age, a supervisor must be a coach and a teacher, not a watchman. A supervisor must continually seek ways to provide his or her people with the tools to do an effective job and the opportunity to use those tools. Continual improvement requires the "catching someone doing something right" mentality, not "catching someone doing something wrong."

Too many management systems today are oriented toward the management of things, not of people. They are oriented toward the management of the outgoing stream of product, not of the management of the people and machines that make the product. Leadership in management should have as its primary goal to motivate people to work to their maximum levels of performance, but too often the management system being used does nothing more than foster mediocrity. Under the system of *management by objective*, the main focus often becomes the job of setting the objective. Just as the engineering specification can be a focus of negotiation to the detriment of quality and productivity, so can the setting of objectives for individual performance. Fear of failure may cause a subordinate to negotiate with his boss to arrive at a lesser objective for his performance for the year. Once the objective is set, however, the only thing that is important from that point is the objective itself and the employee's performance relative to it.

8. Drive out fear, so that everyone may work effectively for the company.

Deming feels that this is perhaps one of the more critical of his 14 points, since it affects directly so many of the other 13. Fear is an all-pervasive quality that touches every corner of the organization in so many different ways. Fear of failure and fear of embarrassment inhibits our ability to capitalize on opportunity. Fear causes us to misinterpret the intentions of others. Fear breeds overreaction, which can only increase variability in the system, thereby increasing waste and inefficiency. To reduce fear requires a fundamental change in the philosophy of management. Implementation of point 7 is an essential prerequisite to the substantial elimination of fear throughout the organization.

9. Break down barriers between departments.

People in all activities within the organization must work as a team on the implementation of the Deming cycle. To accomplish this, point 8 is crucial. But to truly invoke continuous improvement on a company-wide basis requires that everyone come to a common understanding of the needs of the customer and the present capability of the organization to meet those needs. To implement this point requires that the system foster teamwork. This may mean that management systems which destroy teamwork, such as the personnel performance appraisal system, must be changed. In his book, Scherkenbach singles out performance appraisal systems as potentially one of the most significant inhibitors to quality and productivity design and improvement.

10. Eliminate slogans, exhortations, and targets for the work force that ask for zero defects and new levels of productivity.

Personal motivation and improved awareness of needs of the organization will contribute to the reduction of waste and inefficiency. But such are not a substitute for providing training and giving people the tools to do a better job. As we discussed earlier, William Conway, then CEO of the Nashua Corporation, told a story about two employees who were competing for a major company perk. Both were asked to do the same job and both were equally enthusiastic about the reward. Motivation was at a fever pitch. One, however, was given the proper tools by management to do the job while the other was simply told to do the best he could. Clearly, the one who had the proper tools won the perk. Both had the same desire to do a good job but only one was given what was required to make it happen. Everybody wants to do a good job. But management must create an environment that enables people to improve continuously.

11. Eliminate work standards on the factory floor.

Clearly, we have already said much about this point in Chapter 1. Work standards place a cap on productivity improvement and are totally contrary to the concept of never-ending improvement.

12. Remove the barriers that rob employees at all levels in the company of their right to pride of workmanship.

To make this happen, major changes in the way we manage people must take place. Supervisors must emphasize the need for quality, not for getting out the numbers. The systems that evaluate the performance of people on a regular basis must probably be changed. Financial management systems that tend to focus on the short term must change. The use of daily or weekly production performance reports must be abandoned. Such reports encourage the management of things, not people, and often lead to overreaction, which as we have already said, increases the variation and hence the waste in the system.

13. Institute a vigorous program of education and self-improvement.

People must not be treated as a commodity but rather as an asset. People in the organization must believe that management is willing to reinvest in their future, providing them not only with the opportunity for change, but also with the tools to take advantage of the opportunity.

14. Put everybody in the organization to work to accomplish the transformation.

The transformation is everybody's job. Above all, this will require that management seek out and acquire the necessary talent to teach everyone the new philosophy and provide continual coaching and direction on the use of statistical thinking and methods. Such persons must have access to top management and must be provided with the resources to get the job done. To implement the 13 points above requires that management carefully examine each and every system in the organization to determine whether it promotes or inhibits the continual pursuit of quality and productivity design and improvement.

The 14 points are clearly the responsibility of management. They, themselves, define the essential elements of the institutionalization of quality and productivity improvement through statistical thinking and methods. Quality and productivity improvement is everyone's job, but only management can create the structure in the organization that enables the 14 points to be carried out. The 14 points define the elements of a process—a new way of conducting all aspects of the business. As is obvious from the discussion above, they cannot be deployed selectively; success in carrying out any one point depends on carrying out the others. The 14 points provide important guidance in terms of what we must do. It is up to management to redesign the system to achieve the 14 points.

2.3

Traditional View of Quality

Shortcomings of Traditional Definitions of Quality

Before we can really understand what type of a strategy needs to be adopted to move in the direction of enhanced competitive position in the world marketplace, we need to ask some basic questions:

What do we mean by quality products?

What do we mean by improving quality?

What do we mean by quality design?

Quite simply, what do we mean by *quality*?

Over the years many definitions of quality have been offered. A quick look into the first chapter of dozens of books written on the subject over the years provides a variety of definitions of quality:

- Fitness for use/fitness for function.
- The degree to which a specific product satisfies the wants of a specific customer.
- The degree to which a product conforms to design specifications.
- Providing products and services that meet customer expectations over the life of the product or service at a cost that represents customer value.
- The characteristics or attributes that distinguish one item or article from another.
- Conformance to applicable engineering requirements as described in engineering drawings, specifications, and related documents.

While most of the words sound okay, such definitions have not been very useful; that is, they have failed to help us do a better job, to be more competitive. The question is: Why? There may be at least three reasons why the definitions above have not served us well:

1. *They are attribute-based and qualitative in nature.* The precise quantitative definition of quality has not been established through any of the definitions above. These definitions of quality generally prescribe that something is either in one state or another: good or bad, defective or nondefective, within or outside the specifications. As a result, they promote the improvement of quality only up to an acceptable level and therefore are a serious inhibitor to continuous and never-ending improvement.

2. *They are manufacturing rather than design based.* Many definitions of quality attempt to compare the product on completion of manufacturing to the design specifications that have been imposed on the manufacturing process. As a result, they become the final filter that we attempt to push the product through rather than the criteria on which the product design is based. That is, the application of these definitions of quality is taking place at the end rather than at the beginning of the product development life cycle. Over the years, the most common and widely used measure of quality has been "conformance to specifications." However, specifications are generally a product of engineering design, not a design objective. Too often, specifications are driven by the limitations of the manufacturing process. Sometimes, the specification results from a considerable negotiation between design and manufacturing personnel.

3. *They do not clearly establish the proper link between customer wants/needs/expectations and product function.* It is important to make the distinction between customer likes in terms of preferences—color, style, and so on—and customer needs in terms of the performance of a function. A certain customer may prefer a red automobile over a blue one, but once that issue, which is one of subjective preference, is settled, the only issue that remains is that of performance of the function. Function is judged by such things as product life, power consumption, trouble in the field, and production of harmful effects during use. A product may be sold—gain market—by virtue of its subjective characteristics (style, color, etc.) but may lose reputation—market share—by virtue of its function, its quality.

In today's economic age we simply cannot afford to use concepts and measures of quality that do not relate the achievement of function in the field to the engineering design process but instead administer "quality control" through defect detection and containment after manufacturing. Definitions of quality that promote improvement to an acceptable plateau of performance will inhibit the continual pursuit of never-ending improvement and hence have a weak and perhaps opposing relationship to process performance in terms of efficiency/productivity. Rather, to improve competitive position our view or definition of quality should:

- Provide a quantitative basis to move the quality issue upstream to engineering design—a design criterion.
- Promote focus on the process not the product in a manufacturing sense—prevention rather than containment.
- Be strongly tied to the issue of productivity and therefore promote continual pursuit of never-ending improvement.
- Quantify/measure the loss imparted to the customer as a result of poor quality rather than the loss imparted to the producer—consumer versus producer orientation.

Quality and the Engineering Specification

From the above it is clear that the traditional definitions of quality appear to have fallen short in terms of their ability to articulate quality in a way that can foster improvement in competitive position. In particular, the association of quality with *conformance to the engineering specification* puts the measurement of quality on an attribute basis and provides little more than a *shipping criterion* when it is essential that we articulate quality in a manner that enables it to be used as a *design criterion.* The view of quality as conformance to specifications promotes the product control approach to quality control and hence stands as a significant inhibitor to the adoption of a process control approach to manufacturing and the integration of quality and the design process.

Figure 2.5 provides a graphical depiction of the traditional interpretation of the engineering specification. In Fig. 2.5(a) and (b), two representations of a

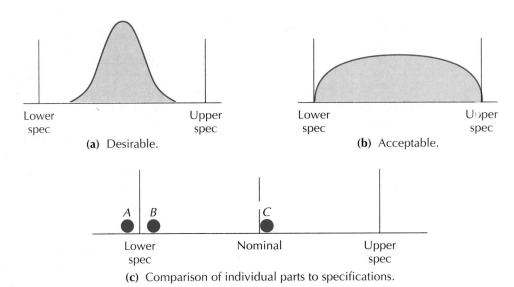

(a) Desirable.   (b) Acceptable.

(c) Comparison of individual parts to specifications.

Figure 2.5 Classical Interpretation of the Engineering Specification

process as a statistical distribution of measurements are shown relative to a certain bilateral specification. Although the bell-shaped process distribution in (a) might be preferred to the more loaf-shaped process distribution in (b), some people would not make much distinction between the two cases in terms of quality; that is, in both cases virtually all of the process/parts are conforming to the specifications. However, when one begins to contemplate Fig. 2.5(c), one begins to realize that the relationship between quality and the engineering specification is not as strong as we may have viewed it in the past. Considering the dots labeled A, B, and C as representing three different manufactured parts, we generally interpret part A to be unacceptable because it is outside the specifications, while parts B and C are considered acceptable because they are within the specifications. The crucial point in this interpretation is the fact that we generally do not make a distinction between parts B and C as far as quality is concerned. It does not take too much thought, however, to come to the realization that as far as performance is concerned, it is unlikely that there is very much difference between parts A and B, although we might well conclude that part C should perform considerably better than either A or B.

If you were a teacher and you were considering the performance of three students on an exam who earned scores of 59, 61, and 95, you would probably view the performance of the two students who got the scores of 59 and 61 as about the same, and much poorer than the student who got 95. Even though 60 might be your lower limit for a passing grade, you would probably not view the two students who got grades of 61 and 95 as being equal from a quality standpoint. Why, then, should we think of manufactured product and its potential for quality performance in the field any differently?

---| **2.4** |---

Taguchi's Definition of Quality

As discussed in Chapter 1, Taguchi's approach to quality engineering has a number of significant strengths. In particular, Taguchi has placed a great deal of emphasis on the importance of minimizing variation as the primary means to improve quality. The idea of designing products whose performance is insensitive to environmental conditions and making this happen at the design stage through the use of design of experiments have been cornerstones of the Taguchi methodology for quality engineering.

The Loss Function Concept

Taguchi offers a view of quality that directly addresses the concerns raised in the preceding section. He suggests that it is important to think of quality in terms of the "loss imparted to society during product use as a result of functional variation and harmful effects." Harmful effects refer to side effects that are realized during the use of the product and are unrelated to product function. Functional variation refers specifically to the deviation of product performance from that intended by design, that is, deviation of performance from the design target, the nominal.

Taguchi uses a loss function concept to quantify quality as "loss due to functional variation." He argues that the loss to society as a whole is minimized when performance is at the design nominal and as performance deviates from the nominal, loss increases. For many applications a quadratic loss function may be appropriate. Figure 2.6 illustrates the loss function idea. The loss function

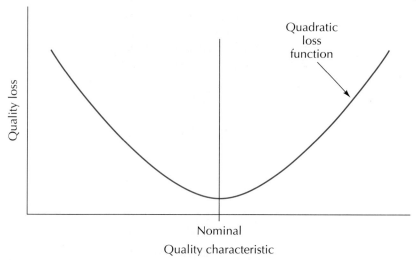

Figure 2.6 Loss Function Representation of Quality

concept as a way to measure quality clearly suggests that the goal of design and manufacturing is to develop products and processes that perform "on target with smallest variation." (Recall the quality/variability relationship we spoke of in Chapter 1 as a motivation to seek continual reduction of variation.) The loss function places tremendous importance on the reduction of variation to achieve the most consistently performing products and processes. To drive performance toward this overall objective, it is important that we learn more about the way in which the important parameters of the product/process influence performance. Over the years the methods of design of experiments have proved to be an effective way to do this.

Figure 2.7 provides an interpretation of the engineering specification when the loss function concept of quality is superimposed. In this case no loss would be realized as long as the quality characteristic in question lies within the lower and upper specifications. Outside of the specifications the product is considered to be unacceptable, and therefore a constant loss would be realized. This loss would probably be measured in terms of scrap cost or rework cost.

Variation Reduction and the Loss Function: A Case Study

The loss function view of quality suggests that there exists a clear economic advantage to reducing variation in the performance of a product. The case study described below clearly indicates how such reduced manufacturing imperfection can lead to a reduction in costs and hence an improvement in competitive position.

Several years ago one of the automobile companies performed a study to compare the manufacturing variations evident in certain transmission components for comparable transmissions made in the United States and in Japan. Random samples of transmissions were selected in each case, the transmissions

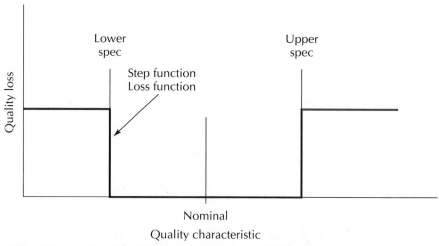

Figure 2.7 Loss Function Interpretation of the Engineering Specification

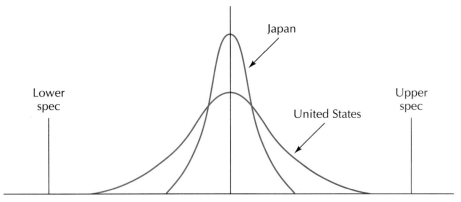

Figure 2.8 Comparison of Critical Dimensions for U.S. and Japanese Transmission Components

were taken apart, and a number of critical dimensions were measured and recorded. Figure 2.8 is representative of the general findings of the study. In particular, it is seen that in the case of U.S. transmissions, the critical dimensions generally consumed about the middle 75% of the tolerance range specified in each case. Normally, one would conclude based on generally accepted standards that the capabilities associated with the manufacturing processes were therefore well within normal expectations. As the figure shows, however, the same critical dimensions for the Japanese counterpart of this transmission consumed only about the middle 25% of the same tolerance range. The question that begs to be asked is: Why would the Japanese strive to make the parts to such tight tolerances? The answer lies at the foundation of what this book is all about.

There are two very good reasons why the Japanese have worked so hard to reduce variation, as shown in Fig. 2.8. The first of these is seen in an examination of Fig. 2.9, where the shaded bars represent the costs associated with warranty claims for the two different transmissions. These warranty cost

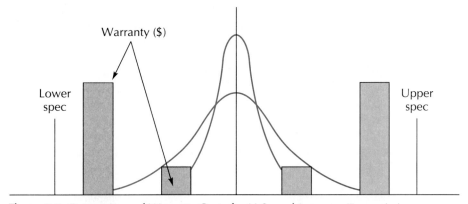

Figure 2.9 Comparison of Warranty Costs for U.S. and Japanese Transmissions

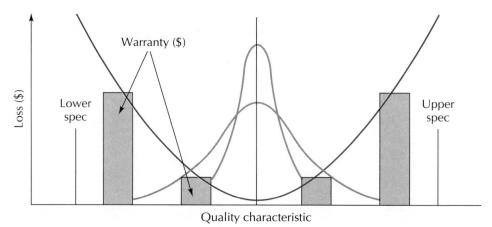

Figure 2.10 Loss Function Interpretation of the Results of the Transmission Study

bars have been plotted at the extremes of the two distributions of the critical dimensions to depict the relative costs associated with variability. It is clear that for U.S. transmissions, the cost associated with repair and replacement is significantly larger than for their Japanese counterparts.

This comparison of the economic data associated with the two transmissions would suggest that the loss function idea is, in fact, at work, whether we realize it or not. Figure 2.10 illustrates the appearance of Fig. 2.9 if one considers that the relative magnitudes of the warranty costs are simply estimating the loss associated with variation in the critical dimensions under study. The figure strongly suggests that there exists a definite relationship between variability and cost in terms of loss incurred due to functional variation. One could think of the warranty cost values as literally mapping out the loss function.

There is still another side to this variation reduction situation and that is the side that suggests that there is an important relationship between variation reduction and reduction in waste and inefficiency. Recently, one of the authors was sitting in a presentation on statistical thinking and listening to the speaker[3] recount an exchange that took place between a manager in a U.S. automobile company plant and a representative of a Japanese firm for which they were making some parts. When the engineering drawings for the parts arrived at the plant, it became immediately obvious that something was wrong. There were no tolerances on any of the dimensions on the parts. The manager called the Japanese firm and asked for a new set of drawings with the specifications indicated. He was promptly told that there were no tolerances specified for the dimensions.

After a brief pause, the manager asked his Japanese friend: "Well, how close do we have to make the parts to the nominal dimension?" "How close

---

[3] Presented by Howard B. Aaron, Lafayette, Indiana, ASQC chapter seminar on "Quality for Today and Tomorrow," March 17, 1988.

can you make them?" was the reply. After some discussion the two agreed upon an allowable amount of variation. The Japanese company representative indicated that this negotiated variation level would apply for one year. Furthermore, he indicated: "After the first year, we will ask you to reduce the amount of variation in the parts you are supplying. We know that to do this you will have to improve your processes and if you are successful you will reduce some of the waste and inefficiency in your operations. As a result, your costs will go down. We want you to share that saving with us and so at the end of the first year when we ask you to reduce the variation, we will also ask you to reduce the price of the parts."

The manager was stunned! How could he possibly be expected simultaneously to tighten up on the specification for the parts and reduce the price? "I did not ask you to tighten up on the specification on the parts," was the reply, "I asked you to improve the consistency of your processes! If you do this by finding the root causes associated with the variation in your processes and take action to remove those problems, the parts will be made closer and closer to the nominal value and the efficiency of your processes will improve. We just want you to share your good fortune with us."

2.5

Cost-of-Quality Issue

The loss function concept may precipitate the thinking that since we have put quality (loss) on economic grounds, we should carefully weigh the cost to achieve quality—reduce variability—against the derived gain using the loss function. Figure 2.11 depicts this scenario graphically. The basic question we

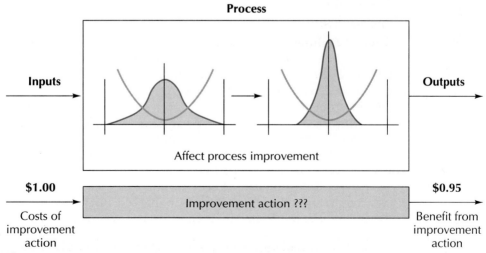

Figure 2.11 Economic Consequences of an Improvement Action

are asking is: Are such continual improvement efforts always cost-effective? Are we reaching a point of diminishing returns?

Some caution may be needed in the quantitative use of the loss function idea. Because the loss function establishes a rigid economic criterion, it may in fact tend to run somewhat contrary to the concept of the continual pursuit of never-ending improvement. In spirit it may be very attractive and its words quite easy to embrace, but its literal translation may be shortsighted, and generally inhibiting, particularly when the question of the cost of quality arises.

In fact, Taguchi and others have given examples to demonstrate the weighting of costs to precipitate, say, a 50% reduction in product variation against the benefit of obtaining such a reduction in terms of functional variation. Such is also depicted in Fig. 2.11. It has been said by some that from an economic standpoint, there is a weakness in Shewhart's theory, that if the cost for control is larger than the profit from the reduction of variation, we should not do anything. In other words, it is the duty of a production department to reduce variation while maintaining the required profit margin. Such thinking could be counterproductive and inhibit us from realizing the full and broad-based benefits of the use of statistical thinking and methods for process improvement.

Perhaps the broader question we ought to ask is: What are the costs of not having quality? What are the costs associated with lost opportunity? Philosophically, we can, and perhaps must, argue that attempts to weigh the input costs continually against the output gain are counterproductive in the long term (i.e., run contrary to the concept of the continual pursuit of never-ending improvement)—that such thinking will continually inhibit our ability to make the right long-term decision.

What is the traditional view of quality costs? Some have suggested that the following quality cost factors should be considered:

Prevention Costs.   maintaining a quality control system.
Appraisal Costs.   maintaining a quality assurance system.
Internal Failures.   manufacturing losses, scrap and rework.
External Failures.   warranty, repair, customer, and product service.

Figure 2.12 shows how these factors are often combined to "optimize" the overall quality requirement. The old concept of AQL—acceptable quality level—is derived from such thinking.

When we ask the question, "What do all of these costs have in common?" the answer is sobering. They are all relatively easy to measure, but they are not representative of the factors that enhance competitive position. The real question we need to be asking is: "What is the cost of not having quality?" The answer to this question can be seen through the following case study.

A comparative study was done by a manufacturer (Sony) of color television sets made in both the United States and Japan.[4] When many sets were examined

---

[4] *The Asahi*, Japanese language newspaper, April 15, 1979. Also reported by Genichi Taguchi during lectures at AT&T Bell Labs in 1980.

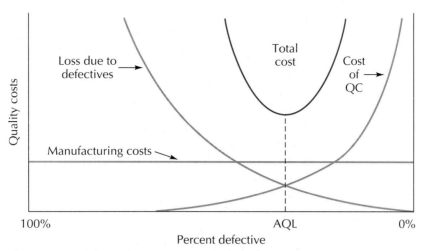

Figure 2.12 Traditional Economic View of Quality Costs

to observe the color-intensity quality characteristic, the results were as depicted in Fig. 2.13. For the U.S. sets, virtually none could be found whose color intensity was outside the specifications. The variation pattern followed the loaf-shaped appearance in Fig. 2.13. For the Japanese sets, a small percentage were found to have color intensities outside of the specifications, but the variation pattern followed the familiar bell-shaped curve in Fig. 2.13. The U.S. variability pattern is unnatural. It should signal to us that something is radically wrong in the process. And what is wrong is probably costing us a great deal of money.

We need to ask ourselves the question, "How could a result such as that for the U.S. sets in Fig. 2.13 be made to occur? Figure 2.14 suggests one answer. In Fig. 2.14 we see that the sets originally manufactured out of the specification are adjusted/reworked after inspection to bring the color intensity within the

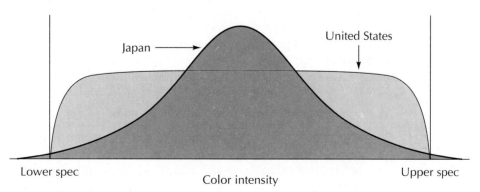

Figure 2.13 Comparison of Quality of TV Sets Manufactured in the United States and in Japan

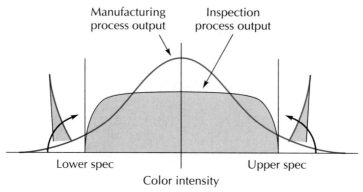

Figure 2.14 Quality "Inspected" into the Process

specifications. This is accomplished only at a tremendous cost, the cost associated with not having quality in the manufacturing process.

Figure 2.15 shows another manifestation of the loaf-shaped process of Fig. 2.13. Although in principle the process in Fig. 2.15 appears quite capable (if stable), its erratic behavior with respect to its mean level causes the color intensity value to be distributed broadly across the specifications. As in Fig. 2.14, if out-of-specification sets are made, inspection will detect this and they will be reworked. Even if all the sets are within the specifications, the erratic process behavior is likely to cause waste and inefficiency.

Figure 2.16 shows still another way that the loaf-shaped process of Fig. 2.13 can arise. Letting the process drift in this manner to "make full use of the wide specifications" can mask the occurrence of other problems that might otherwise show up on a control chart. Further, from a loss function point of view, the cost of not having quality is clear.

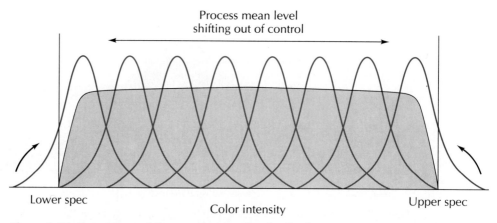

Figure 2.15 Out-of-Control Process Causing Quality Erosion

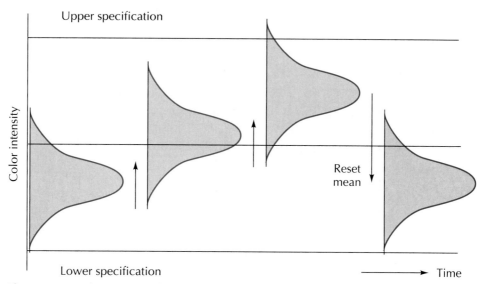

Figure 2.16 Making Use of the Full Range of the Specifications

There is yet another cost of not having quality that is derived from Taguchi's loss function approach to quality. If the loss function is superimposed on the distribution of color intensity, as in Fig. 2.17, then it is clear that total loss, over all sets, is greater for the U.S.-made sets. Furthermore, although the previous examples of the cost of not having quality were manufacturing-related costs, costs on the user end are included in Taguchi's quality definition. For example, costs incurred resulting from having to rent a set while the broken set is being replaced is a cost of not having quality.

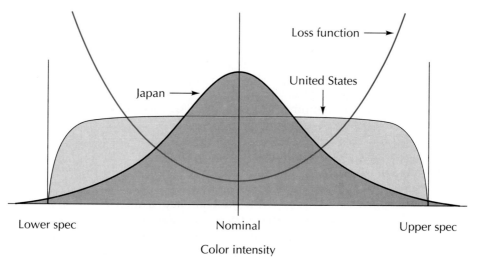

Figure 2.17 Loss Function Assessment of Quality

In considering all of the above, we have not even mentioned some of the more significant benefits of the continual pursuit of quality improvement/variation reduction. We, of course, can also view these as the lost opportunities associated with:

- **Improved Production Management.**   Scheduling/inventory control/MRP.
- **Increased Flexibility/Adaptability.**   Adjusting to changing needs/taking advantage of new market opportunities.
- **Enhanced Environment for Breakthrough.**   Encouraging larger signal-to-noise ratio/ability to more clearly see trouble: easier to study process improvement actions—experimentation.
- **More Positive Employee Attitudes.**   Looking for improvement rather than hiding poor quality; trying to catch someone "doing something right."
- **Identification of Process Flaws.**   Searching for hidden waste and inefficiency.

In summary, we can look at this issue of quality costs in a couple of ways. We can spend our time trying to quantify the apparent costs of achieving a certain level of quality, or we can spend our time trying to understand what the cost of not pursuing quality is and then reap the many, many benefits of the continual pursuit of quality improvement.

Exercises

2.1. What is the purpose of Deming's 14 points? That is, why did Deming feel the need to develop and teach these tenets?

2.2. What is the most fundamental point of departure of the philosophies of product control and process control?

2.3. How does Deming feel that customer–supplier relations ought to be approached?

2.4. Explain the term "hold the gain" in reference to improvement.

2.5. What sort of leadership does Deming feel is necessary in the realm of manufacturing?

2.6. Define "quality."

2.7. Many definitions of the word "quality" have been used in the past. What are some of the reasons why many of them do not serve us well?

2.8. What is the problem with the use of conformance to design specifications as a definition of quality?

2.9. How does the loss function concept differ from the classical definition of the engineering specification?

2.10. What is wrong with the loaf-shaped distribution that is characteristic of the output of processes for so many manufacturers?

2.11. When does the pursuit of quality and productivity improvement become economically inadvisable? Explain your answer.

2.12. What is the real cost-of-quality issue?

2.13. Deming states that it is one of management's responsibilities to provide their workers with the proper tools for the job. To what do you think he was referring here? Can you provide several examples, either real or contrived?

2.14. The elimination of work standards has been one of the more difficult of Deming's points to be accepted by management. Why do you think that is? How would the elimination of work standards positively affect pro-ductivity? Quality?

2.15. Deming's tenth point states that work slogans and targets for the work force should be eliminated. Why? And how will doing so affect quality and productivity?

2.16. Which of Deming's 14 points show the detrimental effect that weekly performance reports have on overall quality? Explain why these inhibit improvement.

2.17. How will staying with single suppliers benefit a company in the long run? When multiple suppliers are used, where does the emphasis tend to be placed? How can a company be sure that its single-source suppliers will not fail to meet demand?

2.18. Which of Deming's 14 points addresses the problems associated with the practice of over-the-wall design that was discussed in Chapter 1? Explain.

2.19. What is the major limitation of the loss function concept in terms of encouraging the continual pursuit of improvements in quality?

2.20. How can a customer benefit from improved quality in a supplier's prod-uct? How will this improved situation affect inventory management for the customer?

C H A P T E R

3

Statistical Methods and Probability Concepts for Data Characterization

---
3.1
Introduction

Throughout our study of statistical methods for quality and productivity improvement, we will be concerned with essentially the same problem: using data obtained from the process to draw conclusions known as inferences about how it is or has been operating. We may be concerned with an injection molding machine that is making truck grilles; or a robot inserting clips into molded instrument panels; or an automatic screw machine making fasteners; or an accounts payable system processing customer invoices; or many other manufacturing processes and service operations. In each case we will find it useful to the improvement of quality and productivity to collect data from the process and model it using statistical concepts and methods. Such will be necessary, since the results of the data collection activity will represent only a portion, sometimes quite a small portion, of the entire process. Therefore, while we will use the results to make inferences about the entire process, these inferences will be subject to error.

3.2
Purpose and Nature of Sampling

The Notion of Sampling

To find out how the process is behaving in terms of an output quality characteristic(s) on the average and in terms of variation, we will observe only part of the output of the process. We will then use these data to infer something about the actual process performance, that is, the performance of all the output that comprises the population. *Population* is a concept that means to imply and include all possible realizations of a process within a certain frame. A population is really defined by a constant system of variation causes characterizable by a succinct set of *parameters* that in an average or collective sense describe the salient features of the variation system.

It is generally neither practical nor necessary to observe the entire population. Rather, we observe only a small subset of it at a time, often doing so periodically. When we do it this way, we are said to be *sampling* the population or process. How much we sample each time, how often and/or when we sample, and how we specifically take the sample are critical issues that must be addressed. But what we do with the data, particularly the information extracted therefrom, will dictate to a great extent the degree to which quality and productivity may be improved.

Various functions of the data may be used to calculate measures, each of which is a reflection of some special feature of the population. These sample measures are called *statistics*, each of which may be used as an estimate of the corresponding population parameter. Together, a set of the parameter estimates is used to characterize the process performance. Because statistics are based on sample data, which represent only part of the population, they are uncertain estimates of the population parameters.

There are many instances in which we need to examine a certain context or environment for the purposes of learning about its makeup or performance. We could be interested in the parts we have on hand, the bills we need to pay, or the character and attributes of our work force. We may wish to know these things because we would like to evaluate the effects of receiving certain types of orders, or evaluating our cash position, or understanding the impact that an early retirement policy might have. The problems at hand here are really "counting" problems: determining how many of "these" and how many of "those" we have.

We often answer the types of questions posed above by sampling—looking at only a fraction of the whole because an exhaustive count would be prohibitive in a cost and time sense. These are the types of problems on which statistical sampling methods and the theory of statistical inference thrive. These are situations where a frame or population is fixed and we are simply trying to observe certain characteristics of it. We are not asking how the parts got into the inventory, why they are there, or what the chances are that the quality of the next ones to join the group might get better or worse. We only want to know what is there right now.

Enumerative Studies

What we have just described are commonly referred to as *enumerative studies*. Over the past half century, W. Edwards Deming has pointed out that it is important to distinguish between the two types of statistical studies: enumerative and analytic.[1] Enumerative studies are statistical analyses of sample observations from a fixed frame, the kinds of things we have just described above. The purpose is to describe a clearly defined population on the basis of sample information. A large body of theory has been developed to deal effectively with these issues. Statistical methods, including significance tests, can be effectively employed to make inferences about the population that has been sampled. In general, the inference becomes more precise as sampling error is reduced, which can be achieved by taking larger samples at random. But even in the case where we are dealing with a fixed frame, there is no guarantee that the assumptions required by the statistical tests employed will be valid. We will study many concepts and techniques in this book that will enable us to check out these assumptions or precipitate an environment in which the assumptions are approximately valid. The control chart is one of these techniques. Certain principles of the design of experiments are others.

In a statistical sense, making a control chart of a day's production is an enumerative study. We collect the data, which is a sampling of the day's production, and we make charts for the purpose of evaluating what took place during that day. We are interested in finding out something about what has already taken place. But why? There can be only two reasons. One reason keeps the problem in the realm of the enumerative study, but the other does not. If our purpose is to determine whether or not we have to take some action concerning the parts we made that day, say, 100% sort them, then the study stays as an enumerative one. In this case any action that we decide to take will be within the frame we are examining. But if our purpose is to learn how to do better tomorrow with an eye toward actions that will precipitate change, then we no longer have an enumerative study, and statistical analysis has done all it can for us. We now are searching for ways to change the system to improve its performance in the future. To bridge the gap between what we did today and what we hope to do tomorrow we will need much more than statistics; we will need knowledge derived from process experience, engineering practice, and perhaps, basic science to solve the problem.

Analytic Studies

When sample results are analyzed for the purpose of understanding the causal formation of the population, Deming calls these studies *analytic*. In an analytic study, no statistical inference could be made about the population, which is undefined or undergoing changes. This is illustrated in Fig. 3.1(a), where the underlying process is changing over time without an identifiable pattern or regularity. The sample information is relevant to the process behavior only at the time the sample is taken. It has no statistical basis to represent either the past or the future of the process. Such sample data can be usefully interpreted by the experimenter only to explain, discover, or predict the dynamics of the

---

[1] W. E. Deming, "On the Use of Judgement-Samples," *Reports of Statistical Application Research, JUSE,* Vol. 23, No. 1, March 1976, pp. 25–31.

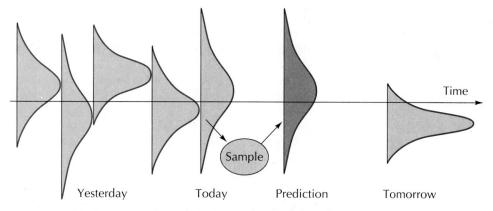

(a) Process not in statistical control: population changes over time.

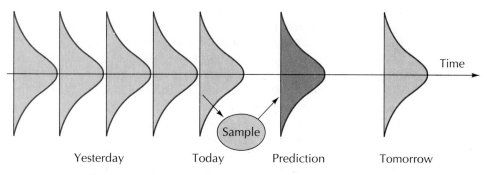

(b) Process in statistical control: one unique causal system.

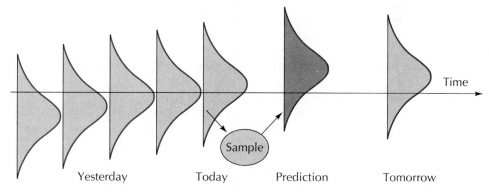

(c) Process in transition: in search of a causal system.

Figure 3.1 Various Processes Under Analytic Study

process using his or her expert knowledge. Note that in these studies, the sample size has very little to do with sampling error, which is undefinable.

Statistical interpretation of the current sample data will be valid for the purpose of explaining or discovering what may have happened in a process within a given frame. But methods such as Shewhart control charts are effective only when the information derived from their patterns, fortified by a statistical interpretation, can be overlaid on our knowledge of the process and its workings to identify possible actions for improvement in the future.

Figure 3.1(b) depicts a process in control. Here control charts are an effective means for analytic studies where the objective is to assess the performance of the process and its capability, and to maintain a process in statistical control. As long as the process remains in a state of statistical control, prediction of future process performance may be inferred statistically from the past data.

Unfortunately, many real-world processes are not stable. For example, in the early development of a product or a process, there is no unique process yet in existence. A stable process is frequently studied by purposeful introduction of changes that will destabilize it to discover opportunities for further improvement. In these situations, usually some knowledge of the process dynamics is available to guide the experimenter in defining a relatively narrow band of uncertainty in his or her studies. In other words, the process as depicted in Fig. 3.1(c), where there is a certain degree of regularity or some inertia of the process dynamics, is more likely to be encountered than the case of Fig. 3.1(a). Again, knowledge derived from process experience, engineering practice, and perhaps, basic science is required to deal with the problem. Statistically analyzed sample information often provides the only basis for actions for these processes.

Statistical Inference in Analytic Studies

In quality design and control, we are concerned primarily with the development and maintenance of statistical control/process stability and the exploration of improvement opportunities of a process such as the ones in Fig. 3.1(b) and (c). Statistical methods can be used effectively in estimating the performance of a process already in control, or for determining that it was not, within a certain frame. For processes in Fig. 3.1(c), however, the method of statistical inference has to be employed with extreme care in drawing conclusions. To improve the validity of the statistical results in these situations, the experiments or sampling method must be planned ahead following certain fundamental principles such as randomization and classification of sources of variation. After sampling, any statistical inferences derived from the sampling results must be confirmed by additional sample results. This is most effectively done in the environment in which the consequences of the action based on the inference are desired to be felt, which may be quite different from the environment in which the inference was developed. It is with these understandings that statistical methods may be used for making inferences about the process performance. Some of these methods are presented in this chapter. Others are presented elsewhere throughout the book.

3.3

Characterizations and Representations of Data

A process engineer is examining an injection-molding process that is producing a certain kind of light fixture for automobile headlights. The purpose of this study is to collect part weight data to help monitor the performance of the molding press as to its state of statistical control and to evaluate its capability in meeting part weight specifications. Table 3.1 gives 44 samples, each of which is a record of five ($n = 5$) part weights in grams of the light fixtures drawn from the press every hour. In addition to the five part weights, the table also lists, in the last two columns, the average weight and the range for each sample.

Measures of Central Tendency and Dispersion

The arithmetic average of the observed values of a sample is defined by

$$\overline{X} = \frac{\sum_{i=1}^{n} X_i}{n}. \tag{3.1}$$

In Table 3.1 the part weights of light fixtures from a molding press are listed five in a row for each hourly sample. The sample mean is calculated using Eq. (3.1). For example,

$$\overline{x}_1 = \frac{167 + 169 + 187 + 160 + 180}{5} = 172.4.$$

Each sample mean is an estimate of the true but unknown process mean, μ_X.

An important measure of the variability in data is the sample variance, which is the average of the squared deviations of the data from their sample mean. The sample variance, denoted by S_X^2, is defined by

$$S_X^2 = \sum_{i=1}^{n} \frac{(X_i - \overline{X})^2}{n - 1}. \tag{3.2}$$

In this equation for the calculation of a sample variance, it is noted that the denominator is $(n - 1)$ instead of n, which is the actual sample size. This is done because in computing S_X^2 using Eq. (3.2), there are n deviations $(X_i - \overline{X})$ but only $(n - 1)$ of them are independent. The positive square root of the sample variance is called the sample standard deviation S_X, which is often more useful because it is expressed in the same units as the data. The sample variance is used as an estimate of the true process variance, which is denoted by σ_X^2. Similarly, the sample standard deviation is an estimate of the process or population standard deviation denoted by σ_X.

Another important measure of the variability is the range, denoted by R, which is the difference between the largest value and the smallest value of the

TABLE 3.1   Light-Fixture Weights (Grams)

Sample	X_1	X_2	X_3	X_4	X_5	Sample Mean, $\overline{X}$	Sample Range, R
1	167	169	187	160	180	172.4	27
2	186	178	172	153	170	171.8	33
3	155	143	171	170	152	158.2	28
4	195	180	168	167	169	175.8	28
5	168	194	169	183	170	176.8	26
6	165	157	148	161	162	158.6	17
7	167	159	174	173	162	167.0	15
8	171	163	181	165	182	172.4	19
9	145	161	171	164	166	161.4	26
10	156	174	170	177	182	171.8	26
11	161	156	171	187	161	167.2	31
12	159	148	157	162	170	159.2	22
13	167	179	166	197	186	179.0	31
14	179	167	166	181	180	173.6	15
15	157	167	166	162	159	162.2	10
16	178	176	164	165	153	167.2	25
17	169	172	180	176	191	177.6	22
18	167	172	173	163	165	168.0	10
19	159	150	172	157	163	160.2	22
20	165	168	161	169	162	165.0	8
21	182	165	193	172	167	175.8	28
22	164	181	165	166	162	167.6	19
23	154	153	162	171	161	160.2	18
24	170	143	168	170	166	163.4	27
25	181	196	169	166	165	175.4	31
26	167	165	168	166	187	170.6	22
27	159	168	167	163	174	166.2	15
28	170	186	167	161	166	170.0	25
29	159	156	186	158	172	166.2	30
30	172	176	177	161	171	171.4	16
31	186	165	169	193	170	176.6	28
32	159	172	162	169	176	167.6	17
33	192	164	167	176	163	172.4	29
34	168	167	166	189	159	169.8	30
35	173	164	155	174	176	168.4	21
36	160	180	181	169	181	173.2	21
37	167	164	170	147	162	162.0	23
38	179	177	174	175	173	175.6	6
39	170	175	193	169	171	175.6	24
40	170	166	167	162	164	165.8	8
41	176	156	157	157	166	162.4	20
42	153	180	170	167	163	166.6	27
43	178	187	164	163	167	171.8	24
44	166	179	181	178	170	173.8	15
						$\overline{\overline{x}} = 169.02$	$\overline{R} = 21.93$

sample. Given the sample measurements from sample 1 in Table 3.1, 167, 169, 187, 160, 180, the range of this sample is

$$R_1 = 187 - 160 = 27.$$

In general,

$$R = X_{\text{largest}} - X_{\text{smallest}}, \tag{3.3}$$

where X_{largest} and X_{smallest} refer to the largest and smallest values in the sample, respectively.

The range R may be used to obtain an estimate of the population standard deviation of the individuals X, σ_X. Under certain assumptions, an exact mathematical relationship between the true mean range and σ_X has been established, namely,

$$\sigma_X = \frac{E(R)}{d_2}, \tag{3.4}$$

where d_2 is a constant that is determined according to the sample size n and $E(R)$, read "the expected value for the range," is a symbolic representation for the true mean range. In practice, an average of many sample ranges denoted by $\overline{R}$ is used to ensure a good estimate of the population standard deviation.

The average range, $\overline{R}$ (read as "R bar"), of the 44 sample ranges of Table 3.1 is 21.93 grams. The value of d_2 for a sample size of 5 is 2.326 from Table A.2 in the Appendix. An estimate of the standard deviation of individual light-fixture weights of the molding process can be obtained using Eq. (3.4) as follows:

$$\hat{\sigma}_X = \frac{21.93}{2.326} = 9.43 \text{ grams.}$$

As a comparison, the sample standard deviation s_X calculated from the 220 individual weights is 10.20 grams.

Both the range and standard deviation are measures of the amount of variation, or the degree of scatter of data. The efficiency of using the range as a variability measure decreases rapidly as the sample size gets very large. But due to its simplicity and the fact that usually small sample sizes are used in quality control applications, the sample range, R, is commonly used.

When the variabilities of two or more sets of data are compared, particularly if their averages are likely to be different, the use of pairs of the measures, such as $\overline{X}$ and S_X, for comparative analysis of the data sets may not be meaningful. For instance, if the variability of the light-fixture weights is compared with that of the instrument panels, another molded part, the standard deviation of the instrument panel weights can be several times greater than that of the light fixtures and still be considered tolerable because the instrument panels are many times heavier. A dimensionless measure of the variability, called the *coefficient of variation*, may be used as defined by

$$CV = \frac{S_X}{\overline{X}}. \tag{3.5}$$

When the CVs in weights of the light fixtures and the instrument panels are compared, they provide a relative measure of the variability of the same injection-molding process in producing two different products. The coefficient of variation can also be used as a measure of quality or productivity of a process over time as either one or both of the mean and the variance of a process are subject to change.

The measures of central tendency and variability may also be combined differently to indicate the overall performance of a process. One such measure, the *signal-to-noise ratio*, could be expressed by

$$S/N = \frac{\overline{X}}{s_X}, \tag{3.6}$$

which is the reciprocal of the CV. As the name suggests and as pointed out in Chapter 1, it is desirable to seek improvement of a process to produce larger and larger values of S/N.

Dot Diagram

To appreciate the manner in which the data are arising and to gain some insights as to the nature of the process, it will be meaningful to display the data pictorially. The simplest form of graphical representation of data is the dot diagram. Each piece of data is represented as a dot on a graph according to its value. When all data are plotted, the diagram, or the graph, presents an overall distribution of the data, such as the one shown in Fig. 3.2. When a sufficient amount of data is available, say, 25 or more measurements, such a simple graphical representation may be quite insightful. However, such a graph becomes tedious to construct with very large sets of data such as Table 3.1 and somewhat difficult to interpret if a data set contains many different values.

Frequency Histogram

With large data sets of, say, 150 data points or more, we will group the data into cells (or intervals) and make note of the frequency of observations falling within each cell. When we do this grouping, all the observations within a cell are considered to have the same value, which is the midpoint value of the cell. The following are guidelines for choosing cell boundaries and the number of cells:

1. Over the full range of the data, somewhere between 10 and 20 cells or intervals are chosen, depending on the size of the set.
2. Equal class intervals, or cell widths, are used in general.
3. Although the measured scale may actually be continuous, measurement or reading limitations allow only a finite number of possible readings to be recorded as data. Cell boundaries are chosen to fall midway between two possible readings.

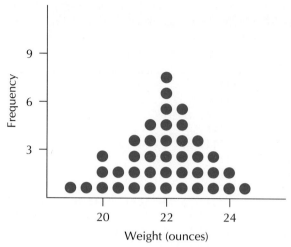

Figure 3.2 Dot Diagram of Part Weights

For our example of the light-fixture data, the observed weights range from 143 to 197 grams. All observations are recorded as integers because the scale used for the measurement reads only to the nearest gram. The data are grouped into 19 cells with a common cell width of 3 grams. The boundaries of the cells will be at xxx.5.

The first step toward constructing a frequency histogram, given that the appropriate class intervals have been chosen, is to tally the data. Table 3.2 is a tally representation for the 220 light-fixture weights, which are grouped into 19 cells. The numerical frequencies f_i for each cell are also listed on the right-hand side in the tally. To convert the tally sheet to a histogram, we allow the number of observations within each cell, the cell frequency, to represent the height of the frequency bars on the ordinate and the difference between adjacent cell boundaries to represent the width of the frequency bars on the abscissa. This is shown in Fig. 3.3.

From the frequency histogram we note three things of importance about the way in which the data arise:

1. The data tend to cluster about a certain value (i.e., more weights arise toward the center of the data).
2. The data are spread out over a broad range (i.e., a wide spectrum of part weights are represented).
3. The histogram has a characteristic shape (i.e., the frequencies fall off rapidly as we move away from the center of the data). Further, the frequencies are approximately symmetric about the center of the data.

The succinct description of these three characteristics of the data—(1) central tendency, (2) dispersion or variability, and (3) distributional shape of frequencies—is central to the successful use of such data as a decision aid.

TABLE 3.2	Tally Sheet for the Data of Table 3.1		
Cell Boundaries	Cell Midpoints	Frequency per Cell	f_i
141.5–144.5	143	11	2
144.5–147.5	146	11	2
147.5–150.5	149	111	3
150.5–153.5	152	11111	5
.	155	11111 11	7
.	158	11111 11111 1111	14
.	161	11111 11111 11111 11111 1	21
.	164	11111 11111 11111 11111 1111	24
.	167	11111 11111 11111 11111 11111 11111 11111 111	38
.	170	11111 11111 11111 11111 11111 11111 1	31
.	173	11111 11111 11111 11	17
.	176	11111 11111 11	12
.	179	11111 11111 1111	14
.	182	11111 11111 1	11
.	185	111	3
.	188	11111 11	7
.	191	11	2
.	194	1111	4
195.5–198.5	197	111	3
			Sum of f_i = 220

To characterize a set of data fully, particularly for probability evaluations and statistical analysis, the shape of the population distribution must be established. Usually, a certain probability distribution is assumed to describe the population either based on theoretical grounds or from the appearance of graphical displays of the sample data. To check the validity of the assumption, a number of methods, including graphical representations and statistical analyses, are often employed. We will discuss some of these methods.

Uniqueness of the Population or the Process "In Control"

Given a set of data, one can always calculate from it a number of sample statistics, such as the sample mean and sample range, or graphically represent it in plots such as the dot diagram and histogram. But these characterizations are meaningless and may provide very misleading information about the product or the process unless all the data are from one unique population or are observations of a process that is subject to a constant system of variability causes.

For a data set consisting of samples collected over time, an effective method to check whether or not every observed sample is from the same unique process is with Shewhart-type control charts. Beginning with Chapter 4, we devote a

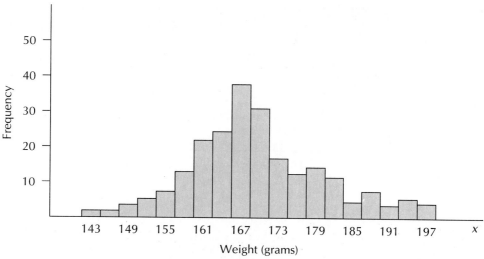

Figure 3.3 Histogram for Individual Light-Fixture Weights

large portion of this book to the use of such control charts. Control charts are graphical displays of statistics plotted according to the order of their observation. If no nonrandom patterns or extreme points are shown on a control chart, then the sample data are considered to have been observed from a process that was subjected to one common cause system during the period of observation. Such a process is said to be in statistical control. The control charts (Fig. 3.4) of the averages and ranges of the data on light-fixture weights (Table 3.1) seem to show that both the average part weight and the weight variability are in a state of statistical control.

3.4

Some Important Concepts of Probability and Probability Distributions

Concepts of Probability

A random experiment is a process that yields any one of several possible outcomes in a trial. The actual occurrence of a particular outcome on any given trial cannot be predicted with certainty. It can only be prescribed in terms of probability. Repeated trials of a random experiment under identical conditions produce a distribution of experimental results. Statistical methods help to analyze and interpret the results of a random experiment. Common examples of random experiments include receipt of invoices in an accounts payable department, sampling inspection of product quality, and number of incoming telephone calls at a switchboard. In each case, the actual outcomes are driven by chance causes.

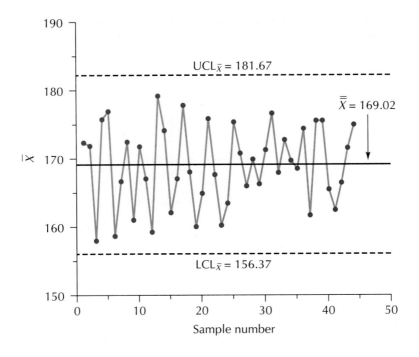

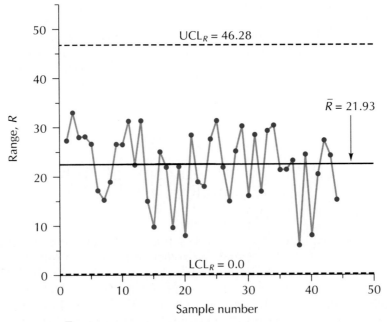

Figure 3.4 $\bar{X}$ and R Charts for the Light-Fixture Data

Suppose that a random experiment has a total of n possible different outcomes, A_i, $i = 1, 2, \ldots, n$. If m trials of the experiment resulted in the outcomes $A_1, A_2, \ldots, A_n$, which occurred $m_1, m_2, \ldots, m_n$ times, respectively, one can estimate the probability of outcome A_i by

$$p_i = \frac{m_i}{m}, \qquad i = 1, 2, \ldots, n.$$

Since $m_i \leq m$,

$$0 \leq p_i \leq 1, \qquad i = 1, 2, \ldots, n \tag{3.7}$$

and $m_1 + m_2 + \cdots + m_n = m$, we have

$$\sum_{i=1}^{n} p_i = 1. \tag{3.8}$$

Equations (3.7) and (3.8) are two important laws of probability. Simply stated, these two laws mean that a probability can only assume values between 0 and 1 and the total probability, or the sum of the probabilities of all possible outcomes, equals exactly 1.

Example. When a fair die is rolled, there are six possible outcomes with equal probability of occurrence:

$$p_1 = p_2 = p_3 = p_4 = p_5 = p_6 = \frac{1}{6}.$$

This example illustrates the notion of *equally likely outcomes*. Not all random experiments produce equally likely outcomes.

Example. A box contains 10 bolts, 20 nuts, and 50 screws. The probability of drawing a bolt from the box at random is $\frac{10}{80}$ and the probability of drawing a screw is $\frac{50}{80}$. Thus these two outcomes are not equally likely to occur in a drawing.

This example shows that the probability of different outcomes may not be the same. It also serves to illustrate the classical approach in assigning probabilities to experimental outcomes.

In the examples above the probability of each possible outcome can be determined a priori, or before experimentation. In most actual situations, one must estimate probabilities of various outcomes by the *frequency approach*, or a posteriori. In these situations the probabilities can be estimated only after sufficient observations of the outcomes have been made.

As part of a summer promotional effort, a rental car company has developed a fleet of 50 cars with the following composition in terms of color and model.

5	White, full-size	[W, F]
10	Red, full-size	[R, F]
10	Blue, full-size	[B, F]
15	Red, luxury	[R, L]
10	Blue, luxury	[B, L].

Patrons pay a special rate and draw keys from a bowl to determine exactly what type of car they will get. For the sake of illustration here we will assume that keys drawn are always from the full population of 50 cars.

(a) If a key taken at random from the bowl is for a car that is red in color, what is the probability that it will be a luxury car?

The required probability is called a *conditional probability* because a condition—color—has been specified, known, or given.

$$P(\text{luxury, given red}) = P(L \mid R)$$
$$= \frac{\text{no. red luxury cars}}{\text{no. red cars}}$$
$$= \frac{15}{25}$$
$$= \frac{3}{5}.$$

(b) If a key is taken at random from the bowl, what is the probability that the car is red and that it is a luxury car?

The required probability is called a *joint probability* because the outcome is a compound outcome that possesses two or more characteristics simultaneously. This joint probability may be obtained by

$$P(R \text{ and } L) = P(R, L)$$
$$= \frac{(\text{no. red luxury cars})}{(\text{total no. of cars})}$$
$$= \frac{15}{50}$$
$$= \frac{3}{10}.$$

This joint probability of two characteristics can also be obtained by

$$P(R, L) = P(L \mid R)\, P(R) \tag{3.9}$$
$$= \left(\frac{3}{5}\right)\left(\frac{25}{50}\right)$$
$$= \frac{3}{10}.$$

Equation (3.9) is the *multiplication law of probability.* The quantity $P(R)$ is known as the marginal probability of the outcome that a key taken at random from the bowl will be red. If we were going to use Eq. (3.9) to find the probability of drawing a car that is blue and luxury, we would get

$$P(B, L) = P(B \mid L) \, P(L) = \left(\frac{10}{25}\right)\left(\frac{25}{50}\right) = \frac{1}{5}.$$

In this case, since $P(B) = \frac{20}{50}$, and $P(B \mid L) = \frac{10}{25}$; the given condition "luxury model" provides no additional information to the estimation of the probability of getting a blue car, and the events B and L are said to be statistically (or probability-wise) independent to each other. For independent events, the multiplication law of probability becomes

$$P(B, L) = P(B) \, P(L), \qquad \text{where } B \text{ and } L \text{ are independent.} \qquad (3.10)$$

In words, the joint probability of independent events is the product of the marginal probabilities of the independent events.

(c) If a key is drawn at random from the bowl, what is the probability that it is either for a red car or for a luxury model?

   The required probability is given by

   P(a car is either red, or luxury, or both red and luxury)

or

$$P(R + L) = P(R) + P(L) - P(R, L) \qquad (3.11)$$
$$= \frac{25}{50} + \frac{25}{50} - \frac{15}{50}$$
$$= \frac{7}{10}.$$

Equation (3.11) is the *addition law of probability.* To use Eq. (3.11) to calculate the probability of getting a car that is either white or luxury would be

$$P(W + L) = P(W) + P(L) - P(W, L).$$

Since there are no white cars of the luxury model, the joint probability $P(W, L) = 0$, we have

$$P(W + L) = P(W) + P(L) \qquad (3.12)$$
$$= \left(\frac{5}{50}\right) + \left(\frac{25}{50}\right)$$
$$= \frac{3}{5}.$$

Equation (3.12) is a special case of the addition law of probability. The fact that the two events, W and L, do not coexist simplifies the calculation of the probability of one of several events that are mutually exclusive.

Random Variables

If a real value is assigned to each of the possible outcomes of an experiment, the set of all assigned values is called a *random variable,* which is different from an ordinary variable (deterministic) in that the values of a random variable can only be determined probabilistically.

Example.   The possible outcomes of tossing two balanced coins are

$$(H, H)   (H, T)   (T, H)   (T, T).$$

If one dollar is paid out for each head that turns up in the tossing, we have

X (dollars)	0	1	2
Outcomes	(T, T)	(T, H) or (H, T)	(H, H)

where X is the random variable, which assumes values 0, 1, and 2 with probabilities 0.25, 0.50, and 0.25, respectively.

Random variables that take on only integer values, or isolated and distinct values, are called *discrete random variables.* Many random variables, however, can have an infinite number of values within a small finite interval, such as time, temperature, and many other dimensional measurements. These are called *continuous random variables.*

Probability Function and Cumulative Distribution Function

There is a probability assigned to each possible value of a discrete random variable. The set of all the probabilities associated with the values of a discrete random variable X is called the *probability function* of the random variable, denoted by $f(x)$. For example, in the problem of tossing two coins, the probability function of X (in dollars) is

x	0	1	2
$f(x)$	0.25	0.50	0.25

The *cumulative distribution function* (CDF) of a random variable X is defined by

$$F(x) = P(X \le x) = P(X \text{ has values up to and including } x)$$
$$= \sum_{t \le x} f(t).$$

For example, the probability of paying out up to 1 dollar in a tossing of two coins is

$$F(1) = P(X \le 1) = \sum_{t \le 1} f(t) = f(t = 0) + f(t = 1)$$

$$= 0.25 + 0.50 = 0.75.$$

Probability Density Function

If a random variable X is continuous, it is meaningful only to consider the probability of X assuming values within an interval. This requires a different function to describe the probability behavior. A function called the *probability density function* (pdf), $f(x)$, is used for a continuous random variable. A pdf must satisfy the following:

$$f(x) \geq 0, \quad \text{all } x, \quad \text{and} \quad \int f(x)\, dx = 1,$$ (3.13)

over the interval for X. The corresponding cumulative distribution function of a continuous random variable with pdf $f(x)$ is given by

$$F(x) = P(X \leq x) = \int_{-\infty}^{x} f(t)\, dt.$$ (3.14)

Example. A random variable X has a pdf $f(x) = 0.5x$, $0 \leq X \leq 2$. The probability of X to be no more than 1.5 is

$$F(1.5) = \int_{0}^{1.5} 0.5t\, dt = 0.25t^2 \Big|_{0}^{1.5} = 0.5625.$$

Similarly, $F(2) = 1$ and $F(1) = 0.25$. The probability of X having values between 1 and 1.5 is given by

$$P(1 \leq X \leq 1.5) = F(1.5) - F(1) = 0.5625 - 0.2500 = 0.3125,$$

which is shown as the shaded area in the sketch of $f(x)$ in Fig. 3.5. This probability is also noted on the CDF in Fig. 3.5.

Mathematical Expectation

The concept of *mathematical expectation* or expected value of a random variable is important in probability and statistics. It is noted that any function of a

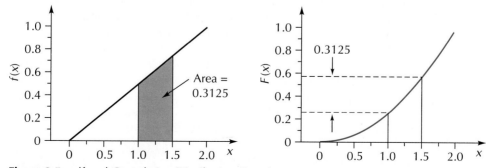

Figure 3.5 pdf and Cumulative Distribution Function

random variable or a function of several random variables is also a random variable. The expectation of a function $g(x)$ of a random variable X is defined as

$$E(g(X)) = \sum g(x)\, f(x), \qquad \text{for discrete } X \tag{3.15}$$

$$= \int g(x)\, f(x)\, dx, \qquad \text{for continuous } X. \tag{3.16}$$

Some important expectations:

1. $g(x) = x$: the expectation of X, $E(X)$, is the true arithmetic mean value of X, usually denoted by μ_X.
2. $g(x) = (x - \mu_X)^2$: the expectation of $(X - \mu_X)^2$ is the true variance of X, which is denoted by σ_X^2.

Example. Referring to the example of the coin-tossing game, the mean and the variance of the random variable X are

$$\mu_X = E(X) = (0)(0.25) + (1)(0.5) + (2)(0.25) = 1$$

$$\sigma_X^2 = \text{Var}(X) = (0 - 1)^2(0.25) + (1 - 1)^2(0.5) + (2 - 1)^2(0.25) = 0.25.$$

Some useful theorems of mathematical expectation include:

1. $E(cX) = cE(X)$, if c is a constant.
2. $E(X \pm Y) = E(X) \pm E(Y)$, where X and Y are two random variables.
3. $\text{Var}(cX) = c^2\, \text{Var}(X)$, if c is a constant.
4. $\text{Var}(X \pm Y) = \text{Var}(X) + \text{Var}(Y)$, if X and Y are independent.

3.5

Normal Distribution

About 200 years ago it was observed that errors of measurement seemed to follow a definite pattern with respect to their relative frequency of occurrence. Repeated measurements of the length of certain objects seemed to arise in a frequency sense according to a bell-shaped distributional curve, symmetric about the mean, which fell off quite rapidly beyond a distance of about one standard deviation from the mean.

This distribution, known as the *normal probability distribution* or *normal distribution*, does an excellent job of approximating the relative frequencies of many natural and man-made phenomena (e.g., heights or weights of people, dimensions of machined parts, strengths of steel specimens, etc.). Mathematically, the normal distribution is defined by the equation

$$f(x) = \frac{1}{\sigma_X \sqrt{2\pi}}\, e^{\,(-1/2)[(x - \mu)/\sigma_X]^2}, \tag{3.17}$$

where the mean μ and the standard deviation σ_X are the two parameters that define the probability density function $f(x)$ as shown above.

Determining Probabilities Using the Normal Curve

The function $f(x)$ has been scaled so that the total area under the curve over the full range of X ($-\infty$ to $+\infty$) equals 1.0. Then the proportion of the whole or the probability that an observation will arise over a given range of values is determined by finding the area under the curve over that range, that is, probability of occurrence between x_1 and x_2 equals the area under curve between x_1 and x_2, as shown in Fig. 3.6.

For the example data, suppose it is desired to find the chances of producing a light fixture that has a weight of 150 grams or less. One method to assess this chance or probability based on the data at hand would be to count the number of fixtures that have weights of 150 grams or less, from among the 220 fixtures measured. The corresponding percentage could then be used as an estimate of the probability. Since there are 7 such fixtures of the 220 that were sampled, one would estimate this probability to be 7/220 = 0.032 or approximately 3%.

A more general way to answer the question posed above would be to use the data and the associated frequency histogram of Fig. 3.3 to develop a mathematical model for the frequency distribution of the phenomenon being considered, light-fixture weights. For example, a normal distribution could be proposed and verified as appropriate using analytical or graphical tests for goodness of fit. We will examine such tests later in the chapter. The area under the normal curve for weights less than or equal to 150 grams will be the estimate of the desired probability. This normal distribution will be completely specified given its mean and standard deviation as seen in Eq. (3.17). The mean and standard deviation may be estimated by the sample mean ($\bar{\bar{x}}$ = 169.02 grams) and standard deviation estimate ($\bar{R}/d_2$ = 9.43 grams). This situation is depicted

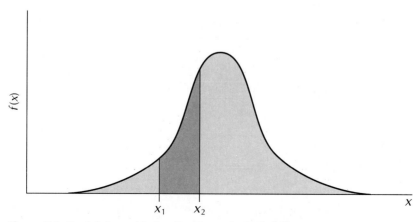

Figure 3.6 Partial Areas Under the Curve As Probabilities

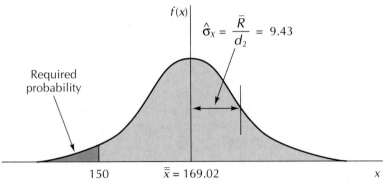

Figure 3.7 Probability of a Fixture Weighing Less Than 150 Grams

graphically in Fig. 3.7. The shaded area under the normal curve shown in Fig. 3.7 is an estimate of the probability of interest. This estimate is a valid prediction of the future chances of obtaining a light fixture weight less than or equal to 150 grams provided the process remains in statistical control at the same mean and standard deviation.

Standard Normal Distribution

Since the mean and standard deviation of the normal distribution can take on many different values from situation to situation, it is convenient to define and work with a standardized normal distribution. Such a standard normal distribution for the random variable Z is defined to have a mean $\mu_Z = 0$ and a standard deviation $\sigma_Z = 1$. Partial areas under this standard normal curve have been calculated and put in a table. It remains for us to establish the relationship between the particular normal distribution we are working with and the standard normal distribution so that we may use it to determine required areas or probabilities. Figure 3.8 illustrates this relationship.

To move from the $f(x)$ normal distribution to the $f(z)$ normal distribution, we use the following transforming equation:

$$Z = \frac{X - \mu_X}{\sigma_X}, \qquad (3.18)$$

so that if x_1 and x_2 are 1 and 2 standard deviations to the right of μ_X, respectively, then

$$z_1 = \frac{x_1 - \mu_X}{\sigma_X} = 1.0$$

$$z_2 = \frac{x_2 - \mu_X}{\sigma_X} = 2.0.$$

In the following section we illustrate use of the normal probability table to obtain probability estimates.

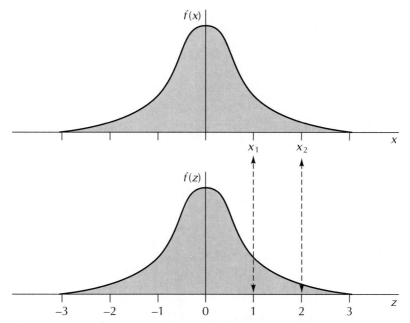

Figure 3.8 Transformation to the Standard Normal Distribution

Using the Standard Normal Table

The diameters of shafts turned on a lathe are considered to have a normal distribution with mean $\mu_X = 1.00$ inch and standard deviation $\sigma_X = 0.01$ inch. Suppose you wish to know the probability that a given shaft will have a diameter between 0.985 and 1.005 inches. We note that the probability $P(0.985 \le X \le 1.005)$ equals the area under the curve from 0.985 to 1.005. Graphically, this is depicted in Fig. 3.9. We may define the equivalent area in the standard normal distribution by determining the number of standard deviations that the boundaries of the area (0.985 inch and 1.005 inch) are from the mean (1.000 inch):

$$z_1 = \frac{x_1 - \mu_X}{\sigma_X} \qquad\qquad z_2 = \frac{x_2 - \mu_X}{\sigma_X}$$

$$= \frac{0.985 - 1.000}{0.01} \qquad\qquad = \frac{1.005 - 1.000}{0.01}$$

$$= \frac{-0.015}{0.01} \qquad\qquad = \frac{+0.005}{0.01}$$

$$= -1.5 \qquad\qquad\qquad = +0.5.$$

Then

$$P(0.985 \le X \le 1.005) = P(-1.5 \le Z \le 0.5).$$

We can use the standard normal distribution table (Table A.1 in the Appendix) to find the required probability.

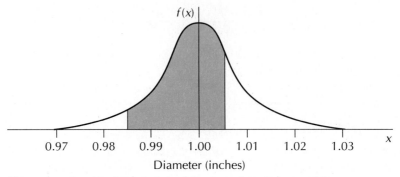

Figure 3.9 Graphical Depiction of Portion of the Process of Interest

1. The area under the curve from $z = -\infty$ to -1.5 is 0.0668.
2. The area under the curve from $z = -\infty$ to 0.5 is 0.6915.
3. Then the area under the curve from $z = -1.5$ to 0.5 is $0.6915 - 0.0668 = 0.6247$.

The probability that a turned shaft will take a value between 0.985 and 1.005 inches is 0.6247.

3.6

Sampling Distribution of $\overline{X}$

We have seen that the sample mean provides us with an estimate of the population mean μ_X. If we sampled a population several times, each time calculating an $\overline{X}$, we would find that the $\overline{X}$ values vary from one another, simply due to sampling variation. This variation is in part a function of the variation of the items in the population and in part a function of the number of items comprising the average. In other words, how good $\overline{X}$ is as an estimate of μ_X; that is, how close we expect $\overline{X}$ to be from μ_X depends on the amount of variation inherent in the process/population σ_X and the size of the samples, n. We need to understand these relationships more specifically; for instance, how the sample means behave with respect to their own mean, standard deviation, and more important, their own sampling distribution.

To characterize the distribution for the sample mean, $\overline{X}$, for a given sample size n, we need to know the mean and standard deviation of the sample means. These parameters can be estimated from the sample data using the general relationships between the mean and variance of $\overline{X}$ and, respectively, the mean and variance of X as follows:

$$\mu_{\overline{X}} = E(\overline{X}) = E\left(\sum_{i=1}^{n} \frac{X_i}{n}\right) = \sum_{i=1}^{n} \frac{E(X_i)}{n} = \frac{n\mu_X}{n} = \mu_X \tag{3.19}$$

$$\sigma_{\overline{X}}^2 = \text{Var}(\overline{X}) = \text{Var}\left(\sum_{i=1}^{n} \frac{X_i}{n}\right) = \sum_{i=1}^{n} \frac{\text{Var}(X_i)}{n^2} = \frac{n\sigma_X^2}{n^2} = \frac{\sigma_X^2}{n}. \tag{3.20}$$

The relationships above hold true for any random variables, provided that the sampling is done independently from one unique population (i.e., a constant system of common causes, a process in statistical control). Note that the mean of the sample means, denoted by $\mu_{\overline{X}}$, is equal to the mean of the individuals, μ_X, while the variance of the sample means $\sigma_{\overline{X}}^2$ is only $(1/n)$th of the variance of individuals, σ_X^2. When the sample size is very large, say, equal to the population size, the variance of the sample mean becomes zero, which should not be surprising because the sample mean is exactly the population mean. Figure 3.10 illustrates the effect of sample size on the variability of sample means. The illustrations show that as the sample size increases, the sample means, $\overline{X}$'s, tend to cluster more about μ_X with reduced spread. This implies that the precision of $\overline{X}$ as an estimate of μ_X increases as sample size increases.

Central Limit Theorem

The illustrations in Fig. 3.10 also show that the sampling distributions of sample means are normal even though the samples may be drawn from a nonnormal population such as a uniform distribution as shown in the figure. A mathematical justification for this relationship is the *Central Limit Theorem*. This theorem states that if $x_1, x_2, \ldots, x_n$ are outcomes of a sample of n independent observations of a random variable X with mean μ_X and variance σ_X^2, the sum of the x's, or equivalently the mean of the x's, will be distributed approximately as a normal distribution with mean μ_X and variance $\sigma_{\overline{X}}^2 = \sigma_X^2/n$. The approximation gets better as n gets larger. In general, the population from which the samples are drawn need not be normally distributed for the sample means to be normal. This constitutes a most remarkable and powerful result.

The following illustration demonstrates the effect of the Central Limit Theorem. Suppose that a particle moves along a line one unit at a time either forward or backward every microsecond with equal probability. Suppose further that each move of the particle is made independent of all previous moves. This is a uniform random variable that can assume two values, -1 and $+1$, each with probability $\frac{1}{2}$. If the particle starts at position 0, it can be at any position between $-n$ and n after n microseconds as depicted in Fig. 3.11. However, the probability that the particle lands at each one of the possible positions will not be the same. Let $X_1, X_2, \ldots, X_n$ be the n independent moves of the random variable X. Each independent move is defined by

$$X_i = \begin{cases} +1 & \text{with probability 0.5, if move forward} \\ -1 & \text{with probability 0.5, if move backward} \end{cases}$$

where $i = 1, 2, \ldots, n$. Then the final position of the particle after n moves is a random variable, the value Y which is given by

$$Y = X_1 + X_2 + \cdots + X_n, \qquad -n \le Y \le n.$$

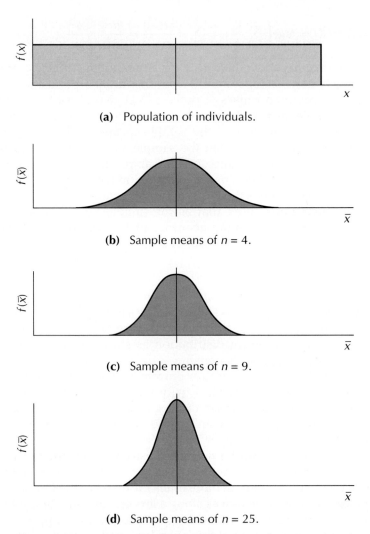

(a)   Population of individuals.

(b)   Sample means of $n = 4$.

(c)   Sample means of $n = 9$.

(d)   Sample means of $n = 25$.

Figure 3.10 Sampling Distributions of Averages from Samples of Various Sizes n

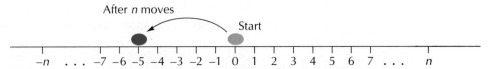

Figure 3.11 Movements of a Uniformly Distributed Random Variable

Y is described by a probability distribution that will approach a normal distribution with a modestly large n. To find the exact distribution of Y, consider the case of $n = 3$:

Possible Outcomes	Move 1	Move 2	Move 3	x_1	x_2	x_3	Sum y	Probability
1	F	F	F	$+1$	$+1$	$+1$	$+3$	1/8
2	F	F	B	$+1$	$+1$	-1	$+1$	1/8
3	F	B	F	$+1$	-1	$+1$	$+1$	1/8
4	B	F	F	-1	$+1$	$+1$	$+1$	1/8
5	F	B	B	$+1$	-1	-1	-1	1/8
6	B	F	B	-1	$+1$	-1	-1	1/8
7	B	B	F	-1	-1	$+1$	-1	1/8
8	B	B	B	-1	-1	-1	-3	1/8

In the table above, F and B denote, respectively, forward and backward moves. The final position of the particle, Y, which is the sum of $n = 3$ moves, can be only one of four possible values, with probabilities given as

y	$P(Y) = f(y)$
-3	1/8
-1	3/8
1	3/8
3	1/8

Similarly, a probability distribution can be constructed for any n. Figure 3.12 shows the plots of these distributions for $n = 3$, $n = 9$, and $n = 12$. In each case the resultant sum of just a few moves tends to behave like a normal distribution, although each individual move has a uniform distribution. If the sum of independent variables (of any distribution) approaches a normal distribution, so do the sample means because they are the sums divided by a constant.

Example. Six batteries taken at random from a large lot were subjected to life tests with the following results:

    19, 22, 18, 16, 25, and 20 hours.

Find the probability that the sample mean of $\overline{x} = 20$ hours, as observed here, or one less than 20 hours could arise if the true mean of the lot of batteries is 21.5 hours, assuming a known variance of $\sigma_X^2 = 10$ hours^2 for the individual batteries in the lot.

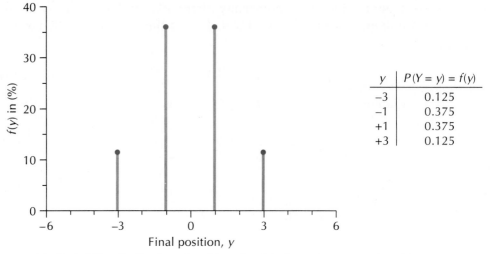

y	$P(Y = y) = f(y)$
−3	0.125
−1	0.375
+1	0.375
+3	0.125

(a) Probability distribution of position after 3 independent moves ($n = 3$).

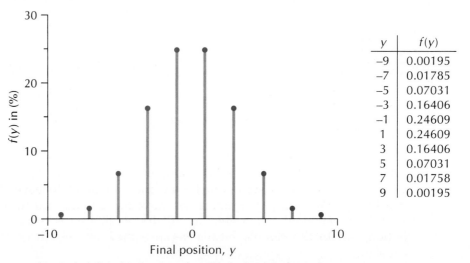

y	$f(y)$
−9	0.00195
−7	0.01785
−5	0.07031
−3	0.16406
−1	0.24609
1	0.24609
3	0.16406
5	0.07031
7	0.01758
9	0.00195

(b) Probability distribution of position after 9 independent moves ($n = 9$).

Figure 3.12 Central Limit Effects on Sample Means

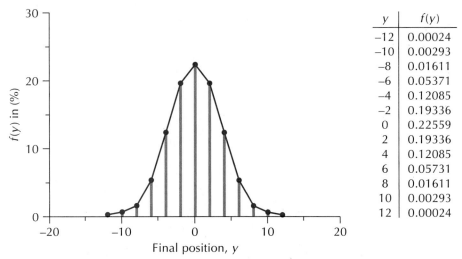

y	f(y)
−12	0.00024
−10	0.00293
−8	0.01611
−6	0.05371
−4	0.12085
−2	0.19336
0	0.22559
2	0.19336
4	0.12085
6	0.05731
8	0.01611
10	0.00293
12	0.00024

(c) Probability distribution of position after 12 independent moves ($n = 12$).

Figure 3.12 *(continued)*

Solution. Given: $n = 6$, $\mu_X = 21.5$, $\sigma_{\overline{X}}^2 = \dfrac{10}{6} = (1.291)^2$.

Using the normal distribution for the sample mean lives, we have

$$Z = \frac{\overline{X} - \mu_X}{\sigma_{\overline{X}}}$$

$$P(\overline{X} \le 20) = P\left(Z \le \frac{20 - 21.5}{1.291} \right)$$
$$= P(Z \le -1.162)$$
$$= 0.123.$$

Figure 3.13 shows the relationships between the distribution of individual battery lives and the distribution of mean lives of samples of size $n = 6$, and between the distribution of $\overline{X}$'s and the standard normal distribution of Z.

3.7

Some Other Useful Graphical Representations of Data

Box-and-Whisker Plot

Another graphical method for displaying data distribution is the box-and-whisker plot. Some commonly used statistics for such a plot are sample quartiles. The first, second, and third quartiles, denoted by Q_1, Q_2, and Q_3, are, respectively, the values of those data points that are positioned in the data sequence

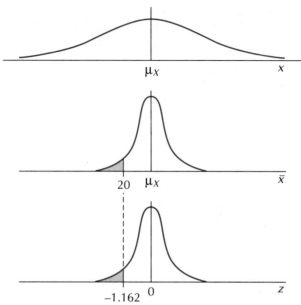

Figure 3.13 Distributions of X, $\bar{X}$, and Z

of ascending order such that 25%, 50%, and 75% of the data are equal to or smaller in magnitude. The second quartile is also known as the *median*, which is the value of the data point in the middle of the ordered sequence of the sample data. When the sample size n is odd, the median is the value of the $[(n + 1)/2]$th point in the ordered sequence. If n is even, the median is estimated by the average of the $(n/2)$th and $[(n/2) + 1]$th data points.

To construct a box-and-whisker plot for any given sample of size n, we record the smallest value, the largest value, and the three quartiles. Then a box is drawn at the first and third quartiles with the median placed inside the box. Finally, the two whiskers extend out from each side of the box with lengths determined according to the smallest and largest values. As an illustration of the construction of a box-and-whisker plot, we list the 44 sample means of Table 3.1 in ascending order in Table 3.3. From the table we have

$$n = 44$$

$$\bar{x}_{(1)} = 158.2, \text{ the smallest value}$$

$$\bar{x}_{(44)} = 179.0, \text{ the largest value}$$

$$Q_1 = x_{(11)} = 165.0, \text{ the first quartile}$$

$$Q_2 = \frac{x_{(22)} + x_{(23)}}{2} = 169.1, \text{ the median}$$

$$Q_3 = x_{(33)} = 173.2, \text{ the third quartile.}$$

TABLE 3.3	Sample Means in Ascending Order, Light-Fixture Weights				
Order No.	$\bar{x}_i$	Order No.	$\bar{x}_i$	Order No.	$\bar{x}_i$
1	158.2	16	167.0	31	172.4
2	158.6	17	167.2	32	172.4
3	159.2	18	167.2	33	173.2
4	160.2	19	167.6	34	173.6
5	160.2	20	167.6	35	173.8
6	161.4	21	168.0	36	175.4
7	162.0	22	168.4	37	175.6
8	162.2	23	169.8	38	175.6
9	162.4	24	170.0	39	175.8
10	163.4	25	170.6	40	175.9
11	165.0	26	171.4	41	176.6
12	165.8	27	171.8	42	176.8
13	166.2	28	171.8	43	177.6
14	166.2	29	171.8	44	179.0
15	166.6	30	172.4		

A box-and-whisker plot of the sample means of the light-fixture data is shown in Fig. 3.14.

When a set of data is perfectly symmetric about the mean, the median will coincide with the mean and the two whiskers will have the same length. Therefore, a box-and-whisker plot presents the degree of symmetry of the data according to the relative positions of Q_1 and Q_3 to Q_2 and the relative lengths of the whiskers. The box-and-whisker plot of Fig. 3.14 shows some asymmetry in the sample means because the median is not in the center of the box and the left-hand whisker is longer than the one on the right-hand side. Overall, the sample means of the light-fixture weights are distributed with heavier clustering on the right but with more skewness to the left tail.

Box-and-whisker plots are often used to compare two or more sets of data. When several data sets are compared for their central tendencies, variabilities, and distribution shapes, all the corresponding boxes and whiskers are plotted, one for each set, to the same scale on one chart. Figure 3.15 shows two such plots of 220 individual light-fixture weights and the 44 sample mean weights

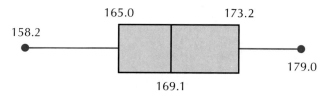

Figure 3.14 Box-and-Whisker Plot of Sample Means of Light-Fixture Weights

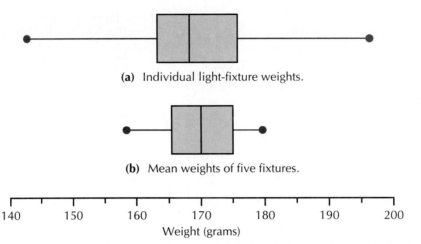

(a) Individual light-fixture weights.

(b) Mean weights of five fixtures.

140   150   160   170   180   190   200

Weight (grams)

Figure 3.15 Box-and-Whisker Plots of Individual and Mean Weights of Light Fixtures

for the data of Table 3.1. It is seen from the figure that the effects of data grouping ($n = 5$) appear to be

1. The variability of the data, or the spread, is drastically reduced.
2. The sample means are distributed more symmetrically about the mean as the median of the sample means is closer to the center of the box.

Box Plots of Percentiles

The method of box plots can be applied to display the sample distribution of data for checking its closeness to an assumed theoretical probability distribution. To do this, one could plot the box based on the percentiles, instead of the quartiles, of the data. The choice of the percentiles can be determined according to the distributional characteristics of the theoretical distribution in question. For instance, if we are interested in checking whether a normal distribution would approximate well the data of light-fixture weights, a box plot of percentiles could be drawn as shown in Fig. 3.16. To better depict the distribution of data for normal fitting, we have used five percentiles, P_{16}, P_{31}, P_{50}, P_{69}, and P_{84} (which corresponds to z values at -1, -0.5, 0, 0.5, and 1), together with the smallest and largest values in the data set. If a normal distribution is a reasonably good fit to the data, then according to the characteristics of a normal distribution illustrated in Fig. 3.17, all the following should hold:

1. P_{50} = mean of the data.
2. The four boxes should have the same width, or $a = b = c = d$, as in Fig. 3.17.
3. The two whiskers should have the same length and each about three times the width of a box.

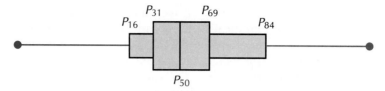

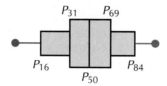

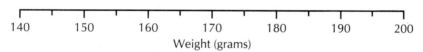

Figure 3.16 Box Plots of Individual and Mean Weights of Light Fixtures

In examining the two plots of Fig. 3.16, the one for the data of individual weights, Fig. 3.16(a), shows somewhat more departure from a normal distribution as the median, P_{50}, is different from the mean, although the other two conditions are not significantly different from their expectations. The other plot, Fig. 3.16(b), appears to be closer to a normal distribution except that the whiskers are too short. For actual data analysis, both of the box plots of Fig. 3.16 would serve as the basis for normal approximation to the data.

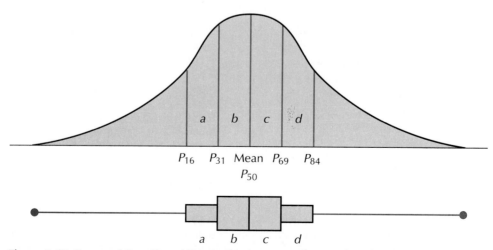

Figure 3.17 Expected Box Plot of Percentiles for Normally Distributed Data

Although the box plots, as well as the histograms discussed earlier, are quite effective in graphically displaying data distributions for large samples, they become less adequate when the sample size is small. Another common concern regarding the use of these plots is the question about the uniqueness of the population from which the sample is obtained. Generally, this means that the sampled process must be in a state of statistical control before any valid interpretation or inference can be made from these plots. If a set of data contains a few extreme points or outliers due to special causes or from other populations, we would like to have our graphical method be able to isolate them so that they are not improperly included for parameter estimation or distributional display. One such method is the use of probability plots, one of which is discussed in the following section.

Normal Probability Plots

Probability plots are a graphical means to judge whether or not a set of data may be reasonably characterized by a specified probability distribution. They may also be used to identify outlying data in a sample. *Normal probability plots*

TABLE 3.4   Calculation Table for the Sampled Data

Index, i	x_i	x_i Values Sorted in Ascending Order	p_i (%)	z_i	q_i
1	9.63	9.34	2.5	−1.96	−1.99
2	9.86	9.51	7.5	−1.44	−1.49
3	10.20	9.63	12.5	−1.15	−1.13
4	10.48	9.69	17.5	−0.94	−0.95
5	9.82	9.75	22.5	−0.76	−0.77
6	10.07	9.82	27.5	−0.60	−0.56
7	10.39	9.86	32.5	−0.46	−0.44
8	10.03	9.89	37.5	−0.32	−0.35
9	9.34	9.96	42.5	−0.19	−0.14
10	10.26	9.98	47.5	−0.06	−0.08
11	9.89	10.03	52.5	0.06	0.07
12	10.67	10.07	57.5	0.19	0.19
13	9.69	10.13	62.5	0.32	0.37
14	10.15	10.15	67.5	0.46	0.43
15	10.32	10.20	72.5	0.60	0.58
16	9.98	10.26	77.5	0.76	0.76
17	9.51	10.32	82.5	0.94	0.94
18	10.13	10.39	87.5	1.15	1.15
19	9.96	10.48	92.5	1.44	1.42
20	9.75	10.67	97.5	1.96	1.98

are a commonly used type of probability plot and are based on the normal probability distribution.

Consider an example in which a sample of $n = 20$ observations is drawn from an unknown distribution. These data are summarized in the second column of Table 3.4. The sample mean and sample standard deviation for this set of data are $\bar{x} = 10.0065$ and $s_X = 0.33427$, respectively.

These data may be graphically displayed using a dot diagram such as that illustrated in Fig. 3.18. As is apparent, for such a small set of data it is difficult to tell the shape of the underlying distribution from the dot diagram.

Another way to graphically display the sampled data is with a cumulative probability plot. To construct such a plot, the members of the sample are sorted in ascending order (column 3 in Table 3.4) and each member, x_i, of the sample is assigned a cumulative probability, p_i. Since it is desired to characterize the shape of the underlying distribution based on the sampled data, we must rely on each sample member to tell us something about the distribution. In this case, each sample member represents $\frac{1}{20}$th of the information available about the distribution. Therefore, we can use the smallest X value to characterize the lowest 5% of the distribution, the next smallest value to represent the next 5% of the distribution, and so on. Since the smallest X value represents the portion of the distribution with cumulative probabilities from 0% to 5%, it may be assigned a cumulative probability value at the midpoint of this interval, that is, $p_i = 2.5\%$. The next smallest X value represents the portion of the distribution with cumulative probabilities from 5% to 10%, so it is assigned a cumulative probability, p_i, of 7.5%, and so on. In general, the cumulative probability associated with a given x_i may be estimated by

$$p_i = \frac{100(i - 0.5)}{n},$$

where p_i is the estimated cumulative probability in percent. The estimated cumulative probabilities for the sample data are shown in column 4 of Table 3.4. The data, (x_i, p_i) pairs, may then be plotted to form the cumulative probability plot. Figure 3.19(a) shows the cumulative probability plot for the sampled data of this example and Fig. 3.19(b) displays the cumulative distribution func-

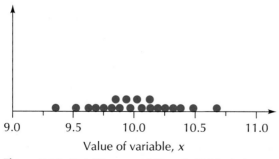

Figure 3.18 Dot Diagram of Data in Table 3.4

tion for a normal distribution having the same mean and standard deviation as the sampled data.

In examining Fig. 3.19(a) and (b), it is evident that the cumulative probability plots of the sampled data and the normal distribution both have "S" shapes. If the S shape associated with the sample is close to the S shape associated with the normal distribution, we may conclude that the sample was drawn from a distribution that is normally, or approximately normally, distributed. Unfortunately, it is not an easy matter to assess the "closeness" of two different S shapes. A normal probability plot solves this problem by using a special scale for the vertical (cumulative probability) axis so that the cumulative distribution function for a normal distribution becomes a line instead of being S shaped. One may visualize the construction of such normal probability paper by imagining a normal cumulative distribution function plotted on graph paper made of thin rubber. The rubber graph paper is stretched along the cumulative probability axis direction in such a way that the normal cumulative distribution function of Fig. 3.19(b) becomes a straight line. The nonlinear scaling of the "stretched" Cartesian scale now defines the scale of the ordinate for normal probability paper.

Commercially available normal probability paper has this specially scaled cumulative probability axis and simplifies the construction of normal probability plots. Figure 3.20 shows a normal probability plot of the sampled data. As is evident, the sampled data in Fig. 3.20 appear to be fairly well described by a straight line. Since the vertical scale for the normal probability plot has been selected so that a normal cumulative distribution function will appear as a line,

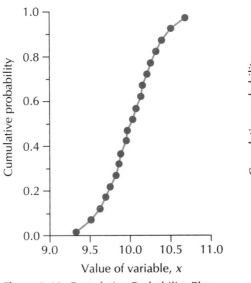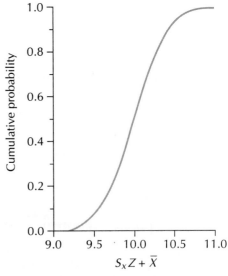

Figure 3.19 Cumulative Probability Plots

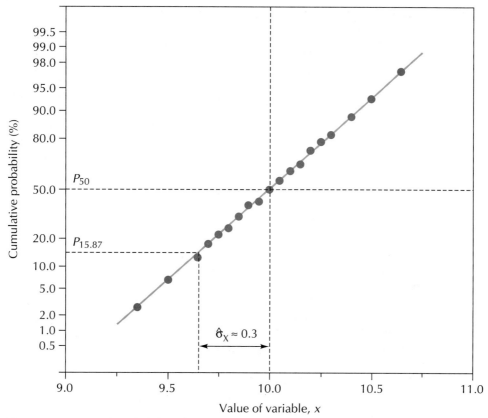

Figure 3.20 Normal Probability Plot for Data in Table 3.4

we may conclude that the distribution from which the sample was drawn is approximately normally distributed.

If all data points appear to be randomly distributed along a line drawn through the data on a normal probability plot, and the line passes on/near the intersection of the mean of X and 50% probability, the fit of the data to the normal distribution is considered good. Conversely, if the points appear to be S-shaped, then the suggestion is that the data are not normally distributed. Sometimes, a line describes a majority of the data but does not characterize a few extreme points. These off-line points are outliers that may have arisen from a different distribution. If the data, part or all, naturally cluster about a straight line that does not go through the mean at 50% probability, then the estimated mean value is probably in error.

We can obtain a graphical estimate, $\hat{\sigma}_X$, of the distribution's standard deviation by finding the difference, measured along the X-axis, between the X values associated with P_{50} and $P_{15.87}$. This estimated standard deviation is shown in Fig. 3.20.

Q-Q Plot

With the help of a normal probability table, the cumulative probabilities, p_i can be converted into their corresponding standardized normal values, $P(Z \leq z_i) = p_i$. These normal values are shown in the fifth column of Table 3.4.

Provided that the mean and variance of the variable X are known, the sampled data may be standardized using the transformation:

$$q_i = \frac{x_i - \mu_X}{\sigma_X}, \qquad \text{for } i = 1, n.$$

Since μ_X and σ_X are generally unknown, we may alternatively use the transformation:

$$q_i = \frac{x_i - \bar{x}}{s_X}.$$

The standardized data values, q_i, are shown in the sixth column of Table 3.4. A plot of the (q_i, z_i) pairs on regular/Cartesian graph paper, called a Q-Q plot, may also be used for judging the normality of a set of data. If the same scale is used for both the q_i and z_i axes, then the line that describes the points (q_i, z_i) should be $45°$ from the abscissa, provided that the sample is indeed from a normal distribution (and good estimates of the mean and standard deviation have been used to standardize the data). Although a Q-Q plot requires several additional calculations compared to a normal probability plot, for a Q-Q plot there is no need to subjectively place a line through the data points on the graph paper. When it is desired to check the normality of several samples of data, a Q-Q plot also makes the relative assessment of goodness-of-fit among the samples much easier. A Q-Q plot using the same data of Figure 3.20 is shown in Fig. 3.21, where it is noted that the same scale is used for both the abscissa and ordinate.

Normal probability and Q-Q plots may be constructed using the following steps.

1. Arrange the data, x_i's, in ascending order.
2. Calculate the estimated cumulative probability, p_i, for each member of the sample based on the sample size n.
3. a. For a normal probability plot, plot the points, (x_i, p_i), on normal probability paper. Attempt to fit a line, usually by eye, to the points on the paper so that it is close to all the data points and they are randomly distributed about the line.
   b. For a Q-Q plot determine $\bar{x}$ and s_X for the data and calculate the standardized normal values, z_i, corresponding to the cumulative probabilities, p_i. Calculate the standardized data values, q_i. Plot the points, (q_i, z_i), on regular graph paper. Put a $45°$ line on the plot.
4. Interpret the plot and draw conclusions.

Figure 3.22(a), (b), and (c) illustrates some typical cases of "poor-fit" evident from Q-Q plots resulting from estimation errors in the mean and variance.

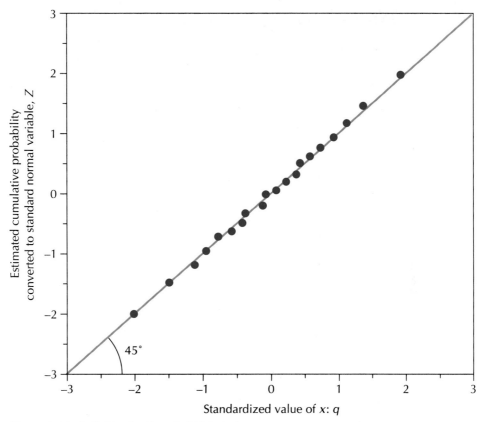

Figure 3.21 Q-Q Plot for Data in Table 3.4

(a) Mean is in error.   (b) Variance is underestimated.   (c) Variance is overestimated.

Figure 3.22 Poor Fits Due to Estimation Errors

| | | Raw Data in | Cumulative | z_i | |
| | Raw | Ascending | Probability, | Associated | Standardized |
Index, i	Data, x_i	Order	p_i (%)	with p_i	Data, q_i
1	172.6	158.2	1.1	−2.29	−1.87
2	171.8	158.6	3.4	−1.83	−1.80
3	158.2	159.2	5.7	−1.58	−1.70
4	175.8	160.2	8.0	−1.41	−1.52
5	176.8	160.2	10.2	−1.27	−1.52
6	158.6	161.4	12.5	−1.15	−1.32
7	167.0	162.0	14.8	−1.05	−1.21
8	172.4	162.2	17.0	−0.95	−1.18
9	161.4	162.4	19.3	−0.87	−1.14
10	171.8	163.4	21.6	−0.79	−0.97
11	167.2	165.0	23.9	−0.71	−0.69
12	159.2	165.8	26.1	−0.64	−0.56
13	179.0	166.2	28.4	−0.57	−0.49
14	174.6	166.2	30.7	−0.51	−0.49
15	162.2	166.6	33.0	−0.44	−0.42
16	167.2	167.0	35.2	−0.38	−0.35
17	177.6	167.2	37.5	−0.32	−0.31
18	168.0	167.2	39.8	−0.26	−0.31
19	160.2	167.6	42.0	−0.20	−0.25
20	165.0	167.6	44.3	−0.14	−0.25
21	175.8	168.0	46.6	−0.09	−0.18
22	167.6	168.4	48.9	−0.03	−0.11
23	160.2	169.8	51.1	0.03	0.13
24	163.4	170.0	53.4	0.09	0.17
25	175.4	170.6	55.7	0.14	0.27
26	170.6	171.4	58.0	0.20	0.41
27	166.2	171.8	60.2	0.26	0.48
28	170.0	171.8	62.5	0.32	0.48
29	166.2	171.8	64.8	0.38	0.48
30	171.4	172.4	67.0	0.44	0.58
31	176.6	172.4	69.3	0.51	0.58
32	167.6	172.6	71.6	0.57	0.62
33	172.4	174.2	73.9	0.64	0.89
34	169.8	174.6	76.1	0.71	0.96
35	168.4	174.8	78.4	0.79	1.00
36	174.2	175.4	80.7	0.87	1.10
37	162.0	175.6	83.0	0.95	1.14
38	175.6	175.6	85.2	1.05	1.14
39	175.6	175.8	87.5	1.15	1.17
40	165.8	175.8	89.8	1.27	1.17
41	162.4	176.6	92.0	1.41	1.31
42	166.6	176.8	94.3	1.58	1.34
43	171.8	177.6	96.6	1.83	1.48
44	174.8	179.0	98.9	2.29	1.72

TABLE 3.5   Calculation Table for the Light-Fixture Weights

Normal Plots of the Light-Fixture Data

Suppose that we would like to check whether a normal distribution describes the sample means of the light-fixture data discussed earlier. Table 3.5 summarizes all the calculations for the normal probability and Q-Q plots for the 44 sample means.

The columns in Table 3.5 contain the following:

Column (1):   The index, i.

Column (2):   The raw data, x_i. The sample mean and sample standard deviation of this data are: $\bar{x} = 169.022$ and $s_X = 5.790$. The sample size is $n = 44$.

Column (3):   The raw data sorted in ascending order.

Column (4):   The cumulative probabilities for a sample of size n, using the following equation: $p_i = 100\,(i - 0.5)/n$.

Example, $p_3 = 100\,(3 - 0.5)/44 = 5.7\%$.

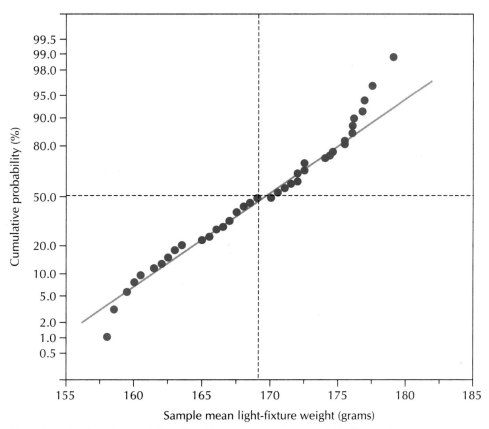

Figure 3.23 Normal Probability Plot of the Sample Means of All Light-Fixture Weights

Column (5):   The standard normal values corresponding to p_i. These are obtained by $p_i = P(Z \leq z_i)$.
Example, $z_5 = -1.27$ for $p_5 = 10.2\%$.

Column (6):   The standardized data values calculated from $q_i = (x_i - \bar{x})/s_X$.
Example, $q_{41} = (176.6 - 169.022)/(5.790) = 1.31$.

The plot of the 44 pairs of (x_i, p_i) data, displayed on normal probability paper, is shown in Fig. 3.23 along with a line drawn through the points. Since the majority of the points are close to the line (except for a few at either end of the line) and it cuts through the intersection of the mean and the 50% probability, it is concluded that the sample mean light-fixture weights, x_i values, approximately follow a normal distribution, as expected based on the Central Limit Theorem.

In general, a Q-Q plot requires more calculations than a normal probability plot, but the former may reveal more of the finer disagreements between a normal distribution and the actual distribution of the data. For instance, the

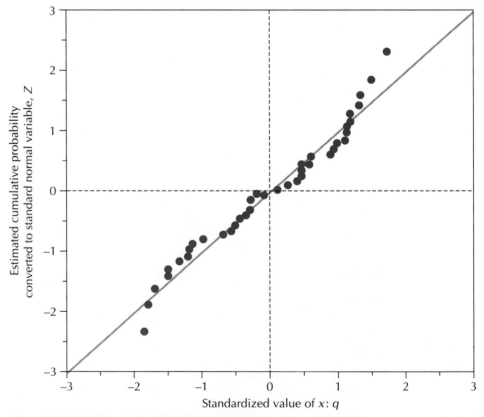

Figure 3.24 Q-Q Plot of the Sample Means of Light-Fixture Weights

Q-Q plot of Fig. 3.24 indicates a possible overestimation of the standard deviation of the raw data. Another interpretation of the Q-Q plot of Fig. 3.24 is that the sample distribution of the data has more spread than would be expected for a normal distribution. A closer examination of both Figs. 3.23 and 3.24 seems to indicate that the data are a bit more uniformly distributed between the mean and about 2 standard deviations than would be expected for a normal distribution.

Exercises

3.1. Your company has received numerous complaints about the life of some of the hand tools you are manufacturing. In investigating these complaints, it is decided to analyze the process used to harden steel wrenches. A few different techniques have been suggested for collecting the hardness numbers that will be used. They include:
   a. Measuring the first three wrenches produced each day and assuming that if they are within the specifications the others that follow will also be satisfactory.
   b. Taking a hardness reading on a wrench drawn from a population at four random times throughout each shift.
   c. Taking samples of four consecutively produced wrenches at approximately half-hour intervals throughout the shift.
   d. Measuring the hardness of each wrench produced.
   What are the benefits and potential problems associated with each of these methods? Which would you recommend?

3.2. The data in the accompanying table represent the weight in ounces of molded instrument display panels. The samples were collected consecutively at half-hour intervals during a two-shift period. Prepare a tally sheet, like that shown in Table 3.2, of the individual measurements and then prepare a frequency histogram of the data, clearly labeling the cell boundaries. Comment on the shape of the distribution.

Sample	X_1	X_2	X_3	X_4	X_5
1	14	15	13	14	13
2	20	18	14	17	8
3	14	17	14	11	14
4	15	16	11	18	14
5	9	17	18	13	12
6	19	15	14	15	16
7	16	13	14	13	17
8	14	17	9	16	15
9	14	14	12	13	13
10	15	13	17	14	16
11	18	18	16	15	11

(continued)

Sample	X_1	X_2	X_3	X_4	X_5
12	20	12	13	17	14
13	11	8	9	12	7
14	12	14	16	14	20
15	18	17	12	19	18
16	19	17	16	16	17
17	14	13	15	16	18
18	14	17	12	16	11
19	18	15	16	15	12
20	15	9	12	13	20

3.3. Given the data in the table, which represent the weld joint strength in ksi from a process known to be in statistical control, make a tally sheet and plot two frequency histograms as follows. For your first histogram use $k = 20$ cells, and for the second use $k = 10$ cells. Clearly show the cell boundaries for each histogram. Comment on the difference between the two histograms.

Sample	X_1	X_2	X_3	X_4
1	23.6	27.2	27.8	24.2
2	23.4	25.2	31.2	23.4
3	29.2	25.8	28.2	29.2
4	26.0	20.8	30.8	29.6
5	29.8	30.8	25.6	24.8
6	25.6	25.8	32.4	33.8
7	33.2	28.2	26.2	31.4
8	29.2	23.2	20.8	26.0
9	23.8	31.4	30.8	28.8
10	32.4	28.6	32.8	25.4
11	24.6	27.2	21.6	24.8
12	31.6	24.4	30.6	28.0
13	30.4	31.2	23.6	20.6
14	28.2	25.6	26.0	33.2
15	15.4	24.4	28.6	27.8
16	30.0	23.8	30.2	28.6
17	22.0	28.2	28.8	23.2
18	21.8	23.0	26.6	24.8
19	26.4	29.8	29.4	25.2
20	25.6	21.4	27.6	31.8
21	33.4	26.2	21.2	26.4
22	22.8	28.2	28.4	27.4
23	28.2	28.2	28.2	31.0
24	27.6	29.8	23.0	23.4
25	28.0	30.6	25.8	26.4

3.4. Given the data in the table, which represent the surface finish in microinches of a cast aluminum part, create a tally sheet of the individual measurements, clearly showing the cell boundaries, and plot a frequency histogram. Also, plot a dot diagram for the individual measurements. What might the shape of the frequency distributions be saying about the process from which the data came?

Sample	X_1	X_2	X_3	X_4	X_5
1	46	46	47	42	41
2	48	44	47	46	52
3	47	45	46	47	42
4	48	37	46	42	45
5	47	43	43	48	46
6	44	45	39	43	49
7	46	41	47	39	45
8	49	50	53	42	45
9	45	39	42	51	46
10	47	41	42	46	46
11	47	45	52	45	45
12	45	44	43	47	46
13	40	43	42	41	46
14	45	43	44	40	46
15	45	39	44	50	47
16	45	39	39	34	41
17	45	37	43	38	47
18	45	42	34	42	34
19	47	43	41	42	44
20	38	35	43	38	43
21	42	40	43	37	36
22	35	36	40	41	39
23	33	44	32	38	33
24	38	44	37	41	43
25	42	38	50	37	47

3.5. You are producing gear blanks that are to have a nominal diameter of 1.060 inches. Samples of size $n = 4$ have been collected once an hour, producing the data shown in the table. Make a tally sheet with clearly labeled cell boundaries, and then plot a frequency histogram and comment on its shape. What might the shape of the distribution be saying about the process from which the data came?

Sample	X_1	X_2	X_3	X_4
1	1.060	1.063	1.053	1.058
2	1.057	1.053	1.048	1.081
3	1.069	1.077	1.038	1.066

(continued)

Sample	X_1	X_2	X_3	X_4
4	1.060	1.050	1.080	1.062
5	1.049	1.048	1.071	1.083
6	1.055	1.064	1.041	1.068
7	1.070	1.080	1.065	1.055
8	1.047	1.040	1.054	1.068
9	1.066	1.025	1.035	1.072
10	1.053	1.064	1.065	1.053
11	1.047	1.050	1.044	1.044
12	1.051	1.084	1.052	1.061
13	1.084	1.058	1.096	1.050
14	1.068	1.035	1.077	1.065
15	1.065	1.074	1.061	1.050
16	1.077	1.064	1.045	1.057
17	1.072	1.062	1.075	1.036
18	1.042	1.094	1.065	1.052
19	1.077	1.052	1.065	1.047
20	1.055	1.036	1.071	1.057
21	1.050	1.066	1.063	1.072
22	1.073	1.059	1.043	1.068
23	1.049	1.048	1.080	1.077
24	1.052	1.049	1.062	1.053
25	1.063	1.074	1.037	1.055
26	1.044	1.083	1.059	1.068
27	1.062	1.046	1.062	1.053
28	1.072	1.057	1.055	1.059
29	1.054	1.076	1.070	1.046
30	1.046	1.065	1.036	1.070

3.6. For Exercises 3.2 and 3.3, prepare dot diagrams for the average values for each sample and compare the dot diagram in each case to the associated histogram of the individual measurements. How do the average value dot diagrams differ from the individual measurement histograms? Does this make sense? Be specific and quantitative in your answer.

3.7. Determine $\bar{x}$, s_X, and R for each sample in
   a. Exercise 3.2, samples 1 to 5.
   b. Exercise 3.3, samples 6 to 10.
   c. Exercise 3.4, samples 11 to 15.

3.8. You have been told to continue monitoring the resistance of an electronic component with the existing Shewhart control charting scheme using the same sampling method as the person on the previous shift. You see in a log that for the samples he collected, $\bar{R}$ was 0.525 ohms, and the estimate of σ_X that he determined from his data was 0.207 ohms. There is, however, no record of the sample size he used. Based on the information available to you, what sample size do you use to continue testing in the same manner?

3.9. You and a co-worker have been asked to analyze a boring process used to increase the inside diameter of a pipe flange. There are two lathes used in the shop to perform the boring operation, and the two of you take 15 samples of four individual measurements of the bore surface roughness from each machine with the results in microinches as given in the tables. Use whatever statistical methods you deem appropriate to compare the output of the two machines using the data given. Be sure to include graphical representations of the data.

Machine 1:

Sample	X_1	X_2	X_3	X_4
1	124.4	121.5	126.0	125.0
2	126.8	127.0	125.9	123.6
3	122.6	123.8	122.3	126.4
4	122.5	125.4	123.0	126.4
5	126.5	128.2	124.3	124.8
6	126.8	124.2	126.7	121.6
7	126.3	126.4	126.4	122.1
8	122.1	124.1	124.2	126.0
9	125.9	123.0	124.2	124.3
10	127.6	126.8	127.2	126.2
11	125.9	123.3	125.5	123.6
12	123.4	125.2	125.2	123.5
13	124.9	123.5	128.1	126.1
14	126.3	123.0	126.3	123.3
15	125.3	126.3	124.5	124.4

Machine 2:

Sample	X_1	X_2	X_3	X_4
1	127.7	129.4	128.2	124.7
2	120.7	125.7	123.8	132.5
3	126.1	125.4	118.1	123.8
4	119.9	123.3	120.4	120.8
5	127.6	126.1	124.7	127.4
6	124.3	121.8	125.1	128.3
7	125.4	119.2	123.1	125.6
8	131.2	126.6	129.6	126.0
9	126.7	125.4	129.3	128.7
10	121.2	121.7	117.7	123.1
11	124.1	126.1	128.9	125.7
12	125.3	120.9	128.3	123.9
13	126.6	128.1	122.2	122.1
14	131.9	126.3	121.0	122.5
15	125.2	125.1	127.0	127.9

3.10. A fastener company is supplying washers to a manufacturer of kitchen appliances who has specified that the outside diameter be 0.500 ± 0.025 inch. Fifty samples of $n = 5$ washers show that the process is in good statistical control, the diameters are normally distributed, and μ_X and σ_X for the process are 0.505 and 0.0065 inch, respectively.

     a. What percent of the washers made by this process are within the specifications? Sketch the appropriate graphs to represent the distribution of the process along with the specifications.

     b. What would happen if the process variability was reduced by about half (i.e., σ_X changed to 0.003 inch)?

3.11. Assuming that the data in Exercise 3.3 can be approximated by the normal distribution, within what range of weld strength would you expect 95% of the welded joints to fall? What range would 99% fall within? Provide a graphical representation of your answer, as well.

3.12. For a process that is in good statistical control and for which the individual measurements follow a normal distribution with $\mu_X = 39.87$ and $\sigma_X = 0.56$, what percent of the process is expected to be in the range 39.34 to 40.52?

3.13. For a process that molds polycarbonate face guards for motorcycle helmets, the thickness of the plastic can be assumed to have a normal distribution with $\mu_X = 1.825$ millimeters and $\sigma_X = 0.026$ millimeter. Below what value will 10% of the thicknesses be?

3.14. A process that drills holes in crankshafts which are subsequently ground has been found to be in good statistical control, with the following statistics used to estimate the mean and the variance of the hole diameter:

$$\bar{\bar{x}} = 15.050 \text{ millimeters}$$
$$\bar{R} = 0.110 \text{ millimeter.}$$

It is also known that 95% of the hole diameters lie within the range 14.945 to 15.155 millimeters. Find the size of the samples employed to produce the ranges R used to estimate the standard deviation of the process. Assume that the hole diameters are normally distributed.

3.15. After taking many samples of size $n = 4$ of the length of a pipe, μ_X and σ_X were determined to be 0.973 and 0.003 meter, respectively. The process is in good statistical control and the individual lengths seem to follow the normal distribution.

     a. What percent of the pipe lengths would fall outside specification limits of 0.965 ± 0.007 meter?

     b. What is the effect on the percent conforming to specifications of centering the process?

     c. What would the effect be if $\mu_X = 0.973$ meter and the process standard deviation were reduced to $\sigma_X = 0.0025$ meter?

Represent each situation above by providing a graphical representation.

3.16. A supplier is producing ball bearings for a major auto producer who has provided a specification on the diameter of 0.250 ± 0.005 inch. Sampling has shown that the distribution of the parts is centered at the nominal, but too many part diameters are outside specifications. What would the standard deviation of the process need to be for 99% of the ball bearings to be within specifications if it were to be assumed that the diameters follow the normal distribution?

3.17. The data in the table represent the average thickness for samples of aluminum sheets after going through a rolling process. Draw a box-and-whisker plot and a box plot of the data.

Sample	Average (inches)	Sample	Average (inches)
1	0.043	21	0.051
2	0.048	22	0.046
3	0.048	23	0.050
4	0.052	24	0.050
5	0.048	25	0.052
6	0.052	26	0.052
7	0.048	27	0.055
8	0.059	28	0.047
9	0.043	29	0.055
10	0.054	30	0.055
11	0.050	31	0.054
12	0.061	32	0.047
13	0.045	33	0.053
14	0.053	34	0.050
15	0.051	35	0.050
16	0.053	36	0.054
17	0.052	37	0.056
18	0.055	38	0.053
19	0.045	39	0.052
20	0.046	40	0.048

3.18. The process for forming steel sheets of gage 8, with a nominal thickness of 0.16440 inch, is in question after many complaints have been received. Your manager would like to see a graphic representation of the mean and spread of the thickness of these sheets. You have decided that a box-and-whisker plot would be best.

   a. Construct the plot using the sample means in the first table. The data have been collected in samples of $n = 5$ at random intervals throughout the day.

Sample	Average (inches)	Sample	Average (inches)
1	0.16425	16	0.16444
2	0.16431	17	0.16447
3	0.16473	18	0.16435
4	0.16459	19	0.16469
5	0.16446	20	0.16420
6	0.16449	21	0.16461
7	0.16433	22	0.16441
8	0.16439	23	0.16449
9	0.16419	24	0.16480
10	0.16454	25	0.16454
11	0.16454	26	0.16419
12	0.16480	27	0.16425
13	0.16513	28	0.16429
14	0.16457	29	0.16439
15	0.16423	30	0.16428

b. After seeing the data, changes were made to improve the process. The data in the second table show the results of a second series of samples collected in a manner similar to that used to obtain the data in part (a). Prepare a box-and-whisker plot of these new data to the same scale as the plot in part (a), and compare the two processes.

Sample	Average (inches)	Sample	Average (inches)
1	0.16439	16	0.16427
2	0.16424	17	0.16441
3	0.16449	18	0.16436
4	0.16450	19	0.16435
5	0.16430	20	0.16432
6	0.16439	21	0.16441
7	0.16444	22	0.16422
8	0.16448	23	0.16446
9	0.16435	24	0.16452
10	0.16448	25	0.16446
11	0.16455	26	0.16433
12	0.16435	27	0.16446
13	0.16448	28	0.16453
14	0.16437	29	0.16443
15	0.16440	30	0.16438

3.19. It has been suggested that a process for drawing copper sheets is no longer in control; the data seem to be more uniformly than normally distributed about a nominal value of 0.180 inch. In the table are averages from samples of $n = 5$ consecutively drawn sheets taken every few minutes throughout the day. Using a box-and-whisker plot, what can be said about the distribution of the averages?

Sample	Average (inches)	Sample	Average (inches)
1	0.183	19	0.192
2	0.189	20	0.164
3	0.183	21	0.175
4	0.185	22	0.174
5	0.179	23	0.178
6	0.174	24	0.190
7	0.177	25	0.185
8	0.181	26	0.179
9	0.189	27	0.175
10	0.168	28	0.175
11	0.181	29	0.192
12	0.177	30	0.190
13	0.182	31	0.166
14	0.190	32	0.173
15	0.196	33	0.197
16	0.183	34	0.177
17	0.206	35	0.170
18	0.181		

3.20. Use the following data obtained from a destructive testing procedure to draw a conclusion as to whether or not the results indicate that the tensile strength values follow the normal distribution. Construct a dot diagram, cumulative probability plot, normal probability plot, and Q-Q plot in doing your analysis. The data have been obtained by subtracting 50,000 psi from the actual measurement and dividing the result by 1000 psi.

−0.980	0.992	1.224	0.921
1.432	1.321	−1.391	1.098
−1.199	−0.987	−0.889	−0.998
1.145	1.042	1.427	

3.21. It is desired to determine if the surface finish obtained in a face milling operation used in machining engine blocks can be approximated by the normal distribution. Use the data in the table to construct a dot diagram, cumulative probability plot, normal probability plot, and Q-Q plot. Draw the appropriate conclusion about the data, given the results of your graphical analysis.

Sample	Surface Roughness (microinches)	Sample	Surface Roughness (microinches)
1	118.4	9	117.1
2	113.4	10	117.9
3	115.0	11	115.2
4	115.4	12	121.6
5	117.7	13	120.0
6	118.5	14	122.5
7	117.3	15	121.1
8	114.4	16	117.6

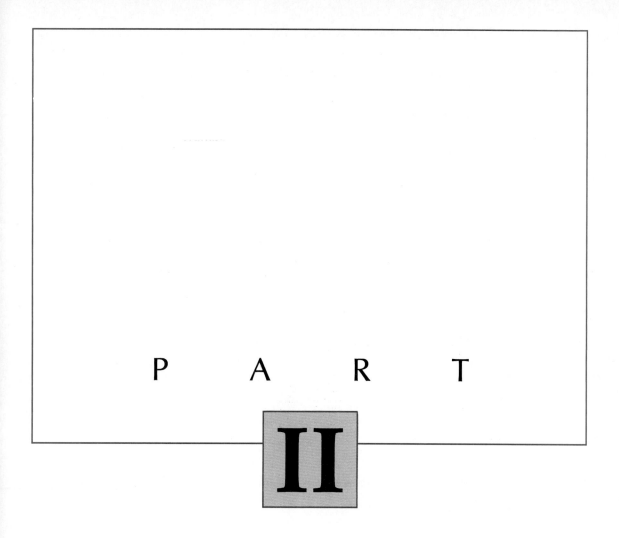

P A R T

II

Process
Control and
Improvement

4

Conceptual Framework for Statistical Process Control

4.1

Introduction

This chapter may well be the most important chapter in this book from the standpoint of successful implementation of statistical process control. This is so because we discuss five fundamental concepts central to the successful use of statistical process control as a tool for quality and productivity improvement:

1. The origins and nature of *variability*.
2. Process evolution over *time*.
3. Shewhart's concept of *statistical control*.
4. *Managing variability* using control charts.
5. The *process* of statistical process control.

One's ability to discard the old philosophy of inspection for product control and embrace the philosophy of never-ending improvement using process control depends entirely on how well the five concepts above are understood. This chapter is devoid of equations, tables of numbers, and other numerical/analytical representations and summaries. Those will follow in great abundance but may well be misinterpreted and misused without a firm understanding of the concepts that follow.

111

4.2

Origins and Characteristic Behavior of Variability

Sources of Functional Variation

To reduce functional variation, that is, increase the consistency of product/process performance, it is essential that we identify the basic sources of functional variation so that appropriate countermeasures can be formulated and implemented. Taguchi suggests that variation in product or process function (he refers to this as *functional variation*) arises from three basic sources.

1. **Outer noise.** External sources or factors that are operating in the environment in which the product is functioning and whose variation is transmitted through the design to the output performance of the product. Examples of outer noise factors include such things as temperature, humidity, contaminants, fluctuations in voltage, vibrations from nearby equipment, and variations in human performance. In Taguchi's definition of outer noise it is important to note that these are sources of noise which are influencing performance as measured during field use under actual operating conditions.
2. **Inner noise.** Internal change in product characteristics such as drift from the nominal over time due to deterioration. Inner noise may be precipitated by such things as mechanical wear and aging.
3. **Variational noise.** Variation in the product parameters from one unit to another as a result of the manufacturing process. For example, the design nominal for a resistor may be 200 ohms, but one manufactured resistor may have a resistance of 202 ohms while another may have a resistance of 197 ohms. Variational noise is sometimes referred to as manufacturing imperfection.

The significance of the recognition of the sources of variation described above becomes evident as one begins to think in terms of the fundamental countermeasures one might invoke to mitigate the forces of these sources of variation. It quickly becomes clear that the forces of outer noise and inner noise can be dealt with upstream effectively only by the engineering design process. Variational noise is a matter of manufacturing imperfection and hence can be dealt with, in part, at the process via techniques such as statistical process control. However, mitigating the forces of variational noise should also be considered to be a product and process design issue as well. In fact, it is likely that variational noise can be dealt with in a more significant and fundamental way if it is thought of as a process and product design problem. Certainly, the concept of design for manufacturability and the current emphasis on the simultaneous engineering of products and processes have a bearing on this issue. Since manufacturing processes are themselves subject to external sources of variation, the concept of robust design applies to them as well as to the products they are manufacturing.

Figure 4.1 illustrates how the sources of variation defined above are transmitted to the quality response of the product. At the product design stage the nominal values of the critical design parameters are selected to produce a prespecified level of performance in the product in terms of one or more quality responses. Sources of variation active in the manufacturing processes cause the design parameters to vary from those values intended by the design process, and hence introduce variation in the quality response. Although these are sources of variational noise or manufacturing imperfection, the extent to which noise in the process is transmitted to the product performance is a matter to be considered at the design stage, using robust design concepts in concert with the principles of design for manufacturability and simultaneous engineering.

Once the product is put into field use, outer noise sources of variation active in the environment in which the product is being asked to function are transmitted through the design and introduce further functional variation in the quality response. For example, variation in temperature may cause product performance to vary.

As time goes on and the use of the product continues, the forces of inner noise are transmitted through the design and introduce still further variation in the quality response. Changes in the mean level of performance may take place as well. For example, as a result of wear a critical design parameter may actually change its mean value. Wear could cause a change in size, resistance, viscosity, and so on, thereby causing a change in quality performance, as is depicted at the bottom of Fig. 4.1.

From Fig. 4.1 it is clear that sources of variation enter the picture and have an impact on the ultimate performance of the product during manufacturing and during field use. Historically, the majority of the efforts at quality control were directed toward manufacturing sources of variation. Only recently have design-based concepts and methods emerged that are aimed at understanding and taking action against the forces of variation arising during field use.

It should be clear after some thought that the only way to take action against the forces of outer noise and inner noise is through the design process. Since the root causes of outer and inner noise cannot be economically reduced or eliminated directly, their influence on the functional performance of the product must be dealt with through improved product design. That is, the design of the product must be such that the transmission of outer and inner noise through the design to the product output performance is suppressed as much as possible. This is the primary objective of Taguchi's method of parameter design.

Manufacturing/Process Variation: Common and Special Causes of Variation

Shewhart, Deming, and Juran all clearly point out in their teachings that quality and productivity problems at the process fall into two basic categories. Shewhart described the variation in the process as arising from either *chance causes* or *assignable causes*. Dr. Deming refers to *system faults/common causes* and *local faults/special causes*; Juran refers to *chronic problems* and *sporadic problems*. Each provides

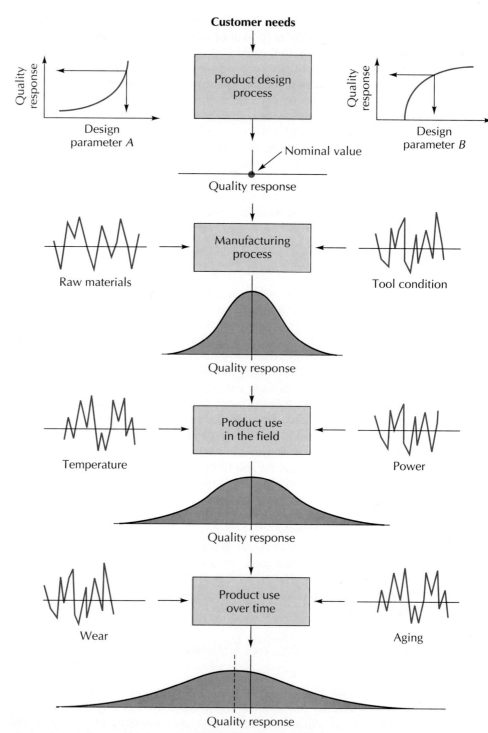

Figure 4.1 Impact of Sources of Variation on Quality Performance

a lucid description of the overall situation. Assignable causes, local faults, or sporadic problems may be thought of as problems that arise periodically in a somewhat unpredictable fashion and can usually be dealt with effectively at the machine or workstation by the immediate level of supervision or the operator. These are the disturbances to the process that interfere with its "routine" operation and cause the economic viability of the process to be jeopardized. It is the primary purpose of statistical process control concepts and methods to identify the presence of these local faults/assignable causes/sporadic problems and then to identify the proper corrective action at the machine.

However, local faults are but a small fraction of the total picture. System faults/chance causes/chronic problems constitute problems with the system itself and are ever present, influencing all of the production until found and removed. System sources of variation are also referred to by Dr. Deming as common causes, since they are common to all of the manufacturing output. It is the attack on the system problems that Dr. Deming stresses, since about 80 to 85% of all the problems we encounter are of this nature, while only 15 to 20% are of a local fault, sporadic problem nature.

Table 4.1 summarizes the basic nature of process variation and provides some examples in each case. In the case of common causes/system faults it is clear by their very nature that management must take the responsibility for their removal. Only management can take the actions necessary to improve the training and supervision of the workers; only management can take the responsibility for the redesign of a poor workstation layout; only management can establish new methods or procedures for the process. Clearly, it is essential that we properly differentiate between the occurrence of these two types of

TABLE 4.1 The Nature of Faults in the Process

All processes are subject to two fundamental
types of faults or problems.

Faults	Local faults	versus	System faults
	Special causes		Common causes
	Sporadic problems		Chronic problems
	Assignable causes		Chance causes
	⋮		⋮
Examples	Broken tools		Poor supervision
	Jammed machine		Poor training
	Material contamination		Inappropriate methods
	Machine-setting drift		Poor workstation design
	⋮		⋮
Action/ by Whom	Correctable locally, at the machine, by the operator or first level of supervision		Requires a change in the system—only management can specify and implement the change

problems so that we assign the proper responsibility for solving the problems—taking corrective action.

Summary: Nature of Variation in Product/Process Performance

- It is essential that we understand clearly where variation comes from—the basic sources of variation—so that we know what actions to take and where to assign the responsibility for taking the actions.
- Quality as loss due to functional variation should be observed in product performance during field use. Deviations in parameter values (deviation from the nominal) observed after manufacturing are but one of several sources of functional variation.
- Although some variation can be dealt with effectively through study of the manufacturing process, the engineering design process plays a strong role in determining the extent to which some variation sources, outer noise and inner noise, influence the amount of functional variation—variation in product/process performance—in the field, under working conditions.

4.3

Process Behavior over Time

A succession of parts emanating from a process under statistical control will exhibit variability in the quality characteristic of interest due to a constant set of common causes. These variable measurements tend to collect themselves into a predictable pattern of variation that can easily be described by a few simple statistical measures, such as a mean, a standard deviation, and a frequency distribution, say, the normal distribution. These measures stand as a model that predicts how the process will behave if subject only to a constant set of common causes. Figure 4.2 illustrates how such a statistical model may emerge. In Fig. 4.2, measurements of thickness, for example, may be recorded for a succession of automobile head gaskets produced by the process and these data may be used to develop a statistical model for the process.

If we can develop a model for the process measurements when subject only to a constant system of common causes, then when a major disturbance affects the process, the ensuing measurements will be seen not to conform to this model; that is, they will stand out clearly in the common-cause variability pattern. This is the case for the situation depicted in Fig. 4.3. For the measured gasket thickness represented by the dot in Fig. 4.3, we ask the question: Could this deviation of this gasket's thickness from the process mean of 0.045 inch be explained by the forces of common-cause variation? If only common-cause forces are at work, the vast majority, say, 99.73% of all measurements, fall between A and B, so our answer will be no, it does not fit into the common cause variability pattern!

The statistical description and interpretation above is useful; however, a

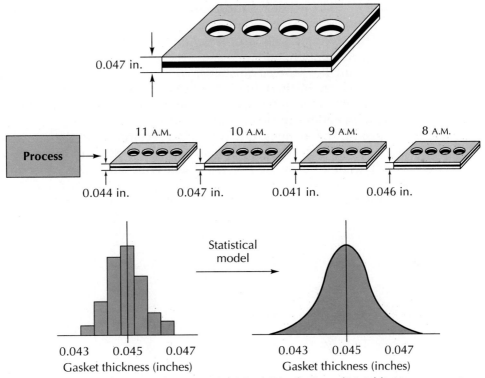

Figure 4.2 Process Representation As a Statistical Distribution of Variable Measurements

very large step has been taken in simply taking a set of measurements over a period of time by sampling the process output and collecting them into the graphical summaries of Figs. 4.2 and 4.3. This very large step has ignored a crucial characteristic of the data: the time order of production! To build statistical models to represent product/process characteristics of importance, we must first determine if the process data indicate consistency over time, statistical control,

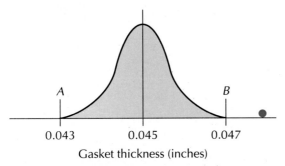

Figure 4.3 Process Behavior Inconsistent with the Common-Cause System

and to do this we must consider the process as it evolves over time of production. Far too often we simply take a set of sampled data, calculate an average, calculate a sample standard deviation, assume a statistical model such as the normal distribution, and then apply that model for some analytic purpose, such as assessing the process capability, without checking to see if the sampled data can reasonably be thought of as emanating from a constant system of common causes. Such abuses are commonplace and unfortunately often lead to gross misinterpretations of process performance.

Figure 4.4 depicts the time-order evolution of the process introduced through Fig. 4.2. In Fig. 4.4(a) the process mean and variation stay constant over time and hence the process is said to be stable (i.e., in statistical control). In Fig. 4.4(b), the process mean begins to shift about erratically at 11:00 A.M. At this point the process is said to have gone out of control. Processes under

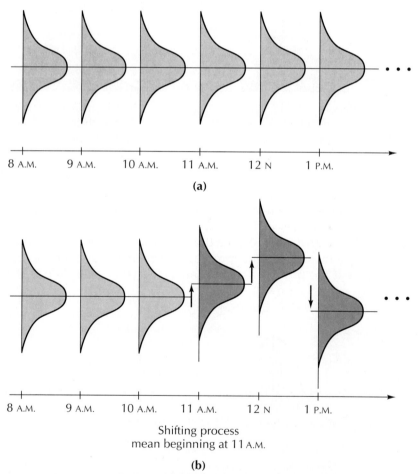

Figure 4.4 Changing Process Behavior over Time

statistical control are driven solely by common causes of variation. Instability, out of control behavior, is driven by special causes of variation.

Over time, a process can be subject to several kinds of disturbances (special causes) that can produce a variety of unstable behaviors with respect to either its mean level or level of variability or both. Figures 4.5 and 4.6 graphically portray this concept of process behavior. In Fig. 4.5 only the mean level of the process is exhibiting a changing nature over time.

In Fig. 4.5(a) the mean level of the process is varying from time to time in an erratic fashion. Figure 4.5(b) shows the appearance of a process for which the mean level has experienced an abrupt and sustained shift. In Fig. 4.5(c) the process mean is changing in a somewhat systematic or gradual fashion. Finally, in Fig. 4.5(d) the mean level of the process is stable; that is, the process is in control with respect to its mean level. In each case except (d), the indication is that a special cause (or causes) has entered the system and is responsible for the change in the process mean level. It is clear that such changes will influence the quality of the product, particularly when we think in terms of the loss function interpretation of quality. The root cause of the mean shift is also likely to create a loss of productivity.

Figure 4.6 illustrates the appearance of a process for which the variation level has changed over time. This, again, signals in each case the presence of special causes that will have deleterious effects on both the quality and productivity of the process. In Fig. 4.6(a) the process has experienced an abrupt and sustained shift in the variation level. In Fig. 4.6(b) the changes in the amount of process variation are occurring in a more erratic fashion. Figure 4.6(c) exemplifies the worst of all worlds,[1] a process that is unstable in terms of both its mean level and amount of variation. For a process to have any chance to operate in an economical fashion it must at a minimum be in a state of statistical control. From the standpoint of the time behavior of the process this means that the process must have the appearance shown in Fig. 4.5(d).

Summary: Time Evolution of Process Behavior

- Data collected over time from a process may be used to develop a statistical model for the process, as long as the data are collected during a period when the process is subject to a constant system of common causes.
- The most common changes in process performance over time are changes in the process mean level and changes in the amount of variation in the process. These changes can take on a variety of characteristic forms (erratic, systematic, etc.), depending on the physical forces that are driving them. These changes indicate an instability of process performance.
- By tracking process performance over time through periodic sampling, it is possible to detect such changes in performance through the characteristic

---

[1] During a supplier seminar for a large off-road construction vehicle manufacturer some years ago, one of the authors, using Fig. 4.6, asked the group, "What is the nature of the situation we have in Fig. 4.6(c)?" A loud voice from the back of the room called out, "Chapter 11!"

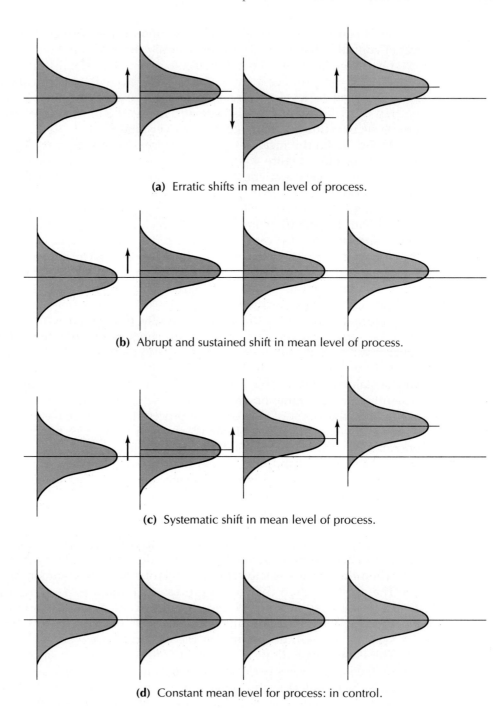

(a) Erratic shifts in mean level of process.

(b) Abrupt and sustained shift in mean level of process.

(c) Systematic shift in mean level of process.

(d) Constant mean level for process: in control.

Time

Figure 4.5 Behavior of Mean Level of the Process over Time

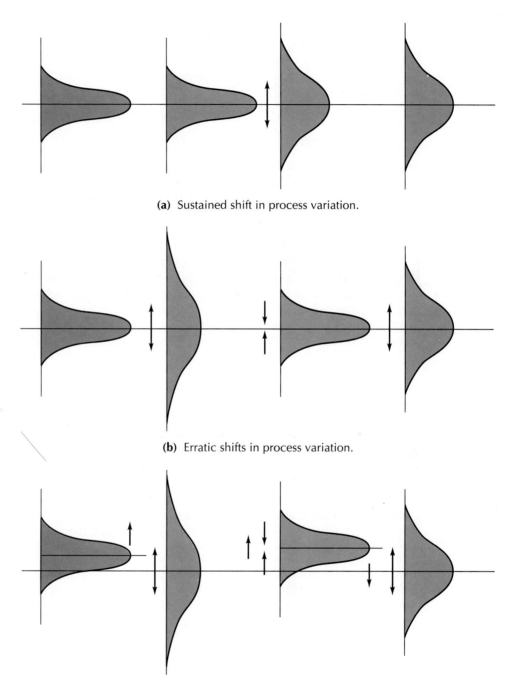

(a) Sustained shift in process variation.

(b) Erratic shifts in process variation.

(c) Erratic shifts in both process mean and process variation.

Time

Figure 4.6 Behavior of the Level of Variation of the Process

changes in the process model, namely, changes in the mean and changes in the standard deviation.

4.4
Shewhart's Concept of Statistical Control
Economic Control of Manufacturing Processes

Collecting production performance data and keeping charts of these data is a common practice in industrial operations. The observation of dimensions, strengths, weights, yield, defects, and other quality characteristics can be of great utility not only in documenting performance relative to requirements, but more important, as a tool to aid the identification of ways to improve performance. Regrettably, such data are often examined in a most cursory manner and then filed away never to see again the light of day. Often, such data are quickly summarized into daily or weekly management reports, and as a result, a great deal of the important information content is lost, *particularly as this information relates to the time history of the process performance.* A basic problem with the use of such production data, even when retained in its raw state, lies with the fact that it is generally not conceived of, modeled, and then interpreted on a statistical basis.

Many years ago Walter Shewhart at Bell Labs showed how such process data could be developed and interpreted through the use of very simple but profound statistical methods. In his benchmark book on the subject, *Economic Control of Quality of Manufactured Product*,[2] he establishes from the very beginning the overarching philosophy that drives the control chart concept. The first paragraph of the preface of this work clearly sets the foundation for what we know today as statistical process control:

> Broadly speaking, the object of industry is to set up economic ways and means of satisfying human wants and in so doing to reduce everything possible to routines requiring a minimum amount of human effort. Through the use of the scientific method, extended to take account of modern statistical concepts, it has been found possible to set up limits within which the results of routine efforts must lie if they are to be economical. Deviations in the results of a routine process outside such limits indicate that the routine has broken down and will no longer be economical until the cause of trouble is removed.

From Shewhart's statement, several things become immediately obvious:

1. The fundamental focus is on the process: "ways and means of satisfying human wants."
2. The overarching objective is economic operation of the process: "Reduce

---

[2] W. A. Shewhart, *Economic Control of Quality of Manufactured Products*, D. Van Nostrand, Princeton, N.J., 1931. Reprinted by the American Society for Quality Control, Milwaukee, Wis.

everything possible to routines requiring a minimum amount of human effort.''

3. During normal operation, process behavior falls within predictable limits of variation: ''It has been found possible to set up limits within which the results of routine efforts must lie if they are to be economical.''

4. Deviations in performance outside these limits signal the presence of problems that are jeopardizing the economic success of the process: ''Deviations in the results of a routine process outside such limits indicate that the routine has broken down and will no longer be economical.''

5. Improvement in quality and productivity requires that attention be directed at the process to find the root cause of the trouble and remove it: ''The routine has broken down and will no longer be economical until the cause of trouble is removed.''

Something else is also evident in Shewhart's statement—there is no mention of the product and the conformance of the product to specifications. The total emphasis that Shewhart placed on the process and its economic operation was to a large extent not clearly understood by many. Hence it was not then nor is it now uncommon to hear questions posed such as the following: ''If the product is by and large meeting the specifications, why is it important to be concerned about the fact that the process is not in a state of statistical control?'' To answer this question, we must understand (1) what we mean by control, and (2) what the implications are of not running a process in a state of control. For these answers we turn again to the teachings of Shewhart:

> A phenomenon will be said to be controlled when, through the use of past experience, we can predict, at least within limits, how the phenomenon may be expected to vary in the future. Here it is understood that prediction within limits means that we can state, at least approximately, the probability that the observed phenomenon will fall within the given limits.

From this definition it is clear that Shewhart is overlaying the statistical characterization of variations on the physical meaning and interpretation of the origins/sources of those variations. This, of course, he made clear in the earlier quote: ''It has been found possible to set up limits within which the results of routine efforts must lie if they are to be economical.'' When a process is not controlled, Shewhart clearly spelled out the consequences:

> Deviations in the results of a routine process outside such limits indicate that the routine has broken down and will no longer be economical until the cause of the trouble is removed.

In short, when a process is not in a controlled state, the productivity or economic success of the process is not guaranteed. We may be making product that meets the specifications but we are not doing so in a manner that will enhance competitive position. In terms of Shewhart's way of thinking, the notion of predictability of the process behavior through the interpretation of the variability pattern in product/process quality characteristics is synonymous with the behavior of a routine process if it is to be economical.

What Shewhart recognized so clearly is simply this. Process instability—the presence of out-of-control behavior—is a signal that forces acting on the process are causing waste and inefficiency (i.e., robbing the process of productivity). To see this fact clearly, we will now consider the important distinction between process capability and process control.

Process Control Versus Process Capability

Perhaps the single most difficult hurdle to clear in coming to a full understanding of the teachings of Shewhart is the separation of the notions of the product's conformance to specifications and statistical control of the process. Not only do these issues need to be totally separated, but the value and importance of achieving both of these goals simultaneously and their essential interrelationship must be completely understood. Since both ideas involve the comparison of variation patterns to a set of preestablished limits, the confusion that can arise is considerable. It is essential that we understand that the origin and interpretation of these two sets of limits are totally different. That the quality characteristics of the product fall within the specification limits (derived during engineering design) implies that the customer expectations are being met. That the performance of the process falls also within the limits of variation for routine operation (derived from the observed pattern of variation in the process) implies that the customer expectations are being met in an economic fashion.

Figure 4.7 describes the separate issues of product conformance and control of the process. In Fig. 4.7(a) product quality characteristic measurements are compared to the specifications. It would appear that the product is meeting the specifications. In Fig. 4.7(b) we see the same quality characteristic but this time in the form of Shewhart $\overline{X}$ and R control charts. It would appear that the process variation pattern is not confined to fall within the predictable limits defined by the common-cause system. The process is not in control in terms of either the mean or the amount of variation.[3]

But as Shewhart has pointed out so clearly, if we fail to control the process in the statistical sense, the process is not performing in an economical fashion. We are spending more money to make the product than we would if the process were in control. In other words, we are not operating the process in the manner in which it was designed to be run. There exist fundamental faults in the process that are robbing us of productivity. The Shewhart control chart is a tool to alert us to the fact that the process routine has in some way broken down and that continued operation under such conditions will be wasteful.

---

[3] During the same supplier seminar referred to in footnote 1, the same author, referring to Fig. 4.7(a), asked, "What can we conclude from the data portrayed in this fashion?" Again, from the back of the room came a loud voice, "Ship em!" The respondent probably didn't realize how right he was . . . that's about all we can conclude.

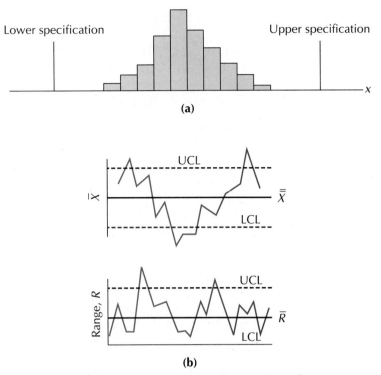

Figure 4.7 Product Conformance Versus Process Control

Summary: Shewhart's Concept of Statistical Control

- "Through the use of the scientific method, extended to take account of modern statistical concepts, it has been found possible to set up limits within which the results of routine efforts must lie if they are to be economical. Deviations in the results of a routine process outside such limits indicate that the routine has broken down and will no longer be economical until the cause of trouble is removed."

- "A phenomenon will be said to be controlled when, through the use of past experience, we can predict, at least within limits, how the phenomenon may be expected to vary in the future. Here it is understood that prediction within limits means that we can state, at least approximately, the probability that the observed phenomenon will fall within the given limits."

- That the quality characteristics of the product fall within the specification limits (derived during engineering design) implies that the customer expectations are being met. That the performance of the process also falls within the statistical limits of variation for routine operation (derived from the observed variation in the process) implies that the customer expectations are being met in an economic fashion.

| 4.5 |

Control Chart Approach to the Management of Process Variation

In the first section of this chapter we have simply referred to the variation evident in process data as arising from common cause/chronic problems and from special cause/sporadic problems. In a subsequent section we saw how Shewhart viewed these sources of variation as arising (1) from the routine operation of the process, and (2) from forces above and beyond the routine operation of the process. We now must be much more specific about the nature of the variability pattern in the data. This is necessary so that we might understand what actions can be taken against the forces of process variation, and more important, who is to be held responsible for formulating and implementing such actions.

Detecting the Presence of Special Causes of Variation

Figure 4.8 illustrates a situation in which a stable process is suddenly subject to erratic shifts in its mean level. As we have just discussed, whatever is causing the process mean to begin to shift after 11:00 A.M. is causing the process efficiency to be less than what it would be if that force were not present. For example, the shifts in the mean level could be resulting from operator overadjustment of some process setting, thereby not only increasing the variability in the process and the product but also reducing the productivity—the process routine has broken down. It is, therefore, important to be able to detect as soon as possible when such a change in the process occurs. We can then look for

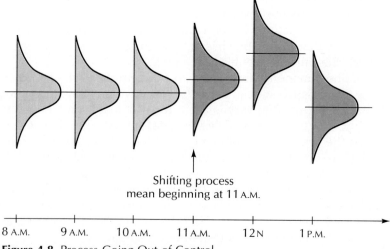

Shifting process
mean beginning at 11 A.M.

8 A.M.   9 A.M.   10 A.M.   11 A.M.   12 N   1 P.M.

Figure 4.8 Process Going Out of Control

the cause of the change and formulate actions that will bring the process back to routine and keep such changes from recurring. This is the job of the Shewhart control chart.

To illustrate the workings of the control chart idea, let's return to the gasket-making process we mentioned briefly above. Suppose that periodically we take small samples of size $n = 5$ gaskets from the production as depicted in the following chart.

	10:00 A.M.	9:30 A.M.	9:00 A.M.	
	0.044 inch	0.046 inch	0.043 inch	
	0.043	0.044	0.044	
Process ...	0.045 ...	0.045 ...	0.044	...
	0.044	0.045	0.042	
	0.044	0.047	0.047	
	$\overline{X} = 0.0440$	$\overline{X} = 0.0454$	$\overline{X} = 0.0440$	
	$R = 0.0020$	$R = 0.0030$	$R = 0.0050$	

Each small sample tells us how the process is behaving at any one time with respect to its mean level and its variation. The summary statistics, namely the average and the range of the sample, provide estimates of two process parameters, the mean and the standard deviation. The question is: Is the behavior of the process changing from time to time, sample to sample? If the process is subject only to common-cause variation, the $\overline{X}$'s and R's should vary randomly within certain probabilistic limits of variation as shown in Fig. 4.9(a). Occasionally, the process may experience some real change in its mean level or amount of variation due to some special causes. When this happens, an $\overline{X}$ or R can indicate this by exceeding the natural variability limits, as shown in Fig. 4.9(b).

Shewhart used the statistical model portrayed above to indicate when the results of a stable/routine process were being disturbed by external forces. In this event the deviations would manifest themselves as variations that could not be explained by or expected to arise within the probabilistic limits of variation of the process when under control. He indicated further that such deviations signaled the presence of forces that were breaking down the routine of the process and causing it no longer to be economical. He referred to such deviations as arising from "assignable" causes, as compared to the "chance" causes driving the routine operation of the process. These assignable causes or special causes signal the occurrence of "local faults" in the process that can be dealt with by the operator or other local authority at the process, provided that these people have the knowledge and understanding regarding the root cause of such faults and the associated action(s) required to remove them.

Figure 4.10 shows an example of $\overline{X}$ and R charts being maintained on a particular process. For a period of time the process appears quite unstable, with numerous indications of out-of-control behavior (i.e., the process is not predictable). The implications of the lack of control of the mean level of the process

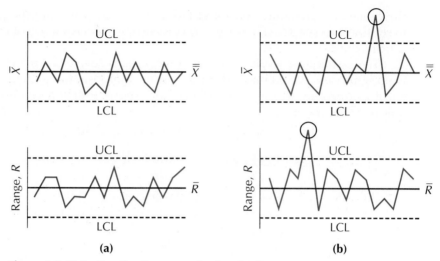

Figure 4.9 Detecting the Presence of a Special Cause

in Fig. 4.10 are serious. First, since the process is not in a controlled state, its results are not predictable. The capability of the process is not guaranteed and, in fact, cannot even be assessed reliably. But the equally serious implication of the lack of control or stability in the process is the fact that it signals that the process is not operating in a routine manner and therefore that its economic

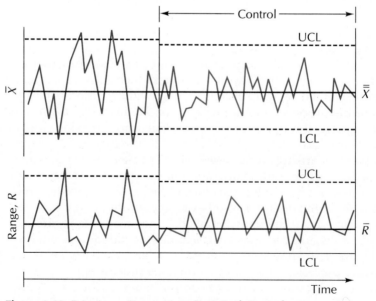

Figure 4.10 Bringing a Process into Statistical Control

feasibility is jeopardized. Until the root cause of the problem is found and action against it is taken, the process will continue to perform in a subpar fashion. However, as seen in Fig. 4.10, it appears that after the source of the trouble has been identified and action has been taken to bring the process into control, its pattern of variation is within the "limits of a routine process." The process is now running as it was designed to run, in a stable fashion, subject only to a constant system of common causes of variation. This movement from instability to control is an essential first step toward competitive manufacturing.

Moving from Control to Improvement

The identification of the presence of special causes and their subsequent elimination brings the process into a state of statistical control. But the presence of control only signals the fact that the process is now operating as it was intended to. We really have not improved the process—we simply have it running like it should! Once control is established, those at the process operation level can only be expected to use the charts to maintain control. To improve the process, we must attack the common-cause system. Such requires a more fundamental examination of the process and can be accomplished only through the involvement of management.

Earlier, Shewhart was quoted as indicating that "it has been found possible to set up limits within which the results of routine efforts must lie if they are to be economical." The suggestion here is that there exists a consistent or homogeneous set of forces, inherent to the process/system, which is driving the variation pattern in the process data. These forces are ever present and cause the process to behave in a stable and predictable, though variable way. Shewhart referred to this state of affairs as indicating that the process was being driven by the forces of chance.

It was the belief of Shewhart that these chance sources were many in number, each contributing only a small amount to the total, but collectively constituting an observable and measurable variation pattern. As a result, it was clear to Shewhart that the laws governing the behavior of random variations applied and hence that limits could be established to predict in probabilistic terms the behavior of these variations. Shewhart further reasoned that if the process operation in terms of its variation could be reduced such that it was to be affected solely by these chance forces, the process operation would be in a sense "doing the best it could." If other forces of variation were present, clearly the opportunity for improvement existed and this opportunity must be seized. The fact that Shewhart's control chart idea has prevailed over more than six decades attests to the validity and simplistic beauty of the foregoing reasoning.

Over the years, Deming, Juran, and others have built on and refined the theories of Shewhart, and in the process a rich and tremendously useful approach to quality and productivity design and improvement has evolved. The set of chance causes referred to by Shewhart have come to be described as the *common-cause system* and constitute those sources of variation that are part of the system. As such, they are ever present and affect all realizations of the process. Of course, it is possible that the common-cause system variability could be excessive, and as a result, process capability is jeopardized. It then follows that to improve the capability, the system must be changed.

The discussion above leaves no confusion as to who is responsible for the improvement of the capability of processes that are already showing good statistical control. Clearly, management has this responsibility. The operator at the machine is helpless against the forces of common-cause variation. Although the operator may have important input to the fault diagnosis process, he or she cannot unilaterally initiate the action(s) required to change the system—only management can do this. Juran has reinforced this notion by referring to the forces of common-cause variation as describing the level of chronic variation in the process. A chronic condition is clearly one that is ever present as a part of the system and requires the attention of a higher level of authority to treat its root cause.

To illustrate the impact of solving a chronic problem at the process, let us suppose that a certain process has been brought into a state of statistical control. However, an assessment of the process capability has revealed that the process is not very capable; that is, a significant amount of the stable production from this process falls outside the quality characteristic specifications. This state of affairs is illustrated in Fig. 4.11, which also shows the impact that a significant reduction in the amount of common-cause variation in the process can have on the capability of the process. Figure 4.11 shows that over the time period T_1 to

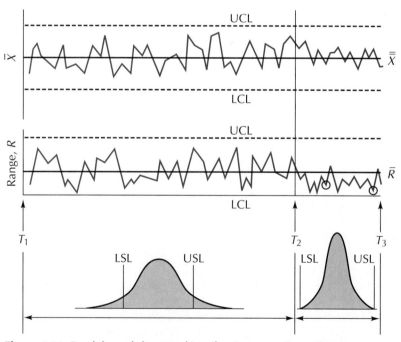

Figure 4.11 Breakthrough by Attacking the Common-Cause System

T_2 the process was running in a state of control but the process output relative to the specifications was not very satisfactory. Something needed to be done—but what and by whom? Now suppose that a team studying the process discovered a problem with the fixturing used and that through redesign, the problem was solved at time T_2. From the appearance of the R chart in Fig. 4.11 one would conclude that this change in the fixturing has caused a major reduction in the amount of common-cause variation in the process. A chronic problem that was compromising the process capability has been solved and the result is clear. As shown in Fig. 4.11, the process capability, as evaluated through process data collected over the time period T_2 to T_3, has been greatly improved. A major "breakthrough" in improving process performance has taken place. It is important to note that a basic change in the process has been made: a change in the fixturing. Such would require a level of authority above the operator. Again, while the operator might have been instrumental in identifying the cause of the trouble, he or she would not be in a position to unilaterally formulate and carry out the actions required to make the change in the process. This is the role and responsibility of management.

Control Chart Approach to Identifying Improvement Opportunities

In developing a discipline of continual improvement, it must be clear what we mean by improvement. Bringing a process into a state of statistical control simply means that we have brought the process back to where it should have been to begin with! Figure 4.12 depicts the basic difference between control and improvement. The act of bringing the process into control, as seen in Fig. 4.12, should be viewed as one of simply properly implementing a system as it was conceived of, planned for, and developed during design. At this point it is then

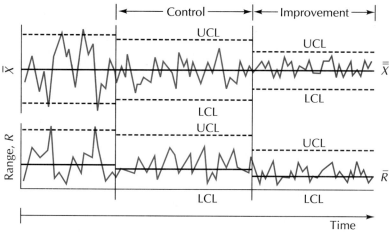

Figure 4.12 Moving from Instability to Control to Improvement

possible to begin to assess rationally the present ability of the process to realize the process potential and to seek improvement opportunities. The process may be failing to realize this potential because the implementation of the process is flawed or because the design of the process itself is flawed. In either case, the root cause(s) of the chronic/common-cause problem must be identified and removed at the system level. Such constitutes a breakthrough in performance; that is, an improvement in the process has taken place. The results of such a breakthrough are depicted in Fig. 4.12.

Summary: Management of Process Variation Using Statistical Methods

- Shewhart used the statistical model for common-cause variation, with associated probability limits, to indicate when the results of a stable/routine process were being disturbed by external forces. In this event the deviations would manifest themselves as variations that could not be explained by the forces of the common-cause system (i.e., expected to arise within the limits of variation of the process, when under control).
- When the source of the trouble has been identified and action has been taken to bring the process into control, that is, its pattern of variation is within the "limits of a routine process," the process is now running as it was designed to run, in a stable fashion, subject only to a constant system of common causes of variation.
- From a statistical standpoint, the fact that the process is subject only to the forces of common causes implies that it is in statistical control. A process in statistical control is "doing the best it can under the present system"; to do better, the system must itself be changed.
- Once a process is brought into statistical control it is possible to begin to assess rationally the present ability of the process to realize the potential it was initially intended to have. The process may be failing to realize this potential because the implementation of the process is flawed or because the design of the process itself is flawed. In either case, the root cause(s) of the chronic/common-cause problem must be identified and removed at the system level. Such constitutes a major breakthrough in performance; that is, an improvement in the process has taken place.

4.6

The *Process* of Statistical Process Control

Purposes of Shewhart Control Charts

There are two fundamental purposes that Shewhart control charts can usefully serve.

1. To serve as a means to assist in the identification of both sporadic and chronic faults in the process and to help provide the basis to formulate improvement actions.

2. To serve as a tool to provide a sound economic basis for making a decision "at the machine" as to whether to look for trouble/adjust the process or leave the process alone.

Although these two purposes may sound essentially the same, they are not. The former is strictly an off-line activity while the latter is an on-line activity. On line, the operator can only recognize and take action against the presence of special causes of variation. If the process, as monitored on-line, is free of special causes (operating within the limits of routine efforts), the operator is doing the best he or she can. In short, if the process is in statistical control, the operator should do nothing at all to disturb the routine behavior of the process.

On the other hand, the fact that the process is in statistical control in no way guarantees that it is a capable process, meeting the customer expectations as they may be communicated through a specification. However, the operator can certainly not be expected to be responsible for this, although he or she often has some valuable input to the problem. A lack of process capability is basically a product/process design and planning problem. To solve the problem may require the efforts of many different individuals, working together off-line to find the basic faults, their root causes, and the actions necessary to remove them. Furthermore, if a process exhibits considerable erratic behavior with little knowledge of the causes of such behavior, the operator will be helpless in his or her efforts to take actions to improve the stability of the process, that is, to remove the forces of special causes.

Perhaps the most fundamental failing of the use of control charts in the past was the failure to recognize them as an important tool to discover off-line the presence of improvement opportunities from both the quality and productivity points of view. Those advocating the use of control charts were often too quick to put them at the machine and place total responsibility for the use of the charts on the operator. As a result, in the past control charts were too often viewed as a tool to be used by operators on the shop floor to keep production tracking within the product specifications. If "points" showed that out-of-control conditions were present but the individual parts were within specifications, often no action was taken. Such misuse of the Shewhart control chart concept was commonplace and eventually often led to the abandonment of the methods.

Closing the Feedback Loop at the Process

For the techniques of statistical process control—in particular, Shewhart control charts—to be employed successfully as an off-line problem identification and problem-solving tool it is essential to keep in mind that we are working with a three-step process:

1. Use statistical signals to find improvement opportunities through identification of the presence of process faults.
2. Use experience, technical expertise, and fault diagnosis methods to find the root cause of the fault that has been identified.

3. Develop an action plan to correct the fault in a manner that will enable any gains that are realized to be held.

This three-step process may be usefully explained through the classical feedback control system perspective shown in Fig. 4.13, which depicts the classical feedback control loop applied to a process. There are five distinct stages in this generic control loop. These five stages facilitate the three-step process outlined above in the following way:

Step 1: using statistical signals   (1) Observation
                                            (2) Evaluation
Step 2: fault diagnosis   (3) Diagnosis
Step 3: action plan   (4) Decision
                                            (5) Implementation.

Each of these stages of the classical feedback control loop will now be discussed in detail.

Observation.   There are two aspects of observation: the physical system aspect of what to observe and how to observe it, and the statistical sampling issue (e.g., how many measurements to take, how often should they be taken, method of defining the sample, etc.). In this book we focus on the latter issue, statistical sampling, in considerable detail in Chapter 7.

Evaluation.   At the evaluation or comparison stage the data are compared to a model that describes how they should look under a predetermined circumstance/state of affairs. The Shewhart control chart is a statistical model that tells us how the data should look/behave if the process is stable/in control; that is, its mean level and level of variability are constant/unchanged over time. It is this comparison stage that provides the basis for step 1 of the three-

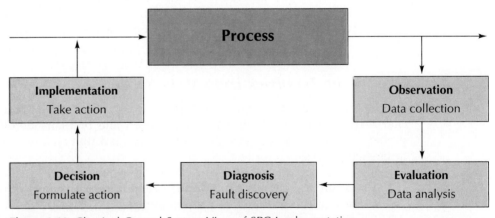

Figure 4.13 Classical Control System View of SPC Implementation

step SPC implementation process, namely, "reading" the statistical signals, which may tell us that a specific opportunity for quality and productivity improvement exists.

Diagnosis.   Diagnosis is perhaps the most crucial stage in the feedback control loop cycle. Being able to move from comparison to diagnosis marks the difference between making charts and solving problems. The key here is to be able to correlate the statistical signals with the physical state(s) of the process, for example, associate a point on a control chart above the upper control limit with a change in a raw material property. Numerous fault diagnosis aids are available from the simple but powerful graphical methods of Pareto charts and cause-and-effect diagrams to decision trees for failure modes effects analysis. If the closed-loop control system is to break open, it is likely that it will happen at the fault diagnosis stage. It is unfortunately altogether too easy to get caught up in making charts and forget why we are doing it.

Decision.   Once a fault is identified at the root cause level, it is necessary to formulate the appropriate action to eliminate the fault.

Implementation.   Finally, once the appropriate correction has been decided upon it is necessary to define the specific means to make the correction. It is important to study carefully the full implications of the action on all elements of the system so that the action taken may be lasting. *Holding the gains* over the long term requires that the system truly accept the total value of the action.

Summary: Process of Statistical Process Control

The fundamental purposes of control charts are

- To serve as a tool to provide a sound economic basis for making a decision "at the machine" as to whether to look for trouble/adjust the process or leave the process alone.
- To serve as a means to assist in the identification of both sporadic and chronic faults in the process and to help provide the basis to formulate improvement actions.

For the techniques of statistical process control, in particular, Shewhart control charts, to be employed successfully as an off-line problem identification and problem-solving tool it is essential to keep in mind that we are working with a three-step process:

- Use statistical signals to find improvement opportunities through identification of the presence of process faults.
- Use experience, technical expertise, and fault diagnosis methods to find the root cause of the fault that has been identified.

- Develop an action plan to correct the fault in a manner that will enable any gains that are realized to be held.

Exercises

4.1. Are chance causes/system faults and assignable causes/local faults useful in describing outer noise? Which causes/faults is it the responsibility of management to address?

4.2. Why is it important to consider the time order of production when describing a set of data, particularly if it is being used to help explain the performance of a process of manufacture?

4.3. What is meant by "statistical control" of a process? Why is the achievement of statistical control important?

4.4. An automobile yields consistent performance for several months in use in and around Peoria, Illinois. However, it experiences a considerable drop in MPG when driven for 3 months at high altitude in Colorado. What is this an example of? In general, how could the performance of this car be improved without restricting it to low altitudes?

4.5. A process is operating under statistical control. Are customer expectations (in terms of conformance to specifications) necessarily being met? Explain.

4.6. A process is manufacturing virtually all of the product within the specifications to satisfy the customer's demand. Is it necessary to monitor this process to see if it is under control? Why or why not?

4.7. Explain two of the most serious implications of running a process that is not in a state of statistical control.

4.8. A process that is making plastic lids for tennis ball containers is finally brought into control after some adjustments were made. Has a fundamental improvement been made? Explain.

4.9. A process manufacturing taillights for a new line of automobiles is seen to be in control, yet it is not meeting customer expectations—too many parts are falling outside the specifications. What should the operator on this line do about the problem? What needs to be done in general?

4.10. Can Shewhart control charts be used off-line, away from the shop floor, in a meaningful way? Explain.

4.11. Why is the diagnosis stage in the feedback control loop cycle so important? Why is it so difficult to make the step from evaluation to diagnosis?

4.12. What type(s) of "noise" are affected by parameter design?

4.13. What group(s) within the organization are responsible for reducing the effect that variational noise has on the quality of products that are being manufactured? What role does the operator/first-level supervisor play in solving problems of excessive variational noise?

4.14. Can a process be considered to be running in an economical fashion if it demonstrates a mean that fluctuates out of control periodically, yet still manages to produce virtually all of parts within the specifications?

4.15. Are erratic shifts in the mean of a process considered special causes or part of the common-cause variability?

4.16. How could bringing a process into statistical control affect a company's production schedule? What benefits can customers see as a result of doing business with a company whose processes are running in a stable, predictable manner?

4.17. If the time order of data collected was not recorded, would it be possible to detect that special causes may have occurred during the production of products over a shift of operation? Would it be possible to develop a track leading back to the root cause of the local faults? Explain.

4.18. What can be said about the variability pattern in a process that is running as it was designed to run? Explain what is meant by the notion "running as it was designed to run."

4.19. A company initiates statistical charting on a process that has never been charted. The operator is trained in making charts and reading the statistical signals. After charting is begun, the process starts to exhibit erratic shifts in variability at random intervals. The operator cannot find any obvious physical cause for this behavior and becomes very frustrated. Are the control charts serving any useful purpose in this situation? What has been done wrong here? Explain.

4.20. What is meant by the term "holding the gains" in reference to SPC implementation? Why is it so important that this be done?

5

Statistical Basis for Shewhart Control Charts for Variable Data

5.1
Introduction

The purpose of this chapter is to provide the reader with the background necessary to understand the statistical foundation on which the Shewhart control chart concept is built. To do this, we need to discuss in some detail the concept of hypothesis testing. We will do this in a somewhat generalized framework and then focus on how this concept is specifically applied and extended for the control chart.

5.2
Concept of Hypothesis Testing

Very often, we are called on to make decisions or draw conclusions on the performance of a process or a product on the basis of sampled data. For example, we might collect data to answer the following questions:

1. Has a new generator design reduced our warranty costs?
2. Is the foam molding process out of statistical control with respect to part weight?

3. Is the concentration of an additive important to the taste of a new food product?
4. Does a new assembly method improve the final product quality?

In making such decisions, we typically make a hypothesis concerning what we believe to be the true state of affairs and then use sampled data to test it. For example, using the scenarios described above, we may hypothesize that:

1. There *is no* difference between the new and old generator designs with respect to the warranty costs.
2. The foam molding process *is* in statistical control.
3. The additive concentration *is not* important to food taste.
4. The new assembly method *does not* affect product quality.

If the data collected to test a hypothesis are inconsistent with it statistically, we must reject the hypothesis. For example, if the data collected for the four scenarios described above are inconsistent with the hypotheses, we must reject them. Therefore, we would conclude that:

1. There *is* a difference between the new and old generator designs with respect to warranty costs.
2. The foam molding process *is not* in control.
3. Additive concentration *is* important to food taste.
4. The new assembly method *does* affect product quality.

In effect, we are rejecting one hypothesis (the one not supported by the data) and accepting another, since the four statements above are also hypotheses. Even if the data collected to test a hypothesis *do* support it, we cannot necessarily accept the hypothesis as the truth. The fact that the data are consistent with the particular hypothesis proposed does not discount the possibility that another hypothesis could be proposed that is also supported by the data. The best that we can say is that the hypothesis has not been disproven. The hypothesis may or may not be true.

Hypothesis testing is very similar to a court of law. In a court of law we assume that the defendant is innocent (we assume that the hypothesis is true) at the outset of the trial. Testimony is presented (data are collected). The jury, or judge, then decides whether the testimony is strong enough to reject the assumption concerning the innocence of the defendant (data do/do not support the hypothesis). If the assumption of innocence is rejected, we conclude that the defendant is guilty. If the assumption of innocence is not disproven, the court does not conclude that the defendant is innocent; rather, it simply concludes that the defendant is not guilty.

In statistical hypothesis testing we generally formulate two hypotheses. The *null hypothesis*, denoted by H_0, will be rejected or nullified if the sample data do not support it. Any hypothesis that is different from H_0 is called an *alternative hypothesis*, denoted by H_1. Whenever an H_0 is rejected, the stated H_1 will be considered accepted. In a court of law the null hypothesis is that the defendant is innocent (H_0) and the alternative hypothesis is that he is guilty (H_1).

In a court of law, guilt must be proven beyond a reasonable doubt. Although it is difficult to quantify a reasonable doubt, we must do so in a statistical test of hypothesis. In all hypothesis tests, the conclusions must be stated with a probability of error. If a sample result differs markedly from that postulated under the null hypothesis (H_0), that is, the chance that it could occur if H_0 is true is some probability α or less, one would begin to think that such a large difference could not have been caused by chance variation alone. In this case, the large difference is said to be statistically significant at a significance level α (the probability of error), and the null hypothesis is rejected. If H_0 is rejected on the basis of data analysis, one accepts H_1 as being true. Because the decision "to reject" or "not to reject" an H_0 cannot be made with absolute surety, one needs to specify the probability of error or risk, α. The process of decision making is called statistical because the determination of significance is probabilistic.

Three Ways of Setting Up the Alternative Hypothesis

Statistical hypothesis testing is a method by which one attempts to prove a point, the alternative hypothesis, H_1, by way of rejecting the null hypothesis, H_0. Therefore, it is important to state the alternative hypothesis correctly. The following example illustrates the subjective nature in formulating the alternative hypothesis.

In the 1970s when several foundries in Milwaukee, Wisconsin, were closing down or moving out of the state, a newspaper reported that the mean wage rate of local foundry workers was $16 an hour. The management of one of the foundries believed that the wage rate was more than $16, while the foundry union believed that its members actually earned less than that. At the same time, a labor economics professor also wanted to dispute the accuracy of the newspaper's figure. In each case, a sample survey could be taken to test the same null hypothesis:

$$H_0: \quad \mu = 16,$$

since all had the same objective, which was to "reject" it (μ represents the true mean value of the wage). But the three parties were to set up three different alternative hypotheses according to their real interests in the event that H_0 were rejected. The three different alternative hypotheses that might be set up are as follows:

Management	Union	Professor
$H_0: \mu = 16$	$H_0: \mu = 16$	$H_0: \mu = 16$
$H_1: \mu > 16$	$H_1: \mu < 16$	$H_1: \mu \neq 16$

The alternative hypotheses above for the management and the union are both called a *one-sided test*; the professor's hypothesis is a *two-sided test*.

Stepwise Approach to Hypothesis Testing

A statistical hypothesis test consists of the following steps:

1. State the null and alternative hypotheses clearly. Define the test statistic used to analyze the situation.
2. Determine a significance level, α, at which the test will be made.
3. Collect the data and calculate the test statistic result.
4. Define the reference distribution for the test statistic.
5. Compare statistically the test statistic and its reference distribution, under H_0. Carry out the necessary analysis of the data.
6. Assess the risk. Weigh the economic or otherwise measurable consequences of saying that H_0 is false when it is true, or failing to recognize that it is false when it is false. This may lead to a redefinition of α. Draw appropriate conclusions.

Hypothesis Testing on the Mean

To test the newspaper claim that the mean wage rate of local foundry workers is $16 an hour, a worker is chosen at random by his union, and his wage rate is found to be $11.50. Historical data suggest that the wage rates follow the normal distribution and that the standard deviation of wage rates is $3. Can the union say that the overall mean wage of Milwaukee foundry workers is less than $16 an hour for $\alpha = 0.05$?

Step 1.
The null and alternative hypotheses are H_0: $\mu = 16$, H_1: $\mu < 16$. The test statistic being used to test H_0 is a single wage rate value, X.

Step 2.
The significance or risk level, α, has been assumed to be 0.05.

Step 3.
$x = 11.50$.

Step 4.
The wage rate values, X's, are known to follow the normal distribution.

Step 5.
Let us compare the specific X value to its corresponding reference distribution under H_0, as shown in Fig. 5.1. Now we must answer the question: Does the observed wage rate value of $11.50 represent a significant departure from the null hypothesis? To answer this question, we can calculate the probability of obtaining a wage rate value less than or equal to $11.50, given that H_0 is true. To perform this probability calculation, we must transform the distribution of

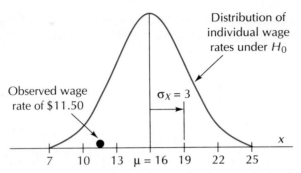

Figure 5.1 Reference Distribution for the Statistic Under H_0

wage rates to the standard normal distribution as shown in Fig. 5.2. The Z value associated with the statistic, $x = 11.50$, may now be calculated:

$$z = \frac{x - \mu_X}{\sigma_X} = \frac{11.50 - 16.00}{3} = -1.50.$$

The standard normal distribution table (Table A.1 in the Appendix) indicates that

$$P(Z \le -1.5) = 0.0668.$$

Therefore,

$$P(X \le 11.50) = 0.0668.$$

Step 6.
Since the alternative hypothesis is concerned only about the wage rate being less than \$16, we should compare the calculated probability (0.0668) with $\alpha = 0.05$. Since 0.0668 is greater than 0.05, we cannot reject the null hypothesis. Therefore, we conclude that there is no reason to doubt the newspaper's claim.

Now let us suppose that 25 foundry workers are randomly surveyed by the professor. He finds the average wage rate for this sample of workers to be

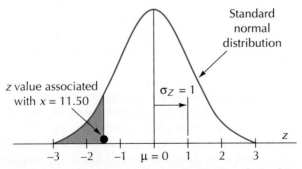

Figure 5.2 Reference Distribution for the Standarized Statistic Under H_0

$14.50. Having analyzed labor data for years, the professor feels confident to assume that the standard deviation of wage rates is $3. Can the professor say that the true mean wage is not $16 an hour at an $\alpha = 0.05$?

Step 1.
H_0: $\mu = 16$ H_1: $\mu \neq 16$. The sample statistic is $\overline{X}$, the sample mean.

Step 2.
Assume that $\alpha = 0.05$.

Step 3.
$\overline{x} = 14.50$.

Step 4.
Regardless of the distribution followed by the individual wage rates, the averages of a number of individual wage rates are approximately normally distributed due to the Central Limit Theorem (Fig. 5.3). The standard deviation of the normally distributed $\overline{X}$'s is given by

$$\sigma_{\overline{X}} = \frac{\sigma_X}{\sqrt{n}} = \frac{3}{\sqrt{25}} = 0.6.$$

Step 5.
We must judge whether we could reasonably expect the observed sample mean to have arisen from the distribution of sample means defined under H_0. To calculate the probability of obtaining an $\overline{X}$ less than or equal to 14.5, we must perform the transformation to the standard normal distribution (Fig. 5.4). The

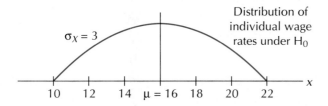

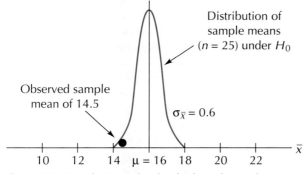

Figure 5.3 Distribution of Individuals and Sample Means Under H_0

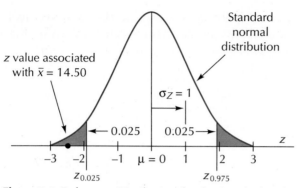

Figure 5.4 Reference Distribution for the Standardized Statistic Under H_0

critical values for Z which place $\alpha/2$ (0.05/2 = 0.025) in each tail of the standard normal distribution (we are concerned with both tails of the distribution since this is a two-sided test) are

$$z_{0.025} = -1.96 \quad \text{and} \quad z_{0.975} = 1.96,$$

where the subscript on z represents the area under the normal curve to the left of that z value. The calculated z value corresponding to the observed sample mean of 14.50 is then

$$z = \frac{\bar{x} - \mu_{\bar{x}}}{\sigma_{\bar{x}}} = \frac{14.50 - 16}{0.60} = -2.50.$$

Step 6.
H_0 is rejected at $\alpha = 0.05$, since z is less than $z_{0.025}$. Equivalently, H_0 may be rejected, since $P(Z \leq -2.5) = 0.0062$ is less than $\alpha/2 = 0.025$. Therefore, the professor has disproved the newspaper's reported wage rate of $16 at a 0.05 significance level.

To carry out a test of the hypothesis as discussed above, the significance level α is interpreted as the probability of rejecting a null hypothesis that is true. This is also referred to as committing a *type I error*. Similarly, a *type II error*

TABLE 5.1 Decision Errors in Hypothesis Testing		
Decision	True H_0	False H_0
H_0 not rejected	No error with probability $1 - \alpha$	Type II error with probability β
H_0 rejected	Type I error with probability α	No error with probability $1 - \beta$

is said to have been committed if H_0 is not rejected at a significance level α when in fact it is false. This type II error, denoted by β, is the probability of failing to reject H_0 given that the true parameter value is at a certain different value, denoted by H_1. Table 5.1 summarizes the errors associated with various conclusions of such a hypothesis test.

5.3

Shewhart Control Chart Model

The Shewhart control chart is a test of hypothesis and therefore also stands much as a jury in a court of law. The information from each sample is judged to determine whether or not it indicates the presence of a special-cause disturbance. Unless the evidence is overwhelmingly in favor of the occurrence of a special-cause disturbance, we enter a verdict: not guilty. In other words, we find no strong reason to believe that forces other than those of common causes are at work. Such an interpretation is, in fact, based on and consistent with the hypothesis-testing approach to making statistical inferences about the process based on information contained in the sample. This was discussed in detail in Section 5.2.

There are two process characteristics that are of general interest in the statistical process control scenario: the mean level of the process and the amount of variation in the process. We are particularly interested in determining when these two characteristics/parameters may have changed in magnitude. We have seen that we could make inferences concerning the mean level of a process through sampling and an associated inferential structure that we called hypothesis testing. The null hypothesis was generally our stated belief about the mean level of the process. Given the sampling distribution of the statistic used to estimate the mean level, we could then evaluate a particular sample result to see if its occurrence was consistent with our belief about the process mean.

Figure 5.5(a) graphically depicts the hypothesis-testing situation we have just presented. If we were to estimate the process mean from the results of a sampling of the process using the arithmetic average $\overline{X}$, the sampling distribution for averages, which is the normal distribution, would provide the probabilistic basis for drawing an inference about the actual process mean. To do all of this, we must have some rational basis for establishing the null hypothesis H_0 in the first place. What do we think the actual process mean is? In the context of the Shewhart control chart method what we do is collect an amount of data (k samples of size n each) that we think is sufficient to allow us to treat the average of all of these data as if it were the actual process mean. Once such a hypothesized mean is established, we can use it as the null hypothesis and conduct the test as illustrated in Fig. 5.5.

In Fig. 5.5(a) we see two sample results ($\overline{X}$ values each from samples of size n), denoted A and B. The reject-H_0 and cannot-reject-H_0 regions are estab-

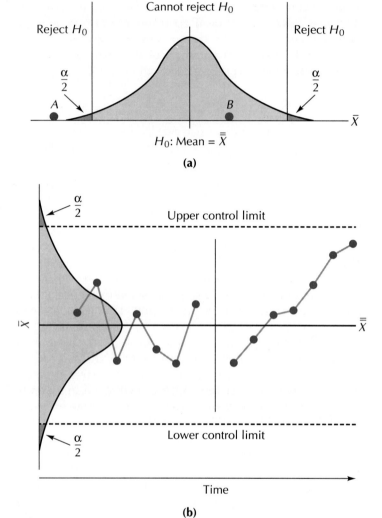

Figure 5.5 Hypothesis-Testing Formulation Extended to Establish the Shewhart Control Chart Model

lished on the basis of the sample size n, the process standard deviation σ_X, and the α risk. Recall that the latter quantity is the risk we are willing to take in concluding that the null hypothesis is false when, in fact, it is really true. Based on sample result A, we would reject the null hypothesis, while the result B leads to the conclusion that we cannot reject H_0.

It is important to note in the formulation above that even if the sample results A and B were obtained from the consecutive sampling of, say, two batches of raw material, they would be used, in the general hypothesis-testing formulation, independent of each other to test the null hypothesis. It was the absence of such serial consideration of sample results that Shewhart recognized when dealing with the results of samples drawn consecutively from an evolving process. His model for the control chart adds this dimension to the problem.

Figure 5.5(b) illustrates how Shewhart extended the hypothesis-testing idea to include the consideration of the time-evolving nature of process behavior. By turning the classical formulation "on its side," considering the horizontal axis to represent time, and extending the reject-H_0, cannot-reject-H_0 lines, he established a model that enables one to track the behavior of the process mean over time. This is done by examining the results of successive samplings to determine in which region they fall and, of equal importance, the relationship that might or might not be evolving when they are considered relative to each other.

Under the assumption that the sample results arise from a constant system of common causes, they should evolve in time in a random fashion. The sequence to the left in Fig. 5.5(b) may be typical of this. On the other hand, if the sample results show evidence of evolving in a nonrandom pattern, such as the sequence to the right, the suggestion is that forces other than those of common causes are at work and some investigation of their origin is warranted. We will see later that a sequence evolving completely within the cannot-reject-H_0 region (what Shewhart called "within the control limits") but having six or more sample results in a row that are continuing to increase provides us with strong evidence that the mean of the process is shifting upward. In doing this, adding a condition(s) on the evolving $\overline{X}$ values for the rejection of H_0 (adding other criteria), we may be increasing the sensitivity of the overall test, but the effective α-risk is increased. This is so since 6 $\overline{X}$ values in a row, continually increasing, could occur due to common causes alone. The value 6 was chosen, so that this risk value is quite small.

To establish Shewhart control charts for the process mean level and process variability, we begin by defining the sampling procedure to obtain the data. For now we will simply say that we will take "snapshots" of the process at periodic intervals by collecting small samples of consecutively manufactured product. How large these samples should be, how often they should be collected, how many should be collected, and finally, the precise manner in which they should be collected are important questions that must be addressed, and will be in Chapter 7.

5.4

Control Chart for Averages, $\overline{X}$

Let us suppose that we select samples of size $n = 5$ every half hour from a process manufacturing gasket material blanks. Each sample will contain five consecutively manufactured gasket blanks. Typical data might be as follows, with the measurements being the thickness of the blanks in inches.

	10:00 A.M.		9:30 A.M.		9:00 A.M.	
	0.044 inch		0.046 inch		0.043 inch	
	0.043		0.044		0.044	
Process ...	0.045	...	0.045	...	0.044	...
	0.044		0.045		0.042	
	0.044		0.047		0.047	
	$\overline{X} = 0.0440$		$\overline{X} = 0.0454$		$\overline{X} = 0.0440$	
	$R = 0.0020$		$R = 0.0030$		$R = 0.0050$	

Each small sample tells us how the process is behaving at any one time in terms of the process mean and variability through the sample mean and the range of the sample. The question is: Is this behavior changing from time to time? If the process is subject only to common-cause variation, the $\overline{X}$'s and R's should be randomly distributed within certain probabilistic limits, as illustrated in Chapter 4. Occasionally, the process may experience some real change in its mean level or amount of variation. When this happens, an $\overline{X}$ or R can indicate this by exceeding the natural variability limits or when a series of either of these values exhibits a nonrandom pattern.

As we saw in Chapter 3, when a series of samples as we defined above are collected on a process in good statistical control, the sample means, $\overline{X}$,

	10:00 A.M.	9:30 A.M.	9:00 A.M.
	$\overline{X} = 0.0440$	$\overline{X} = 0.0454$	$\overline{X} = 0.0440$

will tend to arrange themselves in a frequency sense according to the normal distribution. This is the Central Limit Theorem effect that was discussed in Chapter 3. This tendency for the sample means obtained from rational sampling of a stable process is illustrated again in Fig. 5.6. We will call the limits shown on the right in Fig. 5.6 the *upper control limit* (UCL) and the *lower control limit* (LCL) for the $\overline{X}$ control chart. For the averages $\overline{X}$ of samples of size n varying about a fixed mean $\overline{\overline{X}}$, these limits are defined by

$$\overline{\overline{X}} \pm \frac{Z_{\alpha/2}\, \sigma_X}{\sqrt{n}}.$$

To be precise, $(1 - \alpha)$ 100% of the $\overline{X}$'s should fall within these limits if the process is subject only to common causes of variation. Traditionally, these limits have been established at ± 3 standard deviations from the mean. Hence 99.73% of the $\overline{X}$'s should fall within these limits for a process whose mean level is stable—in statistical control.

It is absolutely essential that we keep in mind that these limits are based on the variation we expect to see in averages, not individual measurements. In

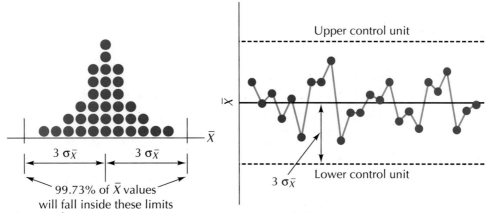

Figure 5.6 Statistical Basis for the Control Limits on the $\overline{X}$ Chart

Chapter 3 the relationship between the distribution of individual measurements and the distribution of averages of several individual measurements was clearly shown. Failure to recognize this important difference is what often leads to an erroneous comparison of control limits and specification limits.

If a process under surveillance by periodic sampling maintains a state of good statistical control, this means that its mean level and level of variability remain constant over time. If such is the case, then, as we have learned, the sample means will have a distribution that follows the normal curve, and therefore nearly all (99.73% to be more exact) of the sample means $\overline{X}$ would fall within the band of ± 3 standard deviations of $\overline{X}$ about the established mean of the process, that we estimate generally by $\overline{\overline{X}}$, the average of a number of averages, $\overline{X}$.

Now suppose that a sample mean $\overline{X}$ falls above the upper control limit. Since the chance of this occurring is so small (about $\frac{1}{8}$ of 1%), if the process mean is at the hypothesized mean of the process $\overline{\overline{X}}$, we must assume that some special cause is present that is influencing this $\overline{X}$ to be larger than expected. In particular, we must assume that the process, at least at that point, cannot reasonably have had the mean level that is represented as the null hypothesis, the hypothesized mean used as the centerline of the chart. Rather, it is likely that an upward shift in the mean has occurred, at least at this point in time. This is the hypothesis-testing interpretation of the data and the one that Shewhart adopted for the control chart. Figure 5.7 illustrates the presence of such a special cause.

Of course, it is possible that no shift in the mean has occurred and therefore we will mistakenly look for a problem that is not there at all. But we know precisely what the chance is that this error will occur. It is the α risk that we defined earlier, the chance that we will reject the null hypothesis that the process mean is the centerline of the chart when, in fact, it really is.

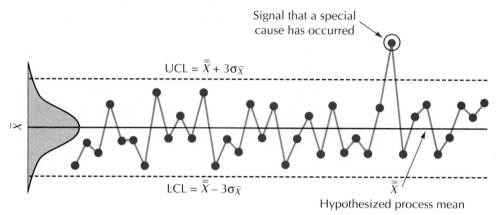

Figure 5.7 Use of Probabilistic Limits to Identify Special Causes

Control Chart Sensitivity

The fact that all points fall within the 3 sigma limits for $\overline{X}$'s and exhibit a random pattern does not mean that no special causes are present. We must also accept the fact that departures from expected process behavior may not always manifest themselves on the control chart immediately, or for that matter, at all. In other words, the mean has shifted—the null hypothesis is now false— but the data do not show this to be the case. This is the β risk we noted in Table 5.1. Although narrower limits (e.g., 2 sigma limits) would enable us to detect the occurrence of special causes easier and faster, such limits would also increase the chances of false alarms, that is, times when the process is actually in good statistical control but a sample mean $\overline{X}$ falls outside the control limits, and therefore we conclude that it is not in control.

The selection of the appropriate placement of the upper and lower control limits, that is, the selection of the α risk, is an economic issue. The intent would be to fix the limits in such a way as to balance the economic consequences of failing to detect a special cause when it does occur—failing to reject the null hypothesis when it is false—and wrongly identifying the presence of a special cause when it has really not occurred—rejecting the null hypothesis when it is really true. Practice over the years has led to the use of 3 sigma limits as a good balance of these two mistakes. But there is nothing sacred about the 3 sigma limit. In the United Kingdom, for example, 99% control limits have been in use for some time.

5.5
Control Chart for Ranges

So far we have confined ourselves to a discussion of the detection of shifts in the process mean. However, sporadic disturbances may also cause shifts or changes in the amount of process variability. In Fig. 5.7 we have represented only one side of the coin. In so doing, we have violated a fundamental rule—that one should never show an $\overline{X}$ chart without also showing the associated range chart. In fact, it is the range chart that is so essential to the Shewhart control chart concept. If the process variation is not stable over time, then the limits, calculated from the unstable process, on the $\overline{X}$ chart do not reflect only the forces of common-cause variation, and therefore the interpretation of the $\overline{X}$ chart could be quite misleading.

A similar hypothesis-testing argument to that above can be put forward for the behavior of a series of ranges of the same samples emanating from a constant system of common causes. The sampling distribution of the ranges then provides the basis for the establishment of the probability limits, which we call the upper and lower control limits when the ranges are plotted in time order of occurrence.

To establish a control chart for the range R, we must be a bit more specific about the distribution of the process (i.e., the quality characteristic of interest X). This is so because the distribution of ranges does not exhibit the same robust behavior as that of the distribution of averages. If we can assume that the individual measurements X are normally distributed, some theory exists that allows us to establish the relationship between the standard deviation of ranges σ_R and the standard deviation of measurements, σ_X.

In general, the distribution of the range may be determined by deriving the distribution of the difference between the largest and smallest values in the sample. By using knowledge of the distribution of the individuals X, such a determination can be made, although the result is much more complicated than that for the $\overline{X}$'s.

For simplicity we will place $+3$ and -3 sigma limits about the average range $\overline{R}$ of a number of sample ranges R to define the upper and lower control limits for R even though the frequency distribution of ranges is not a nice symmetric distribution. Hence the upper and lower control limits for the R chart are given, in principle, by

$$\overline{R} \pm 3\sigma_R.$$

Again, assuming that the quality characteristic X is normally distributed, statistical theory has established the relationship between the standard deviation of the range and the standard deviation of the quality characteristic X:

$$\sigma_R = d_3\sigma_X,$$

where d_3 is a known function of n, the subgroup size. Given the relationship between R and σ_X, we have that

$$\hat{\sigma}_R = \frac{d_3\overline{R}}{d_2}.$$

Therefore, 3 sigma control limits for the ranges are given by

$$\overline{R} \pm 3\left(\frac{d_3}{d_2}\right)\overline{R}.$$

These control limits reduce to forms

$$UCL = D_4\overline{R}$$
$$LCL = D_3\overline{R},$$

where D_4 and D_3 are known functions of n. Values for D_4 and D_3 are tabulated in Table A.2 and their use will be illustrated in Chapter 6. The interpretation of the R chart parallels that of the $\overline{X}$ chart previously discussed, that is, the hypothesis-testing approach enhanced by Shewhart to consider the time order of the data/process.

Summary

Our discussion above constitutes a presentation of the statistical basis for Shewhart control charts in general, and for $\overline{X}$ and R charts in particular. The hypothesis-testing formulation of Section 5.2 is, of course, very general and can be applied to many types of problems. We will examine some of these later in the book. We are now ready to deal with Shewhart control charts in a more mechanical way. In Chapter 6 we establish specific procedures for the construction and interpretation of $\overline{X}$ and R charts.

6

Construction and Interpretation of Shewhart Control Charts for Variable Data

6.1

Introduction

At the end of Chapter 4 we presented the complete process of SPC. In this chapter we focus on two important aspects of this process, *evaluation* and *diagnosis*. We first focus on the construction of $\overline{X}$ and R charts. Then we present some tools that will help greatly in facilitating the fault diagnosis process. In Chapter 7 we will focus on the statistical sampling aspects of *observation*.

Figure 6.1 shows this notion of SPC as a feedback control loop overlaid on the process. A crucial aspect of closing the loop is the ability to convert the statistical signals emanating from the control charts into the root cause(s) of the problem through fault diagnosis. The evaluation stage of the process essentially consists of making control charts from rationally collected data and reading the statistical signals to understand what the charts are telling us about the process behavior in a statistical sense (e.g., behavior of the mean of the process, behavior of the variation of the process). The diagnosis stage is really the key to successful SPC implementation. Once it has been determined that the mean or the range of the process has been influenced by a sporadic/special cause of variation, we must determine the specific physical nature of that cause so that

153

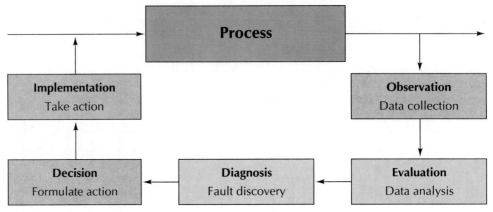

Figure 6.1 Classical Control System View of SPC Implementation

an improvement action can be formulated. At the end of this chapter we examine some useful graphical tools to facilitate fault diagnosis.

6.2
Setting Up $\overline{X}$ and R Control Charts

Once the statistical basis for Shewhart control charts has been established, it remains to define the necessary elements of the charts mathematically and to establish a standard graphical representation. This section deals with the construction of $\overline{X}$ and R control charts. All of the necessary equations and general procedures for the calculation and graphical representation of the basic elements of the charts are presented in this section.

The first step in setting up $\overline{X}$ and R control charts is the selection of the samples. It is important that all samples are *rational samples*, which are groups of measurements whose variation is attributable only to a constant system of common causes. We must strive to choose our samples in such a way as to minimize the occurrence of special causes within the sample. By so doing, we maximize the opportunity to detect special causes when they occur between samples.

Rational sampling is a crucial concept; an entire chapter (Chapter 7) is devoted to this subject. Sampling from different machines, sampling over extended periods of time, and sampling from product combined from several sources are all nonrational sampling methods and must be avoided. As a good rule of thumb, one should select 25 to 50 samples to provide a solid basis for the initiation of the control charts. What this really means is to provide a basis for the estimation of the process mean and process variability with small sampling errors. For reasons to be discussed later, the sample/subgroup size should be relatively small; between $n = 3$ and $n = 6$ is common. A sample size of

$n = 5$ is very commonly used. Once the process is in a state of statistical control, periodic recalculation of the control limits to update the information will be a good practice.

The following simple steps define the procedure necessary to set up $\overline{X}$ and R control charts.

1. From each sample, an average is calculated:

$$\overline{X}_i = \sum_{j=1}^{n} X_{ij}/n, \tag{6.1}$$

where X_{ij} is the jth measurement, $j = 1, \ldots, n$, in the ith sample and where $n = $ sample size.

2. The spread or dispersion within the ith sample is measured by the range R_i:

$$R_i = X_{\text{largest}} - X_{\text{smallest}}. \tag{6.2}$$

Because of its simplicity and ease of calculation, the range, R, is commonly employed as a measure of within-sample variability.

3. The grand average, $\overline{\overline{X}}$, is the arithmetic average of all the available sample averages. This grand average is an estimate of the process mean μ_X and becomes the centerline of the $\overline{X}$ control chart:

$$\overline{\overline{X}} = \frac{\sum_{i=1}^{k} \overline{X}_i}{k}, \tag{6.3}$$

where k is the number of samples being used to set up the control chart.

4. The average of the sample ranges is $\overline{R}$:

$$\overline{R} = \frac{\sum_{i=1}^{k} R_i}{k}. \tag{6.4}$$

5. The true range of samples of size n is related to the standard deviation of the population (process) by the formula

$$\frac{E(R)}{\sigma_X} = d_2, \tag{6.5}$$

where d_2 is a function of the sample size under an assumed normal distribution of X's. Values of d_2 for varying sample sizes n are conveniently tabulated for our use and can be found in Table A.2 in the Appendix. Although this calculation is not needed directly to establish control limits on either the $\overline{X}$ or R charts, it does enter into the formulation as was seen in Chapter 5, and hence is presented here. Since the true mean and true range are estimated by $\overline{\overline{X}}$ and

$\bar{R}$ but these values are assumed to represent μ_X and $\sigma_X d_2$, we emphasize that a minimum of 25, and preferably 40 to 50 subgroups should be available to initiate control charts. If the number of subgroups used is too small, say 10 to 20, the sampling errors associated with $\bar{X}$ and $\bar{R}$ may be large and could lead to misguided conclusions about the state of the process.

    6. We are now ready to calculate the trial control limits for the $\bar{X}$ chart. The limits will be in the form

$$\text{grand average} \pm 3 \text{ sigma of sample averages.} \tag{6.6}$$

The grand average $\bar{\bar{X}}$ is the centerline of the $\bar{X}$ chart. We recall from our previous discussion of basic principles in Chapter 3 that

$$\sigma_{\bar{X}} = \frac{\sigma_X}{\sqrt{n}} \tag{6.7}$$

and

$$\sigma_X = \frac{E(R)}{d_2}. \tag{6.8}$$

The standard deviation of sample means is therefore estimated by

$$\hat{\sigma}_{\bar{X}} = \frac{\bar{R}}{d_2 \sqrt{n}}. \tag{6.9}$$

Thus the formulas for the upper and lower limits of the $\bar{X}$ control chart are

$$\text{UCL} = \bar{\bar{X}} + \frac{3\bar{R}}{d_2\sqrt{n}}$$
$$\tag{6.10}$$
$$\text{LCL} = \bar{\bar{X}} - \frac{3\bar{R}}{d_2\sqrt{n}}.$$

Equation (6.10) can be further simplified. The $3/d_2\sqrt{n}$ term, which is dependent solely on the sample size n, can be combined into a single constant A_2. This gives us

$$\text{UCL} = \bar{\bar{X}} + A_2\bar{R}$$
$$\tag{6.11}$$
$$\text{LCL} = \bar{\bar{X}} - A_2\bar{R}.$$

Values for A_2 for varying sample sizes n are conveniently tabulated for us and are given in Table A.2 in the Appendix.

    7. It should be noted that the distribution for R is not easily determined and is not symmetric about its mean value. However, in the interest of simplicity and ease in establishing the R chart, symmetrical $\pm 3\sigma_R$ limits have been adopted. If the lower limit is less than 0, a lower control limit of 0 is used. As mentioned previously in Chapter 5, it can be shown that σ_R can be estimated as a function of $\bar{R}$. Ultimately, the control limit formulas for the R chart take the form

$$\text{UCL} = D_4\overline{R}, \qquad \text{LCL} = D_3\overline{R}. \tag{6.12}$$

Values for D_3 and D_4 have been tabulated as a function of the sample size n and are also given in Table A.2 in the Appendix.

6.3

Interpretation of Shewhart Control Chart Patterns

General Patterns Indicating Out-of-Control Conditions

The control chart idea is to monitor the process by periodically drawing small samples from production and estimating the process mean and process variability from the sample by the sample mean $\overline{X}$ and the range R of the sample. Such data serve as the basis for establishing Shewhart $\overline{X}$ and R charts.

But before the charts can be used to identify the presence of special causes on-line, the process must be brought into a state of statistical control; that is, the data used to establish the charts and define the control limits must be subject only to a constant system of common causes. Once control is established the charts may be employed to monitor the future behavior of the process. A rigorous examination of the patterns in the data may be quite fruitful in terms of the identification of quality and productivity improvement opportunities.

Too often, $\overline{X}$ and R control charts are examined only for the presence of extreme points (i.e., sample results that fall outside the control limits). However, the Shewhart control chart model suggests that other conditions should also prevail and the absence of these conditions can signal the presence of special causes—chaotic process disturbances.

Since the distribution of $\overline{X}$ is normal, serious departures from normality can signal the presence of special/assignable causes even if all points are within the control limits. Too many points near the control limits, for example, may signal problems with the process, such as an overcontrol of the process mean level. The occurrence of a trend or recurring cycles in the $\overline{X}$'s can indicate special-cause problems that need attention, such as operator fatigue, excessive tool wear, or systematic adjustment of the process.

Runs of points above or below the centerline on either the $\overline{X}$ or R chart may be indicating shifts in mean level or amount of variability, which may not be large enough to manifest themselves quickly through points outside the control limits. It is essential that we strive to associate with each statistical signal of an out-of-control condition the basic process fault that is producing it. Some useful generic conditions to look for might include:

Trends/cyclic behavior. Systematic changes in the process environment, worker fatigue, maintenance schedules, wear conditions, accumulation of waste material, and contamination.

High proportion of points near or beyond control limits. Overcontrol of the process, large differences in incoming raw materials, and/or charting more than one process on a single chart.

Sudden shifts in level.   New machine/die/tooling, new worker, new batch of raw material, change in measurement system, and/or change in production method.

Figure 6.2 shows $\overline{X}$ and R charts constructed from 50 samples of size $n = 5$ measurements, each drawn in succession from a new manufacturing process. Examination of the variability patterns in the charts indicates no reason to suspect that the process was not in good statistical control because:

1. No points exceed the control limits.
2. The points are approximately normally distributed within the limits.
3. The points show no evidence of trends or recurring cycles.
4. The points look quite random with time; that is, no patterns such as runs above or below the centerline are evident.

It is important that we fully appreciate the appearance of a process in good statistical control. We now examine some very specific tests that cover the four items outlined above.

Zone Rules for Control Chart Interpretation

In the discussion above, a number of general control chart patterns have been presented which indicate various types of unnatural process behaviors. Often, these unnatural patterns (e.g., extreme points, trends/cycles, or runs above or below the centerline) are fairly obvious through even the most cursory examination of the charts. However, sometimes special causes produce unnatural

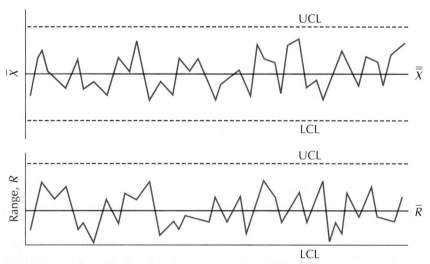

Figure 6.2 Appearance of a Process in Good Statistical Control

Figure 6.3 Control Chart Zones to Aid Chart Interpretation

patterns that are a bit more subtle, and therefore a more rigorous pattern analysis should generally be conducted.

Several useful tests for the presence of unnatural patterns (signals of special causes) can be performed by dividing the distance between the upper and lower control limits on the $\overline{X}$ chart into zones defined by 1, 2, and 3 sigma boundaries, as shown in Fig. 6.3. Such zones are useful, since the statistical distribution of $\overline{X}$ follows a normal distribution if the process is in control, and therefore we expect that certain proportions of the points fall within plus or minus 1 sigma, between 1 and 2 sigma, and so on.

We will now discuss eight specific tests that may be applied to the interpretation of $\overline{X}$ and R control charts.[1] These tests provide the basis for the statistical signals which indicate to us that the process has undergone a change in its mean level, variability level, or both. Some of the eight tests are based specifically on the zones defined in Fig. 6.3 and apply only to the interpretation of the $\overline{X}$-chart patterns. Some of the tests apply to both charts. In these cases the zones are not used. Unless specifically identified to the contrary, the tests/rules apply to the consideration of data to one side of the centerline only.

When a sequence of points on the chart violates one of the rules, the last point in the sequence is circled. This signifies that the evidence is now sufficient to suggest that a special cause has occurred. The issue of when that special cause actually occurred is another matter. A logical estimation of the time of occurrence may be the beginning of the sequence in question. This is the interpretation that will be used here.

Test 1: Extreme Points

Test 1 concerns the identification of extreme points—points beyond the control limits—and therefore applies to both the $\overline{X}$ and R control charts. The specific rule is: The existence of a single point beyond a control limit signals the presence

[1] An excellent article on this subject has been written by L. S. Nelson, "The Shewhart Control Chart: Tests for Special Causes," *Journal of Quality Technology*, Vol. 16, No. 4, 1984, pp. 237–239.

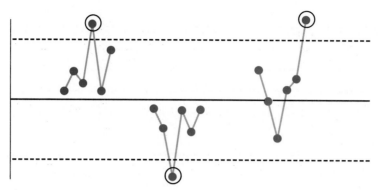

Figure 6.4 Examples of Test 1: Extreme Points

of an out-of-control condition. Figure 6.4 shows several examples of the occurrence of a special cause as indicated by this test.

Test 2: Two Out of Three Points in Zone *A* or Beyond

Test 2 is based on the specific control chart zones and therefore applies only to the $\overline{X}$ chart. The specific rule is: The existence of two of any three successive points in zone *A* or beyond signals the presence of an out-of-control condition. Figure 6.5 shows several examples of how this test is applied.

Test 3: Four Out of Five Points in Zone *B* or Beyond

The probabilistic basis for test 3 is the same, in principle, as that of test 2 and hence the test applies only to the interpretation of the $\overline{X}$ chart. The rule states: Four out of five successive points in zone *B* or beyond signals the presence of an out-of-control condition. Figure 6.6 illustrates the application of this test. As

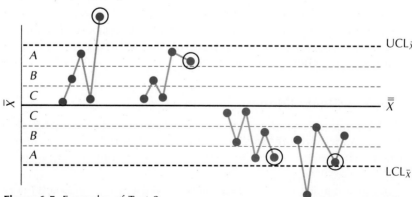

Figure 6.5 Examples of Test 2

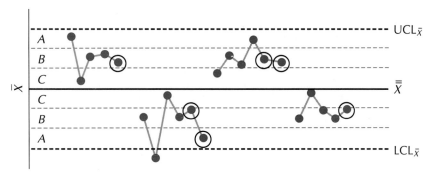

Figure 6.6 Test 3: Four-Out-of-Five Rule

with test 2, this test may help to detect smaller shifts in the process mean which do not give rise to extreme points.

Test 4: Runs Above or Below the Centerline

Test 4 considers long runs (eight or more successive points) either strictly above or strictly below the centerline and therefore applies to both $\overline{X}$ and R charts. The presence of such a run indicates that the evidence is strong that the process mean or variability has shifted from the centerline. Figure 6.7 illustrates the run test. As the run continues beyond eight, each additional point is circled, that is, each point which indicates the endpoint of eight successive points above or below the centerline. This is shown in Fig. 6.7.

Test 5: Linear Trend Identification

When six successive points on either the $\overline{X}$ or the R chart show a continuing increase or decrease, a systematic trend in the process is signaled. Figure 6.8 shows the occurrence of such trends. Neither the zones nor the centerline come into play for this test.

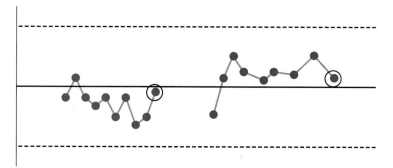

Figure 6.7 Runs Above or Below the Centerline Test

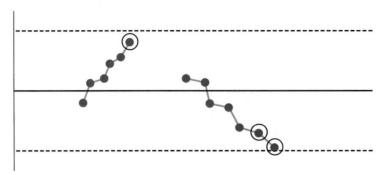

Figure 6.8 Test 5: Upward/Downward Trend Test

Test 6: Oscillatory Trend Identification

In the spirit of test 5, when 14 successive points oscillate up and down on either the $\overline{X}$ or R chart, a systematic trend in the process is signaled. Figure 6.9 shows the application of this test. Again neither the chart centerline nor the zones come into play for this test.

Test 7: Avoidance of Zone C Test

When eight successive points occurring on either side of the centerline avoid zone C (Fig. 6.10), an out-of-control condition is signaled. This rule applies only to the $\overline{X}$ chart. This test could be signaling the occurrence of more than one process being charted on a single chart or perhaps overcontrol of the process. It may also be indicating that improper sampling techniques are being used, in particular, process mixing. This will be discussed later.

Test 8: Run in Zone C Test

When 15 successive points on the $\overline{X}$ chart fall in zone C only, to either side of the centerline, an out-of-control condition is signaled. Such can arise from

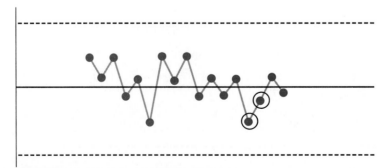

Figure 6.9 Test 6: Oscillatory Trend Test

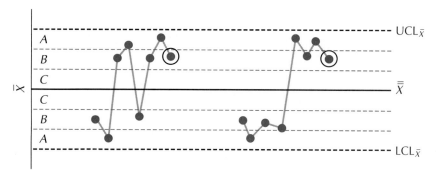

Figure 6.10 Application of Test 7: Mixing/Overcontrol

improper sampling (to be discussed later) or from a change (decrease) in process variability that has not been properly accounted for in the $\overline{X}$-chart control limits. Figure 6.11 illustrates the application of this test.

The series of tests above are to be applied jointly in interpreting the charts. Several rules may be simultaneously violated for a given $\overline{X}$ or R value, and hence that point may be circled more than once. Figure 6.12 illustrates such a situation. In the figure, point A is circled twice because it is the endpoint of a run of eight successive points above the centerline and the endpoint of four of five successive points in zone B or beyond. Another point in this grouping should be circled. Which one, and why? In the second grouping in Fig. 6.12, point B is circled three times because it is the endpoint of:

1. A run of eight successive points below the centerline.
2. Two of three successive points in zone A or beyond.
3. Four of five successive points in zone B or beyond.

Other points in this second grouping should also be circled. Which ones, and why?

The tests discussed above are important to a full interpretation of control

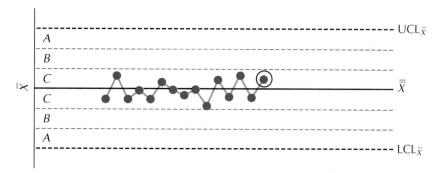

Figure 6.11 Application of Test 8: Stratification/Reduced Variability

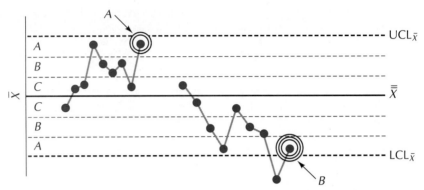

Figure 6.12 Examples of the Simultaneous Application of More Than One Test for Out-of-Control Conditions

chart patterns. These tests may easily be programmed to be carried out by computer. It should be noted that while these tests do improve the sensitivity of the charts, they increase the α-risk as well. Some work has been done recently to quantify the specific nature of the α-risks associated with several of these tests.[2]

6.4

Example: Construction of $\overline{X}$ and R Control Charts

Data were obtained from a process that machines cylinder bores in an engine block. The inside diameter of the cylinder bore was measured following the boring operation. Measurements were made to 1/10,000 of an inch. Samples of size $n = 5$ were taken to obtain some data to initiate $\overline{X}$ and R control charts for this process. The samples were taken roughly every half hour. The sample measurements were all taken on the same cylinder (position-wise) in the block. The results of the first 35 samples are shown in Table 6.1. The actual measurements are of the form 3.5205, 3.5202, 3.5204, and 3.5209 inches, and so on. The table provides only the last three digits in the measurement.

Determination of Trial Control Limits

The sample mean $\overline{X}$ and range R were calculated using Eqs. (6.1) and (6.2) for each of these first 35 samples. The grand average $\overline{\overline{X}}$ and average range $\overline{R}$ were

[2] G. S. Rahn, S. G. Kapoor, and R. E. DeVor, "Development of Operational Characteristics and Diagnostics for $\overline{X}$-Control Charts," Proc. of the Sym. on Sensors, Controls, and Quality Issues in Manuf., ASME, Dec. 1991, Atlanta, Ga.

TABLE 6.1 Cylinder Boring Process Data						
Sample	1	2	3	4	5	$\overline{X}$	R
1	205	202	204	207	205	204.6	5
2	202	196	201	198	202	199.8	6
3	201	202	199	197	196	199.0	6
4	205	203	196	201	197	200.4	9
5	199	196	201	200	195	198.2	6
6	203	198	192	217	196	201.2	25
7	202	202	198	203	202	201.4	5
8	197	196	196	200	204	198.6	8
9	199	200	204	196	202	200.2	8
10	202	196	204	195	197	198.8	9
11	205	204	202	208	205	204.6	6
12	200	201	199	200	201	200.2	2
13	205	196	201	197	198	199.4	9
14	202	199	200	198	200	199.8	4
15	200	200	201	205	201	201.4	5
16	201	187	209	202	200	199.8	22
17	202	202	204	198	203	201.8	6
18	201	198	204	201	201	201.0	6
19	207	206	194	197	201	201.0	13
20	200	204	198	199	199	200.0	6
21	203	200	204	199	200	201.2	5
22	196	203	197	201	194	198.2	7
23	197	199	203	200	196	199.0	7
24	201	197	196	199	207	200.0	10
25	204	196	201	199	197	199.4	5
26	206	206	199	200	203	202.8	7
27	204	203	199	199	197	200.4	7
28	199	201	201	194	200	199.0	6
29	201	196	197	204	200	199.6	8
30	203	206	201	196	201	201.4	10
31	203	197	199	197	201	199.4	6
32	197	194	199	200	199	197.8	6
33	200	201	200	197	200	199.6	4
34	199	199	201	201	201	200.2	2
35	200	204	197	197	199	199.4	7

then calculated using Eqs. (6.3) and (6.4). The control limits for the $\overline{X}$ and R charts were determined using Eqs. (6.11) and (6.12). These calculations are summarized below.

$$\overline{\overline{X}} = 200.25$$
$$\overline{R} = 7.514.$$

For the $\overline{X}$ chart,

$$\text{UCL} = \overline{\overline{X}} + A_2\overline{R} = 200.25 + (0.58)\,(7.514)$$
$$\text{LCL} = \overline{\overline{X}} - A_2\overline{R} = 200.25 - (0.58)\,(7.514)$$
$$\text{UCL} = 204.61, \qquad \text{LCL} = 195.89.$$

For the R chart,

$$\text{UCL} = D_4\overline{R} = (2.11)\,7.514$$
$$\text{LCL} = D_3\overline{R} = (0)\,7.514$$
$$\text{UCL} = 15.85, \qquad \text{LCL} = 0.0.$$

The appropriate values for A_2, D_3, and D_4 are obtained from Table A.2.

   With the calculations above completed we may proceed to construct the $\overline{X}$ and R charts. In plotting the charts, a few simple rules are generally followed:

1. The individual $\overline{X}$ and R values are plotted as solid dots that are connected from sample to sample. This will be helpful in clearly appreciating the patterns in the data.
2. Always plot the R chart directly below the $\overline{X}$ chart using the same horizontal axis scale. This makes it easy to compare $\overline{X}$ and R results for individual samples.
3. Use a heavy solid line to denote the centerline of each chart.
4. Use a heavy dashed line to denote the control limits.
5. Write the specific numerical values for the control limits on the charts as well as for the centerlines.
6. Circle any points that extend beyond the control limits and any points that identify a nonrandom pattern (i.e., that violate any of the tests previously discussed). Points should be circled for each rule that is violated.

   It is essential to start with the R chart. This is necessary since the limits of the $\overline{X}$ chart depend on the magnitude of the common-cause variation that is measured by $\overline{R}$ [recall Eq. (6.11)]. If, initially, some points on the R chart exceed the upper control limit or other indications of special causes are present, the limits in the $\overline{X}$ chart will be inflated (or perhaps deflated) and may therefore project misleading information about the behavior of the process mean.

Interpretation of the Initial Charts

Figure 6.13 shows the $\overline{X}$ and R charts based on the trial limits determined above. Initially, as we examine the R chart, we see that two points exceed the upper control limit. From this we conclude that there are special causes producing an increase in the process variability at least at those points. We now go back and examine these points—samples 6 and 16—to see if we can identify reasons for these special causes. Our records show that at these points, the regular operator was absent, and a relief operator, who was less experienced, was responsible for the boring station in the production line for a short time.

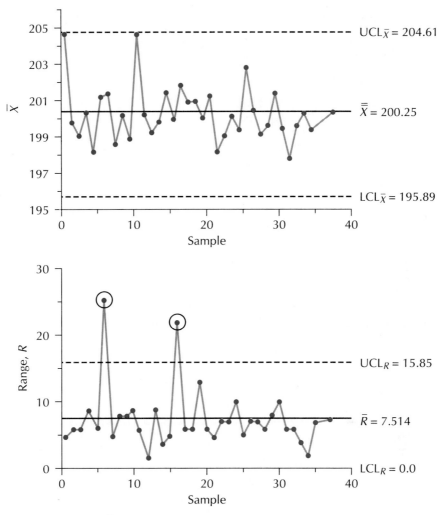

Figure 6.13 Initial $\overline{X}$ and R Charts for Cylinder Boring Example

As a result, samples 6 and 16, taken over the time the relief operator ran the process, exhibited greater variability, perhaps due to inexperience in setting the tools, applying the coolant, and so on.

Since the physical nature of the process driving the special causes present for samples 6 and 16 has been identified, these sample values (both $\overline{X}$ and R) are removed and new centerlines and control limits are calculated. Note that the sample results are removed from both the $\overline{X}$ and R charts (i.e., the entire subgroup is deleted since it has been influenced by a special cause). The calculations for the revised $\overline{X}$ and R charts are

$$\overline{\overline{X}} = 200.24$$
$$\overline{R} = 6.55.$$

For the $\overline{X}$ chart,

$$\text{UCL} = \overline{\overline{X}} + A_2\overline{R} = 200.24 + (0.58)(6.55) = 204.04$$
$$\text{LCL} = \overline{\overline{X}} - A_2\overline{R} = 200.24 - (0.58)(6.55) = 196.44.$$

For the R chart,

$$\text{UCL} = D_4\overline{R} = (2.11)(6.55) = 13.82 \;\cdot$$
$$\text{LCL} = D_3\overline{R} = (0)(6.55) = 0.0.$$

The revised centerlines and control limits for $\overline{X}$ and R charts based on the data after samples 6 and 16 have been deleted are shown in Fig. 6.14.

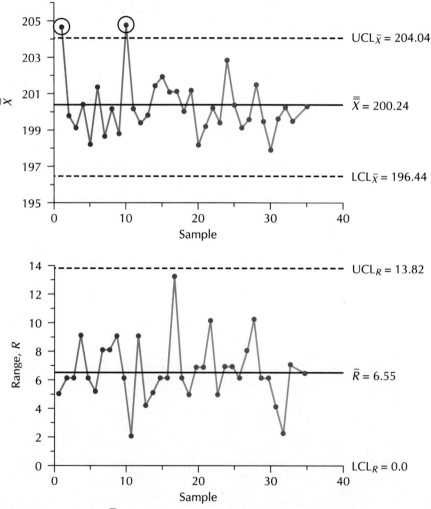

Figure 6.14 Revised $\overline{X}$ and R Charts for the Cylinder Boring Example

Interpretation of Revised Charts

Again, we first examine the R chart. No points are now outside the control limits, and no other unusual patterns of variability appear to be present (i.e., none of the tests discussed previously have been violated). The R chart shows the process to be in good statistical control.

However, when we examine the $\overline{X}$ chart, we see that there are two $\overline{X}$ values above the upper control limit. The investigation of these points reveals the fact that these two samples (1 and 11) occurred at 8:00 A.M. and 1:00 P.M., corresponding roughly to the startup of the boring station in the morning and directly after the lunch hour, when the machine was down for tool changing. It was found that the samples were taken from the first few blocks bored in each case. Once the machine warmed up (approximately 10 minutes) the problem seemed to disappear. It was decided as a policy not to initiate production until 10 minutes after the startup of the machine following any significant period of downtime.

With this special cause identified and the corresponding samples (1 and 11) removed, the new control chart centerlines and control limits are calculated. The calculations are given below. The control charts in Fig. 6.15 now both show good statistical control. We are now ready to extend the limits and begin to monitor the process:

$$\overline{\overline{X}} = 199.95$$
$$\overline{R} = 6.61.$$

For the $\overline{X}$ chart,

$$\text{UCL} = \overline{\overline{X}} + A_2\overline{R} = 199.95 + (0.58)(6.61) = 203.78$$
$$\text{LCL} = \overline{\overline{X}} - A_2\overline{R} = 199.95 - (0.58)(6.61) = 196.12.$$

For the R chart,

$$\text{UCL} = D_4\overline{R} = (2.11)\,6.61 = 13.95$$
$$\text{LCL} = D_3\overline{R} = (0)\,6.61 = 0.0.$$

Summary

This case study points to the importance of maintaining both $\overline{X}$ and R control charts and the significance of first focusing attention on the R chart and establishing its stability. Initially, no points fell outside the $\overline{X}$-chart control limits and we could have been led to believe that this indicates that the process mean exhibited good statistical control from the outset. However, the fact that the R chart was initially not in control has caused the limits on the $\overline{X}$ chart to be somewhat wider because of two inordinately large R values. Once these special causes of variability were removed, the limits on the $\overline{X}$ chart became narrower and two $\overline{X}$ values now fell outside these new limits. Special causes were present

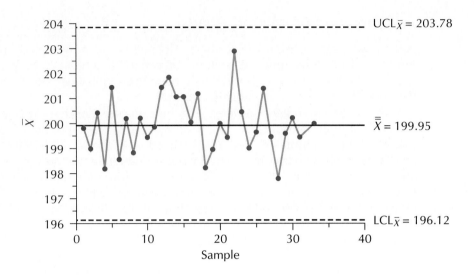

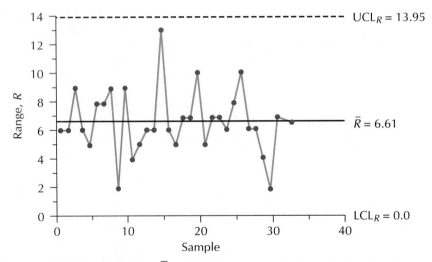

Figure 6.15 Second Revision, $\overline{X}$ and R Charts for the Cylinder Boring Example

in the $\overline{X}$ data but initially were not recognizable because of the excess variability as seen in the R chart.

It is important to note in closing this example that the root causes identified for the out-of-control conditions on both the $\overline{X}$ and R charts are conjecture until verified through continued process monitoring. It is extremely important that this verification take place, and clearly, the $\overline{X}$ and R charts are the appropriate tool to effect this verification.

6.5

Closing the Loop Using the Process of SPC

In previous sections of this chapter we concentrated on only one facet of the complete process of SPC, namely, the evaluation stage of the SPC process discussed at the end of Chapter 4. We have primarily been reading statistical signals to determine whether changes in the process variability and/or the mean might have taken place. We now will present some simple but powerful graphical methods that will help us to take the crucial step from reading statistical signals to diagnosing the cause of the problem. Figure 6.16 again repeats this important graphical depiction of the process of SPC.

Some Useful Graphical Techniques

We now present and discuss three graphical methods that are particularly useful during the fault diagnosis stages of a quality/productivity improvement study: the scatter diagram, the Pareto diagram, and the cause-and-effect diagram. All three of these tools help to focus attention on issues and events relevant to the process at hand and relate effects or symptoms to the factors that are causing them.

Scatter Diagrams

The scatter diagram is a graphical representation aimed at helping to identify correlation that might exist between a quality or productivity performance measure and a factor that might be driving the measure. Figure 6.17 shows one general structure that is commonly used for the scatter diagram. The perfor-

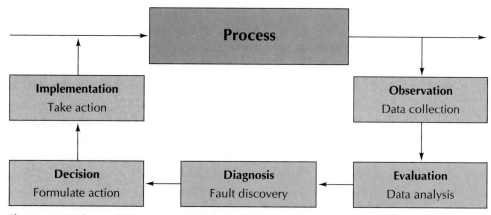

Figure 6.16 Classical Control System View of SPC Implementation

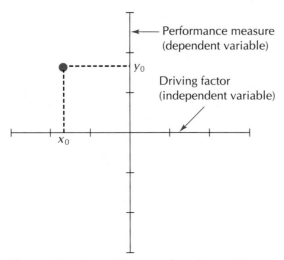

Figure 6.17 General Structure for a Scatter Diagram

mance measure variable (Y) is usually plotted on the vertical axis, while the suspected correlated variable (X) is usually plotted on the horizontal axis. The point where the axes cross is the average of each of the sets of data.

The basic idea of the scatter diagram is to determine if the particular values of the performance measure seem to depend on the values of the factor identified. Although there are a number of more quantitative ways to determine this, the graphical approach is quick, instructive, easy to communicate to others, and generally easy to interpret. Even in the case of the graphical approach, some numbers may be determined, counting the number of occurrences in each quadrant of the plot, for example, to help pin down the presence or absence of correlation. Most often, however, a more qualitative evaluation is made. Figure 6.18 illustrates two hypothetical sets of data plotted on scatter diagrams. In plot (a), the scatter appears quite random and hence we conclude that the two factors seem to have arisen in an unrelated way. In plot (b), however, there appears to be a strong correlation. In particular, the higher the value of X, the higher the value of Y. This is referred to as *positive correlation*. If the plot looked somewhat the same except that the slope of the data was negative, the correlation would be said to be *negative correlation*.

Suppose that $\overline{X}$ and R charts are being kept on the surface finish of a machined part. Further suppose that statistical signals of out-of-control conditions are evident on the $\overline{X}$ chart, while the R chart shows good statistical control. The question is: What is causing the out-of-control conditions on the $\overline{X}$ chart? Suppose that each time a sample was taken for the construction of the charts, observations were also made to record certain key process conditions, such as cutting speed, feed rate, and tool condition. The availability of the latter data provides us with the opportunity to correlate what we see on the charts with what is actually going on in the process.

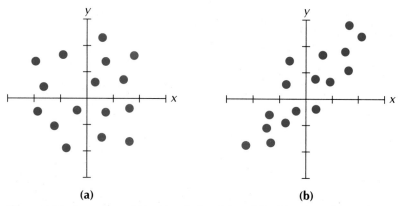

(a)   **(b)**

Figure 6.18 Examples of No Correlation (a) and Positive Correlation (b)

Suppose, for example, that we think that the out-of-control signals on the chart are driven by changes that might be occurring in the tool wear condition. Since the data are observed and recorded in pairs, for example, the part surface roughness is 77 microinches (y_0) and the tool condition is observed to be sharp (x_0), these pairs may be plotted in the manner shown in Fig. 6.19. The data in Table 6.2 are available for 10 samples. These data are plotted as a scatter diagram in Fig. 6.19. Tool condition can take only one of two possible states, while surface roughness is measured along a more continuous scale. From the figure one might conclude that there is some correlation, in particular, that the dull tool condition tends to give rise to a smoother surface (i.e., a lower surface roughness value).

The scatter diagram clearly points to the need to collect data in such a way so as not merely to observe the quality characteristic of interest at points in

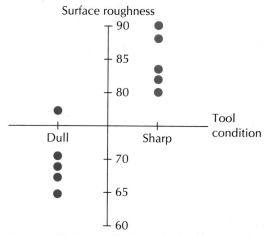

Figure 6.19 Scatter Diagram for Surface Roughness Versus Tool Condition

TABLE 6.2 Tool Condition Data	
Part Surface Finish (microinches)	Tool Condition
80	Sharp
71	Dull
90	Sharp
88	Sharp
69	Dull
82	Sharp
67	Dull
65	Dull
78	Dull
83	Sharp

time during production but also to observe the various states of the process which may be having an impact on the quality characteristic. For example, in monitoring the surface roughness of a machined part, we will, of course, want to sample parts from the production periodically and measure their surface roughness. But at the same time we should note the state of certain factors we have defined as possibly having some bearing on the surface roughness (e.g., tool condition, feed rate, machine operator, etc.). Such information then stands as a basis for us to track down the cause of problems when they occur.

Pareto Diagrams

Pareto diagrams are simply a type of bar chart or histogram which helps us to understand the extent to which a particular phenomenon is occurring relative to the occurrence of others. As such, it helps us to focus on the most frequently occurring problems and prioritize our efforts toward problem solving. This type of diagram is styled after the *Pareto principle,* which says essentially that while the sum total of a certain type of occurrence may arise from quite a number of different sources, it is likely that the vast majority of the occurrences arise from a relatively small subset of these sources. The idea is that if we can identify these few key sources, we can focus our energy on them, thereby hopefully solving most of the problem.

Consider, for example, an injection molding process and the associated defects that can arise on a molded part. These defects could include splay, scratches, flow lines, sinks, and black spots on the surface. We might set up a sampling system whereby we take five successive parts every hour from the production and carefully inspect them for these five types of defects. For each defect we carefully note every incidence of its occurrence. We might be keeping a Shewhart control chart on the number of defects per sample over time.

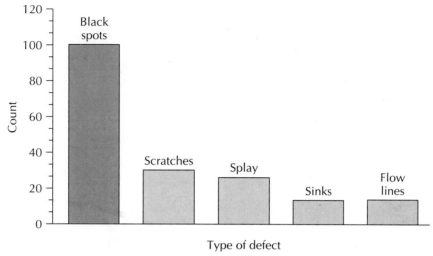

Figure 6.20 Pareto Diagram of Molding Defects

Suppose that the number of defects seems to be fairly stable over time, but the total number is high and we wish to investigate this and perhaps formulate some strategy for the reduction of defects. What should we do first? Which defect or defects should we focus our attention on?

Figure 6.20 is a bar chart of the frequency of occurrence of the various types of defects over a certain period of time. This chart, a Pareto diagram, plots the type of defect along the horizontal axis and the frequency of occurrence of each defect along the vertical axis. The frequencies are plotted in descending order from left to right. The data in Fig. 6.20 are from an actual study conducted in the Saline, Michigan, plant of the Plastic Products Division of the Ford Motor Company.[3] This study is presented and discussed in considerable detail in Chapter 14. From Fig. 6.20 it is clear that over one-half of the total defects are of the black-spots type. Since we generally have limited resources to apply to the problem and we would like to make as much improvement as possible and as rapidly as possible, it seems that we might focus our attention on the black-spots issue. As we will see in Chapter 14, this was the course of action chosen, in this case with considerable success.

A similar Pareto diagram for quality concerns is shown in Fig. 6.21. In this case a quality improvement team was studying an accounts payable process in an administrative area and was collecting data on the various types of problems that can cause a transaction not to be presented for payment. In the figure there are six different types of defects shown that can occur, but the "no purchase order" defect accounts for the vast majority of all the defects that are actually

---

[3] R. J. Deacon, "Reduction of Visual Defects in an Injection Molding Process," *Society of Plastics Engineers, 1983 National Technical Conference Proceedings,* p. 97.

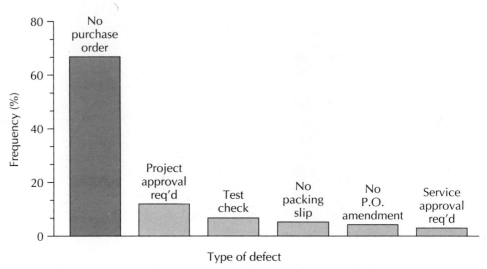

Figure 6.21 Pareto Diagram of Defects Occurring in an Accounts Payable Department

occurring. This case study was also developed within the Ford Motor Company[4] and is discussed in detail in Chapter 14.

Cause-and-Effect Diagrams

If it is the Pareto diagram that helps us to prioritize our efforts and focus attention on the most pressing problem or symptom, it is the cause-and-effect diagram that helps to lead us to the root cause of the problem. The cause-and-effect diagram is a tool that enables the user to set down systematically a graphical representation of the trail that ultimately leads to the root cause of a particular quality concern. The cause-and-effect diagram associates with the symptom the possible factors or causes driving it. This is accomplished in a hierarchical sense, from the broadest of the factor categories to the most specific or lowest-level causal factor under each broad factor. It is best to look at an illustration.

Let us return to the injection molding case study. The symptom being focused on is the "black-spot" defect, and attention now focuses on trying to find the precise cause(s) of black spots. The main factors that could give rise to the problem are the machine, the raw material, the method or process proce-dures, and the human element in the process. The structure of the cause-and-effect diagram, given these factors, is shown in Fig. 6.22. The cause-and-effect diagram is sometimes referred to as a "fishbone" diagram because of its general appearance, shown in Fig. 6.22. To this coarse "skeleton" we begin to add the smaller "bones" by continuing to ask probing questions about the various

[4] R. J. Deacon, "Quality Improvement in an Accounts Payable System," Independent Study Project Report, Eastern Michigan University, College of Business, Ypsilanti, Mich., 1986.

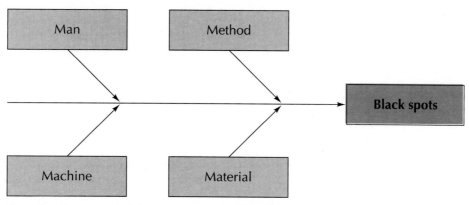

Figure 6.22 Structure of the Cause-and-Effect Diagram

causes of the problem. Why do we think the machine might be the problem? Someone might reply that they think it is a problem with the vent tube. What specifically is the problem with the vent tube? The reply might be that the design is poor, causing material to accumulate, overheat, and then be pushed back into the machine barrel. The root cause of the problem could now be identified; at least we are at a level of specificity that would allow for the development of an action plan.

Figure 6.23 shows the fully developed cause-and-effect diagram for the injection molding black-spot defect concern. Many possible causes of the problem have been identified. It remains to focus on one or two causes for which an improvement action(s) can be developed. Figure 6.24 shows another cause-and-effect diagram, this one for the "no purchase order" defect in the accounts payable case study discussed previously. In this case, the main factors are somewhat different—namely, the buyer, the vendor, the plant involved, and the accounting system—but the idea is still the same.

Cause-and-effect diagrams have several distinct advantages:

- They provide an emphasis that places attention where it ought to be, on looking for solutions to problems at the root cause level rather than on finding ways to treat the symptom.
- They are educational; that is, they summarize the collective wisdom of a group of experts/people who touch the system under study and allow that wisdom to be shared, thereby elevating everyone's understanding of the problem.
- They provide a means to keep the discussion continually focused and converging toward a solution rather than rambling and divergent.
- They provide everyone involved with a quick and clear measure of the level of knowledge currently available on a particular problem. If the diagram does not have very many levels of detail but rather many main factors, this is an indication that either the problem is not well understood or that the right people are not in the room, or both.

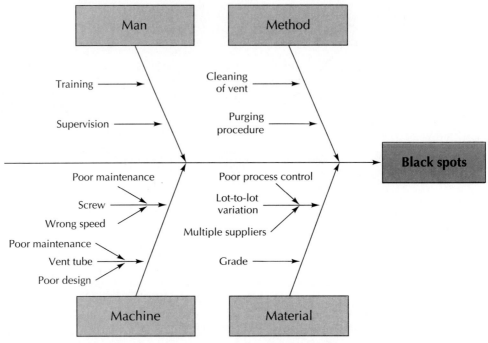

Figure 6.23 Fully Developed Cause-and-Effect Diagram for the Black-Spot Symptom

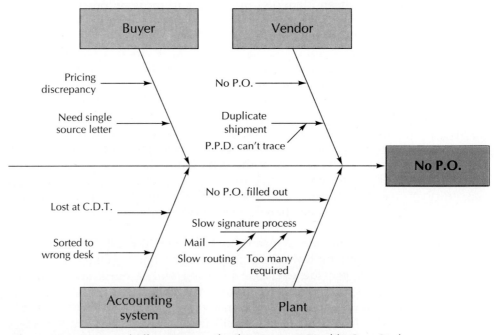

Figure 6.24 Cause-and-Effect Diagram for the Accounts Payable Case Study

Exercises

6.1. What is the purpose of Shewhart control charts? What information can be gained from these charts? How can this information be used to improve the quality and productivity of a manufacturing process?

6.2. Given the table of coating weights shown here for samples of size $n = 4$:
   a. Calculate the $\bar{X}$'s and R's for samples 2 to 8.
   b. Calculate the $\bar{\bar{X}}$ and $\bar{R}$ values based on all eight samples.
   c. Determine the appropriate constants that would be used with these data to construct $\bar{X}$ and R control charts, and calculate the control limits.

Sample	X_1	X_2	X_3	X_4	$\bar{X}$	Range
1	18.5	21.2	19.4	16.5	18.90	4.7
2	17.9	19.0	20.3	21.2		
3	19.6	19.8	20.4	20.5		
4	22.2	21.5	20.8	20.3		
5	19.1	20.6	20.8	21.6		
6	22.8	22.2	23.2	23.0		
7	19.0	20.5	20.3	19.2		
8	20.7	21.0	20.5	19.1		

6.3. Control charts are to be kept on the thickness measurements for a process that rolls 10-gage copper sheets. The current specification on the sheets is 0.1340 ± 0.0010 inch. After collecting 30 rational samples of size $n = 4$ measurements at approximately half-hour intervals, the data were used to determine $\Sigma \bar{X}_i = 4.014$ inches and $\Sigma R_i = 0.027$ inch, $i = 1, 30$. It is desired to set up $\bar{X}$ and R control charts for this process. Use the information above to determine centerlines and upper and lower control limits for the process.

6.4. Averages and ranges of samples of size $n = 5$ are shown in the table. The data pertain to the depth-to-shoulder measurement of glove boxes manufactured by an American car company. The measurements are in inches.

Sample	$\bar{X}$	Range	Sample	$\bar{X}$	Range
1	0.4402	0.015	9	0.4366	0.010
2	0.4390	0.018	10	0.4368	0.011
3	0.4448	0.018	11	0.4360	0.011
4	0.4432	0.006	12	0.4402	0.007
5	0.4428	0.008	13	0.4332	0.008
6	0.4382	0.010	14	0.4356	0.017
7	0.4358	0.011	15	0.4314	0.010
8	0.4440	0.019	16	0.4362	0.015

(continued)

Sample	$\overline{X}$	Range	Sample	$\overline{X}$	Range
17	0.4380	0.019	24	0.4348	0.019
18	0.4350	0.008	25	0.4338	0.015
19	0.4378	0.011	26	0.4366	0.014
20	0.4384	0.009	27	0.4346	0.019
21	0.4392	0.006	28	0.4374	0.015
22	0.4378	0.008	29	0.4339	0.024
23	0.4362	0.016	30	0.4368	0.014

From the first 25 samples, set up and interpret $\overline{X}$ and R control charts. Indicate whether or not control is evident. If the chart shows the process to be in a state of control, plot the next five samples and comment on the results.

6.5. Given the set of data for $\overline{X}$ and R shown in the lower table, plot their values on control charts with the following centerlines and control limits shown in the upper table.

	$\overline{X}$ Chart	R Chart
LCL	9.74	0
CL	10.66	1.59
UCL	11.58	3.35

Sample	$\overline{X}$	Range	Sample	$\overline{X}$	Range
1	10.44	1.8	11	10.56	1.5
2	10.46	1.4	12	9.96	2.4
3	9.98	0.7	13	10.44	2.8
4	9.82	2.6	14	10.96	1.0
5	10.90	2.4	15	11.14	1.5
6	10.36	1.1	16	10.04	0.8
7	11.32	0.8	17	11.44	2.1
8	10.86	1.2	18	11.84	1.0
9	10.58	0.5	19	11.14	2.7
10	9.52	1.9	20	11.44	1.6

   a. Do the charts show good statistical control? Why or why not?
   b. What was the sample/subgroup size used?

6.6. a. Using Table A.2 in the Appendix, find the values for A_2, D_3, and D_4 for samples of size $n = 2$, $n = 5$, $n = 10$. How were the values for A_2 originally determined?
   b. If $\overline{R} = 0.034$ and samples of size $n = 5$ have been collected, what is the corresponding estimate of $\sigma_{\overline{x}}$?
   c. For a certain control chart application the upper control limit on the R chart is 1.840 and the lower control limit on the $\overline{X}$ chart is 4.120. The

centerline for the $\overline{X}$ chart is $\overline{\overline{X}}$ = 4.625. What subgroup size was used for this application?

6.7. An initial set of data has been collected for the construction of control charts to monitor the manufacturing of a spacer with three critical quality characteristics/dimensions: the inside diameter, the outside diameter, and the thickness. Thirty-five rational samples of size n = 6 have been collected for each dimension. For the inside diameter characteristic, $\Sigma \overline{X}$ = 121.423 centimeters and ΣR = 2.714 centimeters. Set up Shewhart control charts for these data using proper convention and clearly labeling all relevant entities on the charts.

6.8. Charts are also to be constructed for the thickness quality characteristic in Exercise 6.7. The quality control engineer left for vacation before finishing the charts. You are given her notes and find that all you know is that the upper control limit for the R chart is 0.0031 centimeter and the upper control limit for the $\overline{X}$ chart is 0.5012 centimeter. The raw data page is missing, but from charts on the other characteristics you learn that the sample size was n = 6. Finish the construction of the charts, again clearly showing the centerlines, control limits, and zone lines and labeling their values.

6.9. A new operator has been working on a foam molding machine that is producing automobile instrument panels. $\overline{X}$ and R control charts are being used to monitor foam thickness. Samples of size n = 4 have been used, but the new operator inadvertently has begun to use a sample size of n = 5.
   a. What impact does this have on the ongoing charting process?
   b. What could be the consequences if the operator had inadvertently changed to a sample size of n = 3?

6.10. Engineers at two sister plants have decided that they would like to compare the variation in performance of two types/manufacturers of end mills used in a machining operation common to both plants. The end mills are procured under the same specifications but have been purchased from different suppliers. Both machines perform the same task, a slotting operation, and the quality characteristic in question is the surface finish of the slot walls. Shewhart control charts have been kept at each plant to monitor the performance, and the engineers have decided to use these to determine which end mill provides for the more consistent surface finish. However, the subgroup sizes used to create the charts are different. The $\overline{X}$ chart for end mill A, established using samples of size n = 4, has an $\overline{\overline{X}}$ value of 85.2 microinches and an upper control limit of 100.3 microinches. End mill B's $\overline{X}$ chart was established using samples of size n = 5 and has an $\overline{\overline{X}}$ value of 84.9 microinches and an upper control limit of 98.5 microinches. Both the $\overline{X}$ and R charts have been shown to be in good statistical control over an extended period of time. Which end mill appears to be producing the most consistent/least variable surface finish?

6.11. Suppose that a co-worker has been using Shewhart control charts to analyze a process but has only been using the upper and lower control limits to determine out-of-control conditions in the form of extreme points without any consideration of the zone rules, runs, and other statistical signals for the $\overline{X}$ chart. When questioning him about this, he responds that we only need to worry about the extreme points, since these are the ones that will indicate problems with lack of conformance of the parts to the specifications. How would you respond to his argument?

6.12. Why is it so important to examine the R chart and interpret its patterns before looking at the $\overline{X}$ chart? Be specific in your discussion.

6.13. Match each of the $\overline{X}$- and R-chart pairs (i) to (vi) shown here with one of the time-varying process behaviors (1) to (10), which could have led to the data on the charts. Clearly identify the location and nature of each of the statistical signals that occur. The time of the process changes and the signals they might produce do not necessarily match precisely across the charts, but the general nature of the patterns can be matched.

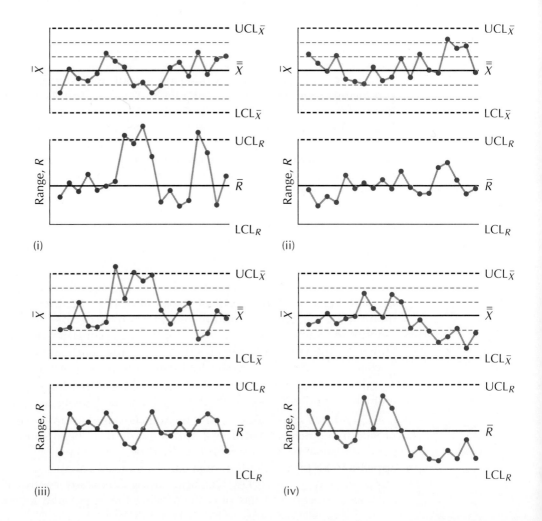

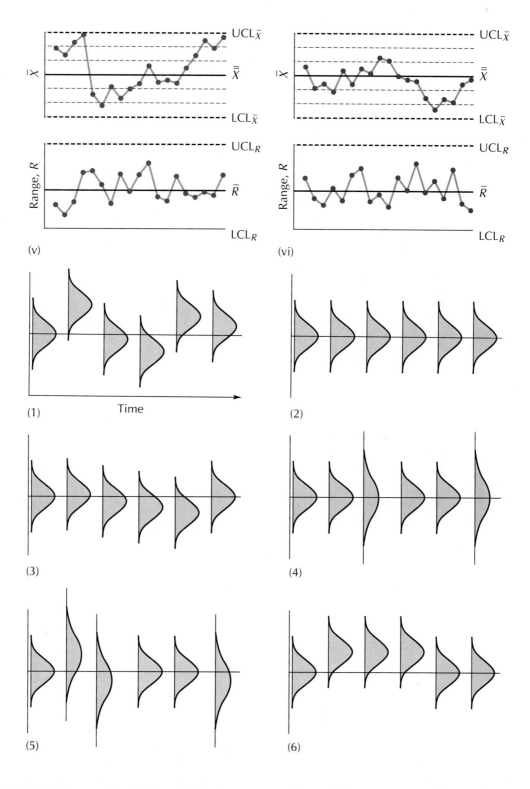

(v)

(vi)

(1)   Time

(2)

(3)

(4)

(5)

(6)

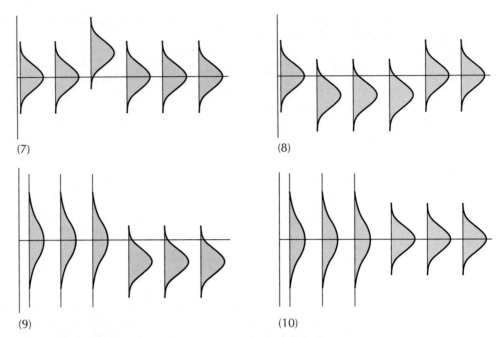

(7)   (8)

(9)   (10)

6.14. Consider charts (i) to (vi), which represent the $\overline{X}$ chart for a certain process. Some of the charts have instances of special cause variation, while others may show good statistical control.

a. For each chart, identify (circle) the points where statistical signals show that special-cause variation is indeed occurring. Indicate the exact rule that applies.

b. Identify from the list below [(1) to (9)] the process behavior or sampling fault that is responsible for the appearance of a special cause on the chart.

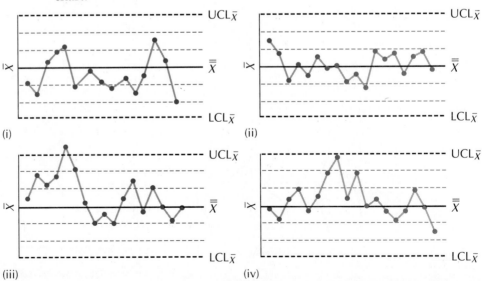

(i)   (ii)

(iii)   (iv)

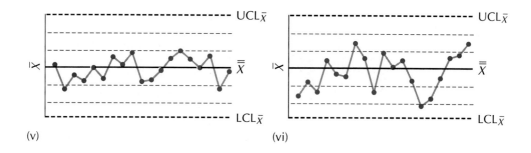

(v)   (vi)

(1) Upward shift in the mean.   (6) Improper sampling.
(2) Downward shift in the mean.   (7) Mixing/overcontrol.
(3) Upward shift in the variance.   (8) Common-cause system only.
(4) Downward shift in the variance.   (9) Trend.
(5) Gradual shift in the mean.

6.15. Use the sketch of the $\overline{X}$ chart shown here as a model in demonstrating the general appearance of the types of process behaviors (a) to (e). On each chart include the occurrence of at least two statistical signals, which could be indicating each type of behavior. Plot 16 points on each chart, clearly identifying the location and type of signals/rule violations.

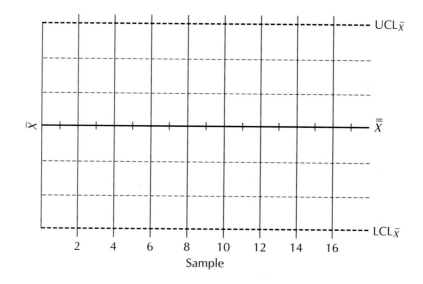

a. An abrupt and sustained upward shift in the mean.
b. Gradual downward shift in the process mean.
c. Overcontrol of the process.
d. Plotting of two processes on a single chart.
e. An abrupt but short-lived downward shift in the process mean.

6.16. Engineers at a manufacturing plant have just been to a course to learn the basics of SPC. They have decided to set up their first charting application on a CNC turning center where a motor shaft is being machined. The quality characteristic that they wish to monitor is the outside diameter, which has a nominal value of 2.125 inches and a tolerance of ± 0.001 inch. Recently, however, they have been feeling pressure from the company to which they are supplying the parts to improve the consistency of the part dimension in question, and the engineers feel that methods of SPC may be able to assist them in reducing variability. Initially, they collect data on the outside diameter in subgroups/samples of five consecutive parts at intervals of approximately $\frac{1}{2}$ hour, with the results shown in the table. The measurements were made in inches to the nearest ten-thousandth of an inch (e.g., 2.1248 and 2.1259 inches, etc.). The data are coded using the last two digits.

Sample	X_1	X_2	X_3	X_4	X_5
1	48	59	42	40	32
2	48	52	51	50	52
3	50	49	60	44	48
4	50	41	48	43	41
5	51	49	60	60	47
6	44	47	54	45	50
7	51	57	53	48	45
8	46	63	63	41	41
9	54	53	53	46	55
10	54	46	55	58	62
11	57	56	53	37	42
12	51	49	67	50	48
13	42	47	48	39	61
14	43	50	44	38	40
15	41	54	59	53	37
16	60	45	52	46	55
17	48	50	45	40	63
18	46	39	43	61	41
19	52	32	54	45	50
20	62	50	47	57	47
21	58	52	57	54	63
22	50	56	44	44	38
23	47	53	42	40	46
24	56	55	44	58	49
25	39	54	54	44	63
26	40	43	59	36	46
27	48	48	34	43	62
28	36	45	43	32	38
29	51	48	24	55	54
30	43	50	40	53	41
31	43	43	49	31	47

(continued)

Sample	X_1	X_2	X_3	X_4	X_5
32	29	42	43	42	43
33	36	43	43	49	48
34	48	42	43	39	46
35	45	51	43	39	29

a. Find $\overline{X}$ and R for each sample.
b. Use all 35 samples to compute the centerlines and control limits for $\overline{X}$ and R charts.
c. Construct the charts, plotting all of the $\overline{X}$ and R values and including all appropriate information on the charts.
d. Apply all the appropriate rules to interpret the charts, indicating every incidence of an out-of-control condition, and identifying the rule that is violated in each case.
e. What can be said about the stability of the process?

6.17. The preliminary results of the investigation in Exercise 6.16 indicated that the cutting tools being used from one of the suppliers did not have the proper edge preparation, causing a more rapid tool wear condition. Once this was determined and verified, the $\overline{X}$ chart seemed to exhibit good statistical control. To continue the charting it was proposed to use control limits determined from the first 25 samples. Twenty-five more samples were collected, again at roughly half-hour intervals, with the resulting data shown in the table. These data are presented in a coded form as in Exercise 6.16, using the last two digits of the actual measurement.

Sample	X_1	X_2	X_3	X_4	X_5	$\overline{X}$	Range
36	50	54	51	45	55	51.0	10
37	54	46	40	53	51	48.8	14
38	60	55	44	47	57	52.6	16
39	46	61	51	56	41	51.0	20
40	39	57	54	56	63	53.8	24
41	52	39	45	48	56	48.0	17
42	49	51	49	54	51	50.8	5
43	51	46	44	50	39	46.0	12
44	54	51	49	48	48	50.0	6
45	54	54	57	50	45	52.0	12
46	54	57	49	47	41	49.6	16
47	46	45	48	51	46	47.2	6
48	49	49	55	53	41	49.4	14
49	45	51	44	55	46	48.2	11
50	54	54	51	47	49	51.0	7
51	43	51	56	52	48	50.0	13
52	51	47	47	56	48	49.8	9
53	50	47	55	59	48	51.8	12
54	48	60	45	49	38	48.0	22

(continued)

Sample	X_1	X_2	X_3	X_4	X_5	$\overline{X}$	Range
55	54	47	49	50	52	50.4	7
56	41	46	46	48	55	47.2	14
57	52	57	47	58	60	54.8	13
58	53	53	55	52	53	53.2	3
59	55	48	52	56	50	52.2	8
60	48	49	43	54	46	48.0	11

a. Determine centerlines and control limits for the $\overline{X}$ and R charts using the data from the first 25 samples from Exercise 6.16. Construct the charts, plotting the first 25 sample results, $\overline{X}$, and R. Verify that the charts to this point show good statistical control.

Note: Read part (b) before you start plotting for part (a).

b. Extend the limits and plot the next 25 sample results (samples 36 to 60).

c. Evaluate the charts using all applicable rules.

d. Comment on what the charts are telling us about the process.

6.18. Our engineer friends from Exercises 6.16 and 6.17 have decided to use their newly acquired quality improvement skills on another problem that has developed. Much rework time has become necessary to correct the depth of a keyway that is being milled into the shaft previously studied. Many of the keyways are being found to be too shallow, causing the key not to fit properly during assembly. A sketch of the part is shown. Measurements are collected in samples of size $n = 4$ consecutive parts at approximately 30-minute intervals over the course of three shifts. The slot depth measurements were made in inches to the nearest ten-thousandths of an inch (e.g., 0.1501 and 0.1494 inch, etc.). The data are coded using the last three digits.

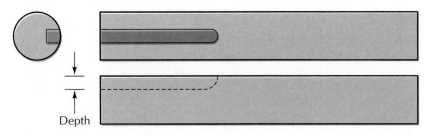

Depth

Sample	X_1	X_2	X_3	X_4	$\overline{X}$	Range
1	501	501	494	496	498.00	7
2	505	493	509	497	501.00	16
3	497	486	503	500	496.50	17
4	507	507	503	494	502.75	13

(continued)

Sample	X_1	X_2	X_3	X_4	$\overline{X}$	Range
5	488	496	496	503	495.75	15
6	505	505	505	488	500.75	16
7	507	496	506	486	498.75	7
8	490	501	494	505	497.50	11
9	490	495	489	505	494.75	15
10	490	493	497	491	492.75	19
11	492	489	485	496	490.50	11
12	496	503	511	499	502.25	15
13	510	507	491	504	503.00	19
14	503	492	496	497	497.00	11
15	503	492	496	497	497.00	10
16	493	500	500	494	496.75	7
17	499	494	512	504	502.25	16
18	505	492	505	493	498.75	13
19	501	505	497	497	500.00	8
20	498	507	496	506	501.75	11
21	502	500	501	501	501.00	2
22	496	492	504	505	499.25	13
23	503	489	501	503	499.00	14
24	490	498	494	509	497.75	19
25	493	502	493	499	496.75	9
26	498	503	490	486	494.25	17
27	500	507	489	491	496.75	18
28	497	502	499	506	501.00	9
29	491	504	506	502	500.75	15
30	502	493	500	501	499.00	9
31	499	504	506	495	501.00	11
32	503	497	502	493	498.75	10
33	493	508	494	496	497.75	15
34	496	495	509	496	499.00	13
35	503	500	508	507	504.50	8

a. Use the data in the table to construct $\overline{X}$ and R charts for the process.

b. Interpret the charts using the appropriate rules. Indicate each incidence of an out-of-control condition.

c. What are the charts telling us about the behavior of the process in terms of its mean and/or standard deviation?

6.19. A study has been initiated on a filling process. The data in the table represent the net weights in pounds (above 20 pounds) for 20-pound bags of dry dog food. The samples of $n = 5$ represent five consecutive bags produced by filling head 2 on a four-head filling machine. The samples were collected at 30-minute intervals. Ultimately, the goal is to fill the bags as close to 20 pounds as possible without going under this nominal weight. Regulations allow only 0.1% of the bags to be below this value. The first step in this study was to examine the filling process from a statistical control point of view.

Sample	Day	Time	X_1	X_2	X_3	X_4	X_5	$\overline{X}$	Range
1	1	7:30	0.92	1.01	0.95	1.04	0.90	0.964	0.14
2	1	8:00	1.15	1.02	0.98	0.94	0.99	1.016	0.21
3	1	8:30	0.94	0.91	1.00	1.05	0.95	0.970	0.14
4	1	9:00	1.11	0.94	0.89	1.11	1.00	1.010	0.22
5	1	9:30	0.95	0.97	0.97	0.98	0.86	0.946	0.12
6	1	10:00	1.02	0.89	0.97	0.95	0.97	0.960	0.13
7	1	10:30	1.18	0.84	0.95	1.39	1.03	1.078	0.55
8	1	11:30	0.94	1.15	1.07	0.99	1.03	1.036	0.21
9	1	12:00	1.03	1.20	1.00	1.10	1.09	1.084	0.20
10	1	12:30	0.98	0.82	0.98	1.02	1.13	0.986	0.31
11	1	1:00	0.98	0.95	0.97	1.04	0.89	0.966	0.15
12	1	1:30	1.10	1.12	1.01	1.12	1.04	1.078	0.11
13	1	2:00	1.10	0.94	0.88	0.92	0.91	0.950	0.22
14	1	2:30	1.01	0.99	1.11	0.96	1.05	1.024	0.15
15	1	3:00	1.17	1.30	1.21	0.69	0.82	1.038	0.61
16	2	7:30	0.97	1.03	1.09	1.04	0.94	1.014	0.15
17	2	8:00	0.92	0.88	0.83	0.94	0.87	0.888	0.11
18	2	8:30	0.99	1.00	0.95	1.00	0.90	0.968	0.10
19	2	9:00	0.88	1.09	1.05	1.05	1.01	1.016	0.21
20	2	9:30	0.87	1.08	0.99	0.97	1.04	0.990	0.21
21	2	10:00	1.08	0.99	1.18	1.02	1.07	1.068	0.19
22	2	10:30	0.60	1.28	0.97	0.84	1.01	0.940	0.68
23	2	11:30	0.89	0.99	1.02	0.95	0.99	0.968	0.13
24	2	12:00	1.01	0.90	0.97	1.09	1.13	1.020	0.23
25	2	12:30	0.95	1.01	1.09	1.10	1.10	1.050	0.15
26	2	1:00	1.10	0.96	1.02	1.03	1.01	1.024	0.14
27	2	1:30	0.92	1.05	1.03	0.99	1.08	1.014	0.16
28	2	2:00	1.00	0.87	1.00	1.05	0.97	0.978	0.18
29	2	2:30	0.96	1.03	1.03	1.11	1.05	1.036	0.15
30	2	3:00	1.15	0.84	1.02	1.18	1.05	1.048	0.34
31	3	7:30	0.91	0.85	0.89	0.82	0.95	0.884	0.13
32	3	8:00	0.95	0.92	0.95	0.84	0.92	0.916	0.11
33	3	8:30	0.98	0.98	1.01	1.12	1.19	1.056	0.21
34	3	9:00	0.89	0.90	1.05	1.05	0.87	0.952	0.18
35	3	9:30	1.16	0.96	0.96	1.06	1.00	1.028	0.20
36	3	10:00	1.10	0.87	0.95	1.05	1.14	1.022	0.27
37	3	10:30	0.83	0.75	1.04	1.25	0.77	0.928	0.50
38	3	11:30	0.98	1.02	1.06	0.87	1.00	0.986	0.19
39	3	12:00	0.95	0.88	0.97	1.01	0.85	0.932	0.16
40	3	12:30	1.04	0.95	1.00	1.14	1.06	1.038	0.19
41	3	1:00	0.98	0.96	1.04	1.09	1.05	1.024	0.13
42	3	1:30	0.80	0.99	0.98	1.03	0.89	0.938	0.23
43	3	2:00	1.04	1.00	0.87	1.02	0.91	0.968	0.17
44	3	2:30	0.94	1.02	0.92	1.00	1.01	0.978	0.10
45	3	3:00	0.71	1.36	1.10	1.24	0.81	1.044	0.65

a. Use the data in the table to establish $\overline{X}$ and R charts for the process.
b. Interpret the charts. If statistical signals/out-of-control conditions are present, assume that the physical cause was identified. Remove the out-of-control subgroups and recalculate the centerlines and control limits.
c. Continue until the charts show good statistical control. Comment.

After these steps were taken a group was formed to study the process further with hopes of reducing the variation in the process. One of the first steps taken was to create a cause-and-effect diagram, as shown.

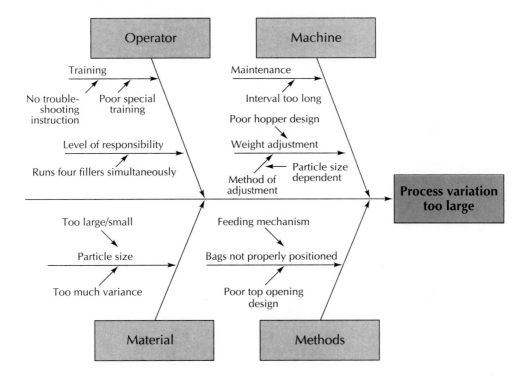

Using this as a guide, it was determined that the most likely cause of variation centered around the fact that adjustments were made assuming a much smaller particle size than actually was present in the dog food. Once this common-cause/chronic problem was addressed, the charting was continued, with the data in the table being collected.

Sample	Day	Time	X_1	X_2	X_3	X_4	X_5	$\overline{X}$	Range
46	4	7:30	0.97	0.99	1.05	0.98	0.96	0.990	0.09
47	4	8:00	1.02	0.99	1.00	1.00	1.00	1.002	0.03
48	4	8:30	0.96	1.05	0.96	1.02	0.97	0.992	0.09

(continued)

Sample	Day	Time	X_1	X_2	X_3	X_4	X_5	$\overline{X}$	Range
49	4	9:00	1.05	1.00	1.00	0.98	0.99	1.004	0.07
50	4	9:30	0.98	1.00	0.95	0.97	0.97	0.974	0.05
51	4	10:00	0.98	0.97	0.98	1.04	0.99	0.992	0.07
52	4	10:30	0.99	1.05	1.03	0.99	0.99	1.010	0.06
53	4	11:30	0.95	0.94	0.99	1.03	1.00	0.982	0.09
54	4	12:00	0.97	1.01	1.01	1.01	1.02	1.004	0.05
55	4	12.30	1.02	0.99	0.97	0.99	1.02	0.998	0.05
56	4	1:00	0.99	1.00	1.01	1.05	1.02	1.014	0.06
57	4	1:30	0.99	0.99	1.00	0.98	1.01	0.994	0.03
58	4	2:00	1.02	1.00	1.01	0.99	0.97	0.998	0.05
59	4	2:30	0.99	0.94	0.98	0.99	0.95	0.970	0.05
60	4	3:00	1.04	1.00	1.01	0.98	0.98	1.002	0.06

    d. Use the charts established in part (c) to continue monitoring this process for day 4. Plot the results of the 15 subgroups above. Comment on the continuing process behavior.

6.20. The manufacturer of exercise weights makes a full range of dumbbells. Recently a new set of molds has been developed for the casting of 6-pound iron dumbbells. However, many of the dumbbells made have been falling outside the specifications of 96 + 1.0 ounce, −0.0 ounces.

    a. Use the data in the table (32 samples of size $n = 4$), collected from one of the new molds roughly at hourly intervals, to establish centerlines and control limits for $\overline{X}$ and R charts. The data are in ounces above 90 ounces.

    b. Plot the values of $\overline{X}$ and R on the control charts. Leave room to plot an additional 12 samples.

    c. What are the charts telling us about the process in terms of the mean and range/variation?

Sample	X_1	X_2	X_3	X_4	$\overline{X}$	Range
1	5.3	4.7	5.1	4.9	5.000	0.6
2	4.7	4.9	5.0	4.9	4.875	0.3
3	5.1	4.5	5.2	5.2	5.000	0.7
4	5.1	5.4	4.6	4.9	5.000	0.8
5	4.9	4.7	4.9	5.2	4.925	0.5
6	5.0	4.7	4.8	4.8	4.825	0.3
7	5.0	4.8	4.7	4.2	4.675	0.8
8	4.6	4.8	5.4	4.7	4.875	0.8
9	5.3	4.7	4.9	4.8	4.925	0.6
10	4.8	5.0	5.4	5.2	5.100	0.6
11	4.7	5.3	4.5	5.2	4.925	0.8
12	5.4	4.8	5.6	4.6	5.100	1.0
13	4.6	5.4	4.8	4.8	4.900	0.8
14	5.4	5.6	5.6	5.3	5.475	0.3

(continued)

Sample	X_1	X_2	X_3	X_4	$\overline{X}$	Range
15	5.0	4.8	4.5	4.3	4.650	0.7
16	5.3	5.6	4.8	5.3	5.250	0.8
17	5.3	4.7	5.1	4.9	5.000	0.6
18	5.3	6.0	5.2	4.9	5.350	1.1
19	5.1	4.4	5.1	4.5	4.775	0.7
20	5.6	4.3	5.2	5.5	5.150	1.3
21	5.6	5.2	4.7	4.7	5.050	0.9
22	5.8	4.8	4.6	4.6	4.950	1.2
23	5.4	5.1	5.0	4.9	5.100	0.5
24	5.6	5.7	4.9	4.8	5.250	0.9
25	4.6	4.8	4.6	5.1	4.775	0.5
26	4.3	4.7	4.8	5.4	4.800	1.1
27	4.9	4.9	5.6	4.8	5.050	0.8
28	4.9	4.7	4.8	4.9	4.825	0.2
29	5.0	5.2	4.7	5.1	5.000	0.5
30	4.9	4.9	4.9	5.3	5.000	0.4
31	5.2	5.1	5.3	4.8	5.100	0.5
32	4.8	5.6	5.3	4.6	5.075	1.0

A team was established to determine the causes associated with having too many of the dumbbells outside the specifications. It became apparent that the consistent occurrence of blows, which produce voids in the casting, was a common-cause/chronic problem. The cause for the blows was found and removed, and the process was continued, with the following data being collected.

Sample	X_1	X_2	X_3	X_4	$\overline{X}$	Range
33	5.0	4.9	5.0	4.8	4.925	0.2
34	5.3	5.2	5.1	5.1	5.175	0.2
35	5.1	5.0	4.9	4.9	4.975	0.2
36	5.2	5.0	5.1	5.1	5.100	0.2
37	4.9	5.0	5.2	5.1	5.050	0.3
38	5.1	4.9	5.1	4.9	5.000	0.2
39	5.2	5.1	4.8	5.2	5.075	0.4
40	4.9	4.9	4.9	5.0	4.925	0.1
41	5.0	5.1	4.8	5.1	5.000	0.3
42	5.1	5.0	5.2	4.9	5.050	0.3
43	4.9	4.9	5.0	5.2	5.000	0.3
44	5.2	4.9	5.1	5.0	5.050	0.3

d. Plot these data on the already existing charts and continue to evaluate the process. Can it be said that the process is continuing in a state of statistical control? Does the team have cause for concern or celebration?

e. Comment on the advisability of continuing with these charts. What might be the next step(s) to take in the charting of this process?

6.21. Data are being collected from a metal stamping process where four presses are simultaneously making small metal clips that will be used in the assembly of automobile instrument panels. The quality characteristic that is important for the clips is an inclination angle that will affect the amount of clamping pressure exerted by the clip. The specifications placed on the angle are 15.0 ± 2.5 degrees. Statistical charting has been initiated on this process, and the averages and ranges of 40 samples are given in the table, along with the associated $\bar{X}$ and R charts. Read the statistical signals and hypothesize as to what may be occurring in this process. Use data/numerical facts to prove your suspicion.

Sample	$\bar{X}$	Range	Sample	$\bar{X}$	Range
1	14.09	10.71	21	15.42	5.98
2	15.81	14.32	22	15.16	9.16
3	13.93	10.46	23	15.21	10.14
4	13.49	7.38	24	15.59	5.32
5	15.29	4.51	25	15.05	10.87
6	16.25	8.49	26	13.97	9.72
7	14.69	5.13	27	15.58	3.95
8	14.88	8.43	28	15.16	9.16
9	15.54	8.86	29	15.62	8.99
10	15.56	8.62	30	13.80	6.48
11	15.56	9.85	31	14.48	7.97
12	14.34	5.69	32	15.67	1.69
13	16.35	5.81	33	14.15	8.99
14	13.48	8.38	34	15.14	8.74
15	14.38	13.84	35	15.27	4.73
16	14.79	3.63	36	13.86	6.45
17	12.29	5.32	37	13.71	10.99
18	15.35	3.25	38	16.10	12.27
19	17.29	10.92	39	14.28	6.14
20	15.20	12.30	40	13.99	4.56

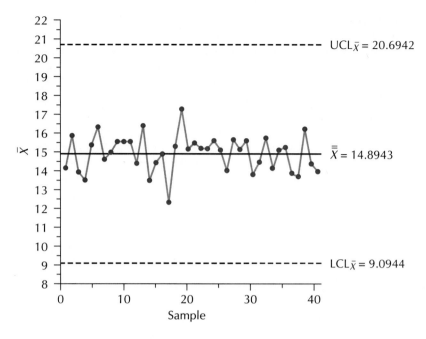

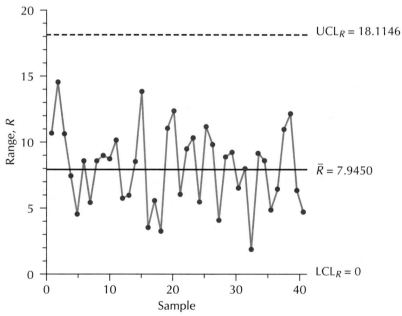

6.22. Explain the Pareto principle as it might apply to studying visual defects in a quality control application.

6.23. A certain grade of polystyrene is being used in an injection molding process used to manufacture a series of toys for children. A study of the process has revealed the defect frequency information shown in the table. Construct a Pareto diagram for the defect information. Comment on the result.

Quality Characteristic	Number of Defects
Scratches	24
Flow lines	189
Flash	17
Voids	96
Short shots	36
Splay	8

6.24. In packaging a board game, many different parts must be included in specific quantities. Recently it has been noticed that occasionally some of the playing pieces, six different-colored plastic parts, have been missing from the completed packages. Sampling over the past week has shown the frequency of missing pieces as given in the table.

Color	Frequency
Black	9
Blue	45
Green	5
Red	8
White	50
Yellow	149

Create a Pareto diagram for the missing pieces.

6.25. Construct a cause-and-effect diagram to show possible causes as to why the crystal glasses shipped from a distribution center by an overnight mail carrier might have arrived at the final destination with some amount of damage.

6.26. An insurance claims agent is preparing a report regarding an accident caused by a car spinning out during a rainstorm. Construct a cause-and-effect diagram outlining possible causes for the accident.

7

Importance of Rational Sampling

7.1

Introduction

We now discuss perhaps the issue most crucial to the successful use of Shewhart control charts: the design and collection of the samples or subgroups. We discuss the concept of rational sampling; talk about sample size, sampling frequency, and sample collection methods; and review some classic misapplications of sampling. Also, we present a number of practical examples of subgroup definition and selection to aid the reader in understanding and implementing this most central aspect of the control chart concept.

7.2

Concept of Rational Sampling

Rational subgroups or samples are collections of individual measurements, whose variation is attributable only to a constant system of common causes. In the development and continuing use of control charts, subgroups or samples should be chosen in a way that provides the maximum opportunity for measurements within each subgroup to be alike (subject only to the forces of common-cause variation) and the maximum chance for the subgroups to differ from one another if special causes arise between subgroups. Figure 7.1 graphically depicts the notion of a rational sample. Within the sample or subgroup only common-cause variation should be present. Special causes/sporadic problems should arise between the selection of one rational sample and another.

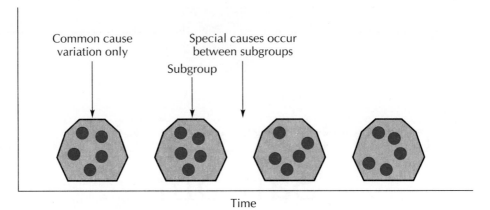

Figure 7.1 Graphical Depiction of the Rational Subgroup

Sample Size Considerations

The size of the rational sample is governed by the following considerations:

- *Subgroups should be subject to common-cause variation.*   The sample size should be small if we wish to achieve this objective. If the sample is allowed to become too large, variation may be introduced within the subgroup due to the occurrence of special causes, such as shifts in the process mean level. If subgroups are large enough to allow this to happen often, primarily because the amount of the process captured within a single subgroup is large, the sensitivity of the control chart will be eroded.
- *Subgroups should ensure the presence of a normal distribution for the sample means.* The construction and interpretation of the $\overline{X}$ chart are based on the assumption that the $\overline{X}$ values follow the normal distribution. In general, the larger the sample size, the better the actual distribution of $\overline{X}$ is approximated by the normal curve. In practice, sample sizes of 4 or more generally ensure a good approximation to normality.
- *Subgroups should ensure good sensitivity to the detection of special/assignable causes.*   The larger the sample size, the more likely we are to detect a shift of a given magnitude.
- *Subgroups should be small enough to be economically appealing from a collection and measurement standpoint.*

When all of the above are taken into consideration, a sample/subgroup size of 4 to 6 is likely to emerge. Five is the most commonly used number, due to the relative ease of further computation, although this is less important today since much of the calculation work is done by computers.

Methods of Subgroup Selection

The most obvious basis of selecting subgroups is the time order of production. One method of selecting subgroups by production time is to sample parts all produced at approximately the same time (i.e., consecutively), for example,

5 consecutive measurements at 9:00 A.M.

5 consecutive measurements at 9:45 A.M.

5 consecutive measurements at 10:30 A.M., etc.

Here the interval between samples is 45 minutes, but we stress that the selection of the sample generally occurs over a much smaller time period within this interval. Taking measurements consecutively, rather than distributed over the interval, permits the minimum chance for other than common-cause variation within subgroups and the maximum chance for special-cause variation arising between the subgroups to be detected. This method is a more sensitive measure of shifts in the process average and variability.

Rational samples or subgroups should be as homogeneous as possible. By this we mean that we wish to obtain a sample where all variation in measurements within the sample is due entirely to one unique common-cause system. To help accomplish this, we wish for our samples to:

1. *Be comprised of measurements that come at approximately the same time*, so that we do not introduce special cause variation due to process changes. This suggests that samples should be selected so that the individual observations are consecutive rather than distributed in a production/time-order sense. More will be said on this later.
2. *Have the time for sampling selected randomly* instead of at exactly the same time every day, thus minimizing the effects of occurrences such as shift change, tool wear, and tool change. This may mean that if samples are collected, on average, about once an hour, varying the actual time by ± 15 minutes might be appropriate.

Sampling Frequency

The question of how frequently samples should be collected is one that needs careful thought. In a good number of the applications of $\overline{X}$ and R charts, samples are selected too infrequently to be of much use in identifying and solving problems. Some considerations in sample frequency determination are

1. *General nature of process stability.* If the process under study has not been scrutinized by control charts before and appears to exhibit somewhat erratic behavior, samples should be taken quite frequently to increase the opportunity to identify improvement opportunities quickly. In some cases, particularly where the application of the process (i.e., the rate of gener-

ation of the quality characteristic in question) is low, all possible process results may need to be observed to obtain enough data to make good charts. In this case individual measurements would be grouped into samples of consecutive observations. As the process exhibits less and less erratic behavior, the sample interval can be lengthened.

2. *Consideration of frequency of process events.* It is important to identify and consider the frequency with which things are happening in the process. This might include, for example, ambient condition fluctuations, raw material changes, process adjustments such as tool changes or wheel dressings, and so on. If the opportunity for special causes to occur over a 15-minute period is good, sampling twice a shift may be of very little value.

3. *Cost of sampling.* Although it is dangerous to overemphasize the cost of sampling in the short term, clearly it cannot be neglected. If destructive testing is required to observe the quality characteristic of interest, frequent sampling is difficult unless the problem is so extreme that economic survival of the process is in serious question anyway. In the past, the cost of sampling was routinely calculated and used as a measure of the overall cost of quality. As discussed in Chapter 2, this approach to quality control is ill-conceived and should be abandoned. What is often not recognized is that the real waste of money due to sampling arises when samples are collected too infrequently, since then the information/value content of the data is so low. There is considerable evidence, as witnessed by these authors on many occasions, to support the conjecture that many companies waste large amounts of money by sampling so infrequently that the resulting charts have no value.

The bottom line in the sampling frequency question is simply that there is no Santa Claus in the world of quality design and improvement. If you wish to use data to find problems and identify improvement opportunities, you have to collect it. Unless processes are very mature in an SPC sense, exhibiting long-term stability and excellent capability, sampling infrequently will not be very useful. On the other hand, intense sampling over a short period of time may lead to considerable insight on the workings of the process.

7.3

Merits of Consecutive Versus Distributed Sampling

The issue of consecutive versus distributed subgroup selection is one that requires some special thought. If a process is subject to abrupt shifts in mean level or variability and those shifts are sustained, consecutive sampling may be preferred. This is so since the shift will be more easily detected from one sample result to another. Figure 7.2 illustrates how the process mean might be experiencing abrupt and sustained shifts over time relative to the consecutive sub-

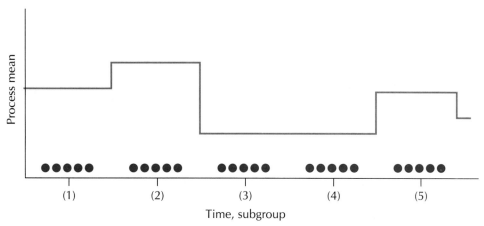

Figure 7.2 Abrupt and Sustained Shifts in the Process Mean and the Use of Consecutive Subgroup Selection

group selection. Figure 7.3 shows what the associated $\bar{X}$ and R charts might look like. However, if a process is subject to frequent abrupt but short-lived shifts in mean level, a distributed sample may be preferred since it provides a better opportunity for detection of a mean level shift through the R chart. Figure 7.4 depicts this type of process behavior and the associated distributed subgroup selection method. Figure 7.5 illustrates how such behavior might be manifest on the $\bar{X}$ and R charts.

If a process is subject to more gradual changes in mean level, this may be detected more readily through distributed sample selection, since a greater opportunity for a large sample range and increased sample mean is present.

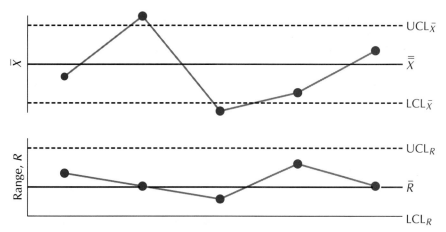

Figure 7.3 Appearance of $\bar{X}$ and R Charts for Conditions of the Nature of Figure 7.2

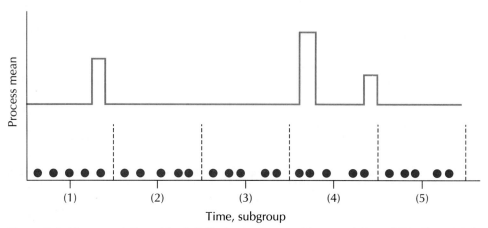

Figure 7.4 Abrupt and Short-Lived Shifts in the Process Mean and Use of Distributed Sub-group Selection

Figure 7.6 shows a gradual shift in the process mean relative to distributed sample selection. Figure 7.7 shows the associated $\overline{X}$ and R control chart patterns that might arise. Here again, a shift in the process mean may be detected on the R chart by forcing the special cause to occur within the subgroup.

   In general, greater attention tends to be focused on the $\overline{X}$ chart and watching for changes in process mean level. Consecutive sampling generally provides greater sensitivity to the detection of mean level shifts since the subgroup variability (and hence the $\overline{X}$-chart control limits) is more likely a reflection of only common-cause variation. Distributed sampling should probably be reserved for those cases where one believes that certain very well known types of mean shifts are likely to occur.

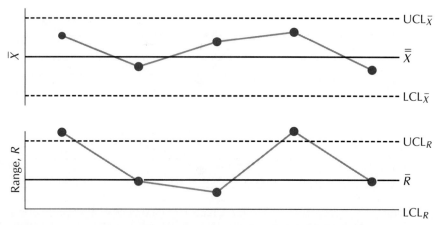

Figure 7.5 Appearance of $\overline{X}$ and R Charts for Conditions of Figure 7.4

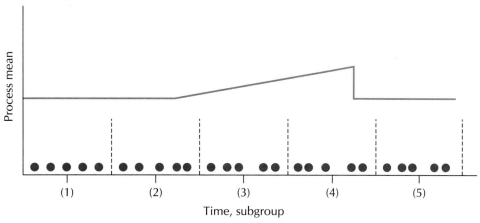

Figure 7.6 Gradual Shift in the Process Mean and the Use of Distributed Subgroup Selection

---

7.4

Some Common Pitfalls in Subgroup Selection

In many situations it is inviting to combine the output of several parallel and assumed to be identical machines into a single sample to be used in maintaining a single control chart for the process. Two variations of this approach can be particularly troublesome: stratification and mixing.

Stratification of the Sample

In the stratification of subgroups, each of several machines, spindles, filling heads, and so on, contributes in the same representative way to each sample,

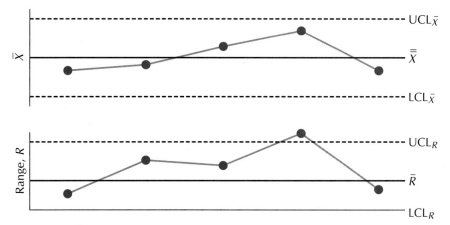

Figure 7.7 $\bar{X}$ and R Chart Patterns for Conditions of the Nature of Figure 7.6

Stratified sampling

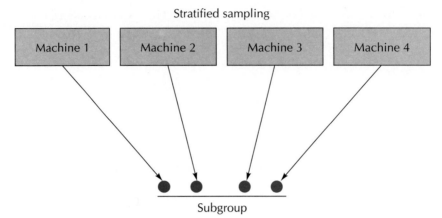

Figure 7.8 Graphical Depiction of Stratified Sample Selection

for example, one measurement each from four parallel machines, yielding a subgroup of $n = 4$, as seen in Fig. 7.8. In this case there will be a tremendous opportunity for any true differences among the machines to occur within subgroups.

When serious problems do arise, say for one particular machine, they will be very hard to detect because of the use of stratified samples. This sampling problem can be detected on the control charts, however, because of the unusual nature of the $\overline{X}$-chart pattern (recall the previous pattern analysis), and therefore rectified, provided that we understand the concepts of rational sampling. The R charts developed from such data will usually show good control (unless the differences in the process/machine means are very large), while the corresponding $\overline{X}$ control chart will tend to show very wide limits relative to the plotted $\overline{X}$'s and hence their control will appear almost too good. The wide limits result from the fact that the variability within subgroups is likely to be subject to more than merely common causes (e.g., real differences in the process means). This is illustrated in Fig. 7.9. Test 8 of the zone rules is specifically directed to this type of sampling problem.

How Stratification of Subgroups Can Occur

Stratification of the subgroup occurs when the subgroup is comprised of representative items that are drawn from similar but distinctly different processes. Stratification occurs primarily for one of three reasons.

1. The process yields a product quality characteristic in a way that gives rise naturally to multiple occurrence of the item. Examples of this would include multiple cavity molds that produce several parts simultaneously, a multihead filling machine that fills several bags, cans, bottles, and so on, simultaneously, or a multiple-orifice extrusion die. In each case the parts

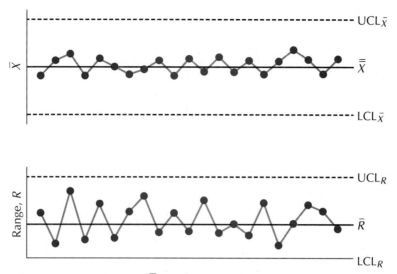

Figure 7.9 Appearance of $\overline{X}$ and R Charts for Stratified Samples

being made are the same, but they arise from distinctly different applications of the general process.

2. There is a conscious effort to distribute the "points of sampling" in an orderly fashion in time or space. Examples of this type of stratification are sampling uniformly across the width of a sheet as it exits a roll mill, taking several measurements around the circumference of an extruded rubber hose, measuring a dimension on a forging several times over a period of time during cooling, and taking measurements of several different bearing journals after a grinding operation on a crankshaft.

3. An effort is made to shortcut the SPC application by taking a representative sample across several like machines, or lines, selecting one item from each as they proceed roughly in unison. One example here might be drawing one item each from four parallel injection molding presses, all making the same part. Another example might be taking one item each from three CNC turning centers, all making the same flange.

   In each of the situations mentioned above, the common denominator is the fact that the items which comprise the subgroup are different for some known and predictable reason, because of an assignable cause. As such, the principle of rational sampling is violated in each case because the variation within the subgroup is not comprised solely of common causes.

   In a roll mill application, sheets exiting the rollers will have a predictable variation in thickness from end to end due to the deflection of the rollers and the associated crowning effect as seen in Fig. 7.10. If four measurements of the sheet thickness are taken across the sheet width as indicated in the figure to comprise the subgroup, the known differences in thickness become part of the

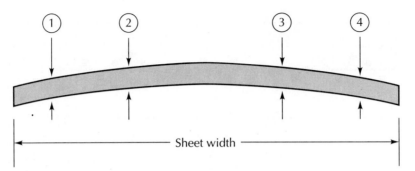

Figure 7.10 Sampling Across a Sheet of Nonuniform Thickness

within-subgroup variation. The subgroup variation will be large and the associated $\bar{X}$ and R charts will have little sensitivity to detect other special causes/ sporadic problems when they occur.

Figure 7.11 shows a schematic of a crankshaft with several bearing journals that undergo a final grinding operation. Due to the nature of the fixturing, the deflections during grinding will be different for different journal locations. Because of grinding wheel differences due to the wheel redress cycle, different grinding forces will prevail for different journals. If a subgroup of $n = 4$ is formed by measuring each of the four journal locations as indicated in the figure, these predictable/assignable differences will become part of the within-subgroup variation, again seriously inflating the subgroup ranges.

In each of the physical situations described above, each of the individual strata (a mold cavity, a filling head, a sheet thickness location, a single journal) is a separate process and should be treated as such in developing statistical charts. In each case, more would probably be learned that would contribute to improvement even if only one of the strata was sampled and charted.

Mixing the Production from Several Machines

Often it is inviting to combine the output of several parallel machines/lines into a single stream of well-mixed product, which is then sampled for the purposes of maintaining control charts. This is illustrated in Fig. 7.12. Often mixing may

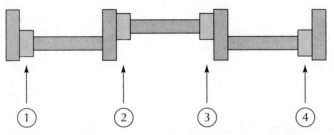

Figure 7.11 Schematic of a Crankshaft Journal Grinding Sampling Situation

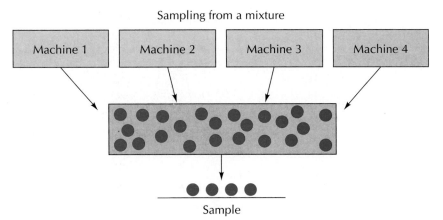

Figure 7.12 Graphical Depiction of Sampling from a Mixture

be inadvertent or occur totally unknown to those sampling the process. We can think of mixing in much the same way that we think about stratification in the sense that it tends to arise when processes are present that have multiple applications in close proximity. Many of the examples cited above for stratification can also easily give rise to mixing. The basic difference is that instead of having a representative sample with all processes being equally represented in each sample, in mixing, the composition of the sample arises through a random sampling of the aggregation of items across all of the processes.

When mixing such as that depicted in Fig. 7.12 occurs, a new distribution of the quality characteristic in question is created such that the control charts often appear in good statistical control. Hence not only will mixing mask problems with individual processes/machines, but it will itself often be difficult to detect. If anything, mixing may tend to produce an appearance in the control charts (particularly the $\overline{X}$ chart) where the points are distributed more toward the control limits than they might otherwise be. This may be more apparent if one (or perhaps two) of the machines have means quite different from the others.

7.5
Examples of the Stratification Phenomenon

Example 1: Filling Process

As an illustration of how stratification during sampling can lead to an unnaturally quiet pattern on the $\overline{X}$ chart, consider the following case: In this situation a sample of size $n = 4$ was taken, one observation from each of four different filling heads on the same liquid filling machine. However, unknown to the quality analyst recording the data, the individual outputs (mean levels) of the four filling heads were somewhat different, as indicated by the four distributions in Fig. 7.13.

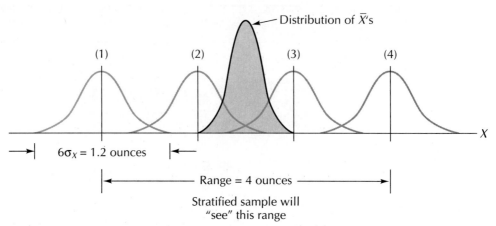

Figure 7.13 Four Parallel Filling Heads on the Same Machine

For this filling head example, given that one measurement comes from each head to comprise the sample, we expect the ranges of the samples to be quite large (i.e., $\bar{R} \approx 4.0$). Then our estimate of the process standard deviation will be

$$\hat{\sigma}_X = \frac{\bar{R}}{d_2} = \frac{4}{2} = 2 \text{ (for } n = 4, d_2 \text{ is approximately 2),}$$

whereas if we consider only a single filling head, as seen in Fig. 7.13, the $6\sigma_X$ spread is 1.2 ounces, so

$$\sigma_X = \frac{1.2}{6} = 0.2 \text{ ounce.}$$

Comparing the actual $\sigma_X = 0.2$ with the estimated value resulting from ranges of stratified samples $\hat{\sigma}_X = 2.0$, we find that they differ by a factor of 10.

Thus, based on the $\bar{R}$ from the stratified sample, the control limits on the $\bar{X}$ chart will be unnaturally wide, and the sample values will hug the chart centerline. There will be an absence of points beyond $\pm 1\sigma_{\bar{X}}$. In fact, the distribution of the averages will be centered approximately at the middle of filling heads 2 and 3, as shown in Fig. 7.13.

Another indication of the representative but inappropriate sample will be a relatively flat histogram of all the individual values, which have come from a very specific form of combined populations—the stratified sample. Other examples of stratified sampling include:

Multispindle screw machine

Multicavity mold

Grinding multiple journals—crankshaft

Stratification is, of course, avoided by taking rational subgroups from each filling head or each machine position and setting up $\overline{X}$ and R charts for each process. Remember, control charts are applicable to one and only one process at a time, and there is no gain in taking a representative rather than a rational sample. In fact, one can create a serious misinterpretation of the process behavior through representative sampling.

Example 2: Process Manufacturing Resistors

A consultant was visiting a plant that was having some quality problems and was, as a matter of course, reviewing each process running in the plant. In one instance control charts were being maintained for a process making electrical resistors, and the consultant was given the charts to study. The process had already been deemed to be showing excellent long-term stability, and the sample measurements were being routinely collected and charted on the floor. The quality characteristic being measured is the resistance, the nominal/target value being 100 ohms, with a specification of ± 2 ohms. The most recent data used to establish the control charts are shown in Table 7.1 (all measurements are in ohms). The consultant was asked to study this process because although it showed good statistical control, the process capability study indicated that the process variation was quite large ($\sigma_X = 2$) and as many as one-third of the resistors were out of the specifications. For some time the resistors were being 100% inspected and many bad ones were found. The quality team assigned to improve the process was looking for common causes of variation but with no success. Once the data had been collected, the limits for the control charts were calculated and the charts were constructed, as shown in Fig. 7.14.

$$\overline{\overline{X}} = \sum \frac{\overline{X}_i}{45} = 100.000 \qquad \overline{R} = \sum \frac{R_i}{45} = 4.013$$

$$\text{UCL}_{\overline{X}} = \overline{\overline{X}} + A_2\overline{R}$$
$$= 100.000 + (0.73)(4.013)$$
$$= 102.929$$
$$\text{LCL}_{\overline{X}} = \overline{\overline{X}} - A_2\overline{R}$$
$$= 100.000 - (0.73)(4.013)$$
$$= 97.071$$
$$\text{UCL}_R = D_4\overline{R}$$
$$= (2.28)(4.013)$$
$$= 9.150$$
$$\text{LCL}_R = D_3\overline{R}$$
$$= (0)(4.013)$$
$$= 0.000.$$

TABLE 7.1   Electrical Resistor Data

Sample	X_1	X_2	X_3	X_4	$\overline{X}$	R
1	98.18	99.85	100.11	101.46	99.90	3.28
2	97.15	100.15	101.53	101.89	100.18	4.74
3	98.24	98.28	100.56	101.35	99.61	3.11
4	97.32	98.68	101.20	102.22	99.85	4.90
5	97.82	98.16	101.46	102.08	99.88	4.26
6	98.33	98.93	100.99	102.12	100.09	3.79
7	98.21	99.19	100.75	102.41	100.14	4.20
8	97.75	98.76	101.39	102.33	100.06	4.58
9	97.59	98.71	100.00	101.61	99.48	4.02
10	98.02	98.90	101.64	101.74	100.07	3.72
11	98.83	99.49	101.89	100.85	100.27	3.06
12	98.89	99.42	100.26	102.78	100.34	3.89
13	98.01	99.04	101.46	101.95	100.12	3.94
14	98.53	98.72	101.36	102.21	100.20	3.68
15	98.06	98.76	101.07	101.91	99.95	3.85
16	97.86	98.51	101.57	102.21	100.04	4.35
17	97.95	98.89	100.14	102.84	99.95	4.89
18	98.12	99.07	101.56	101.45	100.05	3.44
19	97.88	98.56	101.06	101.19	99.67	3.31
20	97.83	98.81	100.99	102.50	100.03	4.67
21	98.99	100.22	101.14	102.11	100.61	3.12
22	98.49	99.82	100.39	101.80	100.13	3.31
23	97.58	99.15	100.99	102.17	99.97	4.59
24	98.51	99.43	100.92	102.72	100.40	4.21
25	98.27	99.04	101.09	101.93	100.08	3.66
26	97.35	98.57	100.78	102.21	99.72	4.86
27	98.10	99.99	101.33	101.78	100.30	3.68
28	97.12	98.50	101.56	102.70	99.97	5.58
29	97.21	99.66	100.47	102.36	99.93	5.15
30	98.24	99.24	100.38	102.50	100.09	4.26
31	97.99	99.18	101.67	102.07	100.23	4.08
32	98.08	99.06	101.21	101.93	100.07	3.85
33	97.89	98.71	100.98	101.31	99.72	3.42
34	97.72	98.19	101.46	101.43	99.70	3.71
35	97.03	99.23	100.85	102.53	99.91	5.50
36	98.03	98.39	100.93	100.99	99.58	2.96
37	97.72	99.30	101.54	101.57	100.03	3.85
38	97.76	99.00	100.96	100.86	99.64	3.20
39	98.36	99.17	101.28	102.55	100.34	4.19
40	97.89	99.10	100.50	101.71	99.80	3.82
41	97.84	98.46	101.43	101.77	99.87	3.93
42	98.28	98.82	101.00	102.45	100.14	4.17
43	97.82	99.92	100.75	101.66	100.04	3.84
44	98.92	98.37	99.99	102.74	100.00	3.82
45	98.16	98.11	101.13	102.25	99.91	4.14

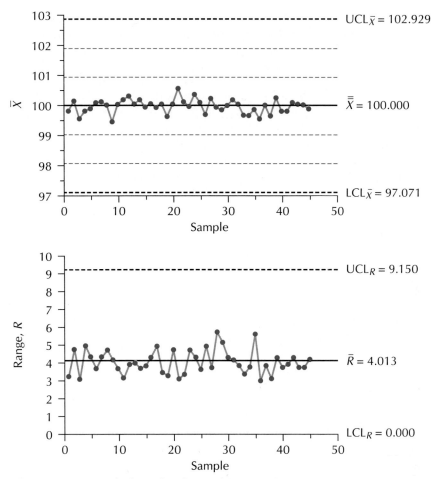

Figure 7.14 Control Charts for Electrical Resistor Data

The zones referred to in Chapter 6 as A, B, and C on the $\overline{X}$ chart can be found by dividing the distance between the centerline and the upper/lower control limit into three equal widths, the one closest to the centerline being zone C, the middle one being the zone B, and the farthest outlying zone being zone A.

Apparently, neither the process operators nor the supervisors had seen problems with this process. The R chart in Fig. 7.14 showed good control, and so did the $\overline{X}$ chart. The $\overline{X}$ chart in Fig. 7.14 shows remarkably good control; it looks almost "too good." All the points are within 1 sigma of the centerline, so immediately it was apparent to the consultant that there was something unusual about the data. The $\overline{X}$ values should arise in a distributional pattern that follows a normal distribution. This means that about one $\overline{X}$ in three should fall beyond $\pm$ 1 sigma from the centerline.

The management of the company was surprised when the consultant decided to explore further the operation of this process; the process seemed to them to be in very good control. The consultant was quick to point out that the unusual pattern on the $\overline{X}$ chart indicated that something was amiss. The manner in which the points on the $\overline{X}$ chart seem to hug the centerline, combined with the fact that the R chart also appears to have an unusually high percentage of points within the middle third of the chart, point to a possible problem with the sampling procedure. The consultant decided to find out exactly how the samples on this line were being collected.

It turns out that the line was set up so that each machine manufacturing the resistors would turn out a stream of parts that would travel to another machine responsible for encasing the parts with an insular coating. This machine was set up to receive one part from each of the four machine streams, coat them, and then eject the four parts simultaneously from the mold. At this point it was convenient to define and physically collect a sample of these four resistors before the parts had a chance to move along a conveyor belt to be color-coded for identification purposes farther down the line. A schematic of the process is shown in Fig. 7.15.

It is clear that a fundamental sampling error known as *stratification* is taking place. Four machines are contributing equally to the composition of the sample, which is then observed for control chart purposes. Because the machines are producing the same product does not necessarily mean that they are identical products. The four machines are in actuality four different processes, so combining the product in this way (i.e., stratifying the data) creates a tremendous opportunity for special causes to occur within subgroups rather than between them. As a result, the within-subgroup range could be quite large, reflecting differences in the mean output of each of the four machines. This means that

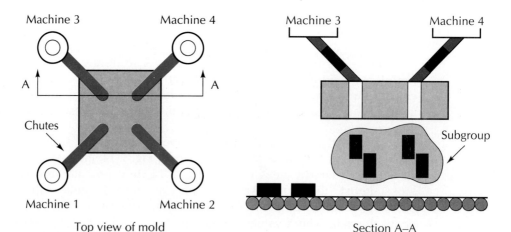

Figure 7.15 Schematic of the Resistor Coating Process

the ranges R will be inflated by such differences, the limits on the $\overline{X}$ chart will be very inflated/wide, and as a result, it becomes very difficult to identify any problems. The process seems to be in control, but the chart is actually very insensitive to changes in the mean level(s) of the individual machines.

The solution to this problem is to make separate control charts for each machine that is producing resistors, so that common causes of variation can be properly observed and any special causes properly identified. In this way, variation and inefficiency in the process can be reduced, making the process stable, more economical, and more capable. Once the error in the sampling method was pointed out, the sampling point was moved upstream, so that sampling could be performed before the parts reached the molding process. Under this system, rational samples could easily be taken from each/any of the

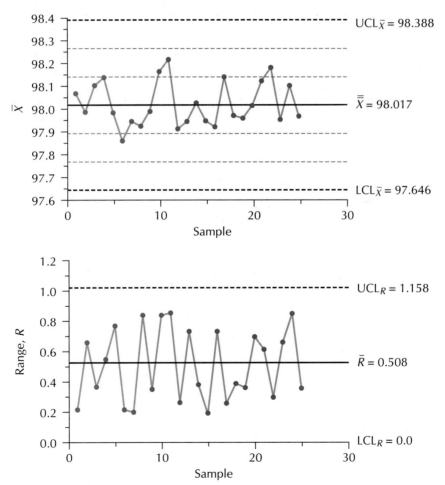

Figure 7.16 Control Charts for Machine 1

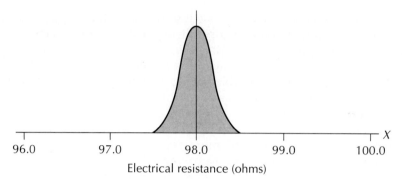

Figure 7.17 Model of Machine 1 Process

four machines, so that four independent sets of control charts could be maintained. Under this new system of sampling, it was discovered that although it appeared previously that the process was well centered at the nominal of 100 ohms, none of the four machines was actually centered at the nominal. The new control charts for machine 1, shown in Fig. 7.16, demonstrate this phenomenon. Although the $\overline{X}$ and R charts show good control, the mean for machine 1 lies at 98.017, or approximately 2 ohms away from the nominal value of 100 ohms. As indicated in Fig. 7.17, the mean of machine 1 is actually about 8 standard deviations from the nominal.

Each of the other three processes shows a similar problem with centering. When plotted as distributions on a common graph (see Fig. 7.18), it is easy to see the wide range of mean values that the processes encompass. Although the means of the four processes when averaged produce a grand average at the nominal, obviously each individual process varies from that by a very significant amount.

It is now clear that the problem was not a common-cause problem at all

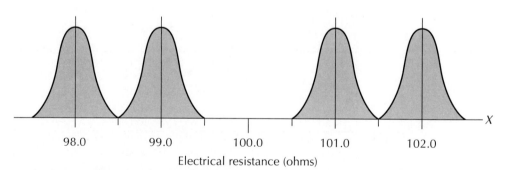

Figure 7.18 Distributions of All Four Machines

but one in which special causes (machine-to-machine differences in mean values) were responsible for the inability of the process, overall, to be capable. By centering all four machines at the nominal with no further reduction in common-cause variation, each machine by itself is 16 sigma capable. This case study clearly demonstrates the problems associated with poor control chart sampling techniques and the wasted time and effort that would result from the misleading information coming from the control charts.

Exercises

7.1. a. When defining and selecting samples/subgroups, do you want special causes to be occurring within the samples/subgroups or between them? Why?

   b. What are the possible dangers involved with taking samples at precisely the same time every day?

7.2. Samples are being taken at 1-hour intervals from an assembly line (600 units per hour). To make up each sample of size $n = 4$, the four units are taken off the line, one by one, at approximately 15-minute intervals (to produce one sample per hour).

   a. Comment on the advisability of this type of sampling to detect:

   (i) Abrupt and short-lived shifts in the mean.

   (ii) Gradual, trendlike shifts in the mean.

   (iii) Abrupt and sustained shifts in the mean.

   b. What sampling scheme would be best for the detection of each type of condition in part (a)?

7.3. a. If you suspect that frequent raw material batch changes are behind special-cause variation in a certain process, would consecutive or distributed sampling be best used in collecting the samples? Why?

   b. If you suspect that accelerated tool wear is behind special-cause variation, would consecutive or distributed sampling be best used in collecting the samples? Why?

   c. If you suspect that very short-lived but somewhat frequent power surges are behind special-cause variation, would consecutive or distributed sampling be best used in selecting the samples? Why?

7.4. For an injection molding process, a four-cavity mold is being used for a certain part. It has been proposed to develop $\overline{X}$ and R charts for part weight where the subgroup/sample is composed of the four parts from a single shot (i.e., one part from each cavity). Comment on the appropriateness of this method of sampling. What impact does it have on the ability of the charts to detect changes in the process? Alternatively, a proposal has been made that after each shot/cycle the four parts are broken off the sprues and runners and put into a rack. Samples would then be randomly drawn from this rack. Again, comment on the appropriateness of this method of sampling and explain the impact it would have on the charts.

7.5. A process for the manufacture of vinyl for use in automobile instrument panels employs a calender-type process to roll-form the vinyl to a pre-specified thickness. The continuous sheets are measured for their thickness several times a shift. Each time this is done by taking three measurements across the width of the sheet: a left-side, center, and right-side measurement. The quality control engineer decides that she will use these data, samples of $n = 3$, to make Shewhart control charts for this process. Comment on the advisability of her approach. Can you suggest an alternative method of charting this process?

7.6. Two identical molding machines producing the same part are being charted separately and both show good stability. The characteristic being monitored is part weight. Both processes are centered at the nominal, and both are 3 sigma capable. Seeing these facts, the quality control engineer decides to start taking 2 parts from each line for every sample ($n = 4$) and only use one set of charts. This way, she feels, she cuts the sampling cost in half while continuing to monitor both machines. Comment on the validity of her reasoning, and explain the impact of this change on the overall scheme of things. How would you modify the sampling process to reduce costs but maintain surveillance of the two machines?

7.7. A process that is making plastic parts for cassette tape cases has never been monitored with SPC charts but has lately been unable to meet production demands, so a decision is made to begin charting the process. The line produces 650 parts per hour, from 8:00 A.M. to 4:00 P.M. each day. Recommend an appropriate sampling scheme, including sample size, time/frequency of sampling, and whether consecutive or distributed sampling should be used. Justify your scheme.

7.8. A process that is producing approximately 10% defects is being reviewed as a candidate for SPC charting. After some consideration, the quality control engineer concludes that it is cheaper to rework the defective parts than to submit the process to the necessary destructive testing to study the characteristic in question properly. If this person were your boss, how would you convince him to do the testing, or would you agree with his decision?

7.9. Why is it so important that sampling schemes be carefully chosen? Give an example of how an unwise selection of a sampling scheme can affect the resulting usefulness of the control charts made?

7.10. Assuming that the R charts associated with each of the $\overline{X}$ charts (i) to (iv) show good control:

     a. Which of the $\overline{X}$ charts might indicate that a problem with stratification might exist? Explain.

     b. Which of the $\overline{X}$ charts may be examples of mixing the production from several machines? Explain.

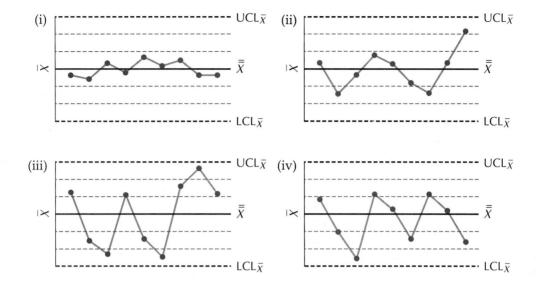

7.11. In this chapter, the practice of taking several measurements around the circumference of an extruded rubber hose is given as an example of stratifying the data. Explain why this is stratification and propose an alternative sampling method.

7.12. At one station in a chrome-plating process for plastic automobile grilles, the parts, hung on 25 individual hooks, are immersed simultaneously for 20 seconds, after which they are removed and held for another 40 seconds before moving to the next station. The level of contaminants on the surface of the part at this point in the process is the quality characteristic of interest here. To measure this, the part must be removed to the lab for testing. Total part cycle time for plating is 20 minutes. To minimize lost production and keep a balanced flow, one part is taken from each successive rack of parts and five consecutive parts are then grouped to form a subgroup for charting. Comment on this scheme. Do you have any suggestions?

7.13. To save money, the quality control engineer at a plant decides to use a sample of size $n = 2$ on a line producing 500 parts per hour. What could you tell him about the advisability of using such a small sample size? What effect will this approach have on the control charts and the information they will yield?

7.14. What is the purpose of collecting all of the observations in a single sample consecutively rather than randomly over the sampling interval?

7.15. What would be some valid reasons for changing from a sample size of $n = 6$ to one of $n = 4$?

7.16. Discuss the considerations/trade-offs involved in selecting the sample size

n that should be used for the purpose of constructing control charts for a certain process.

7.17. What are the considerations that come into play when determining the sampling frequency to be used in collecting data from a process for control charting purposes? Explain.

7.18. A certain process involves punching five holes in sheets of metal. The holes have the same diameter specification. The die contains five punches (i.e., all of the holes are punched in one stroke of the die). It is decided to make control charts for this process to monitor hole diameter. One sheet is taken about every 10 minutes, and the five hole diameters on that sheet are measured to form a subgroup for the chart. Comment on this strategy. Do you have any suggestions?

7.19. A process that fills bottles (throughput = 300 bottles/hour) with a certain expensive chemical uses a two-station filling machine. The first station quickly fills the bottles to nearly the proper level by weight, while at the second precision station the bottle is weighed and then carefully topped off to the proper final weight. Concerned about the throughput of this filling process, charting is begun on the precision filling station. Comment on this situation and recommend how the filling process should be charted to work most effectively toward increasing the throughput of the process.

CHAPTER

8

Interpretation of $\overline{X}$ and R Control Charts: Use of Sampling Experiments

---

8.1

Introduction

This chapter focuses on the interpretation of Shewhart control charts for variable data, in particular, $\overline{X}$ and R charts. To achieve this, the chapter presents the use of sampling experiments to aid in the study of control chart patterns. The results of two sampling experiments are discussed in considerable detail. In the first sampling experiment disturbances are introduced to the data of an initially stable process and chart interpretation rules are invoked, thereby signaling the occurrence of these disturbances. In the second sampling experiment the consequences of one form of nonrational sampling, stratification, are examined in some detail. In presenting these two sampling experiments this chapter is actually demonstrating the applicability of a personal computer–based software package that is used for PC-based computer workshops along with this book. When used in conjunction with this book the workshops provide an excellent environment for students to work in a more realistic way with the charting and analysis methods presented in Chapters 4 to 11.

In Chapter 4 we discussed the nature of variation and, in particular, noted the distinction between the occurrence of special causes of variation and the constant system of common causes that otherwise drive the performance of the process. A process was said to be in control when it was determined to be free of special causes, that is, driven only by the common-cause system. We have seen that when the performance of a process is driven only by such a causal system, data obtained through rational sampling will behave according to certain probabilistic laws. These laws describe how certain functions of the data, such as averages or ranges, will behave when plotted over time or in a frequency sense. These laws allow us to predict in a probability sense the limits of variation in the data. With such a well-defined statistical basis for the behavior of data emanating from a common-cause system, it is therefore possible to determine the precise probability of occurrence of certain characteristic patterns as they arise over time.

8.2

Analysis of Control Chart Patterns: A Sampling Experiment

In this section a sampling experiment is conducted and studied to illustrate the nature of the forces of common-cause variation and the manner in which control charts indicate the presence of special causes/sporadic problems. The sampling experiment illustrates the application of the control chart pattern rules/tests discussed in Chapter 6 toward the identification of the purposeful introduction of changes in the mean and/or the standard deviation of the process.

Value of Sampling Experiments

The use of sampling experiments as a means to illustrate the workings of common-cause variation and the manner in which special causes of variation manifest themselves on control charts is not new. Working and Olds, in their famous World War II War Production Board eight-day courses on quality control using statistical methods, made extensive use of sampling experiments.[1] Previously, Shewhart had introduced such experiments as a vivid way to demonstrate his principles.

In the days of Working and Olds such experiments were conducted using chips and beads, which were either numbered in proportion to the frequencies of various normal distributions or color coded to provide for various proportions of defective and nondefective parts. Burr[2] provides a tabulation of the relative frequencies to be used for numbering chips to obtain normal distributions with varying means and standard deviations. These are given in Table 8.1. These sets of chips may then be placed in a bag or other convenient container, thoroughly mixed, and samples of various sizes may be drawn to simulate sampling from a physical process. These authors have used such sampling experiments on countless occasions and have found them to be tremendously

---

[1] H. Working and E. G. Olds, *Manual for an Introduction to Statistical Methods of Quality Control in Industry: Outline of a Course of Lectures and Exercises,* Office of Production Research and Development, War Production Board, Washington, D.C., 1944.
[2] I. W. Burr, *Statistical Quality Control Methods,* Marcel Dekker, New York, 1976.

TABLE 8.1	Frequency Distributions for Sampling Experiments (Frequencies Approximately Normal)			
X	A	B	C	D
+11				1
+10			1	1
+9			1	1
+8			1	3
+7		1	3	5
+6		3	5	8
+5	1	10	8	12
+4	3	23	12	16
+3	10	39	16	20
+2	23	48	20	22
+1	39	39	22	23
0	48	23	23	22
−1	39	10	22	20
−2	23	3	20	16
−3	10	1	16	12
−4	3		12	8
−5	1		8	5
−6			5	3
−7			3	1
−8			1	1
−9			1	1
−10			1	
Number of chips	200	200	201	201
Mean, μ_X	0.0	2.0	0.0	1.0
Standard deviation, σ_X	1.715	1.715	3.470	3.470

useful. Of course, other normal distributions (different μ_X and σ_X values) may be employed. Today, the use of computers makes it possible to extend and enrich considerably the concept of sampling experiments. In this section we examine the results of one such sampling experiment workshop, conducted using the old technology of bags of chips.

Initial Sampling Results

Table 8.2 gives the results of taking 25 random samples of $n = 5$ observations each from a bag of chips numbered according to one of the normal distributions in Table 8.2 (unknown to the reader at this point). The samples were collected by drawing five chips from the bag, recording the number on each chip, returning the chips to the bag, mixing the chips, drawing another five chips, and so on. For each sample the average $\overline{X}$ and the range R were calculated according to the formulas discussed previously.

TABLE 8.2	Results of the First 25 Samples for the Sampling Experiment						
Sample	X_1	X_2	X_3	X_4	X_5	$\overline{X}$	R
1	3	−1	−1	0	−2	−0.2	5
2	−1	1	1	1	−2	0.0	3
3	3	2	1	−2	0	0.8	5
4	0	0	1	1	−1	0.2	2
5	−1	−3	1	1	2	0.0	5
6	0	0	0	−1	−2	−0.6	2
7	0	0	0	1	1	0.4	1
8	0	2	−2	−2	−3	−1.0	5
9	3	−3	−2	−1	−1	−0.8	6
10	0	−1	2	−2	0	−0.2	4
11	2	2	2	2	−1	1.4	3
12	−1	−1	2	−3	−1	−0.8	5
13	0	0	1	0	2	0.6	2
14	0	−2	−1	1	2	0.0	4
15	1	−3	−1	−1	−1	−1.0	4
16	2	−1	−2	−1	2	0.0	4
17	2	2	−1	−1	3	1.0	4
18	0	3	2	−2	−2	0.2	5
19	0	0	−1	−1	−1	−0.6	1
20	−2	−1	0	−2	0	−1.0	2
21	−1	−1	0	2	3	0.6	4
22	0	1	−1	−1	1	0.0	2
23	1	−1	0	−2	1	−0.2	3
24	0	−2	−1	−1	−1	−1.0	2
25	3	2	−1	−2	0	0.4	5

Determination of Trial Control Limits

By using the results of the 25 samples in Table 8.2, centerlines and trial control limits were determined as shown below for the construction of $\overline{X}$ and R control charts.

R chart:

$$\text{Centerline:} \quad \overline{R} = \sum_{i=1}^{25} \frac{R_i}{25}$$

$$= \frac{88}{25}$$

$$= 3.52$$

$$\text{UCL}_R: \quad D_4\overline{R} = (2.11)(3.52)$$

$$= 7.427$$

$$\text{LCL}_R: \qquad D_3\bar{R} = 0\,(3.52)$$
$$= 0.0$$

$\bar{X}$ chart:

$$\text{Centerline:} \quad \bar{\bar{X}} = \sum_{i=1}^{25} \frac{\bar{X}_i}{25}$$
$$= -0.072$$
$$\text{UCL}_{\bar{X}}: \qquad = \bar{\bar{X}} + A_2\bar{R}$$
$$= 1.970$$
$$\text{LCL}_{\bar{X}}: \qquad = \bar{\bar{X}} - A_2\bar{R}$$
$$= -2.114.$$

By using the plotting conventions discussed previously, $\bar{X}$ and R charts were constructed and are shown in Fig. 8.1.

Interpretation of the Initial Control Charts

To interpret the patterns in the control charts, we always begin with the R chart. This is important, since if the R chart does not exhibit good statistical control, the $\bar{X}$-chart control limits are not appropriate, and therefore the $\bar{X}$ chart should be interpreted with caution. Clearly, if the R chart is not in control, the $\bar{R}$ value is influenced by the presence of special causes, thereby being either inflated or deflated from the value it would take if only common-cause variation sources were present. As a result, the control limits on the $\bar{X}$ chart will be either wider or narrower than they would otherwise be.

The results of examining the pattern of R values on the R chart in Fig. 8.1 are as follows.

- No extreme points are evident.
- There is a reasonable distribution of points within limits. (Roughly two-thirds of the R values should fall in the middle one-third of the chart.)
- No trends or cycles are evident.
- No runs of eight or more are evident.

The conclusion to be drawn from the observations above is that there is no reason to believe that the process is not stable with respect to its variability.

Since the R chart appears to be in statistical control we may continue with the interpretation of the patterns in the $\bar{X}$ chart. As discussed previously, it is useful to establish the zones as shown in Fig. 6.3. With these zones in place, numerous tests, as outlined in Chapter 6, can be applied to the pattern of $\bar{X}$'s. The distribution of $\bar{X}$ values by zones follows.

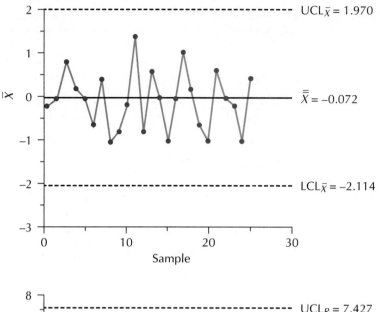

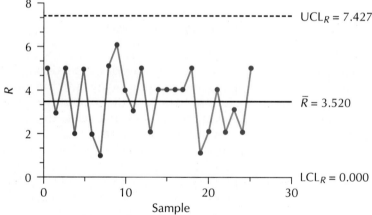

Figure 8.1 $\overline{X}$ and R Charts Based on the 25 Sample Results of Table 8.2

Zone C: $\frac{16}{25}$ = 64%   should be 68.26%

Zone B: $\frac{8}{25}$ = 32%   should be 27.19%

Zone A: $\frac{1}{25}$ = 4%   should be 4.28%.

This distribution of points within the limits appears to follow closely what is expected based on the normal distribution. In addition, the analysis indicates that:

• No extreme points are evident.
• No zone rule tests are violated.

- No trends or cycles or other such nonrandom patterns are evident.
- No runs of eight or more above or below the centerline are evident.

The conclusion that can therefore be drawn is that there is no reason to believe that the process is not stable with respect to its mean value. Since both charts give the appearance of a process in good statistical control, we can extend the centerlines and control limits and continue to sample, using these charts as the model for the way the data should continue to behave if only common causes of variation are at work.

Continuation of the Sampling Experiment

With $\overline{X}$ and R charts in place showing good statistical control, the sampling experiment was continued by selecting an additional five samples of $n = 5$ observations each in the manner discussed previously. Table 8.3 summarizes these sampling results. It should be pointed out that while such charts should, in practice, be continued on an individual sample-to-sample basis as each new sample result becomes available, we have selected the samples in groups of five for the purpose of the presentation here. The $\overline{X}$ and R values for these additional five samples were calculated and plotted as a continuation of the original $\overline{X}$ and R charts, as shown in Fig. 8.2.

In our continuing examination of the evolving patterns on the $\overline{X}$ and R charts, it is important that we adopt a sample-by-sample approach to the interpretation of the charts; that is, we must interpret the charts as if the samples are being taken one at a time, each $\overline{X}$ and R pair interpreted in turn as it arises. This is what we would do if the charts were evolving at the process, with samples taken every hour or half-hour, and so on.

As we, in turn, examine the results of samples 26 to 30, it becomes clear that no unusual patterns are evident. This conclusion comes not from a cursory "eyeballing" of the data but rather by methodically applying each of the rules for chart interpretation that we discussed previously. This is a bit tedious but must be done. Graphical and analytical aids may be developed to assist in this examination. It is clear that the computerization of these rules would be very helpful. The conclusion that we come to in interpreting the patterns in samples 26 to 30 is that there is no reason to suspect that anything has changed from

TABLE 8.3 Results of Samples 26 to 30							
Sample	X_1	X_2	X_3	X_4	X_5	$\overline{X}$	R
26	0	1	−1	3	1	0.8	4
27	0	0	−2	−1	2	−0.2	4
28	−1	1	−1	−3	4	0.0	7
29	0	2	−1	2	1	0.8	3
30	0	−1	0	1	2	0.4	3

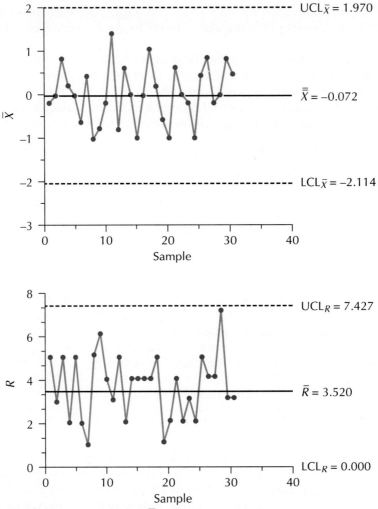

Figure 8.2 Continuation of $\overline{X}$ and R Charts for Samples 26 to 30

what we observed in the first 25 samples. In short, the process appears to be continuing in a state of control or stability—at least we have no evidence to the contrary. Table 8.4 shows the results of taking another five samples. The $\overline{X}$ and R values for samples 31 to 35 were plotted as a continuation of the previous control charts. The results of the continued plotting of the charts are shown in Fig. 8.3.

Applying all the appropriate tests to the interpretation of the evolving patterns in the R chart leaves no reason to suspect that the process is not continuing in a state of statistical control as far as its variation is concerned. However, examination of the $\overline{X}$ chart indicates that an unusual pattern of variation in the $\overline{X}$'s begins to arise at sample 32. Here, again, it is important that we carefully apply all of the appropriate rules, sample by sample.

TABLE 8.4	Results of Samples 31 to 35						
Sample	X_1	X_2	X_3	X_4	X_5	$\overline{X}$	R
31	-1	1	0	-1	-4	-1.0	5
32	-3	-2	-3	-4	0	-2.4	4
33	-3	-2	-2	-2	-1	-2.0	2
34	1	-3	-1	-2	-5	-2.0	6
35	0	-3	-4	-3	-1	-2.2	4

Figure 8.4 focuses on that portion of the $\overline{X}$ chart circled in Fig. 8.3 and shows the results of the proper application of the tests relevant to the $\overline{X}$ chart. The presence of the extreme point at sample 32 is clear evidence that the mean of the process must have shifted downward, at least at that time. This shift

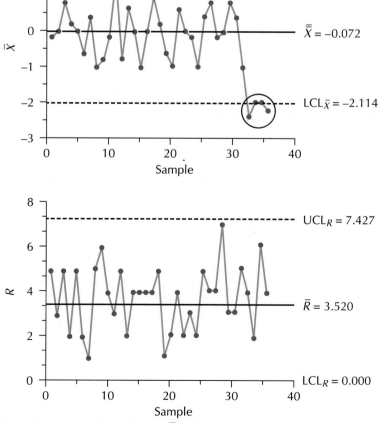

Figure 8.3 Continuation of the $\overline{X}$ and R Charts for Samples 31 to 35

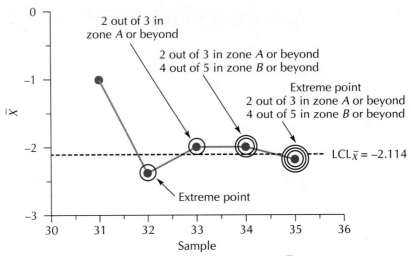

Figure 8.4 Interpretation of Samples 31 to 35 on the $\overline{X}$ Chart

appears to be sustained through sample 35. Sample 33 violates the two-out-of-three samples in zone A or beyond rule. For samples 34 and 35, two and three of the rules are violated, respectively.

Table 8.5 provides the data for the next five samples of this sampling experiment. The $\overline{X}$ and R values are plotted as a continuation of the control charts and are shown in Fig. 8.5. Continuing with our point-by-point examination of the evolving patterns in the $\overline{X}$ and R charts for samples 36 to 40, the following observations can be made:

> The R chart continues to exhibit good statistical control. There is no evidence to suggest that the process variability level has changed.
> The $\overline{X}$ values for samples 36 to 40 all violate more than one rule.
> There continues to be strong evidence of a sustained downward mean shift over samples 32 to 40. It would appear that the process mean has shifted to a level around -2.0.

TABLE 8.5   Results of Samples 36 to 40

Sample	X_1	X_2	X_3	X_4	X_5	$\overline{X}$	R
36	-4	-4	0	-1	-1	-2.0	4
37	0	-2	-4	-1	-1	-1.6	4
38	-3	0	0	-3	-1	-1.4	3
39	-2	-2	-1	-4	-4	-2.6	3
40	-1	-2	0	-3	-3	-1.8	3

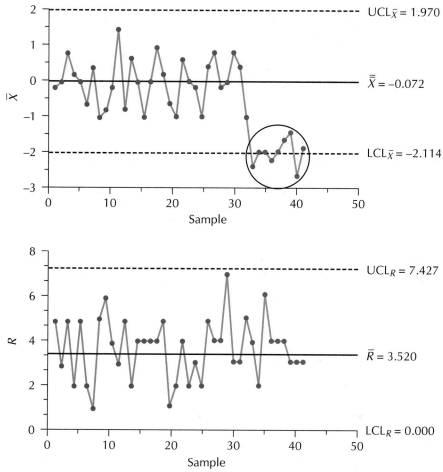

Figure 8.5 Continuation of $\overline{X}$ and R Charts for Samples 36 to 40

Figure 8.6 shows the section of the $\overline{X}$ chart that includes the results of samples 36 to 40. Each time a rule is violated a point is circled. From the figure it is clear that each of the $\overline{X}$ values violates more than one of the rules we have been applying. As an exercise the reader should review exactly which of the rules are violated for each of the $\overline{X}$ values. For example, in the case of sample 40 four rules are violated. Which ones are they?

Table 8.6 gives the results of the sampling experiment for samples 41 to 45. The $\overline{X}$ and R values are plotted as a continuation of the existing $\overline{X}$ and R charts. These updated charts are shown in Fig. 8.7.

Upon continuing with our point-by-point interpretation of the evolving patterns in the $\overline{X}$ and R charts, the following observations are made for samples 41 to 45:

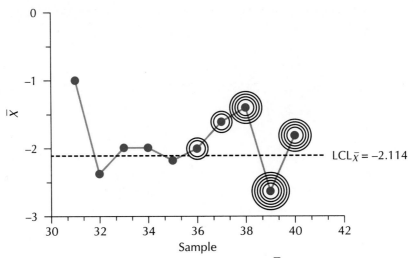

Figure 8.6 Summary of the Rule Violations for the $\overline{X}$ Value from Samples 36 to 40

- On the R chart, samples 41, 42, 44, and 45 are all extreme values (above UCL_R). Therefore, beginning with sample 41 there is a clear indication of an upward shift in process variability over samples 41, 42, 44, and 45.
- Samples 41, 42, and 43 have $\overline{X}$ values that continue to violate at least one of the zone rules. Under the conditions of a stable R chart this would be indicating that there is reason to believe that the shift in the process mean is sustained through sample 42. However, we must be careful at this point since the R chart has shown out-of-control conditions at samples 41 and 42, in particular, an increase in variability. Such would naturally cause the $\overline{X}$ values to exhibit increased variation. Hence we no longer may feel confident that the conditions on the $\overline{X}$ chart at samples 41 and 42 are signaling a shift downward in the process mean. Sample 43 has an $\overline{X}$ value that is now above the original chart centerline but violates the "8 points in

TABLE 8.6   Results of Samples 41 to 45

Sample	X_1	X_2	X_3	X_4	X_5	$\overline{X}$	R
41	0	2	4	−5	−6	−1.0	10
42	−3	−1	−6	3	1	−1.2	9
43	2	4	−1	2	0	1.4	5
44	6	1	−4	3	−3	0.6	10
45	4	−2	−5	−1	−2	−1.2	9

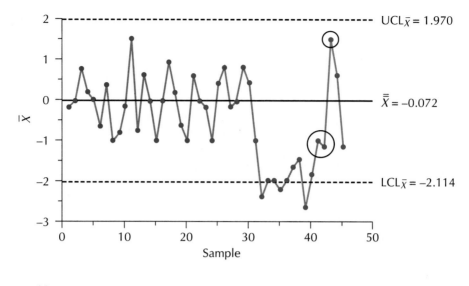

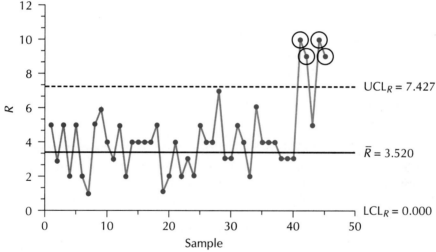

Figure 8.7 Continuation of $\overline{X}$ and R Charts for Samples 41 to 45

a row avoiding zone C″ rule. Samples 44 and 45 have $\overline{X}$ values that do not violate any of the rules.

- It should be noted that the variation in the $\overline{X}$ values for samples 41 to 45 appears to be somewhat larger than that evident in the $\overline{X}$'s up to this point. This should not be too surprising since an upward shift of the process variation as indicated on the range chart should manifest itself on the $\overline{X}$ chart as well. This is so since the variation in $\overline{X}$'s is a function of the variation in the individual observations, σ_X.

In summary, samples 41 to 45 are indicating that the variation in the process has experienced an upward shift. The only exception to this is the range value for sample 43, which does not indicate that a shift in the variation has taken place. Further, it appears that the shift in the process mean that was evident from 32 to 40 is no longer present (from sample 43 to sample 45 no signals are present that would indicate a sustained downward shifts). As indicated above, caution must be exercised in drawing conclusions about samples 41 and 42 because of the increase in variation suggested by the R chart.

Table 8.7 gives the results of the last five samples for this sampling experiment. The associated $\overline{X}$ and R values are plotted as a continuation of the previous $\overline{X}$ and R charts. These updated charts are shown in Fig. 8.8.

Interpretation of the evolving patterns in the $\overline{X}$ and R charts yields the following observations:

- On the R chart sample 46 is an extreme point, while samples 48 and 49 violate the run rule. It should therefore be concluded that the evidence continues to indicate a sustained shift in the process variation. The R value for sample 50 does not violate any of the appropriate rules, and therefore one should conclude that there is no longer any statistical evidence to support the existence of a shift in the variation level of the process.
- The $\overline{X}$ values continue to show a pattern not unlike the pattern evident in the $\overline{X}$'s for the first 30 samples. There is no evidence to suggest that the process mean is anything other than the centerline of the $\overline{X}$ chart.

Revelation of the True Nature of Changes in the Process

It is now time for us to reveal the true nature of the changes that were made to the process mean level and the amount of process variation as the sampling experiment evolved. It is clear from the statistical signals emanating from the $\overline{X}$ and R charts that the distributions from which the samples were drawn were changed from time to time. The statistical signals, in fact, suggested where it appeared that these changes had taken place and what was the specific nature of these changes. The question is: How good a job have the charts really done? A summary of the nature of the statistical distributions used to generate each of the 50 sample results follows.

TABLE 8.7 Results of Samples 46 to 50							
Sample	X_1	X_2	X_3	X_4	X_5	$\overline{X}$	R
46	2	0	−3	6	−5	0.0	11
47	2	−1	1	−2	−1	−0.2	4
48	−3	2	0	−1	2	0.0	5
49	−1	3	3	−1	0	0.8	4
50	0	2	2	−1	0	0.6	3

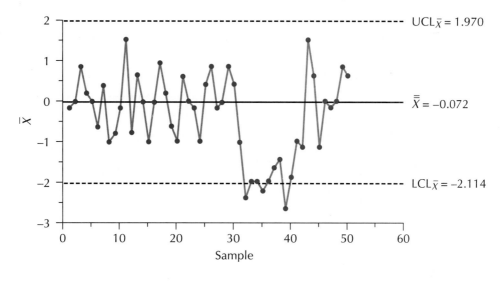

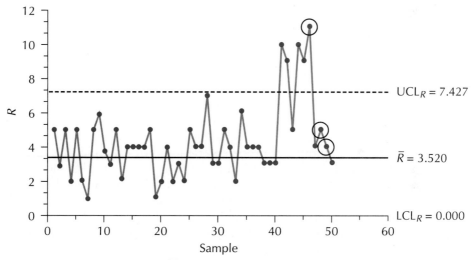

Figure 8.8 Continuation of the $\bar{X}$ and R Charts for Samples 46 to 50

- Samples 1–25: $N(\mu_X = 0, \sigma_X^2 = 1.715^2)$*
- Samples 26–30: $N(\mu_X = 0, \sigma_X^2 = 1.715^2)$
- Samples 31–40: $N(\mu_X = -2, \sigma_X^2 = 1.715^2)$
- Samples 41–46: $N(\mu_X = 0, \sigma_X^2 = 3.47^2)$
- Samples 47–50: $N(\mu_X = 0, \sigma_X^2 = 1.715^2)$

* $N(\mu_X = 0, \sigma_X^2 = 1.715^2)$ denotes a normal distribution with a mean of zero and a standard deviation of 1.715.

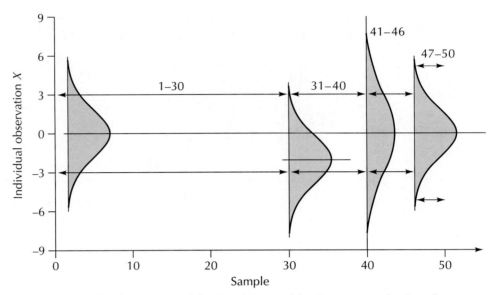

Figure 8.9 Graphical Depiction of the True Nature of the Process over the Sampling Experiment

Figure 8.9 graphically depicts the nature of the evolving process in terms of the statistical distributions outlined above. In comparing the statistical signals emanating from the charts with Fig. 8.9, it would appear that the charts have done an excellent job of signaling the presence of shifts in the process mean and/or the amount of process variation. For example, the mean of the process was actually shifted downward at sample 31, and this fact showed up on the $\overline{X}$ chart at sample 32. The variation of the process was shifted upward at sample 41, and this fact was immediately detected at sample 41 when an R value went above the upper control limit on the R chart. The process mean level was actually shifted back to the original value of 0.0 at sample 41. This fact was revealed by the $\overline{X}$ chart at sample 43, although for samples 41 and 42 we were cautious in saying that the process mean continued to be shifted downward because of the shift in variation detected at samples 41 and 42. Similarly, the process variation was returned to its original level at sample 47. This fact was signaled by the R chart at sample 50.

Sampling experiments of the type we have just reviewed in detail are an excellent tool for learning purposes. In addition to revealing how shifts in the process mean and/or variation can be signaled through the use of $\overline{X}$ and R control charts, these serve as a means to illustrate inappropriate sampling methods such as stratification and mixing and can be used to simulate other common problems, such as overcontrol of the process. We will now make use of these sampling experiments to illustrate the problem of stratification of the subgroup.

8.3

Illustration of Stratification of Subgroups

Simulation of the Stratification Phenomenon

Sampling experiments using several different statistical distributions can be most useful in demonstrating the consequences of poor sampling methods such as stratification. In this section we conduct such sampling experiments to demonstrate the effect that stratification has on the appearance of the associated $\overline{X}$ and R control charts and to demonstrate the loss of chart sensitivity due to stratification of the sample.

Figure 8.10 shows four normal distributions, each with the same standard deviation, but with means of $\mu_X = 2, 5, 5,$ and 7. To simulate a stratified subgrouping, samples of size $n = 4$ were drawn, with each sample consisting of one measurement from each of the four distributions shown in Fig. 8.10. Table 8.8 gives the data for the first 50 of a total of 100 subgroups simulated in this manner and includes the subgroup average and range values. The data in Table 8.8 plus the additional 50 subgroups collected were used to set up $\overline{X}$ and R charts. The calculations are as follows:

$$\overline{\overline{X}} = \frac{\Sigma \overline{X}_i}{100}, \qquad i = 1, 100$$

$$= \frac{481.5}{100}$$

$$= 4.815$$

$$\overline{R} = \frac{\Sigma R_i}{100}, \qquad i = 1, 100$$

$$= \frac{515.5}{100}$$

$$= 5.155$$

$$\text{UCL}_{\overline{X}} = \overline{\overline{X}} + A_2\overline{R}$$

$$= 4.815 + (0.73)(5.155)$$

$$= 8.578$$

$$\text{LCL}_{\overline{X}} = \overline{\overline{X}} - A_2\overline{R}$$

$$= 4.815 - (0.73)(5.155)$$

$$= 1.052$$

$$\text{UCL}_R = D_4\overline{R}$$

$$= (2.28)(5.155)$$

$$= 11.753$$

$$\text{LCL}_R = D_3\overline{R}$$

$$= (0)(5.155)$$

$$= 0.0.$$

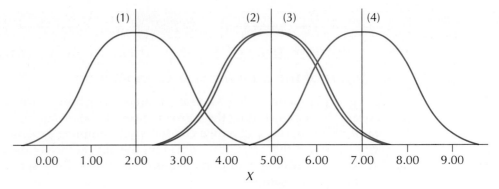

Figure 8.10 Normal Distributions Used to Simulate Stratification

TABLE 8.8	Data for Stratification Example					
Sample	X_1	X_2	X_3	X_4	$\overline{X}$	R
1	6.88	3.32	5.85	3.55	4.900	3.560
2	8.77	0.72	6.08	6.20	5.443	8.050
3	5.84	0.72	5.36	2.50	3.605	5.120
4	7.10	3.77	5.92	1.52	4.578	5.580
5	6.07	2.47	3.10	2.73	3.592	3.600
6	7.00	0.06	6.54	3.12	4.180	6.940
7	7.79	2.24	4.19	4.59	4.702	5.550
8	7.61	3.68	2.60	6.51	5.100	5.010
9	5.21	5.63	5.38	4.57	5.198	1.060
10	7.84	2.29	8.04	5.76	5.983	5.750
11	10.44	0.75	2.59	4.39	4.543	9.690
12	6.76	2.53	1.79	3.17	3.562	4.970
13	7.73	2.81	3.25	5.80	4.898	4.920
14	6.48	3.28	6.21	4.29	5.065	3.200
15	6.24	0.00	5.18	5.45	4.218	6.240
16	4.53	3.11	2.88	6.03	4.138	3.150
17	6.42	0.48	6.04	7.73	5.168	7.250
18	12.09	3.45	6.98	4.76	6.820	8.640
19	7.13	3.47	5.83	3.91	5.085	3.660
20	2.73	2.92	3.29	3.93	3.217	1.200
21	6.28	2.45	5.84	5.63	5.050	3.830
22	4.68	1.32	2.77	5.86	3.658	4.540
23	7.59	5.10	5.10	3.57	5.340	4.020
24	8.64	2.76	3.52	5.92	5.210	5.880
25	9.54	2.60	4.78	5.95	5.718	6.940
26	7.91	1.27	5.71	1.99	4.220	6.640
27	6.28	0.74	4.94	3.86	3.955	5.540
28	6.60	2.39	5.71	4.87	4.892	4.210

TABLE 8.8 *(continued)*					
Sample	X_1	X_2	X_3	X_4	$\overline{X}$	R
29	6.45	0.00	7.14	4.32	4.478	6.450
30	4.26	5.27	4.46	8.57	5.640	4.310
31	6.64	4.05	1.55	5.34	4.395	5.090
32	7.99	0.45	7.36	3.90	4.925	7.540
33	5.17	1.82	3.78	8.93	4.925	7.110
34	5.53	3.68	7.77	4.95	5.483	4.090
35	5.52	0.80	5.18	7.27	4.693	6.470
36	7.18	1.04	3.92	7.65	4.948	6.610
37	8.34	1.92	5.75	3.81	4.955	6.420
38	1.83	2.84	4.74	4.37	3.445	2.910
39	5.95	2.37	5.46	4.39	4.543	3.580
40	7.11	2.13	6.84	4.41	5.122	4.980
41	7.12	0.70	5.81	6.23	4.965	6.420
42	6.95	1.11	9.27	5.23	5.640	8.160
43	5.02	3.38	0.70	7.67	4.193	6.970
44	5.37	3.45	5.26	6.36	5.110	2.910
45	4.69	4.94	5.80	7.32	5.688	2.630
46	9.06	0.00	2.84	5.49	4.349	9.060
47	7.81	4.49	3.08	5.67	5.262	4.730
48	8.51	1.73	7.66	6.61	6.128	6.780
49	6.18	0.05	4.24	4.87	3.835	6.130
50	4.57	3.28	4.77	8.48	5.275	5.200

Figure 8.11 shows the $\overline{X}$ and R charts for the stratified subgroup data of Table 8.8. The R chart looks typical of data arising from a constant system of common causes. No special-cause signals are evident, and the overall appearance of the chart does not reveal anything which would suggest that stratification had taken place. Examination of the $\overline{X}$ chart, however, reveals a somewhat unusual pattern of variability. The $\overline{X}$ values seem to cluster quite closely about $\overline{\overline{X}}$; in fact, 86% of the $\overline{X}$ values are within ± 1 standard deviation of $\overline{\overline{X}}$. Normal distribution theory suggests that we should expect a much smaller percentage: 68%. Furthermore, runs of 15 or more consecutive $\overline{X}$ values within 1 standard deviation of the chart centerline occur twice. These signals indicate that something is wrong—and that something is the sampling method itself.

If samples were drawn separately from each distribution in Fig. 8.10, we would expect the ranges to average about 3.4; that is, $\overline{R}$ should be approximately equal to 3.4 (given that the theoretical standard deviation = 1.715 for each individual distribution, $d_2 = 2.06$ for samples of $n = 4$, and $E(R) = \sigma_X d_2$). However, the observed range is about 50% higher than this expected value. Clearly, this is occurring because we are sampling from several different distri-

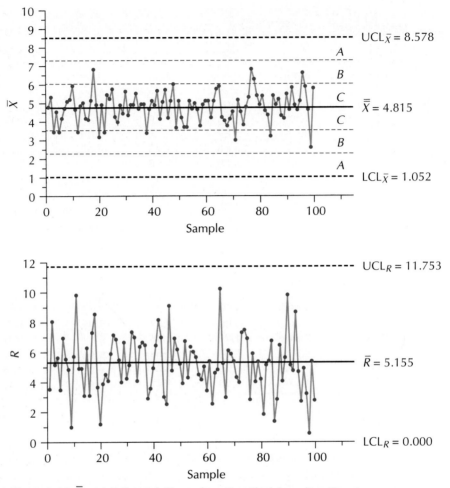

Figure 8.11 $\overline{X}$ and R Control Charts for a Stratified Sampling Situation

butions to make the subgroups, and these distributions have different means. These differences in means are inflating the within-subgroup variation. Fortunately, this characteristic behavior shows up in the $\overline{X}$ chart so that the improper sampling, namely, stratification, is signaled.

Effect of Stratification on Chart Sensitivity

The results of stratification of the sample, as seen in Fig. 8.11, clearly indicate that the special causes due to differences in the process means (refer to Fig. 8.10) that occur within the subgroup significantly inflate the within-subgroup variation. As a result, the limits on the $\overline{X}$ chart seem very wide, and hence one would conclude that the sensitivity of the chart to detect change in process mean level would be greatly reduced. This loss of chart sensitivity can be demonstrated through the use of sampling experiments such as those already conducted in this chapter.

Table 8.9 provides the data for another 50 samples of size $n = 4$ drawn in a stratified way from the four distributions of Fig. 8.10. The data were used to calculate centerlines and control limits for $\overline{X}$ and R control charts. These calculations are given below. The charts are shown in Fig. 8.12.

$$\overline{\overline{X}} = \frac{\Sigma \overline{X}_i}{50}, \quad i = 1, 50$$
$$= 4.7876$$

$$\overline{R} = \frac{\Sigma R_i}{50}, \quad i = 1, 50$$
$$= 5.9201$$

$$\text{UCL}_{\overline{X}} = \overline{\overline{X}} + A_2\overline{R}$$
$$= 4.7876 + (0.73)(5.2901)$$
$$= 9.1093$$

$$\text{LCL}_{\overline{X}} = \overline{\overline{X}} - A_2\overline{R}$$
$$= 4.7876 - (0.73)(5.2901)$$
$$= 0.4659$$

$$\text{UCL}_R = D_4\overline{R}$$
$$= (2.28)(5.9201)$$
$$= 13.4978$$

$$\text{LCL}_R = D_3\overline{R}$$
$$= (0)(5.9201)$$
$$= 0.0.$$

In Fig. 8.12 we see the same characteristic pattern in the $\overline{X}$ chart that was evident before as a result of subgroup stratification. Almost 90% of the $\overline{X}$ values fall within the middle one-third of the chart. Furthermore, a run of 18 in a row within zone C starts at subgroup 5, and another such run of 19 in a row starts at subgroup 24.

Table 8.10 provides the data for 25 additional subgroups chosen in the same fashion as the previous 50 subgroups, with the exception that one of the processes (distributions) has sustained a mean shift of one unit (from 5.00 to

TABLE 8.9	Data for a Stratification Sensitivity Example					
Sample	X_1	X_2	X_3	X_4	$\overline{X}$	R
1	−1.119	8.274	4.858	7.253	4.817	9.400
2	2.694	2.087	4.261	6.866	3.977	4.780
3	0.749	5.444	5.624	5.477	4.323	4.875
4	3.608	8.089	6.711	7.031	6.360	4.481
5	0.258	8.187	4.149	5.148	4.435	7.930
6	3.366	3.299	5.869	4.697	4.308	2.570
7	2.444	2.890	4.869	8.591	4.699	6.147
8	0.282	3.069	4.916	5.161	3.357	4.879
9	3.090	0.399	6.168	5.354	3.753	5.769
10	0.172	8.028	5.488	7.030	5.180	7.856
11	4.159	6.191	2.902	8.671	5.481	5.769
12	0.318	6.882	4.493	9.546	5.310	9.228
13	−0.484	6.642	8.355	5.745	5.064	8.840
14	2.165	3.288	8.148	5.457	4.765	5.983
15	−0.010	8.611	5.293	10.483	6.094	10.494
16	2.265	5.702	6.388	9.930	6.071	7.665
17	−0.503	4.789	6.770	6.146	4.301	7.273
18	3.977	6.757	4.356	5.932	5.256	2.780
19	1.772	5.639	3.531	6.382	4.331	4.601
20	2.107	2.668	4.063	5.357	3.549	3.250
21	1.427	7.736	5.227	6.179	5.142	6.309
22	1.580	2.175	2.842	10.678	4.319	9.098
23	4.376	4.784	8.072	8.694	6.482	4.317
24	1.076	3.764	5.922	9.986	5.187	8.909
25	2.589	4.114	5.833	3.102	3.909	3.244
26	0.091	4.942	2.641	8.053	3.932	7.963
27	4.139	3.490	6.982	7.704	5.579	4.214
28	0.166	3.706	4.584	5.763	3.555	5.597
29	4.308	5.285	4.831	5.593	5.004	1.285
30	2.982	2.832	4.859	4.936	3.902	2.104
31	1.353	1.591	4.309	9.350	4.151	7.997
32	−0.032	3.708	4.615	6.687	3.745	6.718
33	1.153	7.319	4.395	5.018	4.471	6.166
34	1.281	5.643	3.202	8.619	4.686	7.338
35	2.758	5.794	5.294	6.497	5.086	3.739
36	3.163	5.341	7.753	6.166	5.606	4.589
37	−0.625	3.442	5.817	5.162	3.449	6.443
38	2.637	4.516	6.399	6.058	4.903	3.762
39	0.121	2.281	6.907	6.000	3.827	6.787
40	3.727	4.387	6.093	9.402	5.902	5.676
41	3.448	7.121	5.361	6.781	5.678	3.673
42	4.581	8.303	4.821	6.879	6.146	3.721
43	−0.704	1.043	6.987	5.297	3.155	7.691
44	2.704	5.949	7.170	7.320	5.786	4.616

TABLE 8.9	*(continued)*					
Sample	X_1	X_2	X_3	X_4	$\overline{X}$	R
45	−0.029	3.410	6.048	8.099	4.382	8.127
46	4.991	5.533	6.182	9.333	6.510	4.342
47	1.457	5.828	5.394	6.183	4.715	4.726
48	0.076	5.028	3.231	8.947	4.320	8.871
49	2.663	6.572	5.535	5.311	5.020	3.909
50	1.920	3.443	4.823	11.426	5.403	9.505

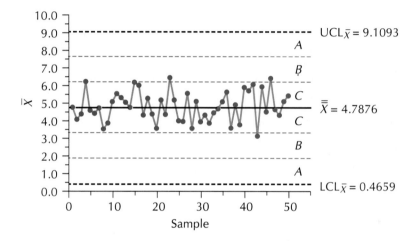

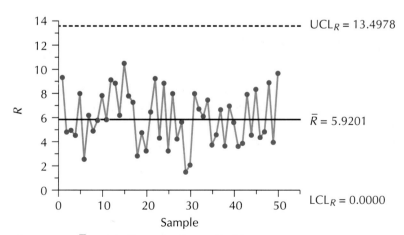

Figure 8.12 $\overline{X}$ and R Charts for Data of Table 8.9

Sample	X_1	X_2	X_3	X_4	$\overline{X}$	R
		TABLE 8.10 Additional 25 Subgroups for Stratification Sensitivity Example				
51	2.935	4.335	5.180	10.002	5.613	7.067
52	2.560	0.626	6.940	7.368	4.373	6.743
53	1.715	5.106	5.252	9.019	5.273	7.303
54	1.410	4.888	6.784	8.758	5.460	7.349
55	3.022	3.272	5.776	7.514	4.896	4.492
56	4.846	4.234	4.720	6.802	5.150	2.568
57	0.828	4.776	0.931	7.534	3.517	6.707
58	1.374	3.542	7.054	5.652	4.405	5.680
59	1.146	5.580	4.159	5.962	4.212	4.816
60	0.063	3.778	5.731	7.148	4.180	7.085
61	1.096	5.837	7.685	4.574	4.798	6.589
62	-0.590	3.633	6.676	7.676	4.349	8.266
63	7.335	2.977	6.634	5.659	5.651	4.358
64	3.520	3.979	3.186	4.492	3.794	1.306
65	0.847	6.205	5.373	6.471	4.724	5.624
66	0.599	7.179	5.936	6.203	4.979	6.579
67	2.197	6.010	3.436	11.114	5.689	8.917
68	2.058	3.275	4.738	7.541	4.403	5.483
69	3.445	7.595	6.185	6.734	5.990	4.150
70	6.716	1.521	5.529	5.481	4.812	5.195
71	3.175	0.715	4.426	4.960	3.319	4.245
72	1.713	3.613	1.759	6.240	3.331	4.526
73	3.838	6.725	5.326	10.457	6.587	6.619
74	0.117	3.668	5.064	7.183	4.008	7.066
75	0.617	3.582	3.062	4.248	2.877	3.632

4.00; again, refer to Fig. 8.10). The limits and centerlines for the $\overline{X}$ and R charts in Fig. 8.12 were extended, and the results of these additional 25 subgroups were plotted on the charts. The updated charts are seen in Fig. 8.13.

In Fig. 8.13 the $\overline{X}$ chart provides no signals that would indicate that the mean of one of the processes had shifted from 5.00 to 4.00. Yet such is clearly a shift away from the chart centerline of 4.7876. In fact, a run of $\overline{X}$ values within zone C that started at subgroup 47 is sustained all the way through subgroup 71, which is an indication that stratification may be occurring. The chart gives no indication that a special cause indicating a shift in the mean actually occurred at sample 51.

To illustrate the result of proper sample selection on chart sensitivity by contrast, still another sampling experiment has been done. In this sampling experiment, 50 subgroups of size $n = 4$ were drawn from a single process (distribution) with a mean of 4.75 (approximately the same mean as the data in Fig. 8.12 that were collected in a stratified way). Table 8.11 provides these new

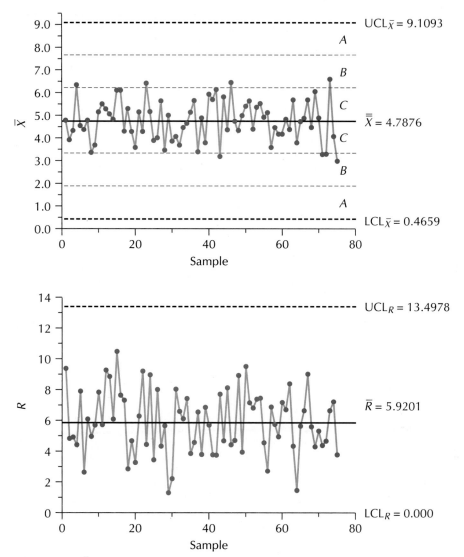

Figure 8.13 $\overline{X}$ and R Charts Extended from Figure 8.12 Using the Data of Table 8.10

Sample	X_1	X_2	X_3	X_4	$\overline{X}$	R
		TABLE 8.11	**50 Subgroups from a Single Distribution/Process**			
1	4.784	3.307	5.980	3.418	4.372	2.673
2	4.891	5.749	3.787	4.314	4.685	1.962
3	5.422	7.241	6.528	7.444	6.659	2.022
4	5.766	3.535	5.313	8.630	5.811	5.094
5	3.460	7.186	3.454	3.473	4.393	3.732
6	3.974	3.053	6.292	6.390	4.927	3.337
7	2.279	4.156	3.654	4.070	3.540	1.877
8	3.950	3.739	4.997	4.902	4.397	1.258
9	2.694	5.347	7.315	6.473	5.457	4.620
10	1.976	5.158	7.685	4.148	4.742	5.709
11	4.279	3.430	2.963	4.984	3.914	2.021
12	6.520	5.643	2.142	6.595	5.225	4.452
13	0.877	3.849	3.205	2.186	2.529	2.972
14	4.935	3.769	3.803	4.590	4.274	1.166
15	4.839	4.448	4.247	5.725	4.815	1.478
16	4.056	2.644	7.048	3.792	4.385	4.044
17	4.342	4.498	4.695	5.328	4.716	0.987
18	3.874	3.735	6.415	2.927	4.238	3.488
19	5.232	5.915	1.274	1.792	3.553	4.641
20	5.253	6.182	5.349	4.266	5.262	1.917
21	3.285	4.017	6.629	8.246	5.544	4.961
22	3.861	6.273	7.191	4.339	5.416	3.329
23	3.799	7.678	4.199	2.427	4.526	5.251
24	4.867	5.053	4.338	7.405	5.416	3.067
25	6.142	6.501	4.877	5.321	5.710	1.624
26	7.901	2.826	0.116	3.213	3.514	7.785
27	3.751	4.587	3.537	4.545	4.105	1.051
28	7.467	4.433	2.255	2.057	4.053	5.411
29	4.880	3.987	5.475	6.754	5.274	2.767
30	3.485	3.460	3.912	4.899	3.939	1.439
31	7.689	4.258	5.857	6.518	6.081	3.430
32	6.011	3.792	5.382	4.220	4.851	2.219
33	4.057	4.850	7.186	7.181	5.818	3.129
34	7.592	4.418	7.402	4.952	6.024	3.443
35	1.496	4.991	5.725	3.356	3.892	4.229
36	2.933	6.691	2.642	1.751	3.504	4.940
37	7.371	1.092	7.974	4.770	5.302	6.882
38	5.416	1.886	5.213	8.230	5.186	6.344
39	4.653	2.907	8.119	3.151	4.707	5.212
40	2.010	4.013	6.614	5.340	4.494	4.604
41	6.872	4.704	6.172	7.031	6.195	2.328
42	5.990	7.138	4.251	2.819	5.050	4.319
43	3.651	4.407	8.336	6.314	5.677	4.685
44	5.554	5.012	6.868	3.844	5.320	3.024
45	4.041	4.330	4.296	7.405	5.018	3.363
46	2.174	5.822	3.613	8.362	4.993	6.188

TABLE 8.11 *(continued)*					
Sample	X_1	X_2	X_3	X_4	$\overline{X}$	R
47	4.963	1.426	6.440	3.423	4.063	5.014
48	3.536	1.945	5.673	7.296	4.613	5.351
49	4.321	6.945	4.897	4.356	5.130	2.624
50	2.362	3.634	3.715	4.814	3.631	2.452

data. $\overline{X}$ and R control charts were developed and are shown in Fig. 8.14. Calculations for the centerlines and control limits are given below.

$$\overline{\overline{X}} = \frac{\Sigma \overline{X}_i}{50}, \quad i = 1, 50$$

$$= 4.7788$$

$$\overline{R} = \frac{\Sigma R_i}{50}, \quad i = 1, 50$$

$$= 3.6055$$

$$\text{UCL}_{\overline{X}} = \overline{\overline{X}} + A_2 \overline{R}$$

$$= 4.7788 + (0.73)(3.6055)$$

$$= 7.4108$$

$$\text{LCL}_{\overline{X}} = \overline{\overline{X}} - A_2 \overline{R}$$

$$= 4.7788 - (0.73)(3.6055)$$

$$= 2.1468$$

$$\text{UCL}_R = D_4 \overline{R}$$

$$= (2.28)(3.6055)$$

$$= 8.2205$$

$$\text{LCL}_R = D_3 \overline{R}$$

$$= (0)(3.6055)$$

$$= 0.0.$$

In Fig. 8.14 it is seen that the charts show good statistical control with no rule violations or unusual patterns of variation. It should also be noted that a 32% reduction in the within-subgroup variation has led to $\overline{X}$ and R charts with much narrower limits than previously was the case with the stratified samples (refer to Fig. 8.13).

An additional 25 subgroups were then selected from a process (distribution) with a mean of 4.00, which would constitute a mean shift of 0.75 from the previous process with a mean of 4.75. These 25 samples were taken from the

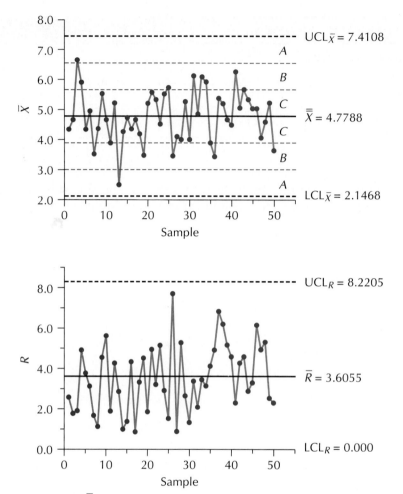

Figure 8.14 $\bar{X}$ and R Control Charts for Data of Table 8.11

same distribution used to introduce a shift in the mean within the stratified subgroup—the 25 samples of Table 8.10. The data for these subgroups are given in Table 8.12, and the $\bar{X}$ and R values for these 25 subgroups are plotted as a continuation of the charts indicated in Fig. 8.14. Figure 8.15 shows charts with extended limits and the additional 25 subgroup results plotted.

   The R chart in Fig. 8.15 continues to show good statistical control, as we would expect, since the process variation has remained constant. The $\bar{X}$ chart, however, provides several signals indicating that the mean has shifted. There is a run below the centerline that starts at subgroup 50, which is 8 long. There is also a signal at subgroup 64, where two of three successive $\bar{X}$ values are 2 or more standard deviations below the centerline. The indications are clear that the mean has shifted downward.

   The sampling experiments above demonstrate the loss of chart sensitivity

TABLE 8.12	Additional 25 Subgroups Selected from a Process with a Mean of 4.00					
Sample	X_1	X_2	X_3	X_4	$\overline{X}$	R
51	2.242	1.735	5.015	1.535	2.632	3.480
52	4.716	0.375	5.118	6.797	4.251	6.422
53	5.564	4.391	3.000	3.781	4.184	2.563
54	3.006	5.879	4.047	3.931	4.216	2.873
55	3.071	1.888	2.705	3.776	2.860	1.888
56	2.119	5.397	3.309	1.928	3.188	3.469
57	3.528	4.775	6.212	1.252	3.941	4.960
58	4.729	3.695	7.341	4.944	5.177	3.646
59	3.040	5.801	6.811	2.677	4.582	4.133
60	5.644	3.025	4.313	4.718	4.425	2.619
61	3.150	5.879	7.466	1.922	4.604	5.544
62	5.004	5.920	2.943	3.837	4.426	2.977
63	4.522	3.013	0.261	2.277	2.518	4.261
64	−0.282	2.216	3.724	5.948	2.901	6.230
65	5.440	2.651	2.473	5.430	3.998	2.967
66	3.644	4.064	5.519	7.478	5.176	3.834
67	5.127	4.296	0.476	4.582	3.620	4.652
68	3.488	1.404	5.015	6.778	4.171	5.374
69	6.562	4.215	3.418	2.989	4.296	3.573
70	4.544	5.273	4.984	6.797	5.399	2.253
71	5.512	3.636	4.731	2.094	3.993	3.417
72	4.221	0.957	4.762	5.964	3.976	5.006
73	2.984	3.821	2.924	4.456	3.546	1.532
74	2.230	4.922	4.862	3.850	3.966	2.692
75	3.893	3.966	4.609	2.221	3.672	2.388

that occurs when samples are stratified by sampling across several different processes. Process mean differences become part of the within-subgroup variation, thereby inflating the range and widening the control limits of the $\overline{X}$ chart. The comparison of the $\overline{X}$ charts of Figs. 8.14 and 8.15 clearly show the sensitivity issue for the same process shifts in mean level.

Computer Workshop 1

Introduction

In this workshop, a sampling experiment will be performed with the aid of two computer programs. One computer program, PROSIM, will be used to generate data samples collected from a fictional process. The second computer program, XRCHRT, will be used to create $\overline{X}$ and R control charts for the data generated

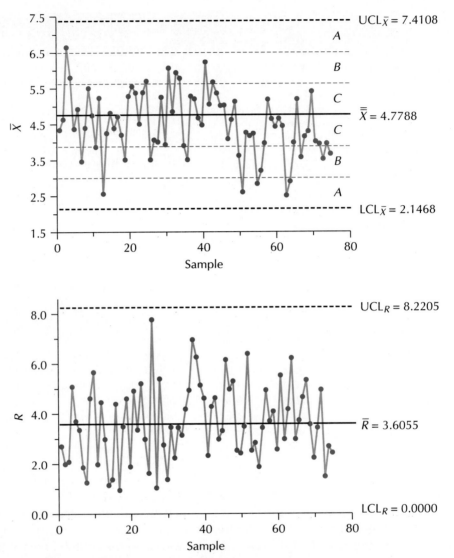

Figure 8.15 $\overline{X}$ and R Charts Extended from Figure 8.14 Using the Data From Table 8.12

by PROSIM. The objective of this computer workshop is to read the charts and identify the specific statistical signals that indicate that special/sporadic causes of variation are present in the data/process generated by PROSIM.

Description of the Computer-Based Process Simulator, PROSIM

PROSIM is a simulation program that introduces special causes into a fictional process, and generates and stores data for control charting in a user-defined file on the computer. These data may subsequently be analyzed with $\overline{X}$ and R

charts using the computer program XRCHRT. This workshop permits a user to gain experience in the interpretation of $\overline{X}$ and R charts and to observe the effects of process changes on the charts. PROSIM simulates a process by drawing from one or more of the following distributions: $N(0, 2^2)$, $N(2, 2^2)$, $N(-2, 2^2)$, $N(0, 1^2)$, $N(0, 4^2)$, $N(-1, 1^2)$, and $N(1, 1^2)$. These distributions are illustrated in Figs. 8.16 and 8.17.

When the program PROSIM is started, it randomly selects a process history from one of 30 stored in a process history library. A process history is a chronological description of the distributions to be used for the series of subgroups to be collected. For example, a process history could be

Samples 1–27:	$N(0, 2^2)$
Samples 28–36:	$N(-2, 2^2)$
Samples 37–43:	$N(0, 4^2)$
etc.	

PROSIM simulates the selected process history by drawing 125 subgroups of size 5 from the statistical distributions specified by the process history. PROSIM begins the simulation by automatically generating at least 25 samples of size $n = 5$ from the distribution $N(0, 2^2)$.

It is also possible that the process history selected by PROSIM may contain one or both of two common but inappropriate methods of collecting samples: stratification and mixing. For stratification, subgroups of size $n = 5$ are formed by selecting exactly one individual measurement from each of the five distributions illustrated in Fig. 8.18. Stratification arises in practice when samples are collected by drawing from each of several processes: machines, filling heads, spindles, and so on. Mixing is simulated by forming subgroups of size $n = 5$, randomly selecting the individual measurements from among the five distributions of Fig. 8.18. Mixing arises in practice when the output from several processes is first thoroughly mixed and then random samples are drawn from

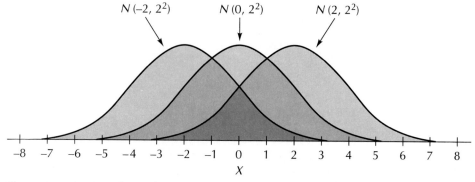

Figure 8.16 Statistical Distributions with Standard Deviations of 2 and Various Means

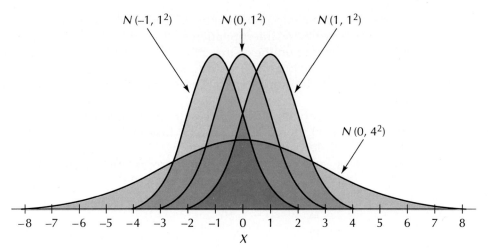

Figure 8.17 Statistical Distributions with Various Standard Deviations and Means

the mixture. Both of these types of sampling violate the fundamental rule of rational sampling—that within-subgroup variation should be solely attributable to one common-cause system.

In addition to PROSIM having the ability to simulate such improper sampling methods as stratification and mixing, it may also simulate a common process fault, overcontrol! Overcontrol is often introduced into a process by a well-meaning operator or controller who/that recognizes any appreciable deviation of a sample mean from the target value as a special cause. In this case the operator is wrongly viewing common-cause variation as arising due to local faults in the process. Overcontrol is simulated through changes to the mean of a normal distribution with a standard deviation of 2 using the following rules:

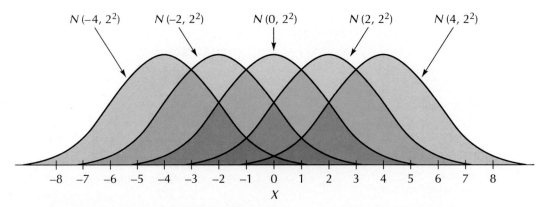

Figure 8.18 Statistical Distributions Used to Generate Stratification and Mixing

1. If the sample mean, $\overline{X}$, of the most recently collected sample is within a narrow range about the target value of zero, no action is taken.
2. If the $\overline{X}$ is above the target value (and outside the narrow range), the true mean for the next sample is reduced.
3. If the $\overline{X}$ is below the target value (and outside the narrow range), the true mean for the next sample is increased.

There are three rules that PROSIM uses when generating the process data:

1. At least the first 25 subgroups are taken from the distribution $N(0, 2^2)$.
2. Shifts in the mean and/or variation are sustained for at least five consecutive subgroups.
3. Stratification, mixing, and overcontrol are sustained for at least 20 consecutive subgroups.

When analyzing the data generated by PROSIM with $\overline{X}$ and R control charts, there are several things to keep in mind:

1. We actually know more here than we would normally know in a real-world situation (i.e., we know the means and standard deviations of each distribution). This information should prove to be very useful.
2. The control chart construction software, XRCHRT, provides you with the flexibility of basing control chart limits on all 125 samples or on just some part of the data, say the first 25 samples only. After some thought, it should be clear why one might wish to construct the control limits based on the first 25 samples alone.
3. The control chart construction software, XRCHRT, allows you to examine only portions of the total record of 125 samples. For example, after generating an initial set of charts, you may wish to go back and determine limits for samples 36 to 60.
4. It is essential that you carefully consider the summary information that precedes the control charts produced by XRCHRT. This information includes all violations of the rules that govern statistical control/process stability. It will be useful to indicate on the charts exactly where unusual behavior is taking place.
5. Based on the ways that stratification, mixing, and overcontrol are created, you should be able to make some calculations to help you identify the presence of these phenomena.

Using PROSIM

The computer software used with this book contains four computer programs: PROSIM, XRCHRT, PROCAP, and TURNSIM. A short description on how these programs are used accompanies the software. PROSIM is the process simulation program associated with this workshop. XRCHRT is a program that may be used for the construction and analysis of $\overline{X}$ and R control charts. This program will be described shortly. PROCAP is a program for the determination of process capability, and its use is described at the end of Chapter 10. TURNSIM is the process simulation program used in the computer workshop described at the end of Chapter 10.

PROSIM was written for use on IBM PCs (and compatibles). The process simulation program for this workshop, PROSIM, may be started simply by typing in "d:PROSIM" at the DOS prompt, where d is the drive that contains the program disk. PROSIM will ask the user to enter a filename into which the data associated with the 125 subgroups of size $n = 5$ will be stored. Although any valid DOS filename may be used, it is suggested that the filename be given an extension of "DAT". The filename "d:PROSIM.DAT", for example, indicates that the file contains data generated by program PROSIM. In defining the filename, d is the drive onto which the file is to be stored. Following the entry of the filename, PROSIM will automatically generate 125 samples of size $n = 5$ and store them in the file. Once the data have been generated and stored in a file, the user may proceed to the construction of $\overline{X}$ and R control charts.

Using XRCHRT

XRCHRT was written for use on IBM PCs (and compatibles). The control chart construction and analysis program, XRCHRT, may be started simply by typing in "d:XRCHRT" at the DOS prompt, where d is the drive that contains the program disk. XRCHRT will ask the user to enter the name of the file that contains the data associated with the 125 subgroups of size $n = 5$. The user is also asked to enter the date and his or her name (this information will subsequently be displayed on the program output).

At this point, XRCHRT will ask the user to enter the number of data blocks to be separately studied. XRCHRT performs a separate control chart analysis for each data block, and the control chart limits for each data block may be different. If, for example, a user believes that the process mean was approximately zero for samples 1 to 32, and approximately 2 for samples 38 to 74, he or she may wish to specify that there are two data blocks to be studied, one block being samples 1 to 32 and the other being samples 38 to 74.

Once the number of data blocks has been specified, the user is prompted for the beginning and ending sample numbers for which the control chart limits are to be calculated for each block. Additionally, the user is asked to specify the beginning and ending sample numbers for data plotting and chart interpretation for each data block.

At this point users are asked if they wish to skip the data dump. The data dump gives a complete listing of all the individual measurements for each sample, as well as the $\overline{X}$ and R values for each sample. It is recommended that the data dump not be skipped the first time XRCHRT is used so that the user has a complete record of the results from all the samples. Users are then asked if they wish to skip the plot (i.e., control chart plotting). The plot should rarely, if ever, be skipped. Finally, users are asked if they wish to invalidate any samples. This option permits a user to remove specific points from control chart limit calculations or from control chart interpretation.

Finally, users are asked where they would like to direct the output of program XRCHRT: to the printer or to a file. If the printer is selected, XRCHRT attempts to open the printer and send output directly to it (this may not be a feasible alternative for a system in which several computers are networked to a printer). If it is desired to send the output to a file, XRCHRT will then ask the user to specify a filename into which the output will be placed. Once again, although any valid DOS filename may be used, it is suggested that the filename be given an extension of "REP" or "OUT", indicating that the file contains a "report" or "output." When the user leaves program XRCHRT, the output stored in a file may be printed using the suitable DOS or system command. Once the user has specified where the output is to be directed, all user input is completed, and XRCHRT will begin constructing and interpreting the control charts.

Assignment

At least four tasks must be completed for this assignment:

1. Run program PROSIM to simulate the collection of 125 samples of size $n = 5$ from the fictional process.
2. Run program XRCHRT to construct and interpret control charts for the data collected from the fictional process (it may be necessary to run XRCHRT several times).
3. Study the charts and determine where the statistical signals are and what they are saying concerning shifts in the mean and/or the standard deviation of the process.
4. Prepare a report that summarizes what you have learned about the process.

The report, describing what you have learned about the process, must include at least the following:

1. An executive summary, which in approximately one page summarizes the contents and conclusions of your report.
2. A body of the report, which includes the following components:
   a. An introductory section, which briefly describes the concepts and methods relevant to the interpretation of statistical signals on control charts and indicates the methodology employed in solving the problem.
   b. A procedures section, which describes the methodology used to study the process. Remember that the identification procedure you are undertaking involves two basic steps: identifying statistical signals indicating the presence of a special cause, and identifying the specific nature of that cause (i.e., how the process is being affected). The logic used in

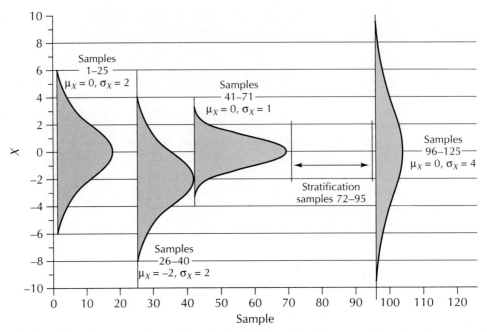

Figure 8.19 Typical Summary of the Special Causes That Were Introduced to the Fictional Process

deciding what to do at each step should be spelled out carefully. You should clearly state how you arrived at any conclusions. In this section it will clearly be necessary to refer to the control charts themselves, which should be contained in the appendix to the report.

c. A conclusions section, which summarizes the special-cause nature and time period over which you believe each special cause was present. Remember, you can conclude that a special cause is present only when the control charts so indicate. Your conclusions may be nicely summarized with a figure like that shown in Fig. 8.19.

3. An appendix section, which contains all the control charts that were generated during the course of your investigation. Note that these control charts should clearly indicate (with circled points) all the statistical signals on ·the charts.

9

Process Capability Assessment

9.1
Introduction

In this chapter we present both traditional and more modern views of process capability and how it is assessed. The presentation stresses the relationship between the stability of a process and its capability. The clear distinction between the engineering specification and the statistical control limit in terms of use and interpretation is emphasized in this chapter. By using the traditional engineering specifications to view process capability, this chapter illustrates the consequences of excessive common-cause variation in terms of the manner and extent to which a process produces parts that meet design intent. Because of the importance of the engineering specification in the traditional approach to process capability and the need to begin to think of engineering specifications with a loss function point of view, this chapter also examines the statistical assessment and assignment of tolerances. In particular, the relationship between statistically determined specifications and the achievement of their intent at the process via a statistically stable manufacturing environment is discussed. Finally, the loss function approach to the articulation of process capability is presented. In this regard the consequences of failure to ensure statistical control of the process, as well as poor process centering, are demonstrated in a quantitative fashion.

9.2

Process Capability Versus Process Control

There are two separate but vitally important issues that we must address when considering the performance of a process: the ability of the process to produce parts that conform to engineering specifications, and the ability of the process to maintain a state of good statistical control. Since most process capability assessments are made on statistical grounds, these two process characteristics are linked in the sense that it is both inappropriate and statistically invalid to assess process capability with respect to conformance to specifications without being reasonably assured of having good statistical control. Although control certainly does not imply conformance, it is a necessary prerequisite to the proper assessment of conformance.

In a statistical sense, conformance to specifications involves the process as a whole and therefore we are focusing our attention on the distribution of individual measurements. In dealing with statistical control we employ summary statistics from each sample, for example, $\overline{X}$ and R, and therefore are dealing with the distributions of these statistics, not individual measurements. Because of the distinction between populations and samples, it is crucial not to compare or confuse specifications associated with quality characteristics and control limits. In fact, we should never place tolerance/specification limits on an $\overline{X}$ control chart. This is so because the $\overline{X}$ control chart is based on the variation in the sample means, $\overline{X}$, while it is the individual measurements in the sample that should be compared to specifications. By placing specifications on the control chart, we may get the mistaken impression that good conformance exists when, in fact, it does not. Furthermore, it is philosophically wrong to place specifications on a control chart, since such would suggest that the purpose of the chart is to track a quality characteristic relative to specifications.

A process may produce a large number of pieces that do not meet the specified tolerances, even though the process itself is in a state of good statistical control (i.e., all the points on the $\overline{X}$ and R charts are within the 3 sigma limits and vary in a random manner). This may be because the process is not centered properly; in other words, the actual mean value of the parts being produced may be significantly different from the specified nominal value of the part. If this is the case, an adjustment of the machine to move the mean closer to the nominal value may solve the problem. Another possible reason for lack of conformance to specifications is that a statistically stable process may be producing parts with an unacceptably high level of common-cause variation, although it may be centered at the nominal value.

In summary, if a process is in statistical control but not capable of meeting the tolerances, the problem may be one of the following:

- The process is off-center from the nominal.
- The process variability is too large relative to the tolerances.
- The process is off-center and has large variation.

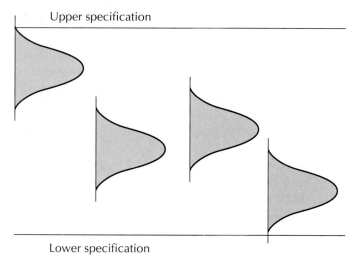

Upper specification

Lower specification

(a) Process variation is small relative to the specifications so that the process mean can shift about without causing the process capability to be jeopardized.

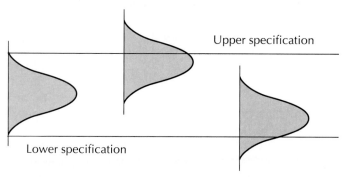

Upper specification

Lower specification

(b) Process variation is large relative to the specifications such that the process must remain well centered for the process capability to be maintained.

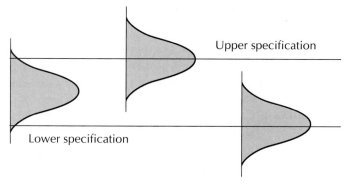

Upper specification

Lower specification

(c) Process variation is large relative to the specifications so that the process cannot be considered capable regardless of the process centering.

Figure 9.1 Relationship Between Process Variability and Product Specifications

Figure 9.1 illustrates the possible relationships that could exist between the process variation and the specifications. Of particular importance may be attention to the uppermost case in the figure. Here the process variation is small relative to the specifications. In such a situation it is easy to forget about the issue of statistical control since the specifications are so wide. In fact, statistically based modified control chart limits have even been suggested to take advantage of this happy state of affairs. Such are probably inadvisable since they mask the presence of improvement opportunities as a result of sporadic behaviors that rob the process of productivity as well as quality. Again, it is essential that we separate the issues of control and conformance.

9.3

Statistical Assessment of Process Capability: A Case Study Revisited

Recall the machined cylinder bore example of Chapter 6. Measurements of the inside diameter of the bore were made after the machining process. The specifications for the bore are 3.5199 ± 0.0004 inches. Many blocks were being rejected on 100% inspection using a go/no-go gage because they failed to meet these tolerances. It was decided to study the capability of the process using $\overline{X}$ and R charts, so the data given in Table 9.1 were collected. These are the same data previously shown in Table 6.1 with samples 1, 6, 11, and 16 deleted. These data were taken from the same bore, positionwise, in the block, and at the rate of about one sample per half hour. The table provides 31 samples of size $n = 5$ each. Recall that this data was presented in coded form, being the last three digits of measurements of the form 3.5205, 3.5202, 3.5204, and so on.

Checking for Statistical Control

The first step in the process capability assessment should be to examine the state of statistical control (or lack of it) of the process. To do this, we will construct $\overline{X}$ and R control charts. This, of course, requires that the data be collected in a manner (samples collected over a period of time) that is consistent with control charting, as was the case here. Recall that in Section 6.4 these data (Table 9.1) were thoroughly studied and the process, after the removal of four samples that contained special causes, showed good statistical control. The necessary calculations are repeated as follows.

We first estimate the mean of the process by $\overline{\overline{X}}$:

$$\overline{\overline{X}} = \sum_{i=1}^{k=31} \frac{\overline{X}_i}{31}$$

$$= 199.95.$$

TABLE 9.1   Cylinder Boring Process Data—In Control

Sample	1	2	3	4	5	$\overline{X}$	R
2	202	196	201	198	202	199.8	6
3	201	202	199	197	196	199.0	6
4	205	203	196	201	197	200.4	9
5	199	196	201	200	195	198.2	6
7	202	202	198	203	202	201.4	5
8	197	196	196	200	204	198.6	8
9	199	200	204	196	202	200.2	8
10	202	196	204	195	197	198.8	9
12	200	201	199	200	201	200.2	2
13	205	196	201	197	198	199.4	9
14	202	199	200	198	200	199.8	4
15	200	200	201	205	201	201.4	5
17	202	202	204	198	203	201.8	6
18	201	198	204	201	201	201.0	6
19	207	206	194	197	201	201.0	13
20	200	204	198	199	199	200.0	6
21	203	200	204	199	200	201.2	5
22	196	203	197	201	194	198.2	7
23	197	199	203	200	196	199.0	7
24	201	197	196	199	207	200.0	10
25	204	196	201	199	197	199.4	5
26	206	206	199	200	203	202.8	7
27	204	203	199	199	197	200.4	7
28	199	201	201	194	200	199.0	6
29	201	196	197	204	200	199.6	8
30	203	206	201	196	201	201.4	10
31	203	197	199	197	201	199.4	6
32	197	194	199	200	199	197.8	6
33	200	201	200	197	200	199.6	4
34	199	199	201	201	201	200.2	2
35	200	204	197	197	199	199.4	7

We will also need to estimate the range of the process:

$$\overline{R} = \sum_{i=1}^{k=31} \frac{R_i}{31}$$

$$= 6.61.$$

The control limits for $\overline{X}$ and R charts are (using the constants for $n = 5$)

$$\text{UCL}_{\overline{X}}: \quad \overline{\overline{X}} + A_2\overline{R} = 199.95 + (0.58)(6.61)$$

$$= 203.78$$

$$\text{LCL}_{\overline{X}}: \quad \overline{\overline{X}} - A_2\overline{R} = 199.95 - (0.58)(6.61)$$

$$= 196.12$$

$$\text{UCL}_R: \quad D_4\overline{R} \quad = (2.11)(6.61)$$
$$= 13.95$$
$$\text{LCL}_R: \quad D_3\overline{R} \quad = (0)(6.61)$$
$$= 0.00.$$

Now we can construct $\overline{X}$ and R control charts. These were previously shown in Fig. 6.15 and are repeated here in Fig. 9.2.

Both the $\overline{X}$ and R charts seem to indicate good statistical control with no points exceeding the 3 sigma limits, a reasonably normal distribution of points

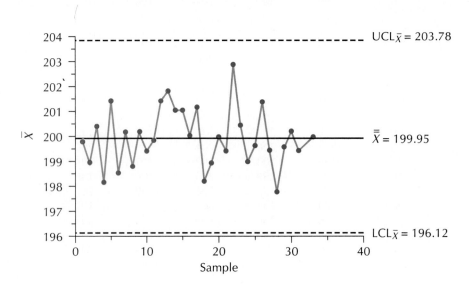

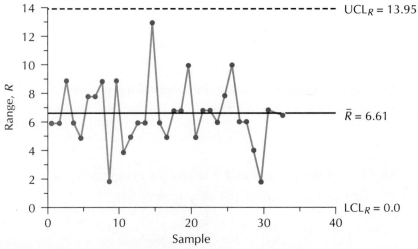

Figure 9.2 $\overline{X}$ and R Charts for Process Capability Case Study Data

between the limits, and no discernible trends, cycles, zone rule violations, and so on. The good control of the R chart indicates that our estimate of the process variation,

$$\hat{\sigma}_X = \frac{\bar{R}}{d_2}$$

$$= \frac{6.61}{2.326}$$

$$= 2.8418,$$

is a valid estimate of common-cause variation. With a stable process demonstrated through the $\bar{X}$ and R charts, we may now proceed to evaluate the process with respect to its conformance to specifications.

Statistical Assessment of Process Capability

To get a clear picture of the statistical nature of the data from an individual measurements point of view, a frequency histogram was plotted. This is shown in Fig. 9.3. The histogram seems to exhibit the shape of the normal distribution, but the mean appears to be a little higher than the nominal value of 199.

Using the knowledge above about the distribution of the process and our estimates of the process mean $\bar{X}$ and process standard deviation $\hat{\sigma}_X$, we can sketch the process as a normal distribution, as shown in Fig. 9.4. The specifications are also shown in the figure. The shaded area under the normal curve represents the probability of obtaining a bore that does not meet the specifications.

To compute the probability of a bore falling outside of the specifications, we calculate Z and use the standard normal curve table (since n is quite large)

$$Z = \frac{X - \bar{X}}{\hat{\sigma}_X},$$

where X is the value of either the lower or upper specification, $\bar{X}$ is our estimate for the process mean, and $\hat{\sigma}_X$ is our estimate of the process standard deviation. To find the probability of a measurement below the lower specification limit, LSL = 195, with $\bar{X} = 199.95$ and $\hat{\sigma}_X = 2.8418$,

$$Z_L = \frac{195 - 199.95}{2.8418}$$

$$= -1.742.$$

Looking this value up in the normal curve table (Table A.1), we find that the probability of $X \leq 195$ is 0.0407. This means that we have a 4.07% chance of an individual bore measurement falling below the specified lower specification limit.

To find the probability of an individual bore measurement falling above the upper specification limit, USL = 203, with $\bar{X} = 199.95$ and $\hat{\sigma}_X = 2.8418$,

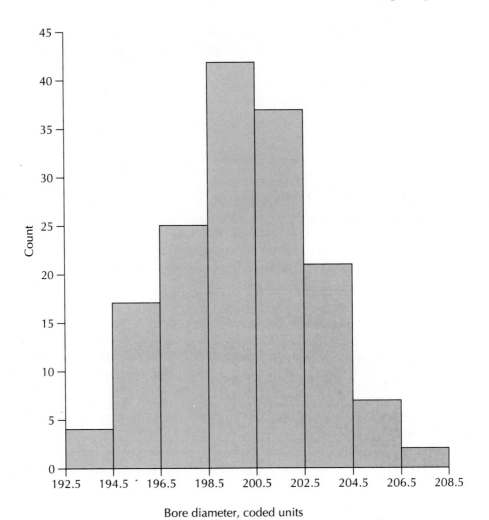

Figure 9.3 Histogram of Case Study Data

$$Z_U = \frac{203 - 199.95}{2.8418}$$

$$= 1.073.$$

The value in the normal curve table for $Z = 1.073$ is 0.8586, but this includes all of the area to the left of the USL. Thus the probability of a bore being above the upper specification limit is equal to $1.0 - 0.8586 = 0.1414$. The percentage of product that does not meet the specifications is therefore 4.07% + 14.14% = 18.21%.

Now we might ask. Would centering the process at our nominal value of 199 help? To check, we construct a new normal curve, centered at 199. The

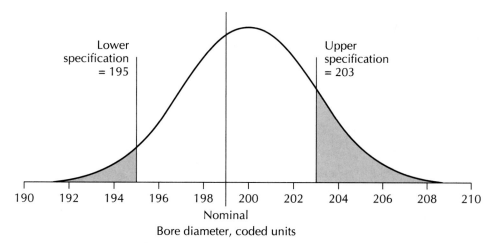

Figure 9.4 Normal Distribution Model for Process Capability Case Study Data

new probability of a point below the lower specification is found by computing Z using the nominal value as the process mean:

$$Z_L = \frac{195 - 199}{2.8418}$$
$$= -1.408.$$

Looking up the area for this Z value in the normal curve table, we get a value of 0.0793, which is the probability of getting a value below the lower specification limit. The probability of getting a value above the upper specification is found by

$$Z_U = \frac{203 - 199}{2.8418}$$
$$= 1.408,$$

which from the tables gives an area = 0.0793. The total new probability of a bore not meeting specifications is the sum of these, or 0.0793 + 0.0793 = 0.1586.

We can see that recentering the process will not help very much. The process is in control—no special causes of variability were indicated. However, over 18% of the part measurements fall outside the specifications. Possible remedies may include:

1. Continue to sort by 100% inspection.
2. Widen the tolerances (e.g., 3.5199 ± 0.0009).
3. Use a more precise process to reduce process variation.
4. Use statistical methods to identify variation reduction opportunities for the existing process.

Too often, strategy 2 is used (or at least urged). Clearly, we suggest that stronger consideration be given to strategy 4.

Summary: Comparison of Specification Limits and Control Limits

It is important to differentiate clearly between specification limits and control limits. In particular, we must note that specification limits or tolerances for a product quality characteristic are

- Characteristic of the part/item in question.
- Based on functional considerations.
- Related to/compared with an individual part measurement.
- Used to establish a part's conformability to design intent.

Control limits on a control chart are

- Characteristic of the process in question.
- Based on the process mean and variability.
- Dependent on sampling parameters, namely, sample size and the α-risk.
- Used to identify the presence/absence of special-cause variation in the process.

Control limits and specification limits must never be compared numerically and should not appear together on the same graph. Tolerances are limits on individual measurements and as such may be compared against the process as a whole as represented by a statistical distribution, as was done in the previous example to assess overall process capability. Figure 9.5, based on the data of the previous process capability study, depicts the difference between specification and control limits, and clearly indicates the danger of putting specification limits on the control chart. The larger solid dots represent the individual measurements from which the sample means ($\overline{X}$'s) shown in the figure were determined. It is clear that while the $\overline{X}$'s are well within the control limits, many (4 out of 25) individual measurements fall on or outside the specifications.

9.4

Some Common Indices of Process Capability

Process Capability Index, C_p

It is common to measure process capability in units of process standard deviations. In particular, it is common to look at the relationship between the process standard deviation and the range between the upper and lower specification limits. This measure is known as C_p:

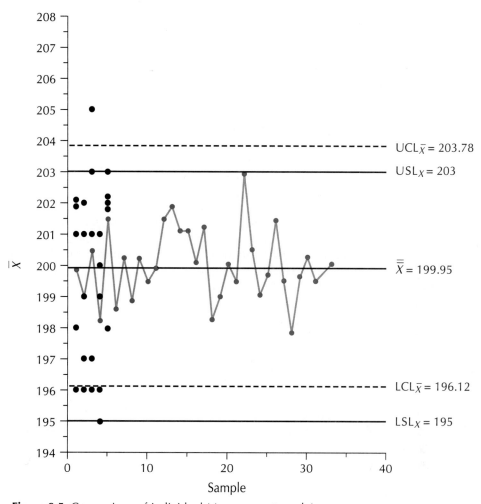

Figure 9.5 Comparison of Individual Measurements and Averages

$$C_p = \frac{\text{USL} - \text{LSL}}{6\,\sigma_X} \tag{9.1}$$

The minimum acceptable value for C_p is considered to be 1.

Process Capability Index, C_{pk}

Recently, many companies have begun to use a capability index referred to as C_{pk}. The primary reason for this is that C_{pk} relates the process mean to the nominal value of the specification. For bilateral specifications, C_{pk} is defined in the following manner. First, we determine the relationship between the process mean μ_X and the specification limits in units of standard deviations:

$$Z_{\text{USL}} = \frac{\text{USL} - \mu_X}{\sigma_X}, \qquad Z_{\text{LSL}} = \frac{\text{LSL} - \mu_X}{\sigma_X}. \tag{9.2}$$

Then we select the minimum of these two values:

$$Z_{\min} = \min[Z_{\text{USL}}, -Z_{\text{LSL}}]. \tag{9.3}$$

The C_{pk} index is then found by dividing this minimum value by 3:

$$C_{pk} = \frac{Z_{\min}}{3}. \tag{9.4}$$

Commonly, C_{pk} should be ≥ 1.00 for the process capability to be considered acceptable.

The idea of using the minimum Z value is to promote adherence of the process mean to the nominal value. Figure 9.6 demonstrates this apparent motivation for being on target. Consider initially that the process mean is located at 130, which is somewhat away from the nominal of 145, with a standard deviation of $\sigma_X = 10$. Then

$$Z_{\text{USL}} = \frac{190 - 130}{10}$$
$$= 6$$
$$Z_{\text{LSL}} = \frac{100 - 130}{10}$$
$$= -3$$
$$Z_{\min} = \min[6, -(-3)]$$
$$= 3$$
$$C_{pk} = \frac{3}{3}$$
$$= 1.00.$$

If the process mean were centered at the nominal of 145, then

$$Z_{\text{USL}} = \frac{190 - 145}{10}$$
$$= 4.5$$
$$Z_{\text{LSL}} = \frac{100 - 145}{10}$$
$$= -4.5$$
$$Z_{\min} = \min[4.5, -(-4.5)]$$
$$= 4.5$$
$$C_{pk} = \frac{4.5}{3}$$
$$= 1.50.$$

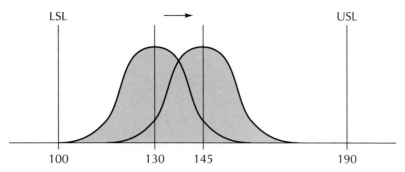

Figure 9.6 Motivation via C_{pk} to Be on Target

Therefore, by centering the process, the value of C_{pk} is increased by 50%.

The foregoing apparent motivation via C_{pk} to adhere to the nominal value must be viewed, however, with some caution. It is possible, with the use of C_{pk}, to use reduction in variation as a means to shift the process mean away from the nominal and still get an increase in C_{pk}, or at least not experience a decrease in C_{pk}. Returning to our previous example, suppose that the process is running at the nominal with a standard deviation of 15. C_{pk} under this circumstance is equal to 1.00. However, by reducing the variation to a standard deviation of 10, the process mean can be shifted well away from the nominal and C_{pk} is not reduced. Figure 9.7 illustrates this problem. The parameter values for each case are shown below.

Case 1: $\mu_X = 145$ and $\sigma_X = 15$, so $C_{pk} = 1.00$.
Case 2: $\mu_X = 130$ and $\sigma_X = 10$, so $C_{pk} = 1.00$.

Therefore, in this case, there is no penalty for purposely moving away from the target. Today, other measures of process capability are being considered to avoid the problem illustrated above.

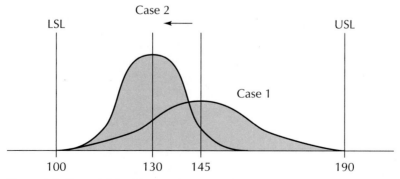

Figure 9.7 Potential Problem with C_{pk}

9.5

Statistical Assessment and Assignment of Tolerances

Suppose that a process is found to be in good statistical control but a subsequent process capability study indicates that the process does not meet the customer needs in terms of conformance to the specifications. One of two possible problems is evident. Either the process variation is too large or the specifications have been set too narrow. Our emphasis in this book seems to favor strong attention to the first possibility—that the process variation is too large and that efforts ought to be made to find and remove the major sources of common-cause variation. It is possible, however, that the specifications have been determined in an improper fashion, and as we will soon see, a common improper approach to the determination of specifications can lead to narrower specification limits than are otherwise required when the statistical methods are used.

Previously, we have emphasized the importance of pushing the quality issue upstream, into engineering design. We have briefly introduced the tolerance design concept based on the loss function, which addresses the economics of tolerance assignment based on loss due to functional variation. This approach also assumes that the process may be represented by a statistical distribution, and therefore that the process is in statistical control.

Statistical Determination of Assembly Tolerances: An Example

Consider the three-part assembly shown in Fig. 9.8. The dimension in question is the length of the assembly and is equal to the sum of the three component lengths. We will assume that the processes producing each of the three components are in statistical control, the processes are centered at the nominal value, and the process capability for each is about $\pm 4\sigma_X$; that is, the specifications for each part are $8\sigma_X$ wide.

In examining Fig. 9.8, the question that might be asked is: What is the tolerance that should be assigned to the assembly dimension based on the tolerances that have been assigned to each of the individual part dimensions

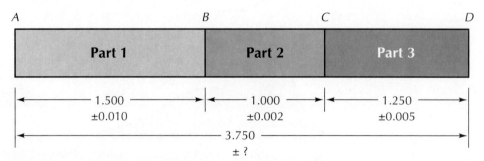

Figure 9.8 Tolerance Determination for a Three-Part Assembly

as specified in the figure? An altogether too common answer to this question is that the assembly tolerance is simply equal to the sum of the individual part tolerances. This would result in an assembly tolerance of 0.010 inch + 0.002 inch + 0.005 inch = 0.017 inch. However, under the conditions of the preceding paragraph, one would find that the actual assemblies would come out much better than this, that is, would have a smaller natural spread. It is important that we understand exactly why this is so.

There are two important forces at work that govern this problem:

1. Statistical distribution of component dimensions.
2. Random assembly.

When in statistical control, the individual processes produce parts that vary over some range due to common-cause variation. This variation tends to arise in a frequency sense according to a statistical distribution and therefore is confined to some range of occurrence by probabilistic limits. If we assumed a normal distribution for each part's individual measurements, the state of affairs would be as shown in Fig. 9.9, with the process capabilities as specified above.

Now, considering random assembly, that is, the parts are drawn randomly from processes 1, 2, and 3, respectively, the assembly deviation from nominal will behave in a predictable fashion based on precise probability laws. Because of this statistical way of thinking about the dimensions of the individual parts emanating from their respective processes, the chances of randomly selecting three parts, all in the far-left tail of their respective distributions (at the 4 sigma distance from the mean), is very, very small. This is depicted graphically in Fig. 9.10.

The probability of getting any one part measurement 4 or more standard deviations below the nominal is about 0.00003. The probability of getting all three parts simultaneously at −4 sigma or less under random assembly is $(0.00003)^3 = 0.000000000000027$! It is clear from this that using simple additivity

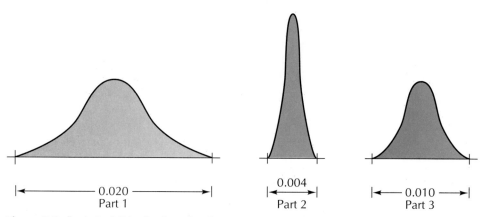

Figure 9.9 Statistical Distributions for the Dimensions of Each Part

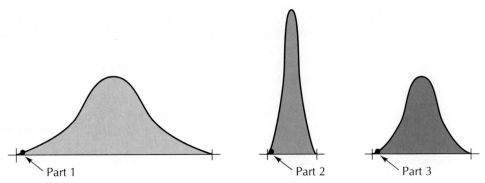

Figure 9.10 Simultaneous Selection of Parts from the Extreme Left of Their Respective Distributions

of the individual part tolerances is an unrealistic way to combine them to obtain the assembly tolerance. Then how shall we proceed to determine the assembly tolerance more realistically? Suppose that for the three parts in question the specifications were set at $\pm 4\sigma_{process}$; then

> *Part 1 tolerance:* ± 0.010 inch implies that $\sigma_1 = 0.0025$.
> *Part 2 tolerance:* ± 0.002 inch implies that $\sigma_2 = 0.0005$.
> *Part 3 tolerance:* ± 0.005 inch implies that $\sigma_3 = 0.00125$.

Now there is a third force at work here, the *additive law of variances:*

$$\sigma^2_{assembly} = \sigma^2_1 + \sigma^2_2 + \sigma^2_3. \tag{9.5}$$

Using the additive law of variances to combine the tolerances produces the following result in terms of the standard deviation of the assembly:

$$\sigma_{assembly} = \sqrt{(0.0025\ \text{inch})^2 + (0.0005\ \text{inch})^2 + (0.00125\ \text{inch})^2}$$
$$= 0.00284\ \text{inch}.$$

If the assembly tolerance is also set at $\pm 4\sigma_{assembly}$, virtually all assemblies will fall within $\pm 4\,(0.00284\ \text{inch}) = \pm 0.01136$ inch of the nominal assembly dimension, not ± 0.017 inch as obtained by adding part tolerances. In fact, 99.73% of all assemblies will fall within ± 0.0085 inch of the nominal assembly dimension, *one-half* of the value obtained by adding part tolerances. In summary:

- *By adding tolerances for individual parts:*   assembly tolerance $= \pm 0.017$ inch.
- *By combining tolerances statistically:*   assembly tolerance $= \pm 0.01136$ inch.

Statistical Assignment of Component Tolerances

In practice the problem above is not the one we are generally faced with; rather, it is the reverse problem. Based on functional grounds, it is likely that the assembly tolerance is specified first. The question that we are then faced with is: How do we partition the assembly tolerance; that is, how do we assign tolerances back to the individual parts? To answer this question, we consider another example.

Consider a certain assembly made up of three identical parts as shown in Fig. 9.11. The assembly tolerance is ±0.009 inch. How do we distribute this tolerance among the individual parts that comprise the assembly? By invoking additivity of tolerances, we would obtain the following answer:

Tolerance of the assembly = tolerance of part 1 + tolerance of part 2

+ tolerance of part 3

± 0.009 inch = ±0.003 inch + ±0.003 inch + ±0.003 inch.

But we already know that this is not the correct way to combine the individual part tolerances. What we now understand is that this places tolerance requirements on the parts which are much tighter than they need to be to guarantee the required assembly tolerance. From what we have seen already, it would appear that the individual part tolerances could exceed ± 0.003 inch and still the assembly tolerance could be ±0.009 inch.

To determine precisely what the individual part tolerance should be, let us again invoke:

1. Random assembly.
2. Statistical (and in this case normal) distribution of part dimensions.
3. The additive law of variances.

Further, let us assume:

4. The processes making the parts will be ±4σ_{part} capable.
5. The part standard deviations are equal (made by same or similar processes).

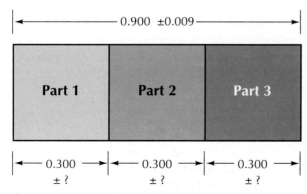

Figure 9.11 Part Tolerance Assignment for an Assembly

With the assumptions above we can easily determine the individual part tolerances:

$$4\sigma_{assembly} = 0.009 \text{ inch}$$

$$\sigma_{assembly} = 0.00225 \text{ inch}.$$

Furthermore,

$$\sigma_{assembly} = \sqrt{\sigma_1^2 + \sigma_2^2 + \sigma_3^2}$$

and we will assume that

$$\sigma_1^2 = \sigma_2^2 = \sigma_3^2 = \sigma_{part}^2.$$

Therefore,

$$\sigma_{assembly} = \sqrt{3\sigma_{part}^2}.$$

But $\sigma_{assembly} = 0.00225$ inch, as calculated above, so

$$\sigma_{part} = 0.0013 \text{ inch}.$$

Now if the individual part tolerances are set at $\pm 4\sigma_{part}$, then a ± 0.0052 inch tolerance for each part ensures that virtually all of the assemblies will fall within ± 0.0090 inch. This is in contrast to a much tighter ± 0.003 inch tolerance as would be specified by using simple addition of part tolerances.

In summary:

Simple division: part tolerance $= \pm 0.0030$ inch.

Statistical division: part tolerance $= \pm 0.0052$ inch.

An important implication of the statistical tolerance assignment situation that we saw above is that by inappropriately setting the part tolerance using additivity/division, you could end up with a situation in which you could be scrapping/reworking perfectly good parts. This fact is seen in Fig. 9.12, using the numbers of the previous example.

It is important that in the above we have the proper view of what we really mean by σ_{part} and setting tolerances, for example, $\pm 4\sigma_{part}$. This notion has meaning only when we think of σ_{part} as a process-generated variation. But the tolerance is based on part function requirements. What the design engineer is doing when he or she sets the tolerance is setting the capability requirement for the process. This, of course, requires that the process is in control. Conformance to specifications cannot be interpreted otherwise.

Importance of Process Centering in Maintaining Capability: Individual Component Versus Assembly Tolerances

Suppose we are concerned with a part that is really a three-piece construction using three identical parts, as shown in Fig. 9.13. Based on design intent the assembly nominal is 0.390 inch and the tolerance is set at ± 0.050 inch. Given the functional needs above, the tolerances for the individual parts are to be set.

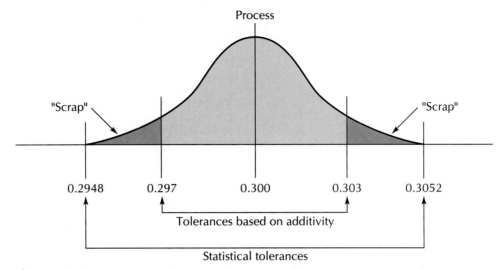

Figure 9.12 Consequences of Improper Tolerance Assignment

Using the additive law of variances, assuming random assembly and assuming that the assembly process is to be ±3 sigma capable, we can determine the required part tolerance.

$$\sigma_{assembly} = \frac{0.050 \text{ inch}}{3}$$

$$= 0.017 \text{ inch.}$$

But as we have seen previously,

$$\sigma_{assembly} = \sqrt{\sigma_1^2 + \sigma_2^2 + \sigma_3^2} = \sqrt{3\sigma_{part}^2}$$

since the three parts are identical. Therefore,

$$0.017 \text{ inch} = \sqrt{3\sigma_{part}^2}$$

$$\sigma_{part} = 0.010.$$

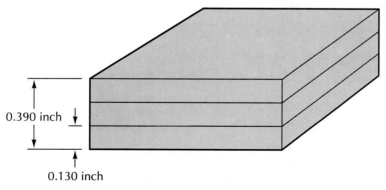

Figure 9.13 Three-Part Assembly

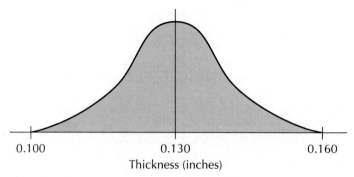

Figure 9.14 Process for the Component Parts

It seems logical, therefore, to set the individual part requirements at 0.130 ± 0.030 inch (lower specification = 0.100 inch, upper specification = 0.160 inch), as shown in Fig. 9.14.

Suppose that a stable process for the individual part is established as seen in Fig. 9.15 and that through a series of process improvements, the process common-cause variation has been halved. Now $\sigma_{process}$ or σ_{part} = 0.005 inch. Let us further suppose that as a result of this process improvement there is a change in the operating policy for the process. Since the process variability has been reduced, it is decided to adjust the process mean for "economic" reasons, so that the process is now operating as seen to the right in Fig. 9.15.

Now it is important that we examine how this change in the component part process operation will affect the product, which is the assembly. Using the additive law of variances,

$$\sigma^2_{assembly} = 0.005^2 \text{ inch}^2 + 0.005^2 \text{ inch}^2 + 0.005^2 \text{ inch}^2$$
$$= 0.000075 \text{ inch}^2.$$

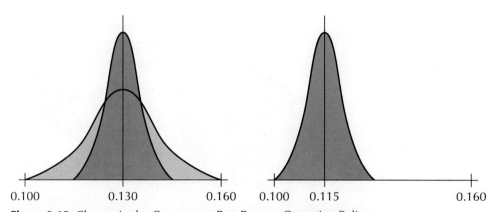

Figure 9.15 Change in the Component Part Process Operating Policy

Therefore,

$$\sigma_{assembly} = 0.0087 \text{ inch.}$$

The actual assembly mean value will be (3)(0.115 inch) = 0.345 inch because of the change in operating policy for the process. For the assemblies, 99.73% will now be within 0.345 ± 0.026 inch.

Recall that based on design intent, assemblies must be 0.390 ± 0.050 inch. But the result now seems unbelievable. The average of the actual assemblies is now within 0.005 inch of the lower specification limit. Figure 9.16 shows the consequences of the change in component-part processing policy. As a result of the change in the individual component process mean, one of every six assemblies will be out of the specification. It is clear from this example that merely operating within the bilateral specification is not sufficient to guarantee that assemblies of individual components will meet the specifications developed by design based on function. The relationship between design intent and its realization through the process is a strong one and must not be ignored. In the next section this fact is reemphasized.

9.6

Statistical Process Control and the Statistical Tolerance Model

As we continue to examine many of the systems that we must deal with in both the product/process design arena and in manufacturing, we will find that the concept and achievement of statistical process control is an ever-present central issue. One of the important issues discussed throughout this book is

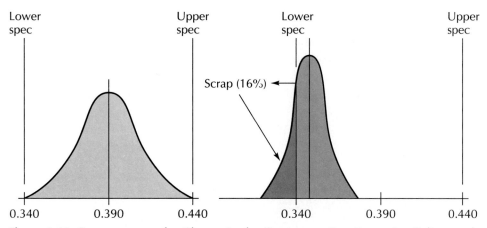

Figure 9.16 Consequences of a Change in the Component Part Processing Policy on the Assembly Conformance to Specification

the relationship between the design process and the manufacturing process. Both parties must realize together the common goals they share and the importance of developing a joint strategy to meet those goals.

The issue of setting tolerances and, in particular, the statistical assignment and assessment of tolerances is an excellent example of the need for design and manufacturing to understand what each other is doing and why. The design process can go unfulfilled if the manufacturing process is not operated in a manner totally consistent with design intent. To appreciate the marriage of thinking that must exist between the design and manufacturing worlds, we will now reexamine some of the basic assumptions of setting tolerances that we discussed earlier and their relationship to the manufacturing process. What we are about to see clearly points to the importance of statistical process control relative to the issue of process capability.

The key concepts in statistical tolerancing are

1. The use of a statistical distribution to represent the design characteristic and hence the process output for the product in question relative to the design specifications.
2. The notion of random assembly (i.e., random part selection from these part process distributions when more than one part is being considered in an assembly).
3. The additive law of variances as a means to determine the relationship between the variability in individual parts and that for the assembly.

To assume that the parts can be represented by a statistical distribution of measurements (and for the assumption to hold in reality), the processes must be in a state of statistical control.

Example: Importance of Statistical Control to Tolerance Fulfillment

To illustrate the importance of statistical process control in achieving design intent in a tolerancing problem, we will consider the following example. Figure 9.17 shows drawings for two simple parts, a plate with a hole and a pin that

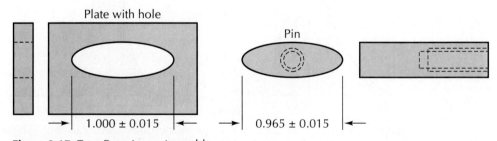

Figure 9.17 Two Parts in an Assembly

will ultimately be assembled to a third part but must pass through the hole in the plate. For the assembly, it is necessary for its function to be served that the clearance between the plate hole and the pin be at least 0.015 inch but no more than 0.055 inch.

To achieve the design requirement stated above, the nominal values and tolerances for the plate hole and pin were statistically derived and are given in Fig. 9.18. In arriving at these tolerances, it was assumed that:

1. The parts would be manufactured by processes that behaved according to the normal distribution, with the same process variation.
2. The process capabilities would be at least 6 sigma, the processes would be centered at the nominal values given in Fig. 9.17, *and the processes would be maintained in a state of statistical control.*
3. Random assembly would prevail.

If these assumptions are met, the processes for the two parts, and hence the clearance associated with assembled parts, would be as shown in Fig. 9.18, and therefore design intent would be met at the 6 sigma level.

Let us suppose now that despite the assumptions made and the tolerances thus derived, the processes used to manufacture the pin and plate hole were not maintained in good statistical control. As a result, when examined, the parts actually more nearly follow a uniform/rectangular distribution within the specifications, as shown in Fig. 9.19. Such could have arisen as a result of sorting or rework of a more variable process(es), in which case the results are doubly distressing—poor-fitting assemblies and increased cost to the system due to sorting and rework.

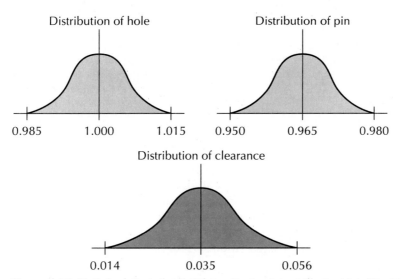

Figure 9.18 Statistical Basis for Satisfying Design Intent for the Hole/Pin Clearance

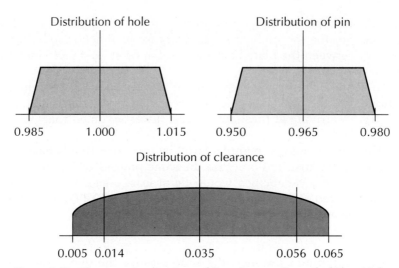

Figure 9.19 Clearance Implications of Poor Process Control of Plate Hole and Pin Dimension

Figure 9.19 shows the distribution of the clearance if the hole and pin dimensions follow the uniform distribution within the specifications. The additive law of variances has been used to derive the variation in the clearance distribution, but this time assuming the uniform distribution for the individual part processes. The distribution of the clearance will not be uniform, but it will not be near-normal either. It is important to note that now many assemblies will fit quite tightly and may later bind if foreign matter, for example, gets into the gap, while others will fit together with much larger clearance than desired. The overall range within which we expect the assemblies to lie is now 0.060 with more uniform probability of occurrence.

The problem here is not a design problem. The plate hole and pin tolerances have been derived using sound statistical methods. However, if the processes are not in good statistical control and hence are not capable of meeting the assumptions made during design, poor-quality assemblies will follow. It should be noted that the altogether too common process appearance of a uniform distribution of measurements within the specifications can arise in several different ways:

1. It can arise from processes that have good potential, in terms of variation, but are not kept in good statistical control.
2. It can arise from unstable and/or large variation processes which require sorting/rework to meet the specifications.
3. It can arise from processes that are allowed purposely to vary over the full range of the specifications to take advantage locally of wide specifications relative to the process variation.

In all three cases mentioned above, additional costs will be incurred and product quality will be eroded. This was the message of the final section in

Chapter 2. Clearly, statistical process control is crucial to the issue of setting tolerances in engineering design. Taguchi's loss function model, which is an essential element in tolerance design, assumes similarly that quality characteristics can be represented by a statistical distribution of measurements. Again, for this assumption to be met at the process and hence in the ultimate product in the field, the manufacturing processes must be maintained in a state of statistical control.

9.7

Loss Function Approach to Capability Assessment

The process capability study discussed earlier in this chapter measured the overall conformance to specifications, which may be useful for comparisons between different processes. To pursue the notion of being on target with smallest variation as introduced in Chapter 1, and perhaps to begin to direct our thinking toward a more global and economic interpretation of the engineering specification, the concept of the loss function advocated by Taguchi and used by economists for decades will now be examined. To see how the concept of the loss function is applied in the evaluation of process capability relative to the engineering specifications, we will use some of the illustrations from an article written by Peter T. Jessup of the Ford Motor Company.[1]

Defining a Loss Function

The quality characteristic of interest in the Jessup case study is the lever effort on a heater control. Very low efforts would cause customer complaints of flimsy feel or rattle, while very high efforts would also cause customer complaints of stiffness. And in each case, the more extreme the condition, the more likely it would prompt a customer complaint. The total complaint rate is the sum of these two individual kinds of complaints. Between the extremes the net complaints from the two conditions drop to a minimum. This is illustrated in Fig. 9.20.

By choosing the point with the minimum loss on the net loss curve as the target and setting it to zero, we establish a continuous loss function of the quality characteristic as shown in Fig. 9.21. In general, a loss function may be defined by evaluating the expected losses for several values of the quality characteristic. Figure 9.22 shows calculations for one such point.

The calculations for the loss incurred at two particular values for the quality characteristic are as follows.

1. Identify a most desirable or target value:

$$x = 0, \text{ at the target;} \quad L(x = 0) = 0.$$

---

[1] P. T. Jessup, "The Value of Continuing Improvement," *Proc. IEEE International Communications Conference, ICC 85,* 1985.

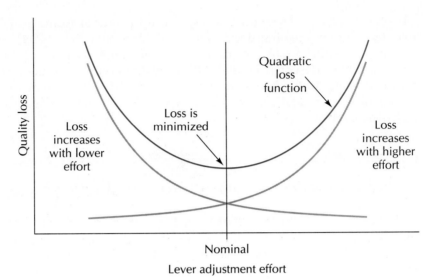

Figure 9.20 Composition of a Net Loss Function

2. Expected cost of a particular failure type (in the field):

$x = 1.5$, and

at $x = 1.5$, the cost of a complaint $= \$45$; the chance that this complaint occurs $= 5\%$; loss $= L(1.5) = 0.05 \times \$45 = \$2.25$.

If it is assumed that a quadratic equation closely approximates the true loss function, then the loss function can be derived from the expected losses esti-

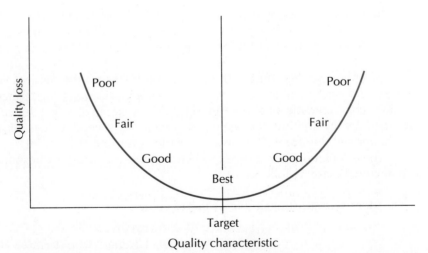

Figure 9.21 Loss Function Representation of Quality

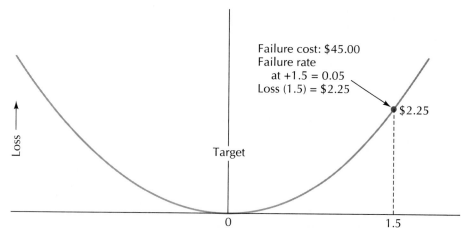

Figure 9.22 Construction of the Loss Function

mated from any two points, such as the loss at the target and the loss at one other point. The quadratic loss function can be written as

$$L(x) = k(x - m)^2,   \tag{9.6}$$

where $L(x)$ is the loss when the quality characteristic is x units away from the origin, m is the target, and k is the loss coefficient. The loss function above can be rewritten as

$$k = \frac{L(x)}{(x - m)^2}.$$

Using the data evaluated earlier, we have

$$k = \frac{L(1.5)}{(1.5 - 0)^2} = \frac{\$2.25}{2.25} = \$1.0.$$

Loss Function Interpretation of the Engineering Specification

It is possible to use the continuous loss function to give an economic interpretation to a specification limit. This is illustrated in Fig. 9.23. With reference to the figure, one has a choice of either

1. Letting the part pass on to the customer with an expected loss as determined by the loss function, or
2. Reworking or replacing the part at a certain cost.

To minimize the total costs, the alternative with the lower cost should be chosen. The point of indifference between the two alternatives determines the specification limit.

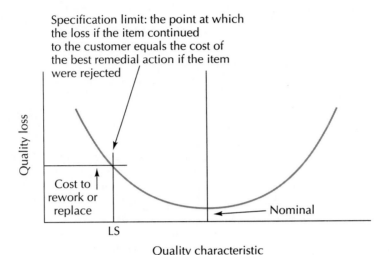

Specification limit: the point at which
the loss if the item continued
to the customer equals the cost of
the best remedial action if the item
were rejected

Quality loss

Cost to
rework or
replace

LS

Nominal

Quality characteristic

Figure 9.23 Loss Function Interpretation of an Engineering Specification

Suppose, in the example here, the replacement/rework cost is $1.00. Since the loss coefficient k = $1.00, then the points of indifference—the lower and upper specifications—would be ± 1.0.

Evaluation of Expected Process Loss

Although a loss function can be used for the economic establishment of specifications, it can be put to an even more illuminating task—evaluating the aggregate expected loss from a process, based on the distribution of actual performance of the quality characteristic in question. Hence the loss function constitutes a new way of assessing the capability of a process. This approach to assessing process capability takes into account the process distribution relative to the specifications in terms of the economic loss to the user due to deviation from the nominal. This average (expected) realized loss is found by weighting the loss function by the probability density function (statistical distribution) of the actual process output. In general this can be written as

$$E[L(X)] = \int_{R_X} L(t)\,f(t)\,dt \qquad\qquad (9.7)$$

where

$L(x)$ is the loss function for the variable X
$f(x)$ is the p.d.f. for the variable X

For a quadratic loss function (Eq. 9.6), the expected loss is simply the loss

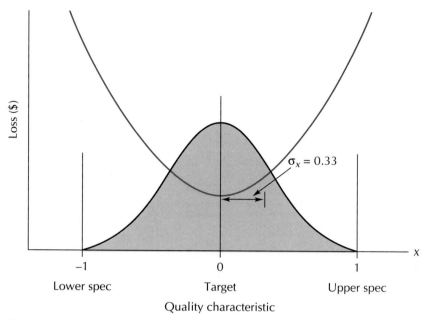

Figure 9.24 Evaluation of Quality Loss of a Process That Is Normally Distributed

coefficient k multiplied by the average of the squared deviations of the individual measurements from the target. The squared deviations can be decomposed into two elements, the process variance plus the square of the offset of the process mean from the target, the process bias:

Expected loss per piece = k(process variance + process mean offset^2)
$$E[L(X)] = k[\sigma_X^2 + (\mu_X - m)^2].$$

To illustrate this use of the loss function, consider a situation defined by the following parameters: $m = 0$, $k = \$1.0$, LSL $= -1.0$, USL $= 1.0$, X is normally distributed with mean 0, and $\sigma_X = 0.33$, as shown in Fig. 9.24. The expected loss is therefore given by

$$E[L(X)] = \$1.00[0.33^2 + (0 - 0)^2]$$
$$= \$0.11.$$

The process in the example above would typically be considered capable in a traditional sense because the specification limits constitute a spread of 6 sigma. Applying our derived quadratic loss function, however, we find that this \$1.00 part (the replacement cost at manufacturing) carries with it an additional loss of about \$0.11. This is a hidden loss above and beyond direct costs of materials, labor, and processing incurred so far. This incremental cost is experienced downstream by the customers, in terms of additional costs they will incur due to decreased utility of the product.

Now suppose that, instead, X is distributed uniformly between -1 and 1.

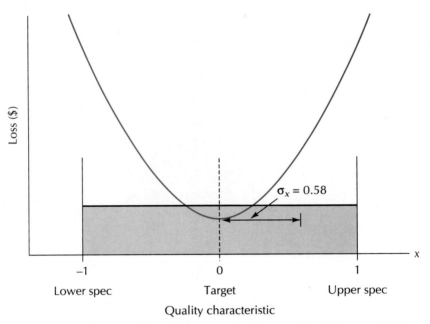

Figure 9.25 Evaluation of the Quality Loss for a Process That Follows a Uniform Distribution

As shown in Fig. 9.25, the standard deviation of the process would then be σ_X = 0.58.[2] Consider that the other data are the same as in the previous example. The expected or average loss per part is now

$$E[L(X)] = \$1.00[0.58^2 + (0 - 0)^2]$$
$$= \$0.33.$$

The effect of a change in the process distribution is to increase the loss threefold. The important question is: What would make the process follow a uniform distribution? The likely answer is that the process is exhibiting instability—shifting of its mean value, for example.

Figure 9.26 shows the effects of reducing the variation of a normally distributed process that remains centered on the target. The process has a standard deviation of 0.10, which constitutes a reduction to about one-third of the previous value. The associated expected loss,

$$E[L(X)] = \$1.00[0.10^2 + (0 - 0)^2]$$
$$= \$0.01,$$

is reduced by a factor of 10. Traditional manufacturing wisdom values such better-than-necessary performance primarily because it can absorb some shocks from special causes of variation without generating nonconforming output. This example, however, shows that additional benefits can be gained from reduced

---

[2] For the uniform distribution, $\sigma_X^2 = [(b - a)^2]/12$ for X varying on the interval $[a, b]$.

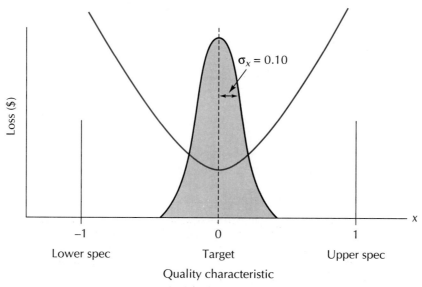

Figure 9.26 Loss Reduction Due to Reduced Process Variation

variation even when the process is stable and centered on the target. The loss function as a measure of process capability takes into account the loss incurred by the customer as a result of reduced utility in field use. As such, it clearly demonstrates the presence and importance of this incremental cost from the standpoint of the producer as well. Such loss incurred by the customer is directly transmitted back to the producer in terms of warranty/replacement cost, lost business cost, and so on.

Process Not Centered on the Target

When a process shows a great deal more than just marginal capability, one seemingly economically attractive way of capitalizing on this potential is to seek short-term benefit by, for instance, running the process toward the side of the specifications that minimizes, perhaps, material usage or cycle time. Figure 9.27 shows the results of such an approach, taking the preceding example ($\sigma_X = 0.10$) and locating the process mean as far to the low side as possible, consistent with remaining 3 standard deviations away from the specification limit. The loss under this process operating condition is now

$$E[L(X)] = \$1.00[0.10^2 + (-0.7 - 0)^2]$$
$$= \$0.50.$$

This process, which previously generated only $0.01 in hidden losses when centered on the target, now causes $0.50 in customer losses when run close to the lower specification limit. Very few pieces do not conform to specifications,

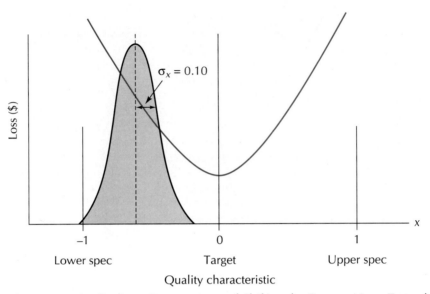

$\sigma_x = 0.10$

Loss ($)

−1
Lower spec

0
Target

1
Upper spec

x

Quality characteristic

Figure 9.27 Quality Loss Consequences of Shifting the Process Mean Toward the Lower Specification

and such a process would meet the explicit requirements of the bilateral specification. The loss arises because the output is consistently mediocre in a design intent sense—and the customer (and therefore the producer as well) suffers. If running the process at the shifted mean saved $0.10 in direct (manufacturing) costs, there would still be $0.40 in net waste, forever lost, as a result of lost utility of the product during field use. Taguchi describes this operating strategy as "more immoral than the actions of a thief."

Processes with Linear Drift

Another common way that we often attempt to take advantage of small process variation relative to the specifications is to reduce reset or tooling costs by allowing the process to drift across the specification range. A tool-change discipline that involves setting the process 3 standard deviations away from one specification limit, then letting the tool wear continue until the process is 3 standard deviations away from the other specification limits seems rational. However, these apparent savings are not achieved without risk of hidden customer loss.

The normally distributed process with a standard deviation of 0.10, discussed previously, may be allowed to drift to its maximum extent, with its centers moving from −0.7 to +0.7. Figure 9.28 shows the process distributions at these extreme centers, as well as the resulting product quality characteristic distribution of a complete drift cycle. As can be seen from the figure, this net distribution is no longer bell-shaped but more loaf-shaped.

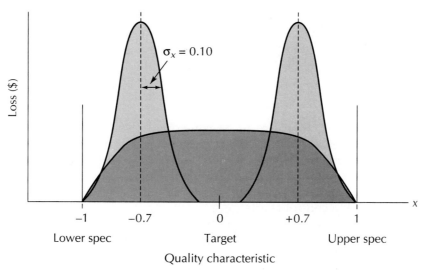

Figure 9.28 Consequences of a Process with Linear Drift

The total hidden loss in this example is now $0.33, coming mainly because the lengthy tool change interval causes marginally acceptable products to be produced near the beginning and end of each cycle. The loss calculation results from the fact that what we essentially have as a result of this operating policy is roughly equivalent to the case of Fig. 9.25 (i.e., the quality characteristic follows a uniform distribution). If the tool-change interval was reduced by 30%, for instance, with set and change points at -0.5 and $+0.5$, the hidden loss would be reduced to almost one-fourth of the loss calculated above. It is clear that tool-change policy should really balance the economies of longer change intervals against the downstream/customer-incurred losses to achieve the lowest total cost.

It is also apparent that when a process with very small variations is allowed to meander within the relatively wide specifications, rather than being maintained in statistical control near the target, this lack of shop discipline is not cost-free, but comes at the expense of the customer.

Exercises

9.1. In your own words, explain the difference(s) between process capability and statistical control. In particular:
  a. What important relationship exists between these two concepts?
  b. What are the basic goals/objectives of each?
  c. Can a process be in control but not capable? Can a process be capable but not in control? Explain.
9.2. Consider a process that continues to produce many parts that are not

meeting the required specifications. However, the $\overline{X}$ and R control charts being maintained for the process show it to be in a state of statistical control. Is this possible? Explain why this state of affairs may be occurring. What needs to be done to improve the process capability?

9.3. What are the steps involved in performing a process capability study? Be specific in outlining the data collection, analysis, and evaluation procedures.

9.4. Control charts are to be constructed for a process turning the outside diameter of a cylinder. The diameter has specifications of 4.500 ± 0.050 centimeters.

a. If $\Sigma \overline{X} = 157.85175$ centimeters and $\Sigma R = 2.18750$ centimeters for 35 samples of size $n = 5$, calculate the centerlines and control limits for the control charts.

b. Assume that the charts show that the process is in good statistical control. What percent of the parts produced will be within tolerances if this process continues under a constant system of common causes and can be approximated by the normal distribution?

c. What percent will be within the specifications if the process were to be centered at the prescribed nominal?

d. Graphically show the situations in parts (b) and (c) in terms of the distributions of the individual measurements.

9.5. Two different sizes of a certain type of fastener are being manufactured by a cold heading process. The nominal outside diameters of the fastener heads, with associated specifications, are

>   *Fastener A:* 0.600 ± 0.020 centimeter
>   *Fastener B:* 0.800 ± 0.020 centimeter.

SPC studies show the processes to be in good statistical control. The individual measurements follow the normal distribution. After collecting 50 samples of size $n = 4$:

>   *Fastener A:* $\Sigma \overline{X} = 30.310$
> $\Sigma R = 0.654$
>   *Fastener B:* $\Sigma \overline{X} = 40.011$
> $\Sigma R = 1.034$.

a. For each fastener dimension of interest, find the percent conforming.

b. If the process mean for fastener A could be brought to the nominal, what would be the percent conforming?

c. If the variation in the process for fastener B could be reduced by a factor of one-third, what would be the percent conforming?

9.6. In Exercise 6.17 you were given data that were collected from a process (which machined the outer diameter of motor shafts) after it was brought into statistical control. For that exercise, the nominal value was 2.1250 inches with a tolerance of ±0.0010 inch. The data provided (page 187) were coded to be the last two digits of the measurements. Use those data to determine the percent conforming to specifications. If you assume that the data are normally distributed, verify your assumption with the appropriate data analysis.

9.7. One of your employees has come to you with the data shown in the table and with associated control charts to show that improvements have been made on a metal casting process after an initial study of its capability. The data are based on samples of size $n = 5$. The specifications for the quality characteristic of interest are 3.50 ± 0.10 millimeters. Based on the charts, he tells you that the process is now capable and that further improvements are no longer necessary.

a. As an informed manager, what are your opinions regarding the process, the approach used to study it, and the conclusion?

b. If you feel that the situation should have been handled differently, what would you do? Use the data to develop your presentation and conclusions.

Sample	$\overline{X}$	Range	Sample	$\overline{X}$	Range
1	3.49	0.13	26	3.55	0.07
2	3.49	0.13	27	3.55	0.07
3	3.49	0.16	28	3.54	0.13
4	3.50	0.12	29	3.53	0.13
5	3.50	0.16	30	3.54	0.08
6	3.53	0.05	31	3.53	0.07
7	3.49	0.08	32	3.57	0.16
8	3.53	0.21	33	3.50	0.15
9	3.46	0.09	34	3.51	0.14
10	3.45	0.07	35	3.53	0.09
11	3.48	0.12	36	3.51	0.18
12	3.52	0.16	37	3.50	0.13
13	3.47	0.22	38	3.46	0.13
14	3.50	0.09	39	3.54	0.08
15	3.51	0.09	40	3.51	0.11
16	3.50	0.20	41	3.51	0.10
17	3.52	0.17	42	3.53	0.09
18	3.49	0.07	43	3.57	0.09
19	3.50	0.14	44	3.57	0.12
20	3.49	0.08	45	3.53	0.12
21	3.49	0.18	46	3.53	0.08
22	3.49	0.03	47	3.53	0.08
23	3.53	0.11	48	3.48	0.15
24	3.52	0.12	49	3.49	0.08
25	3.51	0.14	50	3.48	0.09

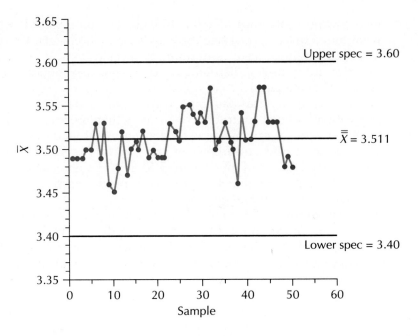

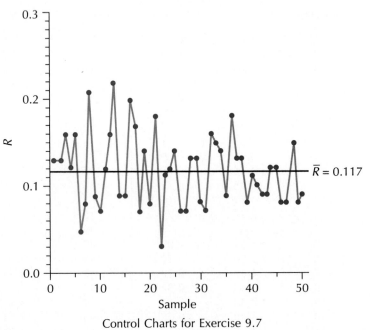

Control Charts for Exercise 9.7

9.8. You need to make some decisions regarding the procurement of bolts from three different suppliers. You have requested that the shank diameter (above the thread) have specifications of 1.500 ± 0.009 inches. The SPC studies done by the suppliers have indicated that their processes are behaving consistently in statistical control with the following process parameters. The individual measurements follow the normal distribution.

> Supplier 1: $\overline{\overline{X}}$ = 1.500 inches, σ_X = 0.003 inch
> Supplier 2: $\overline{\overline{X}}$ = 1.500 inches, σ_X = 0.0022 inch
> Supplier 3: $\overline{\overline{X}}$ = 1.4950 inches, σ_X = 0.0015 inch.

Which supplier would you purchase from? Why? Explain your logic and show calculations and graphical evidence to back it up.

9.9. The specifications of the thickness of a low-alloy steel gear blank are 0.3000 ± 0.0015 centimeter.

a. If it is desired that the value of C_{pk} be at least 1.25, what would be the minimum σ_X of a centered process?

b. If the specifications change to 0.3000 ± 0.0010 centimeter, what would be the minimum σ_X for a centered process?

9.10. Consider again the steel gear blank in Exercise 9.9. The specifications are 0.3000 ± 0.0015 centimeter.

a. If it is desired for the process to have a C_p value of at least 1.25, what is the minimum σ_X if $\overline{\overline{X}}$ = 0.301 centimeter? What if $\overline{\overline{X}}$ = 0.298 centimeter?

b. What does this say about using C_p as a measure of capability?

9.11. The maker of grandfather clocks needs to be very careful about the weight of the swinging pendulum, which is essential to the time-keeping precision. It is desirable for 99.9% of the pendulums to have a weight of 1.0000 ± 0.0002 kilogram.

a. If the process is in statistical control, centered, and has a normal distribution, what value of σ_X would be required to meet the criterion?

b. If the process is centered, what values of C_p and C_{pk} would the value of σ_X found in part (a) produce?

9.12. A die-casting process is being used by a supplier to fabricate a cast housing that will be supplied to a major client who will perform further machining operations on the pieces. It has been determined that monitoring part weight is a good indication of quality process performance. The casting should have a nominal weight of 775 grams and a tolerance of ±20 grams. Use the data given in the table (coded as the last two digits, such that 79 = 779 grams, etc.) to perform a capability study. Determine the percent conforming to specifications, C_p, and C_{pk}. If you assume that the data are normally distributed, verify your assumption with the appropriate data analysis.

Sample	X_1	X_2	X_3	X_4	X_5
1	79	82	71	71	73
2	75	63	75	79	74
3	72	82	70	90	75
4	79	76	73	80	79
5	71	63	79	76	87
6	81	67	72	83	71
7	88	77	70	65	59
8	80	77	78	86	80
9	65	72	75	84	85
10	72	73	75	75	65
11	73	74	72	76	77
12	73	70	83	78	85
13	84	74	78	79	76
14	88	70	79	88	78
15	64	71	65	78	82
16	86	73	71	78	71
17	84	67	71	59	87
18	76	81	66	74	73
19	75	69	83	79	71
20	79	75	73	75	74
21	79	86	85	87	84
22	76	78	88	75	89
23	81	88	84	70	82
24	81	74	72	78	60
25	84	65	65	77	79
26	85	72	76	72	78
27	83	72	72	81	66
28	74	78	74	82	93
29	77	73	74	71	83
30	59	71	85	74	80
31	82	73	76	74	85
32	66	74	79	81	73
33	77	73	77	86	74
34	89	83	71	75	82
35	82	78	66	69	67
36	73	82	67	73	70
37	83	69	77	77	61
38	82	82	80	70	72
39	82	70	70	74	75
40	77	69	80	68	79

9.13. To determine the present status of a piston-turning operation where the surface roughness is an important characteristic, it has been decided that a process capability study should be performed. The surface roughness has specifications of 105 ± 10 microinches. Use the data shown in the table, collected in samples/subgroups of size $n = 5$, to perform this study.

Sample	$\overline{X}$	Range	Sample	$\overline{X}$	Range
1	108.2	13	16	103.6	8
2	105.8	27	17	115.4	24
3	104.0	24	18	103.2	34
4	104.6	16	19	108.0	71
5	106.6	22	20	108.2	57
6	111.4	32	21	100.6	26
7	104.6	12	22	98.2	73
8	102.2	23	23	96.4	47
9	111.0	22	24	101.4	26
10	113.4	18	25	105.6	49
11	102.2	23	26	91.4	68
12	108.0	26	27	116.6	52
13	111.2	22	28	113.6	53
14	103.2	31	29	112.4	42
15	100.4	6	30	99.6	37

9.14. A decision needs to be made regarding the possible purchase of some new casting equipment. To aid in this process, a capability study is being performed to assess the performance of the current equipment. An important part that will consume much of the run time of the equipment is to be used for the study. The width of the cast part at a crucial spot should be 3.000 ± 0.025 centimeters, and rational sampling has produced the data listed in the table. Perform the capability study on this process using the samples below and comment. The samples, of size $n = 5$, were collected every half hour over two continuous shifts.

Sample	$\overline{X}$	Range	Sample	$\overline{X}$	Range
1	2.991	0.038	17	3.005	0.039
2	3.004	0.027	18	3.008	0.054
3	2.997	0.049	19	3.001	0.009
4	3.004	0.033	20	3.000	0.062
5	2.999	0.043	21	2.993	0.034
6	3.000	0.070	22	2.982	0.012
7	2.988	0.044	23	2.990	0.028
8	2.986	0.050	24	3.003	0.091
9	2.991	0.064	25	3.007	0.049
10	2.994	0.048	26	2.998	0.052
11	3.017	0.070	27	3.012	0.077
12	2.987	0.051	28	2.996	0.040
13	3.013	0.060	29	2.997	0.018
14	2.993	0.019	30	3.004	0.020
15	3.009	0.044	31	2.987	0.063
16	2.990	0.043	32	3.002	0.028

9.15. An assembly, composed of three parts, has a total nominal length of 1.500 inches with a tolerance of ±0.050 inch. The three parts are essentially identical in terms of the processes manufacturing each. The processes making the parts (and the assembly) should have a C_p value of at least 1.5.

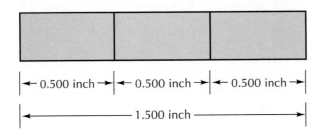

- 0.500 inch → ← 0.500 inch → ← 0.500 inch →

- 1.500 inch

a. What should be the standard deviation for the processes making the individual parts to meet the capability criterion above, and what would the individual tolerances be for these parts?
b. If the tolerances had been determined incorrectly by simple division of the assembly tolerance, what must the individual process values for σ_X be to meet the $C_p = 1.5$ requirement?
c. If the processes for making the individual parts were such that the average was at 0.5002 inch and the standard deviation was 0.0094 inch, what percent would have been incorrectly rejected if simple division was used to set the tolerances?

9.16. A hollow cylinder is made up of two semicylindrical parts as shown in the figure. The assembly is held together by an epoxy layer that must be between 0.002 and 0.005 inch in thickness. The two cylindrical halves are made from the same extrusion. If the outer diameter of the assembly is to be 2.000 inches with tolerances of ±0.010 inch, and a 5 sigma assembly is required, what would be the required nominal value and tolerance for the dimension A of the individual parts?

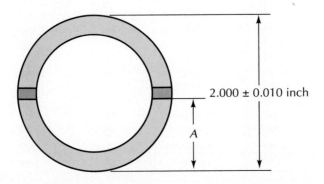

2.000 ± 0.010 inch

A

9.17. An assembly is being produced, as shown in the figure, that is comprised of two aluminum plates with a hard rubber insulation in between. The aluminum plates have a nominal thickness value of 0.09375 inch and the insulation is to be 0.0625 inch thick. Because of the resilience of the hard rubber material, it is deemed necessary that the standard deviation of the process making this material be one-half that of the process making the aluminum. If the overall assembly has specifications of 0.2500 ± 0.0040 inch, what are the required specifications on the individual parts? Assume a ±4 sigma assembly.

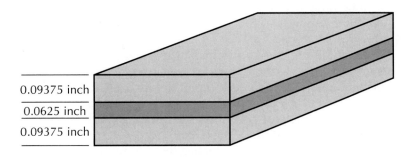

| 0.09375 inch |
| 0.0625 inch |
| 0.09375 inch |

9.18. A woodworking company is shipping flat, plain doors to another firm, which will be using them for cabinets. The thickness of these doors is 0.75 inch with a tolerance of ±0.03 inch. When the doors are shipped they are put in crates, with each crate holding 10 doors. A piece of cardboard is placed on either side of each door. The cardboard is 0.050 inch thick, with a tolerance of ±0.005 inch. Statistical charting has shown that the process for making the doors is ±4 sigma capable, and the process for making the cardboard is ±2 sigma capable. The shipper wants to be 99.5% certain that 10 doors will fit into each crate, with the assembly at or below the cover level. What should the minimum depth of the crates be?

9.19. A company is making 10-ohm resistors to be used in electric thermostats. The resistors have specifications of 10.00 ± 0.10 ohms. It has been determined that if the resistor's actual resistance is 10.15 ohms, a failure of the thermostat will occur 15% of the time with a cost of $45. A quadratic loss function is assumed to approximate the loss.

   a. Use these data to find the value of k in the loss function.

   b. Sampling performed on resistors shows that $\overline{\overline{X}} = 10.01$ ohms and that σ_X (approximated by $\overline{R}/d_2$) is 0.025 ohm. Use this information to find the average loss per part made.

   c. What is the average loss per part if the process in part (b) is centered?

   d. What is the average loss per part if the process in part (b) is centered and the variance is reduced by a factor of 2? (See p. 296 for part (e).)

e. Assuming a process centered on the nominal value, if the replacement/ manufacturing cost of these resistors is $4.32, what should the tolerance on the resistors be?

9.20. The company from Exercise 9.19 has been required to have a minimum C_{pk} value of 1.3 for the resistors. Due to the way the process that makes the resistors works, it is less expensive by $0.05 per resistor to manufacture the resistors at a nominal of 9.95 ohms instead of at 10.00 ohms. Through adjustments it has been possible to reduce the standard deviation to half of its original value, to 0.0125, and it has been deemed appropriate to adjust the mean of the process to 9.95 ohms.

a. What happens to the C_{pk} value if this change in the variation and adjustment to the mean occurs?

b. What happens to the average loss per piece if this process mean shift is implemented?

c. Evaluate the decision to make the proposed process change.

9.21. Two parts in an assembly are in constant contact and have frequent sliding movements relative to each other. Because of this, it is necessary to have a certain surface roughness so that the pieces are able to move easily upon application of the proper lubricant. Therefore, in the machining operation that produces these parts, the surface roughness needs to be between 110 and 140 microinches, with a nominal of 125 microinches.

a. If the loss for parts made with a roughness of 106 microinches is $3.29, find the value of k for the loss function. Assume a quadratic loss function.

b. What would be the loss for a part with a surface roughness of 150 microinches?

c. If the process is in control, approximately normally distributed, and centered at the nominal with a C_{pk} value of 1.2, what is the average loss incurred per assembly?

9.22. Consider that the data collected during an SPC study indicate that a normal distribution is occurring in the measurement of the depth of a slot machined in an aluminum fixture. The slot should have a depth of 2.750 ± 0.010 centimeters. It is known that for these pieces if the depth is 2.765 centimeters, a cost of $100 is incurred by the user to replace the entire fixture due to improper fit.

a. Use the information above to find the value for k, assuming a quadratic loss function.

b. If the process machining the slot is centered at 2.752 with a standard deviation of $\sigma_X = 0.003$ centimeters, calculate the average loss per part. What effect does centering the process have on the loss?

9.23. There are some important differences in the way the various capability measures respond to changes in the process as it is represented by a statistical distribution. Consider a process producing a product with an important quality characteristic that has a nominal value of 74, an upper specification of 82, and a lower specification of 66. For the following process situations, graphically depict each of the three cases, assuming the process to be in control, and find the values for C_p, C_{pk}, and expected $L(x)$. For the loss function, leave k as a constant.

a. $\overline{\overline{X}} = 74$, $\hat{\sigma}_X = 2.67$.

b. $\overline{\overline{X}} = 78$, $\hat{\sigma}_X = 2.67$.

c. $\overline{\overline{X}} = 82$, $\hat{\sigma}_X = 2.67$.

d. Comment on the trends that occur for each index.

9.24. In Chapter 2 we saw the example of the loaf-shaped distribution that occurred in the color-intensity quality characteristic of TV sets produced in the United States. In the television example, it was noted that the quality characteristic of the TV sets produced by the Japanese company followed a normal distribution, which nearly stayed completely within the specifications. Compare these two situations by finding the expected $L(x)$ values for the process approximated by the uniform distribution and by the normal distribution. Both processes are centered at the nominal and are 6 sigma capable.

Statistical Thinking for Process Study: A Case Study

Introduction

The concept of bringing a process into a state of statistical control and then maintaining that state of control is often difficult to appreciate completely because of the tremendous emphasis we have historically placed on the engineering specification as a measure of quality. As a result, the primary focus of production is to maintain the existence of a process that is producing parts which conform to specifications without a great deal of attention given to the notion of achieving and maintaining a stable, consistently performing process. Central to the product control mentality is the notion that the engineering specification is an appropriate measure of quality. If it is, then of course we have put quality on an attribute basis, categorizing finished parts as either good or bad, defective or nondefective, which makes little sense from a functional point of view.

It is important to recognize that the specification is truly an inhibitor to the promotion of the concepts and methods of statistical control of processes. Until rejected as a measure of quality, the specification will continue to stand between us and our ability to make full use of and receive the broad-based benefits of statistical process control.

To convince yourself that the problem posed above is a real one, try the following: Ask someone who is aware of but not familiar with the concept of

SPC this simple question: "If you had a process that was making parts which were consistently meeting the design specifications, would you think it necessary or even advantageous to chart this process and bring it into a state of statistical control?" Chances are you may quickly get the response, "No." This alone should be sufficient to convince you that there is a real problem with fully embracing SPC when the concept of the engineering specification is tied so closely with our apparent understanding of the meaning of quality.

Much of what we have discussed in the first nine chapters has been directed toward overcoming the misconceptions cited above. We have been introduced to the concept of quality as loss due to functional variation and have come to recognize the poor correlation that actually exists, from the customer's point of view, between quality and conformance to the engineering specification, particularly when the specification is misused by manufacturing. This has helped us to see why the pursuit and maintenance of a process in good statistical control is important and how this can affect our ability to identify opportunities for the continual reduction of variability. Furthermore, Shewhart and later Deming have so clearly pointed out to us the important relationship between variation reduction and increase in productivity—variation as an indication of the presence of waste and inefficiency. We have seen that there are still other forces driving the importance of statistical process control. We reexamine all these issues in this chapter as we review in detail an actual case study of the use of statistical thinking and methods for quality and productivity design and improvement.

10.2
Overview of the Case Study

This case study chronicles the activities of a company which, over a period of months, learned and applied the techniques of statistical process control to one of its key processing operations. The case study clearly demonstrates the power of and the broad-based benefits to be derived from the use of statistical thinking and methods for quality and productivity design and improvement. The story is real and the results are dramatic. This case study will call on virtually all the concepts and methods of data analysis and interpretation that we have discussed in the first nine chapters. It points out many of the misconceptions that we have examined as well.

The Dilemma.   A certain manufacturer of gaskets for automotive and off-road vehicle applications awoke one morning to find that a major customer had significantly tightened the bilateral specification on the overall thickness of a hard gasket used in automotive engines. Although the current specification was by and large being met, the product did not come close to meeting the new specification. The first reaction of the gasket manufacturer was to negotiate with the customer to obtain relaxation of the new specification. When efforts in this regard did not bear fruit, the relationship became somewhat strained. At first the gasket manufacturer thought that if he waited long enough, the automotive company would eventually be forced to loosen the requirements and purchase the existing product. However, as time went on it became clear that some positive steps would have to be taken to improve the quality. But what should be done? And by whom?

The Product.   Figure 10.1 shows the product in question, a hard gasket. A hard gasket is a three-piece construction comprised of two outer layers of soft gasket material and an inner layer consisting of a perforated piece of sheet metal. These three pieces are assembled, and some blanking and punching operations follow, after which metal rings are installed around the inside of the cylinder bore clearance holes and the entire outside periphery of the gasket. The quality characteristic of interest in this case study is the assembly thickness.

The Process.   An initial study by the staff engineers revealed that the variation of the thickness of soft gasket material—the two outer layers of the hard gasket—was large and undoubtedly responsible for a large portion of the total variability in the final product. Figure 10.2 shows the roll mill process, which fabricates the sheets of soft gasket material from which the two outer layers of the hard gasket are made. To manufacture a sheet of soft gasket material, an operator adds raw material, in a soft pelletlike form, to the gap, called the knip between the two rolls. The larger roll rotates about its axis with no lateral movement, while the smaller roll rotates and moves back and forth laterally to change the size of the roll gap. As the operator adds more and more material to the gap, the sheet, with the application of a starter material, is formed around the larger roll. When the smaller roll reaches a preset destination (final gap/sheet thickness) a bell rings and a red light goes on telling the operator to stop adding raw material. The operator stops the rolls and cuts the sheet horizontally along the larger roll so that it may be pulled off the roll. The finished sheet,

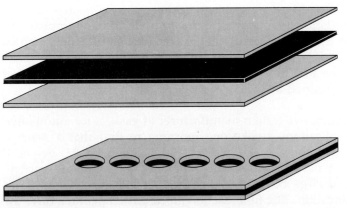

Figure 10.1 The Product: A Hard Gasket of Three-Piece Construction for Automotive Applications

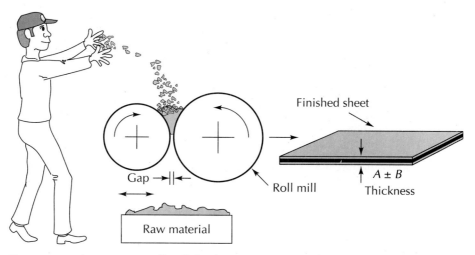

Figure 10.2 The Process: Roll Mill for the Manufacture of Soft Gasket Material

called a pull, is pulled onto a table where the operator checks its thickness with a micrometer. As we will see later, the operator can adjust the roll gap if he or she feels that the sheets are coming out too thick or too thin relative to the prescribed nominal value.

10.3
Product Versus Process Control: Management Strategy

While the staff engineers were studying the problem to formulate an appropriate action plan, something had to be done to make the necessary production requirements for hard gaskets within the specifications. A management decision was made to intensify product inspection and, in particular, to grade each piece of material according to thickness so that the wide variation in thickness could be balanced out at the assembly process. Figure 10.3 shows the selective assembly process that was put in place to provide a solution to the problem. Extra inspectors were used to grade each piece of soft gasket material. The sheets of the same thickness were shipped in separate bundles on pallets to a sister plant for assembly. Thick and thin sheets were selected to make a hard gasket to balance out the highs and the lows. The process worked pretty well and there was some discussion about making it permanent. However, some felt it was too costly and did not get at the root cause of the problem.

Meanwhile, the staff engineers in the company were continuing to study the problem more carefully and came to the conclusion that the existing roll mill process equipment for making the soft gasket sheets simply was not capable of meeting the new specifications. This conclusion was reached as a result of the examination of production data and scrap logs over the past several months.

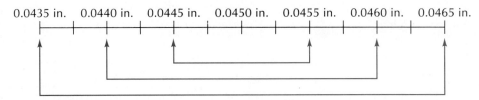

0.0435 in.   0.0440 in.   0.0445 in.   0.0450 in.   0.0455 in.   0.0460 in.   0.0465 in.

Figure 10.3 Selective Assembly Process Used to Assemble Hard Gaskets That Meet the Specifications

They had researched some new equipment that had a track record for very good sheet-to-sheet consistency and had decided to write a project to replace the existing roll mills with this new equipment. To help the approval process their boss asked them to get some data on the poor process capability of the existing equipment. They therefore set out to conduct a "statistical process capability study."

Initial Process Capability Study

The engineers, confident that the equipment was not capable, selected what they thought was the best operator and the best roll mill (the plant has several roll mill lines) and took careful measurements of the thickness of each sheet made on an 8-hour shift. During that particular shift a total of 72 sheets/pulls were made. This was considered quite acceptable since the work standard for the process is 70 sheets per shift. The measurements of the sheet thickness (in the order of manufacture) for the 72 sheets made on that shift are given in Table 10.1. The engineers set out to use these data to conduct a statistical process capability study.

Calling upon a statistical methods course that one of the engineers had in college, the group decided to construct a statistical distribution from the data and use it to estimate the percentage of the measurements/process that falls within the specifications. The engineers first estimated the mean and standard deviation of the 72 sheet thicknesses, using the arithmetic average and the sample standard deviation as defined in Chapter 3. These calculations are

$$\overline{X} = \frac{\Sigma X_i}{72}$$
$$= 0.0449$$

$$s_X = \left[\frac{\Sigma(X_i - \overline{X})^2}{72 - 1} \right]^{1/2}$$
$$= 0.0014.$$

At this point the engineers would normally simply assume that the normal/bell-shaped curve would adequately represent the frequency distribution of the data. But to be more rigorous about their analysis, they decided to construct a frequency histogram of the data to verify this. The following procedure, as outlined in Chapter 3, was followed to construct the frequency histogram:

TABLE 10.1	Measurements of Sheet Thickness						
Sheet	Thickness (in.)	Sheet	Thickness (in.)	Sheet	Thickness (in.)	Sheet	Thickness (in.)
1	0.0440	19	0.0472	37	0.0459	55	0.0425
2	0.0446	20	0.0477	38	0.0468	56	0.0442
3	0.0437	21	0.0452	39	0.0452	57	0.0432
4	0.0438	22	0.0457	40	0.0456	58	0.0429
5	0.0425	23	0.0459	41	0.0471	59	0.0447
6	0.0443	24	0.0472	42	0.0450	60	0.0450
7	0.0453	25	0.0464	43	0.0472	61	0.0443
8	0.0428	26	0.0457	44	0.0465	62	0.0441
9	0.0433	27	0.0447	45	0.0461	63	0.0450
10	0.0451	28	0.0451	46	0.0462	64	0.0443
11	0.0441	29	0.0447	47	0.0463	65	0.0423
12	0.0434	30	0.0457	48	0.0471	66	0.0447
13	0.0459	31	0.0456	49	0.0427	67	0.0429
14	0.0466	32	0.0455	50	0.0437	68	0.0427
15	0.0476	33	0.0445	51	0.0445	69	0.0464
16	0.0449	34	0.0448	52	0.0431	70	0.0448
17	0.0471	35	0.0423	53	0.0448	71	0.0451
18	0.0451	36	0.0442	54	0.0429	72	0.0428

1. The range of the data was determined by subtracting the smallest value from the largest (i.e., $0.0477 - 0.0423 = 0.0054$).
2. It was decided to use 12 cells of width 0.0005, with midpoints at 0.0XX5 and 0.0XX0.
3. Therefore, the boundaries were chosen such that the first cell goes from 0.04225 to 0.04275, the second from 0.04275 to 0.04325, and so on. This ensured that each thickness measurement would fall within a specific cell.
4. Next, a tally sheet of the points was made:

Cell Midpoint	Frequency			Total
0.0425	11111	1		6
0.0430	11111	11		7
0.0435	1111			4
0.0440	11111	1		6
0.0445	11111	11111		10
0.0450	11111	11111	111	13
0.0455	11111	11		7
0.0460	11111			5
0.0465	11111			5
0.0470	11111	11		7
0.0475	11			2

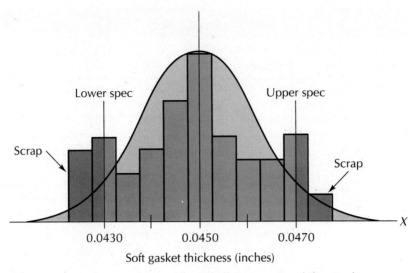

Figure 10.4 Histogram of Data from Initial Process Capability Study

5. Finally, the histogram was plotted, as shown in Fig. 10.4. Also shown in the figure are the product nominal value and the upper and lower specification values. The dark shaded part of the histogram represents the amount of the product that lies outside the specifications.

It is immediately apparent from Fig. 10.4 that a large proportion of the process does not meet the individual sheet specification. To quantify the precise nature of the problem, the engineers assumed that the data could be represented by the normal distribution curve as seen overlaid in Fig. 10.4. Despite the somewhat irregular appearance of the frequency histogram when compared to the normal distribution, they contended that the normal distribution was a reasonably good approximation for the process and proceeded to determine the area under the curve that fell inside the specifications. Figure 10.5 shows the

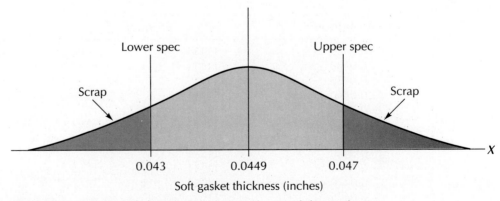

Figure 10.5 Statistical Model for Process Data for Capability Evaluation

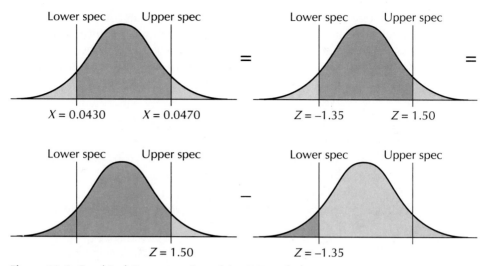

Figure 10.6 Graphical Representation of the Z Transformation and Probability Calculation

process, as represented by the normal distribution, with the product specifications overlaid on the model.

To determine the process capability in terms of percent conforming to the specification, the following calculations were made:

$$P \text{ (a part within specs)} = P \text{ } (0.04300 \leq X \leq 0.0470).$$

If we use the standard normal or Z distribution, the required probabilities can be determined as shown in Chapter 3 and as shown graphically in Fig. 10.6 with σ_X estimated by s_X to be 0.0014 inch, and μ estimated by $\overline{X}$ to be 0.0449 inch:

$$Z = \frac{X - \mu}{\sigma_X}$$

$$Z_{LSL} = \frac{0.0430 - 0.0449}{0.0014}$$

$$= -1.35$$

$$Z_{USL} = \frac{0.0470 - 0.0449}{0.0014}$$

$$= 1.50.$$

The Z table in the Appendix (Table A.1) can then be used to find the following:

$$P \text{ (a part within specs)} = P \text{ } (Z \leq Z_{USL}) - P \text{ } (Z \leq Z_{LSL})]$$

$$= P \text{ } (Z \leq 1.50) - P \text{ } (Z \leq -1.35)]$$

$$= 0.9332 - 0.0885$$

$$= 0.8447.$$

Therefore, the process capability in terms of percent conforming to the specifications is estimated to be 85%. Such a capability is clearly unacceptable.

In terms of the process capability index C_p, the process capability was found to be $C_p = 0.477$, which by all measures is totally unacceptable.

$$C_P = \frac{USL_X - LSL_X}{6\,\sigma_X}$$

$$= \frac{0.0470 - 0.0430}{0.0084}$$

$$= 0.477.$$

Calculations were also performed to find the value of the index C_{pk}:

$$Z_{USL} = \frac{USL - \overline{X}}{\sigma_X} \qquad\qquad Z_{LSL} = \frac{LSL - \overline{X}}{\sigma_X}$$

$$= \frac{0.0470 - 0.0449}{0.0014} \qquad\qquad = \frac{0.0430 - 0.0449}{0.0014}$$

$$= 1.5 \qquad\qquad\qquad\qquad = -1.36$$

$$Z_{min} = min[Z_{USL},\, -Z_{LSL}]$$

$$= min[1.5, 1.36]$$

$$= 1.36$$

$$C_{pk} = \frac{Z_{min}}{3}$$

$$= \frac{1.36}{3}$$

$$= 0.45.$$

This value for C_{pk} is also far from the acceptable value for C_{pk} of at least 1.00. Clearly, the results of this process capability study show that by any measure the process capability is very poor indeed. This was precisely the answer the engineers were looking for. Armed with their statistical process capability study results, the engineers went to their boss to seek the go-ahead to finalize their report, which would then be presented to the division top management to obtain approval to procure the new roll mill equipment. Their boss was pleased with their work and asked them to prepare the project documentation and presentation materials.

Review of Progress to Date

It is worthwhile to pause here for a moment and reflect back on what has transpired to this point. In particular, we may wish to examine some fundamental problems with the way this situation has been handled.

1. Product Control Way of Thinking

It is important to note that in all regards the actions of management and the engineering staff have been dictated by the product control way of thinking. Management devised a procedure to attempt to work with the existing level of quality by screening, sorting, and selective assembly. This strategy is consistent with the "detect and contain" approach to quality control. Engineering focused on the notion of process capability purely from the "conformance to specifications" point of view. Their approach left them with a yes/no answer to the question, "Is the process capable?" but with little direction in terms of where to look for improvement other than the drastic action of replacing the existing equipment. At no time to this point was there any attempt to consider what improvements might be made to the existing process to make it more capable.

2. Tunnel Vision of the Engineers

Initially, the engineers rather hastily concluded that the major source of variation in gasket sheet thickness was the equipment—the roll mill. This is only one of several possible sources of variation, which could include the raw materials (properties and preparation), the methods (procedures for setup, machine operation, etc.), the machine condition (operating settings, maintenance condition), and the operator (training, supervision, technique). With little concern for these issues and no initial data to provide direction the engineers have quickly focused on inherent equipment capability as the problem. It is often the case that we focus quickly on the equipment and propose solutions that call for the purchase of new machines. Dr. Deming has, of course, provided us with some insight in this regard. He has cautioned us that *there is too much talk of the need for new machinery and automation—most people have not learned to use what they have.*

3. Selective Assembly Process

The mere thought that such an approach might be an acceptable solution to the problem fails to recognize the fundamental fallacies of product control as a way of doing business. Such an approach can never cause both quality and productivity to move together in the desired direction and therefore will not enhance competitive position. Such can only be accomplished through process control—attacking the process to find the root causes of the problem.

4. Statistical Control/Process Capability Relationship

The engineers have treated the notion of process capability in a classical but totally inappropriate fashion. When they took the data from an entire shift and assumed that in a frequency sense it could be represented by a statistical distribution—in this case the normal distribution—they did not realize that the implicit assumption being made is that the process is in a state of statistical control. This must be the case if the process is to be represented over time by a single statistical distribution, which has a fixed mean and a constant level of variation. To estimate the mean, standard deviation, assume the normal distribution, and then use this model for analytical purposes, the assumption we must make is that the data have arisen from a constant system of common causes. The engineers, however, did not examine the process data over time and therefore could neither confirm nor deny the process condition in terms of statistical control.

Figure 10.7 illustrates the appearance of a process, represented as a statistical distribution of measurements, when its mean level is changing over time. The bottom half of the figure shows the time-oriented progression of the process behavior and illustrates the inability of the process to operate in a stable fashion in terms of its mean value. The top part of the figure shows what this means in terms of the process appearance over time when represented by a statistical distribution of measurements relative to the specifications. It is clear from Fig. 10.7 that no single process (distribution) exists over this period of time—rather, many different processes are present. Which one should we use to assess the

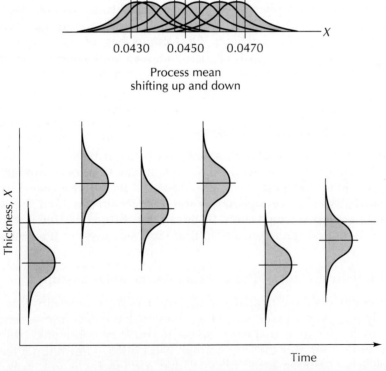

Figure 10.7 Time Behavior of a Process That Is Changing Mean Level in a Sporadic Fashion

process capability? Hopefully, the inadvisability of "pushing" all of these processes together by compressing Fig. 10.7 along the time dimension and thereby creating the illusion of the existence of a single, stable process is clear. Furthermore, the results of the capability study of the engineers may not only be totally misleading, but also, their analysis has little diagnostic value—the process is incapable, but why? What are the major sources of variation contributing to the lack of capability relative to the specifications? If a process capability study is to be a meaningful and valid representation of the ability of the process to manufacture product consistently at a certain level of conformance, it must be preceded by statistical charting which confirms that the process is in a state of statistical control.

10.4

Finding Improvement Opportunities Through Charting
A Different Approach to the Problem

During the time that the study of the staff engineers was taking place, the people in the plant were being introduced to Dr. Deming's concepts and the philosophies and methods of statistical charting for fault diagnosis and process improvement. The plant manager decided to form a team to study the problem. In addition to himself, he comprised the team of the production manager, the quality manager, a process engineer, a product engineer, a production supervisor, and an operator. He arranged for a consultant to meet with the group about once every two weeks to review their progress and make suggestions.

The group began to meet frequently to plan a course of action. As a first step the group made a detailed process flowchart, carefully defining the complete sequence of processing steps from raw material handling, the blending of the gasket material for rolling, to the roll mill operation, and on to the cutting/sizing operation, and finally, the blanking operation. At each step along the way they discussed the possible factors that could be contributing to the variation in the final product.

After studying and discussing the complete process flow they concluded that raw material blending and the roll mill process were the two most likely areas to seek and find improvement opportunities. Since the raw material blending operation involves several steps and is somewhat complicated, they decided to focus their attention first on the roll mill process. It was felt that this process was simpler and easier to obtain data from and was therefore more manageable for their first attempt at tackling a process problem using statistical thinking and methods.

The next step for the group was to begin to do statistical charting of the roll mill process using $\overline{X}$ and R control charts. In employing this technique, it was their hope to learn more about the nature of the variability pattern of the process data, so that they could identify the root causes of some of the major variation sources. Actions could then be formulated to remove these sources of variability, thereby making the process operate more consistently.

Method of Process Operation

Before we proceed to examine the first set of statistical charts that the group developed, it is important that we understand a bit more about how the process is actually operated. Such information is essential to the fault diagnosis process. Statistical process control chart interpretation should be thought of as a two-step process. First, we read the statistical signals to determine what they are telling us about the behavior of the process mean and process variability. Then we use all available data about the operating conditions of the process; the events that may have taken place at the process during the time the data were collected; and our general knowledge about, and our experience with, the process to determine what is causing the process mean and/or variation to behave in the manner that has been detected. It is our ability to embrace this second step that determines whether we are simply making charts to hang on the wall or whether we are using charts to find improvement opportunities.

Investigation revealed that the operator runs the process in the following way. After each sheet is made, the operator measures the thickness with a micrometer. The thickness values for three consecutive sheets are averaged and that average is plotted on a piece of graph paper that at the start of the shift has only a solid horizontal line drawn on it to indicate the target gage thickness value for the particular soft gasket sheet the operator is making. Periodically, the operator is to review these evolving data and make a decision as to whether or not the process mean—the sheet thickness—needs to be adjusted. This can be accomplished by stopping the machine, loosening some clamps on the small roll, and jogging the small roll laterally in or out by a few thousandths of an inch—whatever the operator feels is needed. The clamps are tightened, the gap is checked with a taper gage, and if adjusted properly, the operator begins to make sheets again. Typically, this process takes about 10 to 15 minutes. The questions of when to make such adjustments and how much to change the roll gap for each adjustment are completely at the operator's discretion, based on the evolving sheet measurement averages.

Figure 10.8 shows a series of plots that detail the history of one particular shift over which the operator made several process adjustments based on the data. These data actually come from the shift that the staff engineers used to collect the data for their process capability study. Figure 10.8(a) shows the process data after the first 12 sheets have been made—four averages of three successive sheet thicknesses. At this point the operator judged that the data were telling her that the process was running below the target, so she stopped the process and made an adjustment to increase the final roll gap slightly. She then proceeded to make more sheets. Figure 10.8(b) shows the state of the process somewhat later. Now it appeared to the operator that the sheets were coming out too thick, so she stopped and made another adjustment. As shown in Fig. 10.8(c), the process seemed to run good for a while, but then an average somewhat below the target led her to believe that another adjustment was necessary. Figure 10.8(d) and (e) show points in time where other adjustments were made. Figure 10.8(f) shows the complete history of the shift. A total of 24 × 3 or 72 sheets were made during this shift. When asked, the operator

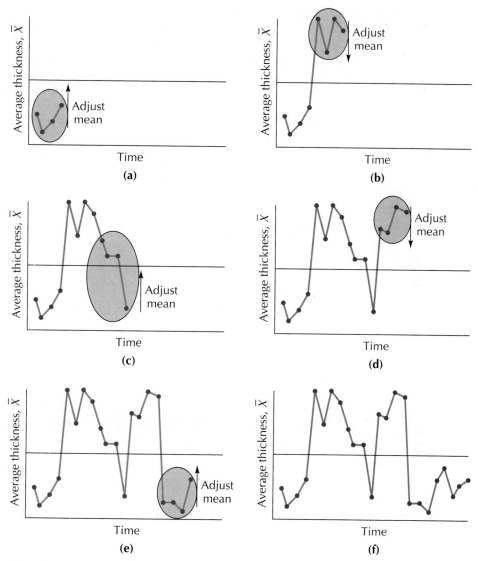

Figure 10.8 Process Adjustment History over One Shift

indicated that the history of this shift was quite typical of what happens on a day-to-day basis.

Statistical Charting of the Process

The roll mill operator has actually been collecting data on sheet thickness in a manner such that the data could be used to construct $\overline{X}$ and R control charts. In fact, it was suspected by those in the plant that the peculiar procedure of averaging three successive sheet thickness values and plotting that average on the graph paper, as witnessed in Fig. 10.8, might actually be the remnants of a once-existing SPC charting procedure for this process.

The improvement team in the plant decided to view the process in a series of samples/subgroups of three consecutive measurements. This would enable them to assess the process centering and amount of variation using an average and a range for each sample at many points in time throughout the day. As an initial charting effort they decided to take all of the sheets from one entire shift and, in fact, were able to use the same data that the staff engineers had collected for their "process capability study." These data are shown again in Table 10.2, but this time it is tabulated in the subgroups that were used. The values of $\overline{X}$ and R for each subgroup are also summarized in Table 10.2. A sample calculation is shown below.

Calculations for subgroup 1:

$$\overline{X}_1 = \frac{X_{11} + X_{12} + X_{13}}{3}$$

$$= \frac{0.0440 + 0.0446 + 0.0437}{3}$$

$$= 0.0441 \text{ inch}$$

$$R_1 = X_{1\,max} - X_{1\,min}$$

$$= 0.0446 - 0.0437$$

$$= 0.0009 \text{ inch.}$$

Using the data above, the plant group proceeded to determine the centerlines and the upper and lower control limits for the $\overline{X}$ and R control charts. These calculations are provided below.

Centerlines for the $\overline{X}$ and R charts:

$$\overline{\overline{X}} = \frac{\Sigma \overline{X}_i}{24}, \qquad i = 1, 24$$

$$= 0.0449 \text{ inch}$$

$$\overline{R} = \frac{\Sigma R_i}{24}, \qquad i = 1, 24$$

$$= 0.0018 \text{ inch.}$$

Control limits for the $\overline{X}$ and R charts:

$$\text{UCL}_{\overline{X}} = \overline{\overline{X}} + A_2 \overline{R}$$

$$= 0.0449 + (1.02)(0.0018)$$

$$= 0.0467 \text{ inch}$$

$$\text{LCL}_{\overline{X}} = \overline{\overline{X}} - A_2 \overline{R}$$

$$= 0.0449 - (1.02)(0.0018)$$

$$= 0.0431 \text{ inch}$$

TABLE 10.2	Original Data Grouped into Subgroups of Size $n = 3$ (inches)				
Subgroup	X_1	X_2	X_3	$\overline{X}$	R
1	0.0440	0.0446	0.0437	0.0441	0.0009
2	0.0438	0.0425	0.0443	0.0435	0.0018
3	0.0453	0.0428	0.0433	0.0438	0.0025
4	0.0451	0.0441	0.0434	0.0442	0.0017
5	0.0459	0.0466	0.0476	0.0467	0.0017
6	0.0449	0.0471	0.0451	0.0457	0.0022
7	0.0472	0.0477	0.0452	0.0467	0.0025
8	0.0457	0.0459	0.0472	0.0463	0.0015
9	0.0464	0.0457	0.0447	0.0456	0.0017
10	0.0451	0.0447	0.0457	0.0452	0.0010
11	0.0456	0.0455	0.0445	0.0452	0.0011
12	0.0448	0.0423	0.0442	0.0438	0.0025
13	0.0459	0.0468	0.0452	0.0460	0.0016
14	0.0456	0.0471	0.0450	0.0459	0.0021
15	0.0472	0.0465	0.0461	0.0466	0.0011
16	0.0462	0.0463	0.0471	0.0465	0.0009
17	0.0427	0.0437	0.0445	0.0436	0.0018
18	0.0431	0.0448	0.0429	0.0436	0.0019
19	0.0425	0.0442	0.0432	0.0433	0.0017
20	0.0429	0.0447	0.0450	0.0442	0.0021
21	0.0443	0.0441	0.0450	0.0445	0.0009
22	0.0443	0.0423	0.0447	0.0438	0.0024
23	0.0429	0.0427	0.0464	0.0440	0.0037
24	0.0448	0.0451	0.0428	0.0442	0.0023

$$\begin{aligned}
\text{UCL}_R &= D_4\overline{R} \\
&= (2.57)(0.0018) \\
&= 0.0046 \text{ inch} \\
\text{LCL}_R &= D_3\overline{R} \\
&= (0)(0.0018) \\
&= 0.
\end{aligned}$$

Recall that the values of A_2, D_3, and D_4 are given in Table A.2.

Figure 10.9 shows the $\overline{X}$ and R charts for the roll mill process. Examination of the R chart, which must show good statistical control before the $\overline{X}$ chart can be fully interpreted, indicates no reason to suspect that the process is not in control with respect to its variability. Examining the general patterns and applying the rules discussed in Chapters 6 and 7, the R chart does not contain any statistical signals indicating that the process variation is unstable (i.e., there is no indication of any out-of-control conditions on the R chart).

Interpretation of the $\overline{X}$ chart in Fig. 10.9 shows a much different story.

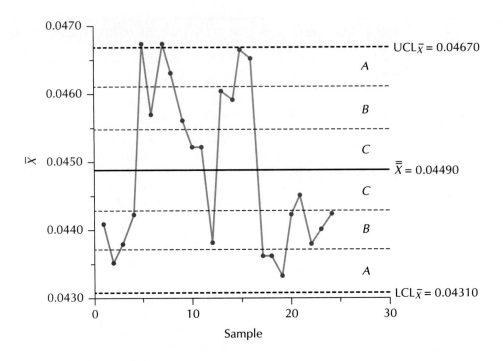

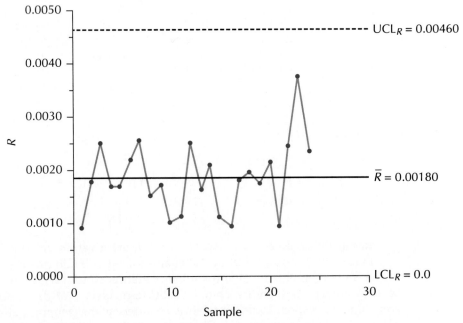

Figure 10.9 $\overline{X}$ and R Control Charts for the Roll Mill Process

Subgroup	Rule(s) Violated
TABLE 10.3 Zone Violations from Fig. 10.10, $\overline{X}$ Chart	
4	1. Four out of five in zone B or beyond
5	1. Extreme point
7	1. Extreme point
	2. Two out of three in zone A or beyond
8	1. Two out of three in zone A or beyond
	2. Four out of five in zone B or beyond
	3. Eight in a row outside zone C
9	1. Eight in a row outside zone C
	2. Four out of five in zone B or beyond.
16	1. Two out of three in zone A or beyond
	2. Four out of five in zone B or beyond
18	1. Two out of three in zone A or beyond
19	1. Two out of three in zone A or beyond
	2. Eight in a row outside zone C
20	1. Four out of five in zone B or beyond
	2. Eight in a row outside zone C
22	1. Four out of five in zone B or beyond
23	1. Four out of five in zone B or beyond
24	1. Four out of five in zone B or beyond
	2. Run of eight below the centerline

Two of the $\overline{X}$ values are extreme points (at or outside of the control limits), in many cases the $\overline{X}$ values violate zone rules, and in one case a run below the centerline occurs. The summary in Table 10.3 identifies each point that constitutes a signal of an out-of-control condition and indicates which rule(s) are being violated. Figure 10.10 shows the control charts with all the points appropriately circled that violate rules.

Overcontrol of the Process

In Fig. 10.10 the statistical signals are clear, but what is causing the process mean to shift up and down in a somewhat sporadic fashion? A review of the operating procedures for the process begins to shed some light on this question. A review of Fig. 10.8 and its associated discussion indicates that the statistical signals in Fig. 10.10 correlate quite well with the points in time at which operator adjustments were made. It appears that the operator may have been overcontrolling the process. Why is this occurring? What are the consequences of overcontrol? What can be done to avoid this mistake? The answers to these questions lie at the very foundation of statistical process control.

The operator is overadjusting the process because she cannot distinguish between the forces of common-cause variation and special-cause variation (i.e., between chronic variability sources and sporadic variability sources). In fact,

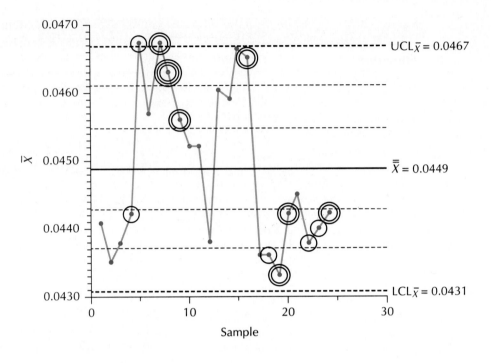

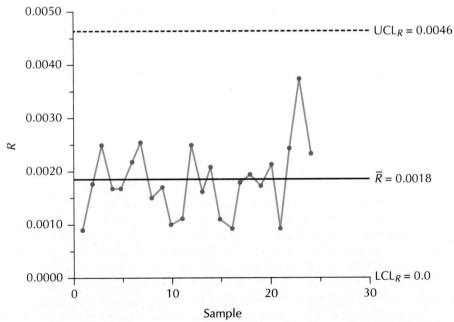

Figure 10.10 $\bar{X}$ and R Control Charts of Figure 10.9 Showing Out-of-Control Points

she is reacting to patterns in the data that can be explained as coming from the common-cause system as if they were indicating the presence of sporadic disturbances in the process. The consequences of this mistake are devastating. Each time an adjustment is made when it is not warranted, a source of variation is introduced that would otherwise not be there. As a result, quality loss is increased. Furthermore, each time the operator stops to make an unnecessary adjustment, productivity is lost.

It should also be noted that even though the operator is charting the process of making the sheets over time, the mentality is still one of product control. Her primary concern is to keep the sheet thickness as close to the nominal as possible—conformance to the specification. The concept of sporadic and chronic sources of variation is not present. Further, the concept that the nature of the variability pattern could reveal opportunities for productivity improvement is not present; rather, quite the contrary. The entire operation, from management strategy for inspection to the operator's method for making sheets, revolves around a product control way of thinking about quality.

Although we cannot totally avoid making the costly mistake of overcontrol, we can minimize its occurrence by using statistical charting to help distinguish between the presence of only common causes of variation and the occurrence of special causes. If the operator were using charts such as those in Fig. 10.9 instead of the ad hoc graphing of gasket thickness averages as depicted in Fig. 10.8, she would be able to avoid the mistake of process overadjustment.

Figure 10.11 is similar to Fig. 10.4 and shows the histogram of the individual measurements for the sheet thickness values obtained by the staff engineers during their process capability study. A more careful examination of these data indicated some reason to be suspicious from the very beginning. The high percentage of measurements in the tails of the histogram seem to be suggesting

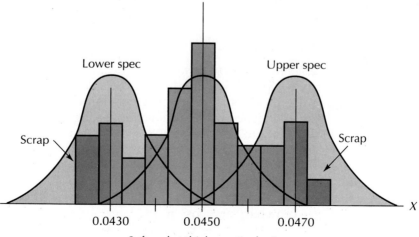

Figure 10.11 Consequences of Overcontrol on Process Performance

that more than one process might actually be represented in the data (i.e., the overcontrol of the process might be recognized by the trained eye, so to speak). The three different normal curves superimposed over the histogram in Fig. 10.11 could well be a plausible result that could easily arise from the type of overcontrol situation which is now believed to be present.

Productivity/Variability Relationship

In Chapter 2 it has been emphasized that Dr. W. Edwards Deming has, at every conceivable opportunity, pointed out to the world that sources of variability are sources of waste and inefficiency. Each time we identify and eliminate a source of variation, productivity must improve/increase. At the same time, Dr. Deming points out that elimination of a source of variation means that quality must improve as well. This is absolutely consistent with Taguchi's definition of quality and the loss function.

The importance of the role of statistical process control in improving productivity can be easily illustrated via the problem of overcontrol of a process. Overcontrol arises when operators or other process personnel or electronic/automatic controllers make adjustments to the process in the interest of causing a change in performance when no such adjustments are called for; that is, the process is stable if left alone. Overcontrol clearly affects the process in a negative way in two regards. First, it increases the process variability so that quality is eroded. Second, it decreases productivity since when the process is being adjusted, no product is being made. The result is a costly mistake from both a quality and a productivity point of view. We have seen this clearly in the gasket-making case study. Overcontrol of a process cannot be totally eliminated, but the frequency of making this mistake can be greatly reduced if statistical signals from control charts are used as a basis for deciding whether to adjust a process or leave it alone. Statistical process control provides an informed basis for this decision-making activity.

Process Improvement Action

As a first process improvement action it was decided by the group to train the operator in the concepts of SPC charting; in the simple calculation of $\overline{X}$ and R, the plotting of these values, and the interpretation of $\overline{X}$ and R control charts. Procedures were established by the production manager and the quality control manager for charting and the operator was given this new tool to help her to decide when to adjust the process, and more important, when not to adjust the process.

Since the initial control charts developed for the roll mill process showed good statistical control for the variability (the R chart was in control) and since the $\overline{X}$ chart was centered very close to the target value for the process, it was decided to initiate control at the machine simply by extending the centerlines and control limits for these charts. The operator followed the same general procedures as before except that she stopped to adjust the process mean level only if the $\overline{X}$ chart showed that the mean had shifted. The measurements that were obtained for the first shift run in this manner are summarized in Table 10.4.

Figure 10.12 shows the $\overline{X}$ and R control charts for the shift for which the process was run under SPC-based operating guidance. Remember that over this shift, the operator was to make adjustments to the roll gap—process mean—only when the $\overline{X}$ chart gave a signal that the mean had shifted. It turned out that no adjustments were required. The charts in Fig. 10.12 show good statistical control, indicating that if left alone, the process appears to operate in a stable fashion. As can be imagined, the consequences of all of this are quite dramatic.

TABLE 10.4	Process Measurements (inches) for a Shift Run Under SPC Guidance				
Subgroup	X_1	X_2	X_3	$\overline{X}$	R
1	0.0445	0.0455	0.0457	0.0452	0.0012
2	0.0435	0.0453	0.0450	0.0446	0.0018
3	0.0438	0.0459	0.0428	0.0442	0.0031
4	0.0449	0.0449	0.0467	0.0455	0.0018
5	0.0433	0.0461	0.0451	0.0448	0.0028
6	0.0455	0.0454	0.0461	0.0457	0.0007
7	0.0455	0.0458	0.0445	0.0453	0.0013
8	0.0445	0.0451	0.0436	0.0444	0.0015
9	0.0443	0.0450	0.0441	0.0445	0.0009
10	0.0449	0.0448	0.0467	0.0455	0.0019
11	0.0465	0.0449	0.0448	0.0454	0.0017
12	0.0461	0.0439	0.0452	0.0451	0.0022
13	0.0443	0.0434	0.0454	0.0444	0.0020
14	0.0456	0.0459	0.0452	0.0456	0.0007
15	0.0447	0.0442	0.0457	0.0449	0.0015
16	0.0454	0.0445	0.0451	0.0450	0.0009
17	0.0445	0.0471	0.0465	0.0460	0.0026
18	0.0438	0.0445	0.0472	0.0452	0.0034
19	0.0453	0.0444	0.0451	0.0449	0.0009
20	0.0455	0.0435	0.0443	0.0444	0.0020
21	0.0440	0.0438	0.0444	0.0441	0.0006
22	0.0444	0.0450	0.0467	0.0454	0.0023
23	0.0445	0.0447	0.0461	0.0451	0.0016
24	0.0450	0.0463	0.0456	0.0456	0.0013
25	0.0442	0.0451	0.0450	0.0448	0.0009
26	0.0450	0.0471	0.0451	0.0457	0.0021
27	0.0457	0.0432	0.0449	0.0446	0.0025
28	0.0460	0.0453	0.0452	0.0455	0.0008

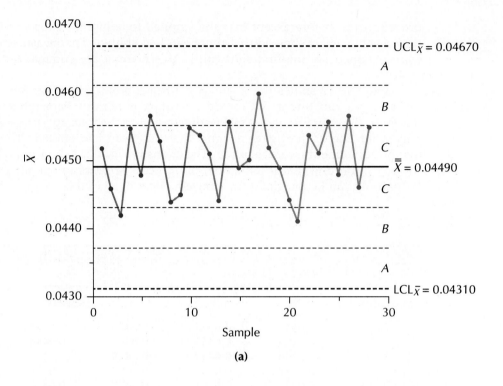

(a)

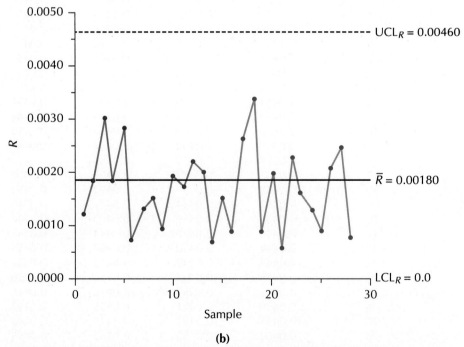

(b)

Figure 10.12 $\overline{X}$ and R Control Charts for the Roll Mill Process Under SPC Process Operation

10.5

Process Capability Versus Process Control

Reassessment of Process Capability

Since the process now exhibits good stability, it is valid and meaningful to conduct a process capability study. To determine if the process could be modeled by the normal distribution, two graphical techniques were employed, the histogram and the normal probability plot. First, a frequency histogram of the data in Table 10.4 was constructed. The steps used for the first set of data were followed once again.

1. The range of the data was determined by subtracting the smallest value from the largest: $0.0472 - 0.0428 = 0.0044$.
2. It was decided to use nine cells of width 0.0005, with midpoints at 0.0XX5 and 0.0XX0.
3. Therefore, the boundaries were chosen such that the first cell goes from 0.04275 to 0.04325, the second from 0.04325 to 0.04375, and so on.
4. A tally sheet of the points was made.

Cell Midpoint	Frequency	Total
0.0430	11	2
0.0435	11111	5
0.0440	11111 111	8
0.0445	11111 11111 11111	15
0.0450	11111 11111 11111 11111 11	22
0.0455	11111 11111 11111	15
0.0460	11111 111	8
0.0465	11111 1	6
0.0470	111	3

5. The histogram itself was constructed, as shown in Fig. 10.13. This histogram has very much the appearance of the bell-shaped normal distribution.

To investigate further the possible normality of the data, a normal probability plot was made of the measurement points in the first 10 subgroups. Table 10.5 was constructed following the guidelines established in Chapter 3. The plot is shown in Fig. 10.14. Both the frequency histogram and the normal probability plot seem to indicate clearly that the individual gasket thickness measurements closely follow a normal distribution curve.

Figure 10.15 is a graphical depiction of the process capability for the stabilized process and emphasizes the fact that over the period of time that the process was sampled, it has been established that the process was in statistical control. It appears that when examined relative to the specifications the process

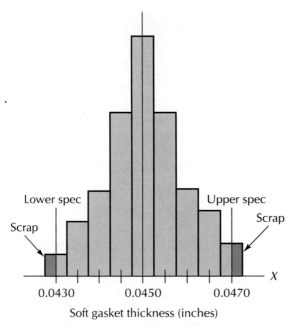

Figure 10.13 Histogram of the Gasket Measurements in Table 10.4

variation is greatly reduced. About 95% of the process is now operating within the specifications.

To establish the improved process capability more precisely, the normal distribution was employed as a model for the process. The calculations are as follows, using the data from Table 10.4:

$$\overline{\overline{X}} = \frac{\sum \overline{X}_i}{28}, \quad i = 1, 28$$

$$= 0.04505$$

$$\overline{R} = \frac{\sum R_i}{28}, \quad i = 1, 28$$

$$= 0.00175$$

$$P(\text{within specs}) = P(0.0430 \le X \le 0.0470)$$
$$= [P(X \le 0.0470) - P(X \le 0.0430)]$$

$$Z = \frac{X - \mu}{\sigma_X}$$

$$\hat{\sigma}_X = \frac{\overline{R}}{d_2}.$$

For $n = 3$, $d_2 = 1.693$,

TABLE 10.5 Data for Normal Probability Plot of Fig. 10.14					
Order	Measure-ment	Estimated Cumulative Probability (%)	Order	Measure-ment	Estimated Cumulative Probability (%)
1	0.0428	2	16	0.0450	52
2	0.0433	5	17	0.0451	55
3	0.0435	8	18	0.0451	58
4	0.0436	12	19	0.0453	62
5	0.0437	15	20	0.0454	65
6	0.0441	18	21	0.0455	68
7	0.0443	22	22	0.0455	72
8	0.0445	25	23	0.0455	75
9	0.0445	28	24	0.0457	78
10	0.0445	32	25	0.0458	82
11	0.0448	35	26	0.0459	85
12	0.0449	38	27	0.0461	88
13	0.0449	42	28	0.0461	92
14	0.0449	45	29	0.0467	95
15	0.0450	48	30	0.0468	98

$$\hat{\sigma}_X = \frac{0.00175}{1.693}$$
$$= 0.00103$$
$$Z_{LSL} = \frac{0.0430 - 0.04505}{0.00103}$$
$$= -1.99$$
$$Z_{USL} = \frac{0.0470 - 0.04505}{0.00103}$$
$$= 1.89.$$

The Z table (Table A.1) can then be used to find the required probability:

$$P \text{ (within specs)} = [P(Z \leq Z_{USL}) - P(Z \leq Z_{LSL})]$$
$$= [P(Z \leq 1.89) - P(Z \leq -1.99)]$$
$$= 0.9706 - 0.0233$$
$$= 0.9473.$$

Compared to the previous determination of the process capability (which, strictly speaking, was not a valid use of the normal distribution), it appears that the process capability has been greatly improved. Previously, by a simple count of the actual number of sheets that were nonconforming (in Table 10.1), the process capability might have been crudely estimated to be about 73%; at

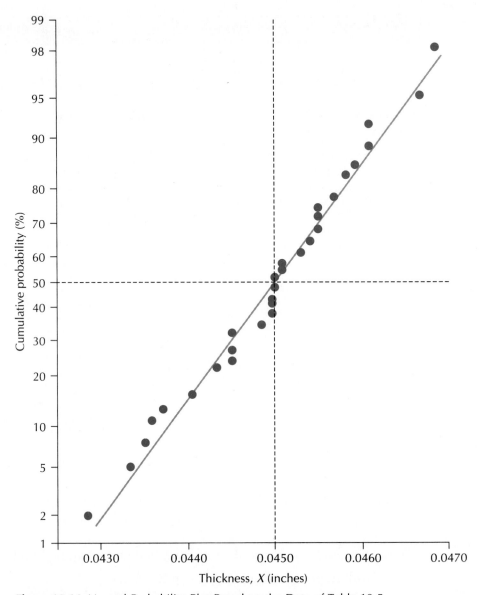

Figure 10.14 Normal Probability Plot Based on the Data of Table 10.5

least this is the percentage of those 72 sheets that were within the specifications. Now that the process has been stabilized by removing the overcontrol operating fault, the process capability is 95%.

In terms of the C_p index, the process capability is now equal to 0.645, as calculated below. While the process capability is still not acceptable by usual

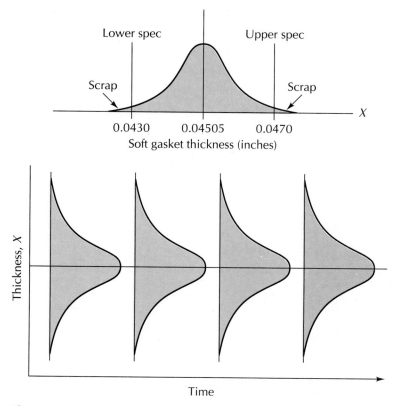

Figure 10.15 Process Capability Analysis Based on the Process Data of Table 10.4

standards ($C_p \geq 1$ is generally required), as indicated above, scrap has been considerably reduced.

$$C_p = \frac{\text{USL}_X - \text{LSL}_X}{6\,\sigma_X}$$

$$= \frac{\text{USL}_X - \text{LSL}_X}{6\,\overline{R}/d_2}$$

$$= \frac{0.0470 - 0.0430}{6(0.00175/1.693)}$$

$$= 0.645.$$

Calculations were also performed to find the value of C_{pk}:

$$Z_{\text{USL}} = \frac{\text{USL} - \overline{\overline{X}}}{\sigma_X}$$

$$= \frac{\text{USL} - \overline{\overline{X}}}{\overline{R}/d_2}$$

$$= \frac{0.0470 - 0.04505}{0.00175/1.693}$$

$$= 1.89$$

$$Z_{LSL} = \frac{LSL - \overline{\overline{X}}}{\sigma_X}$$

$$= \frac{LSL - \overline{\overline{X}}}{\overline{R}/d_2}$$

$$= \frac{0.0430 - 0.04505}{0.00175/1.693}$$

$$= -1.99$$

$$Z_{min} = \min[Z_{USL}, -Z_{LSL}]$$

$$= \min[1.89, 1.99]$$

$$= 1.89$$

$$C_{pk} = \frac{Z_{min}}{3}$$

$$= \frac{1.89}{3}$$

$$= 0.63.$$

But improved product quality—reduced variability—is only one-half of the good news of Fig. 10.15. Since no process roll gap adjustments were made, productivity must also have improved. A review of the production data reveals that 84 sheets were made for this shift, compared with the 72 sheets previously manufactured under the original method of process operation. This constitutes a 17% increase in productivity! It is now becoming clearer that when we seek improvements at the process through SPC concepts and methods, we can simultaneously realize improvements in both quality and productivity.

Quality/Variability Relationship

Since one major source of functional variation, as discussed in Chapter 4, is manufacturing imperfection, the methods of statistical process control should stand as an important tool for quality improvement. In this case study operator overcontrol has been identified as a major source of manufacturing imperfection/variability. This source of variation has led not only to a serious loss in productivity but has also greatly contributed to the poor process capability.

As outlined in Chapter 2 and discussed in Chapter 9, the loss function approach to defining quality provides a clear motivation for seeking continually to reduce functional variation in product/process performance. Figure 10.16 depicts the change in the process that has been precipitated by the removal of the overcontrol source of variation. Since the loss is a function of the process standard deviation as was demonstrated in Chapter 9, removal of every source

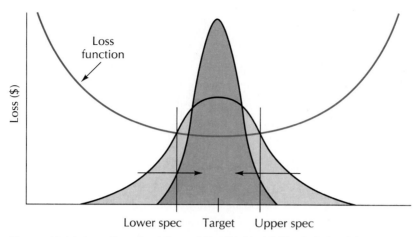

Figure 10.16 Loss Function View of Quality/Variability Relationship

of variation must lead accordingly to a reduction in loss due to functional variation and therefore an improvement in quality.

Holding the Gains

It was made clear from the discussion in Chapter 9 that the loss function provides a quantitative means to assess product/process performance relative to the nominal value. This assessment, however, requires that we be able to represent the process in terms of a statistical distribution, for example, the normal distribution. However, such a representation is valid only if the process is in a state of statistical control. If we do not, therefore, strive to keep processes in statistical control, the loss function analysis used for design purposes will go unachieved at the production stage. In general, this will mean that the actual loss encountered by the customer in the field will be greater than what we believe it to be based on the design process, a truly regrettable state of affairs.

The argument above can also be applied to "holding the gains" precipitated by control charting of the process. In the case study at hand both quality and productivity gains have been realized in stabilizing the process through removal of the overcontrol source of variation. However, failure to maintain this state of statistical control can lead to the loss of all that has been gained! This loss, generated at the process, can be illustrated by overlaying the statistical process control and loss function concepts.

Suppose that Fig. 10.17 represents what we can achieve if the process potential is consistently realized; that is, the process, as shown in the figure, is in statistical control with the indicated level of variability. Now if the process is not kept in a state of statistical control (e.g., its mean value is allowed to shift about as depicted in Fig. 10.18), the customer would experience an increased loss due to functional variation. This increase in loss, as measured by the loss

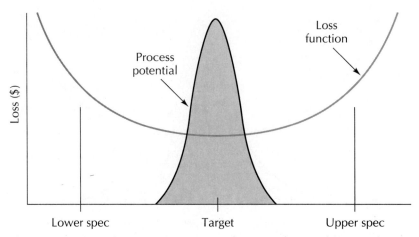

Figure 10.17 Loss If Process Is in a State of Statistical Control Centered at the Nominal

function, results from the fact that over the long term the lack of control/stability may give rise to a distribution of measurements of the appearance of Fig. 10.19, a more flat, loaf-shaped appearance.

We could actually calculate the increase in loss as a result of this lack of statistical control of the process by assuming, for example, that the process of Fig. 10.19 can be approximated by a uniform distribution with endpoints equal to the lower and upper specification limits. Then, using the formula for the calculation of the standard deviation of the uniform distribution, we could calculate the loss and compare this value to that when the process is performing consistently (in control) at the nominal according to its position and statistical

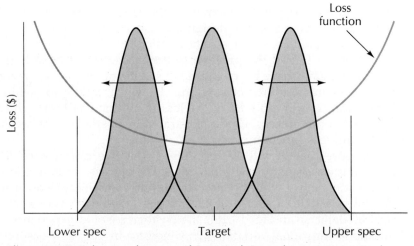

Figure 10.18 Behavior of Process If Statistical Control Is Not Maintained

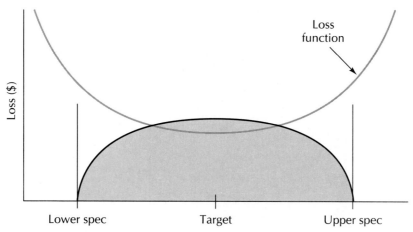

Figure 10.19 Loss If Statistical Control Is Not Maintained

distribution as shown in Fig. 10.17. It is clear that statistical control is an important concept with significant economic implications from both the producer and consumer points of view.

10.6

Seeking Further Improvement Through Design of Experiments

Encouraged by the marked improvement in process performance, the group in the plant continued to hold their meetings and continued with the charting activity in the hope of finding still other improvement opportunities. At one of the meetings the process engineer suggested that two machine conditions, roll speed and roll temperature (cooling lines in the large roll), might have better settings (i.e., ones that would improve sheet consistency). To investigate this possibility it was decided to run a simple two-level factorial design experiment. This experiment is depicted graphically in Fig. 10.20. A total of $2^2 = 4$ tests are required (graphically, the corners of the square in Fig. 10.20).

It was decided to run the four tests of the experimental design over a period of one shift, each test being comprised of about 2 hours of production. During the tests, the $\overline{X}$ and R charts would be continued using the previously established values for $\overline{\overline{X}}$, $\overline{R}$, and the control limits, and the R chart would be examined to see if changes in roll speed and/or roll temperature were causing a change in the sheet-to-sheet consistency, that is, any change in the process variability.

Figure 10.21 shows a portion of the data plotted on the R chart maintained during the time that the experiment was being conducted. Examination of Fig.

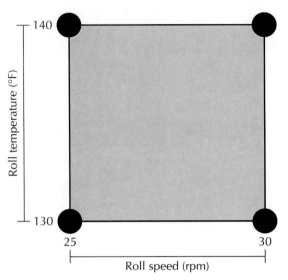

Figure 10.20 Two-Level Factorial Design to Study Roll Mill Process

10.21 indicates that changes in roll speed and/or roll temperature do not appear to be influencing the amount of variation in the process. In fact, if the 2-hour segment boundaries are removed (the dashed vertical lines in Fig. 10.21) and the range values are "connected," the chart clearly shows that the process variation is in good statistical control. It appears, therefore, that changes in roll speed and/or roll temperature do not influence process variation.

The last sentence of the preceding paragraph needs to be considered more carefully. The truth is that we cannot draw this kind of conclusion about the results of the experiment. The fact that no effects of roll speed and/or roll temperature can be detected from the experiment (Fig. 10.21) can result from either of two distinct possibilities:

1. These variables truly have no effect on process variation.
2. These variables have some effect on process variation, but these effects cannot be detected because the noise level of the process is too great (i.e., the true effects are masked/hidden by the noise).

Too often, possibility 2 is really the factor driving the experimental design results and we will therefore either end up mistakenly overlooking improvement opportunities that are there but hidden by the noise in the process, or we will conclude that the design of experiment techniques themselves are flawed or do not apply to what we are doing. In either case the results are regrettable.

A Major Breakthrough

Disappointed with the results of the factorial experiment but still encouraged by the progress to date, the group in the plant continued to meet to discuss

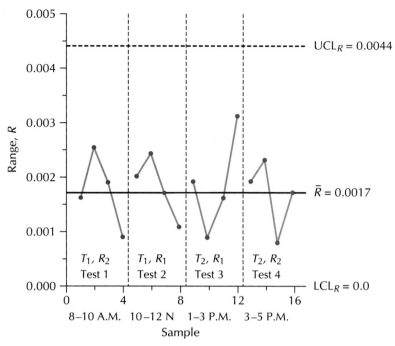

Figure 10.21 Range Chart for the Process over the Period of the Factorial Experiment

possible improvement actions. One Friday, discussion turned to the amount of material that was added to the roll gap to make each sheet. The operator stated that she had been told years ago simply to add the material one handful after another, evenly distributing it along the roll gap until the bell rang and the light went on, indicating that the smaller roll had reached its final destination (i.e., the sheet was now completed). Clearly, each sheet was not receiving the same amount of material, by weight, for the specified target moisture content. Yet each sheet is constrained by the roll diameter to be the same length and each sheet is constrained by stops on the rolls to be of the same width. How could we, then, expect to control closely the sheet thickness and the sheet density, which was another quality characteristic under study?

To solve the problem above, which is clearly a common-cause/chronic variability problem, it was suggested that a fixed amount of material, by weight, be metered into each sheet's construction. A long discussion about how this would be done ensued and a lot of disagreement followed. A primary concern seemed to be not whether the procedure being used now was a problem but rather, how this metering problem might be accomplished. Certain members of the group thought that to meter the material into the roll gap more carefully would be too costly in terms of retrofitting a metering device to the roll mill. The quality control and production managers, however, felt that logistics were getting in the way of common sense and reason, so they decided to prepare for an experiment over the weekend.

On Saturday morning they took it upon themselves to measure many sheets of recently manufactured gasket material to determine the proper sheet weight for the nominal thickness. They then took raw material blended the day before that was ready for use on Monday and carefully measured out portions just enough for one sheet, based on the weight/thickness relationship they had established, and placed the metered raw material in plastic bags. They prepared enough of these bags of material for one shift of production. On Monday morning they asked the operator to run all day using the material from each bag to make just one sheet, all to be added for the construction of each sheet, no more, no less. The $\overline{X}$ and R charts were continued. The results are shown in Fig. 10.22 and are dramatic. From the figure it is clear that the effect of metering a precise amount of material into every sheet has been to cause a significant reduction in the common-cause variability in the process. From the R chart we can see that the common-cause variability has been cut almost in half. It should be clear that this result will have a tremendous positive impact on both sheet quality and process productivity.

Figure 10.23 is a graphical representation of the results of a process capability study done on data for which the sheet material metering was in effect. It is obvious from the figure that the process capability has been greatly improved in terms of the capability indices C_p and C_{pk}. The process now has a capability of $C_p = 1.253$. Calculations for C_p and C_{pk} are as follows:

$$C_p = \frac{\text{USL}_X - \text{LSL}_X}{6\,\sigma_X}$$

$$= \frac{\text{USL}_X - \text{LSL}_X}{6\,\overline{R}/d_2}$$

$$= \frac{0.0470 - 0.0430}{6(0.0009/1.693)}$$

$$= 1.253$$

$$Z_{\text{USL}} = \frac{\text{USL} - \overline{\overline{X}}}{\sigma_X}$$

$$= \frac{\text{USL} - \overline{\overline{X}}}{\overline{R}/d_2}$$

$$= \frac{0.0470 - 0.0449}{0.0009/1.693}$$

$$= 3.95$$

$$Z_{\text{LSL}} = \frac{\text{LSL} - \overline{\overline{X}}}{\sigma_X}$$

$$= \frac{\text{LSL} - \overline{\overline{X}}}{\overline{R}/d_2}$$

$$= \frac{0.0430 - 0.0449}{0.0009/1.693}$$

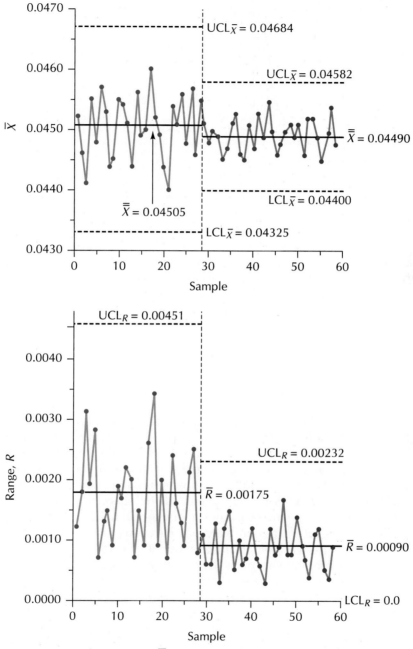

Figure 10.22 Continuation of $\overline{X}$ and R Charts During Sheet Construction Under Raw Material Metering

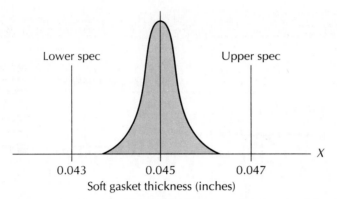

Figure 10.23 Process Capability After Material Metering Was Put into Effect

$$= -3.57$$

$$Z_{min} = \min[Z_{USL}, -Z_{LSL}]$$

$$= \min[3.95, 3.57]$$

$$= 3.57$$

$$C_{pk} = \frac{Z_{min}}{3}$$

$$= \frac{3.57}{3}$$

$$= 1.19.$$

In terms of productivity it may at first be more difficult to understand how material metering might lead to a significant improvement. However, when the operator was questioned about this she indicated that in the past she had often stopped the machine during the construction of a sheet to check the progress in terms of thickness to determine whether she should add a bit more or less material from that point to the end of the sheet construction. In the case of material metering, since the amount of material to be added was predetermined, such process interruptions were not necessary. As a result, she was not surprised that she was able to make about 90 sheets per shift. This is now a 25% increase in productivity from the outset of the study.

Once the clear advantage of material metering had been proved through the data, interest within the project group centered on finding an efficient engineering solution to the material metering problem. A very simple and cost-effective solution was identified rather quickly. Scales were placed under the track that supported the raw material tote pans at the machine. The digital readout for the scale could be set to the prescribed weight and the operator merely had to remove material from the tote pan until the digital readout went to zero. At that point she would know that she had added the correct amount of material to the sheet. This engineering solution was implemented and sometime later the return on investment was calculated to be a mere 13 weeks in terms of both quality and productivity improvement.

It should also be noted that since the raw material metering problem was common to all of the roll mill lines in the plant, the foregoing solution was applied to each line. This is an important aspect of the SPC approach to improvement, that some of the problems discovered through the charting of one machine/process are common to all similar processes. Therefore, solutions may be usefully applied across all machines.

Experimental Design Revisited

It is clear from Fig. 10.22 that the amount of variability of the process has been greatly reduced (i.e., the signal-to-noise ratio has been increased). With this in mind and still with the firm conviction that either roll speed or roll temperature or both might have a real impact on process consistency, the group decided to rerun the same two-level factorial experiment that was conducted previously (see Fig. 10.20). The $\overline{X}$ and R charts employed at this point were based on data obtained from process monitoring after the metering of the raw material was installed as an operating method. Hence with the significant reduction in variability, the charts should be a much more sensitive indicator of process changes when they occur.

Figure 10.24 shows the results of this second running of the factorial experiment. At first the results are quite surprising since now the range chart

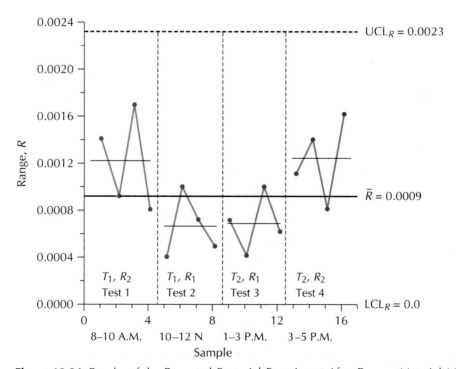

Figure 10.24 Results of the Repeated Factorial Experiment After Process Material Metering Was Initiated

shows clearly that roll speed appears to have a strong influence on process consistency. In particular, the lower level of roll speed seems to give rise to a significant reduction in process variability. It is important in Fig. 10.24 to note that the vertical scale is approximately half that of Fig. 10.21, which shows the results when the experiments were run earlier. It is now apparent that the excessive noise in the system (due to the variation in the amount of material added to each sheet) was masking the roll speed effect.

Based on the results of the second experiment, the roll speed was reduced and a significant improvement in the process capability resulted. Over a period of the next several weeks the group continued to study the process and evaluate other potential improvement actions. At one point the operator noted that excessive blacklash in the gears often made it difficult to set the roll gap when adjustments were required or when she changed from one gage sheet to another. When the maintenance department made an inspection of the gearing, they found excessive wear and replaced several gears. This greatly increased the probability of arriving at the proper roll gap through a single adjustment. Of course, this led to a further reduction in product variability and a further increase in process productivity. Figure 10.25 graphically depicts the process capability after the roll speed change and the maintenance actions. It is clear that the process is becoming extremely consistent.

Statistical Process Control/Design of Experiments Relationship

From the result above it is clear that the power of design of experiments can be greatly enhanced if the environment in which the experiments are conducted has been affected through variation reduction methods such as statistical process control. Although statistical control of the process is not necessarily a prerequisite to drawing valid conclusions from the results of designed experiment, it can greatly enhance the ability to detect the presence of real variable

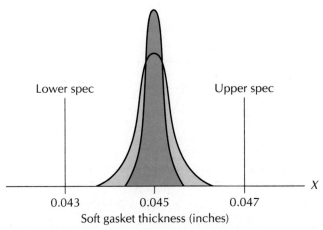

Figure 10.25 Process Capability After Additional Improvement Actions

effects. As we will see later, there are certain fundamental design countermeasures that we can take to ensure experimental validity. SPC is an important countermeasure that we must consider whenever possible to improve experiment sensitivity. Furthermore, in cases where statistical control of a process has been established, subsequent experimentation and associated improvement actions taken are more likely to be realized in future operation of the process. Under statistical control the future is more predictable!

It should also be clear from the above that design of experiments stands as a powerful technique to reveal basic opportunities for improvement during an SPC study. In Chapter 4 we discussed the feedback control loop interpretation of the process control concept. It was indicated then that stage 3 in this loop—fault diagnosis—was critical and that techniques such as design of experiments could be useful for this purpose.

A Postscript.   Several months later the consultant who worked with the people in the plant returned to review the current status of their statistical implementation program. He was greeted by the plant manager, who reported enthusiastically that over the past six months they had saved over $500,000 in material costs alone. When the consultant inquired as to exactly how this was accomplished, the plant manager replied that since the process capability had been increased to over 12 sigma, it was possible to recenter the process very close to the lower specification, thereby greatly reducing the amount of material required to make a sheet. At the same time he could satisfy the customers' defined product quality requirement in terms of the bilateral specification. Figure 10.26 graphically depicts the action of the plant manager.

Hopefully, by now your understanding of the loss function definition of quality and the notions of statistical tolerance design makes clear the serious implications of such actions. One way to think of this action at the process is that it is equivalent to making a design change in the product, changing the

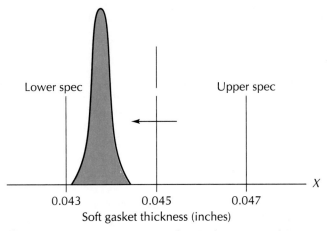

Figure 10.26 Process Recentered to Reduce Material Requirements

nominal value for the gasket material thickness to a value quite near the lower specification limit. Is it reasonable to think that this can be done without having any effect on product quality?

Summary

Through the case study above a number of extremely important concepts and methods central to reducing variability—increasing quality and productivity—have been discussed. What is perhaps most important to note is that today, by using the same machines, same people, and same materials, but in a much more effective fashion, the problem that this company originally faced has been solved. By applying simple but powerful techniques of statistical charting and fault diagnosis, the company avoided perhaps needlessly spending a large amount of money on new equipment that has been shown to be totally unnecessary. Such results are being experienced by many others as they overlay statistical thinking and methods on their engineering knowledge and experience.

Computer Workshop 2

Introduction

In this workshop we study, with the aid of three computer programs, a process that machines a bronze bushing. One computer program, TURNSIM, will be used to collect data from the process, while the other two programs, XRCHRT and PROCAP, will be used to construct control charts and perform a process capability analysis, respectively. The objectives of the workshop are

1. Read the charts to determine what are the statistical signals.
2. Relate these statistical signals to the physical conditions of the process.
3. Take action on the causes to improve the stability and capability of the process.

Background

The part shown in Fig. 10.27 (a bronze bushing) is being machined on its outside-diameter surface to meet functional requirements on its diameter dimension and surface roughness. The process in question is a single-point lathe-turning operation. Recently, inspection of the bushings has shown that many of the parts produced by the process frequently have a surface roughness (R_a value in microinches) that falls outside the specifications of 70 ± 15 microinches. Ad hoc changes have been initiated at the process on several occasions, but no noticeable improvement has been seen as parts have been inspected periodically.

The parts are being made on three different machines, all of which must

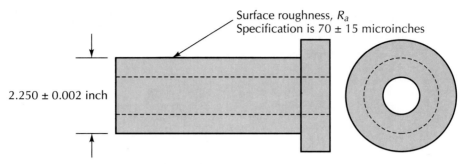

Surface roughness, R_a
Specification is 70 ± 15 microinches

2.250 ± 0.002 inch

Figure 10.27 Part of Interest in the Turning Process Investigation with Dimensional and Surface Roughness Specifications Given

operate in parallel at all times to produce the required daily output. After a series of discussions among those associated with the process, a list was drawn up of 10 factors that can vary somewhat during production. It is believed that the variation of some of these factors may be what is causing the poor product conformance to the functional requirements. One of the individuals involved in the discussions has suggested that these 10 factors actually represent a list of potential special causes. A description of these 10 factors and a description of how they vary during production are shown in Table 10.6.

Description of the Turning Process Simulator, TURNSIM

A computer simulation program called TURNSIM may be used to simulate the data collection activity for the turning process that produces the bronze bushings. The program may be used to collect k subgroups of size $n = 5$, suitable for the construction of $\overline{X}$ and R control charts. The investigator inputs the number of samples, k, to be collected. It happens that only three of the 10 candidate special causes are truly active in the process simulation program (i.e., produce unusual patterns of variability on the $\overline{X}$ and R control charts). The other seven, as they vary, have no special cause impact on the process. It is desired to find out which of the three factors are, in fact, the special causes. Each special cause must be identified and verified separately. Additionally, the corrective actions taken to remove the special causes and bring the process into a state of statistical control must all be specified each time, as successive actions are taken.

For each subgroup simulated by the program TURNSIM, the surface roughness measurements for the sample of size $n = 5$ are output to a user-specified file. Additionally, to diagnose the special causes, TURNSIM also generates a listing of the levels for the 10 candidate special causes for that sample and stores that in a file. Of course, the surface roughness data may be input into program XRCHRT (described in the workshop at the end of Chapter 8) to generate $\overline{X}$ and R control charts. By correlating the levels (process conditions) for the 10 factors with the patterns in the charts, the special causes can be

TABLE 10.6 Potential Special Causes of Variation	
Cause	**Nature of the Variation**
Cutting speed	Has been observed to vary from 1000 to 1100 feet per minute.
Feed rate	Has been observed to vary from 0.0080 to 0.0089 inch per revolution.
Operator	The regular operator is sometimes replaced by a substitute.
Setup person	Two people perform the setup procedure, Mr. Samuel and Mr. Ricard.
Tool type	Tools are purchased from three different sources: Nork-V, Cutgo-T, or Roved Cube.
Tool condition	The cutting tool has been observed to be relatively sharp or dull.
Depth-to-shoulder measurement	This dimension, measured from the outside diameter of the flange, can vary from 0.0997 to 0.1003 inch.
Lathe used	As noted, three lathes are required to meet the production requirements: Nacirema, Rex, and LeLathe.
Surface measuring device	Two profilometers are used to make the R_a value measurement: Talymeas 5 or Surfchek 3.
Tool back rake angle	Two different tool geometries are commonly used with back rake angles of $+5°$ and $+10°$.

identified. It may be noted that the three active special-cause factors produce variations in the surface roughness that can be explained physically, based on general knowledge concerning the single-point lathe-turning process. You should develop this physical explanation to aid your fault diagnosis analysis and to justify your actions.

If you think you have identified a special cause, and you wish to collect additional samples from the turning process, you may reenter program TURN-SIM and tell it that you have found a special cause and that corrective action at the process has taken place. If you have identified a bona fide special cause, TURNSIM will "disarm" it (i.e., the effect of that special cause on future samples collected will go away). You should continue to sample the process to observe the effect (if any) of the corrective actions you are taking. If you identify a factor that is not one of the true special causes, nothing will happen to change the future data collected. Each time you enter program TURNSIM to collect samples, you must identify all previously identified special causes (if verified as being bona fide special causes).

Once all three special causes have been identified and verified, both through physical reasoning and through the evidence provided by the control charts, a process capability study may be performed. To perform the process capability study, given a stable process, the computer program PROCAP may be used. The computer program PROCAP will generate the following information:

1. A histogram of the individual measurements.
2. A normal probability plot of the individual measurements.
3. Calculations of the process capability (percent conforming, C_p, and C_{pk}) given the part specifications and assuming that the individual measurements are normally distributed.

If you find that the process is not capable (i.e, the level of common-cause variability is too large, for example), you should offer some corrective actions that might reduce the common-cause variability or otherwise improve the process capability.

Using TURNSIM

TURNSIM was written for use on IBM PCs (and compatibles). The process simulation program for this workshop, TURNSIM, may be started simply by typing in "d:TURNSIM" at the DOS prompt, where d is the drive that contains the program disk. TURNSIM will ask the user to enter a filename into which the surface roughness data (subgroups of size $n = 5$) will be stored. Although any valid DOS filename may be used, it is suggested that the filename be given an extension of "DAT". The filename "d:TURNSIM.DAT", for example, indicates that the file contains data generated by program TURNSIM. In defining the filename, d is the drive onto which the file is to be stored. TURNSIM will also ask the user to specify a filename into which the process conditions (levels of the 10 candidate special causes) will be stored. It is suggested that this filename be given the extension of "CON", indicative of the fact that the file contains process conditions.

With filenames specified for the surface roughness data and the process conditions, TURNSIM then presents a list of the 10 candidate special causes and asks users if they have identified one or more of them as special causes. If users indicate that they have indeed isolated one or more special causes, they are prompted to specify which factors are the special causes. Once the identified special causes have been indicated, the users are asked how many additional subgroups of size $n = 5$ they wish to collect. At this point, TURNSIM collects the specified number of samples, having disarmed the active special causes that have been identified as such, and stores the data in the two user-specified files.

Following the data collection activity of program TURNSIM, users may proceed to use the program XRCHRT. As described in the workshop in Chapter 8, XRCHRT may be used to construct and analyze $\overline{X}$ and R control charts. Following construction of the control charts, the user may correlate signals/patterns on the control charts with the levels of the 10 factors for the corresponding subgroups.

Using PROCAP

PROCAP was written for use on IBM PCs (and compatibles). The process capability analysis program, PROCAP, may be started simply by typing in "d:PROCAP" at the DOS prompt, where d is the drive that contains the program disk. PROCAP will ask the user to enter the name of the file that contains the data associated with the subgroups of size $n = 5$. The user is also asked to enter the date and his or her name (this information will subsequently be displayed on the program output).

At this point, PROCAP prompts the user to specify the specification limits for the data under analysis. Users are then asked if they wish to skip the data dump. The data dump gives a complete listing of all the individual measurements for each sample. Then users are asked if they wish to skip the plots (i.e., the histogram and normal probability plot of the individual measurements). These plots should rarely, if ever, be skipped. Finally, users are asked if they wish to invalidate any samples. This option permits a user to remove specific subgroups from the capability analysis.

Finally, users are asked where they would like to direct the output of program PROCAP: to the printer or to a file. If the printer is selected, PROCAP attempts to open the printer and send output to it directly (this may not be a feasible alternative for a system in which several computers are networked to a printer). If it is desired to send the output to a file, PROCAP will then ask the user to specify a filename into which the output will be placed. Once again, while any valid DOS filename may be used, it is suggested that the filename be given an extension of "REP" or "OUT", indicating that the file contains a "report" or "output." When the user leaves program PROCAP, the output stored in a file may be printed using the suitable DOS or system command. Once the user has specified where the output is to be directed, all user input is completed, and then PROCAP will begin performing the process capability analysis.

Assignment

To complete the assignment, the following tasks must be undertaken:

1. Run program TURNSIM to collect samples of size $n = 5$ from the turning process. The number of samples you collect each time (k) is your decision.
2. Run program XRCHRT to construct and analyze control charts for the data collected from the turning process (i.e., identify the statistical signals).
3. Correlate the statistical signals/patterns on the control charts with the levels for the 10 candidate special causes.
4. If you believe that a special cause has been identified, move to step 1 to collect additional data. These data should also be charted to verify that the identified special cause truly is a special cause and also to collect additional data to identify remaining special causes.

5. When all three special causes have been separately identified and verified, program PROCAP may be run to assess the capability of the process.
6. Prepare a report that details the findings of your study.

The report, describing what you have learned about the process, must include at least the following:

1. An executive summary, which summarizes the contents and conclusions of your report in approximately one page.
2. A body of the report, which includes the following components:
   a. An introductory section, which briefly describes the problem at hand and indicates the methodology to be employed in solving the problem.
   b. A procedures section, which describes the step-by-step methodology used to study the process. The logic used in deciding what to do at each step should be spelled out carefully. You should state clearly how you arrived at any conclusions. Clearly indicate the data and reasoning that led you to select the special causes from the candidate list of 10 variables. For each special cause that is identified, you should have a physical explanation that describes why the special cause manifested itself on one/both of the charts. In this section it clearly will be necessary to refer to the control charts themselves, which should be contained in the appendix to the report. The results of the process capability study should also be reported. You will need to refer to the output of program PROCAP, which should also be placed in the appendix.
   c. A conclusions section, which summarizes special causes that were identified and the clues that led you to their identification. If the process is not capable, indicate what should be done next (i.e., where additional attention should be focused).
3. An appendix section, which contains all the control charts that were generated during the course of your investigation. Note that these control charts should clearly indicate (with circled points) all the statistical signals on the charts. The appendix should also contain the process capability analysis.

| Exercises |

10.1. Construct a cause-and-effect (fishbone) diagram for the gasket sheet roll mill process. Include all aspects of the process that were mentioned in the chapter as well as factors you think may also have existed that were affecting the final product thickness.
10.2. On page 299, the dilemma facing the gasket manufacturer is outlined. It is stated that the gasket manufacturer realized that "some positive steps would have to be taken to improve the quality [of the gaskets]."

What do you think he had in mind when he said that? Would his ideas differ from your concept of quality and quality improvement? Explain.

10.3. Describe the benefits associated with including the process operator in the team studying the problem with the gasket assembly. In general, what should the composition of such teams be like? Why?

10.4. C_p and C_{pk} values are routinely calculated as measures of the quality of a process. Explain how the widespread use of these indices may inhibit the promotion of the concepts and methods of statistical control of processes.

10.5. Explain why the selective assembly process used by the company for the gasket assembly process was not a good solution to their quality problems although it did allow them to meet the product specifications imposed on them by the customer.

10.6. What types of costs will the company incur to implement the proposed changes in the roll mill process? Comment on the significance of these costs compared with other alternatives that have been suggested for solving the problem.

10.7. What was the effect of metering the raw material in the roll mill process? Would you consider this change in the process as precipitating a major breakthrough or contributing more to the increased stability of the process? Explain.

10.8. How might regular/preventive maintenance on the roll mill machines affect the process as a whole? How would you justify this added cost? Explain.

10.9. At the end of the chapter it was discovered that the process has been recentered below the nominal in order to save on raw material costs. List the reasons why this is a bad practice. Can the savings in excess of $500,000 that they have realized justify the decision to make this change in the process centering? Why or why not? What important concept was missing here?

10.10. From reading about the incident described in Exercise 10.9, describe the notion of quality that might have existed here. How does this idea fit with the notions of Taguchi related to quality?

10.11. Address the issue raised in the question posed in the last sentence of the chapter (just prior to the Summary).

10.12. Discuss the relationship between process capability and process control. Initially, the staff engineers did a capability study to "prove" that the process needed radical change (i.e., new roll mill equipment). Discuss the dangers inherent in their approach.

10.13. It could be argued that the people in the plant were very lucky. They set out to reduce variation through statistical charting, but each time they did, the productivity of the process went up as well. Explain their good fortune.

10.14. Initially, the operator of the roll mill made adjustments in the gap thickness each time she felt, based on her measurements, that such was warranted. What impact did this approach to maintaining the product quality have on the product and the process?

10.15. Why did the initial design of experiments conducted on the roll speed and roll temperature fail to show any effects? What is this saying about the DOE/SPC relationship?

10.16. In this case study, major sources of variation were discovered and acted upon. Categorize these into special-cause and common-cause variation sources. Explain why each is categorized as it is.

10.17. Make a list of the key concepts of quality design and improvement stressed in this book and addressed specifically by this case study. Indicate how each concept was reinforced through this case study.

10.18. Where in the case study is the concept of increasing the signal-to-noise ratio to enhance process performance clearly demonstrated?

10.19. Who do you think played the most important role as a participant in this case study? Why? What is this telling us?

10.20. Sometimes, frustrated by the apparent complexity of implementing a solution to a problem, we tend to attempt to draw attention away from the problem or downplay its importance. Did that happen in this case study? Where?

CHAPTER

11

Some Control Chart Methods for Individual Measurements

11.1
Introduction

In previous chapters dealing with control charts for variable measurements, the notion of the rational subgroup was central to data collection, chart construction, and chart interpretation. The ability to observe the forces of common-cause variation alone is critical to establishing the model for $\overline{X}$ and R control charts and providing a sensitive indication of the presence of special causes when they arise.

In certain situations the notion of taking several measurements to form a rational sample of size greater than one simply does not make sense because only one measurement is available or meaningful at each time that samples are to be taken. For example, certain process characteristics, such as oven temperature, number or dollar value of accounts receivable, suspended air particulate count, and machine downtime, will not vary in close proximity in time or space during the time that such sampling normally occurs because either the characteristic is only determined at discrete and infrequent points in time (dollar amount of accounts receivable) or the medium being sampled is quite homogeneous in nature (oven temperature). The apparently different values from multiple observations of such processes at each sampling are results of reading and measurement system errors rather than reflections of true process variability. In these situations, special control charts may be used. Presented in this chapter are two types of control charts to be used for process control with individual measurements, namely X and R_m charts and exponentially weighted moving average (EWMA) charts.

11.2

X and R_m Control Charts

The X and moving range charts are perhaps the most simple types of control charts that can be used for the study of individual measurements. The construction of X and R_m charts is similar to that of $\overline{X}$ and R charts except that now X is the value of the individual measurement and R_m, the moving range, is the range of a group of n consecutive individual measurements combined artificially to form a subgroup of size n. Most often the moving range is calculated from two or three consecutive individual measurements. The moving ranges are calculated as shown below for the case of using three consecutive measurements to form the artificial samples of size $n = 3$.

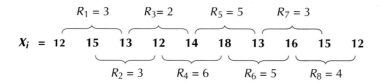

Since the moving range R_m is calculated primarily for the purpose of estimating the common-cause variability of the process, the artificial samples that are formed from successive measurements must be very small in size—taken as close together in time, space, and so on—to minimize the chance of including data in the subgroup that come from out-of-control conditions. If special causes occur within the subgroup, the estimate of the common-cause variability will be inflated and the sensitivity of the chart will be eroded. It is noted that X and R_m are not independent of each other and successive sample R_m values come from subgroups that are overlapping. Considerable caution and good judgment must be exercised in the use of these charts and in the selection of the artificial subgroup. This is particularly so since the nature of the circumstances that lead to the consideration of these charts is such that the time between the observation of individual measurements is often large. This means that the opportunity for special causes to contaminate the subgroup and produce inflated estimates of the level of common-cause variation is larger than it is for the $\overline{X}$ and R chart situation.

X and R_m charts are perhaps the most misused (often, abused) of all of the charts in common use today. The reason is very simple—they seem to suggest that few data need be taken to make a valid control chart. Often, individual measurements are taken and used in cases where natural subgroups can and should be used. The result is (1) a control chart that often has inflated within subgroup variability because too much time has transpired from one measurement to another, and (2) a control chart constructed with a small number of points and hence subject to large sampling errors. Such charts may be worse than using no charts at all.

Construction of X and R_m Control Charts: A Step-by-Step Procedure

Let us assume that k consecutive individual observations of a process have been obtained and that these observations are at least approximately normally distributed. In general, one may set up X and R_m control charts to analyze the process for any appropriate artificial sample size n. However, as noted above, the sample size should be small ($n = 2$ or 3) to minimize the opportunity for special causes to arise within the artificial samples and hence inflate the moving ranges. The following step-by-step procedure provides a basis for the construction of X and R_m charts.

1. Let the k consecutive individual measurements be denoted by $X_1, X_2, \ldots, X_k$, where k should be 25 or more (preferably 40 to 50). Calculate from the data a sample mean by

$$\overline{X} = \frac{\sum_{i=1}^{k} X_i}{k}.$$

   This sample mean will serve as the centerline for the X chart.

2. Compute sample moving ranges for an artificial sample of size n starting with R_{m1}, which is the difference between the largest and the smallest value in the first artificial sample ($X_1, X_2, \ldots, X_n$). Repeat this computation for each succeeding moving sample of artificial size n as follows:

   compute R_{mi} from $(X_i, \ldots, X_{i+n-1})$ for $i = 1, 2, \ldots, k - n + 1$.

3. Calculate an average sample moving range $\overline{R}_m$ from the $(k - n + 1)$ sample moving ranges,

$$\overline{R}_m = \frac{\sum_{i=1}^{k-n+1} R_{mi}}{k - n + 1}.$$

   This average moving range will serve as the centerline for the R_m chart.

4. Plot the X's and R_m's using the standard control chart conventions described previously.

5. Use $\overline{R}_m$ as an estimate of the true process range. Calculate the control limits for the R_m chart using $\overline{R}_m$ and the constants D_3 and D_4 in Table A.2 associated with the sample size n.

6. Divide $\overline{R}_m$ by d_2 (value corresponding to sample size n in Table A.2) for estimating the process standard deviation. Determine control limits for the X control chart using $\overline{X} \pm 3\hat{\sigma}_X$. It may be noted that the A_2 value is not used in calculating the control limits for the X chart. This is so because the sample size for the X chart is always 1.

In examining X and R_m control charts, it should be noted that since the artificial subgroups are overlapping, the interpretation of X and R_m charts are not independent. Sometimes only the X chart is analyzed for out-of-control signals, while the R_m chart is checked for proper construction of the X chart.

11.3

Case Study: White Millbase Dispersion Process

An automotive paint plant was studying the batch processing of white millbase used in the manufacture of topcoats. The basic process begins by charging a sandgrinder premix tank with resin and pigment along with other additives. The premix is agitated until a homogeneous slurry is obtained and then a small portion of the slurry is pumped through the sandgrinder. The grinder output is evaluated to check for fineness and gloss via an ash test that measures pigment content. A batch may require adjustments by adding pigment or resin to achieve acceptable gloss.

The plant was experiencing a situation where too many batches required two or three adjustments or required additional processing time to achieve an acceptable level of gloss. Each time an adjustment was made, about 4 hours would elapse before the resulting gloss could be evaluated. No quantitative method of calculating pigment and resin additions was available, so a rule of thumb was used. Sometimes after numerous adjustments, the premix tank was so full that further batch adjustments were impossible.

To identify potential areas of quality and productivity improvement, it was decided to begin studying the process using X and R_m control charts. The quality characteristic selected for study in this case was the weight per gallon of the batch. The rationale behind the selection of this quality characteristic was that if a batch was overcharged with pigment (high density), the weight of the batch tended to be large, while if it was overcharged with resin (low density), the weight tended to be small. The weight (pounds per gallon) of 30 consecutive batches from the millbase process were collected and are displayed in Table 11.1.

To determine the R_m values, an artificial sample size of $n = 2$ was selected. The R_m values are therefore given by

$$R_{mi} = |X_{i+1} - X_i|, \qquad i = 1, \ldots, 29,$$

that is,

$$R_{m1} = |X_2 - X_1|, \quad R_{m2} = |X_3 - X_2|, \quad \text{and so on.}$$

TABLE 11.1	Millbase Data for Samples/Batches 1 to 30 (pounds per gallon)		
Batch	X	R_m	Moving Range Calculations
1	14.56		
2	13.88	0.68	$= R_{m1} = 14.56 - 13.88 = 0.68$
3	13.98	0.10	$= R_{m2} = 13.98 - 13.88 = 0.10$
4	14.50	0.52	$= R_{m3} = 14.50 - 13.98 = 0.52$
5	14.22	0.28	
6	14.36	0.14	
7	14.46	0.10	
8	14.32	0.14	
9	14.27	0.05	
10	14.20	0.07	
11	14.35	0.15	
12	13.84	0.51	
13	14.43	0.59	
14	14.72	0.29	.
15	14.75	0.03	.
16	14.27	0.48	.
17	14.23	0.04	
18	14.60	0.37	
19	14.25	0.35	
20	14.12	0.13	
21	14.13	0.01	
22	14.47	0.34	
23	14.73	0.26	
24	14.02	0.71	
25	14.11	0.09	
26	14.48	0.37	
27	14.36	0.12	
28	13.70	0.66	
29	14.36	0.66	
30	14.66	0.30	$= R_{m29} = 14.66 - 14.36 = 0.30$

Table 11.1 displays the 29 calculated moving range, R_m, values that were determined from the data using an artificial sample size of $n = 2$.

Construction of the X and R_m control charts may be accomplished using the step-by-step procedure outlined previously. For the X chart, the centerline is

$$\overline{X} = \frac{\sum\limits_{i=1}^{30} X_i}{30} = \frac{429.33}{30}$$

$$= 14.311.$$

For the R_m chart, the centerline is

$$\overline{R}_m = \frac{\sum\limits_{i=1}^{29} R_{mi}}{29} = \frac{8.54}{29}$$

$$= 0.294.$$

In the calculation of averages above, $\overline{X}$ is an average of all 30 individual measurements, and $\overline{R}_m$ is an average of $30 - 1 = 29$ R_m values because there are only 29 moving ranges for a subgroup size of $n = 2$. If the artificial samples were of size $n = 3$, there would have been only $30 - 2 = 28$ moving ranges.

Once the $\overline{X}$ and $\overline{R}_m$ values are calculated as the centerline values of X and R_m charts, respectively, the calculation of upper and lower control limits for the R_m chart can proceed in the same way as for a regular R chart, using the artificial sample size n to determine D_3 and D_4 values. However, the upper and lower control limits for the X chart should always be based on a sample size of 1 [i.e., setting the limits at $\pm 3 \hat{\sigma}_X$ or $\pm 3(\overline{R}_m/d_2)$]. These calculations are shown below for the millbase data.

1. For the R_m chart: For $n = 2$, $D_3 = 0$ and $D_4 = 3.27$,

$$\text{UCL}_{R_m} = D_4 \overline{R}_m = (3.27)(0.294) = 0.963$$
$$\text{LCL}_{R_m} = D_3 \overline{R}_m = (0.0)(0.294) = 0.0.$$

2. For the X chart: To set the control limits for the X chart, we need an estimate of the standard deviation of X. This estimate is obtained from the relationship between the range and the standard deviation, $\hat{\sigma}_X = \overline{R}_m/d_2$. The constant d_2 is found from Table A.2 using $n = 2$. From the table, $d_2 = 1.128$ and therefore

$$\hat{\sigma}_X = \frac{0.294}{1.128} = 0.261.$$

The X-chart control limits are, therefore,

$$\text{UCL}_X = \overline{X} + 3\hat{\sigma}_X = 14.311 + (3)(0.261) = 15.094$$
$$\text{LCL}_X = \overline{X} - 3\hat{\sigma}_X = 14.311 - (3)(0.261) = 13.528.$$

The resulting X and R_m charts are shown in Fig. 11.1.

When the individual X values follow a normal distribution, the patterns on the X chart are analyzed in the same manner as for a Shewhart $\overline{X}$ chart. A sample X is circled as an out-of-control signal if it falls outside a control limit, if it is the endpoint of a sequence that violates any of the zone rules discussed in Chapter 6, or if it simply indicates a nonrandom sequence. Generally, the zone rule tests should be used with caution on X charts since, unlike $\overline{X}$ charts, the distribution of the individuals may have considerable departure from the normal distribution.

To check if the white millbase data can be assumed to have a normal distribution, a normal probability plot was constructed from the data in Table

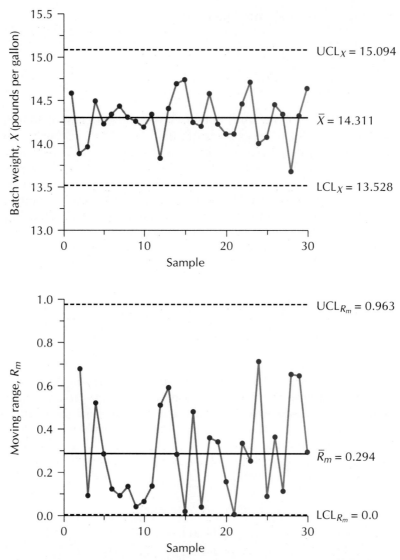

Figure 11.1 X and R_m Control Charts for the Millbase Process, Batches 1 to 30

11.1 and is shown in Fig. 11.2. This set of data, with the exception of one or two points, appears to be well described by the indicated line, and thus we may conclude that the data follow a normal distribution. Therefore, in examining the X chart of Fig. 11.1, in addition to looking for points outside the limits and nonrandom patterns, we may also apply the zone rules that are based on the normal distribution. It should be noted here, however, that if the X chart as initially viewed showed considerable lack of control, the normal probability plot of the individual measurements would probably show the data to be nonnor-

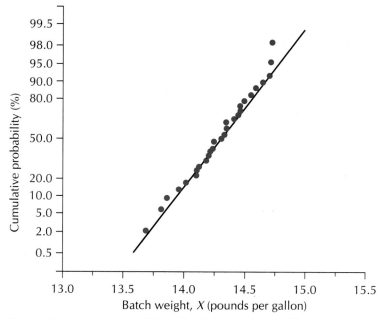

Figure 11.2 Normal Probability Plot for the White Millbase Data

mal, limiting additional chart interpretation via the zone rules. Since the charts of Fig. 11.1 are fairly well behaved, the normal plot is helpful here in learning the extent to which we could use the zone rules for fine-tuning the chart interpretation.

In examining the R_m chart in Fig. 11.1, it may be seen that there are no points beyond the limits. The chart appears to show no other statistical signals. An examination of the X chart shows no signals.

Change in the Millbase Dispersion Processing System

Since the process appeared to be in good control, the engineers involved in the millbase study decided to review the entire system and its operating procedures for possible areas of improvement (i.e., look for system faults). A cause-and-effect diagram was made to help focus on a few areas for more detailed study. An analysis of the production data indicated the need for a better method of determining the pigment concentration in the millbase. Several ash tests were performed to ascertain the balance between resin and pigment levels. These tests also verified the expected correlation between the weight per gallon and the ash test reading. Through curve fitting to the test results, a batch adjustment chart was developed using weight per gallon to determine the amount of resin or pigment to be added to bring the millbase to the target level of ash content. More important, a theoretical target batch weight was determined. The adjustment chart indicated that a weight of 13.77 pounds per gallon should produce

TABLE 11.2 Millbase Data for Samples/Batches 30 to 57 (pounds per gallon)			
Batch	X	R_m	Moving Range Calculations
30	14.66		*Copied from Table 11.1*
31	14.04	0.62	$= R_{m30} = 14.66 - 14.04 = 0.62$
32	13.94	0.10	$= R_{m31} = 14.04 - 13.94 = 0.10$
33	13.82	0.12	$= R_{m32} = 13.94 - 13.82 = 0.12$
34	14.11	0.29	$= R_{m33} = 14.11 - 13.82 = 0.29$
35	13.86	0.25	
36	13.62	0.24	
37	13.66	0.04	
38	13.85	0.19	
39	13.67	0.18	
40	13.80	0.13	
41	13.84	0.04	
42	13.98	0.14	
43	13.40	0.58	.
44	13.60	0.20	.
45	13.80	0.20	.
46	13.66	0.14	
47	13.93	0.27	
48	13.45	0.48	
49	13.90	0.45	
50	13.83	0.07	
51	13.64	0.19	
52	13.62	0.02	
53	13.97	0.35	
54	13.80	0.17	
55	13.70	0.10	
56	13.71	0.01	
57	13.67	0.04	$= R_{m56} = 13.71 - 13.67 = 0.04$

the target ash content. The weights for 27 new batches using the batch adjustment chart (along with batch 30 from Table 11.1) are shown in Table 11.2. It is noted here that while some time had actually passed between batch 30 and the batch identified as 31, for the purpose of charting the process the data are considered here as if they are consecutive.

To demonstrate the effect of using the batch adjustment chart on the process mean and variability, the X and R_m values were plotted on a continuation of the charts displayed in Fig. 11.1. These new charts that display the data for all 57 batches, with control limits based only on the first 30 batches, are shown in Fig. 11.3. As is readily apparent from an examination of Fig. 11.3, there are a number of statistical signals on the X and R_m charts for batches 31 to 57, indicating that both the process mean and variability have been reduced.

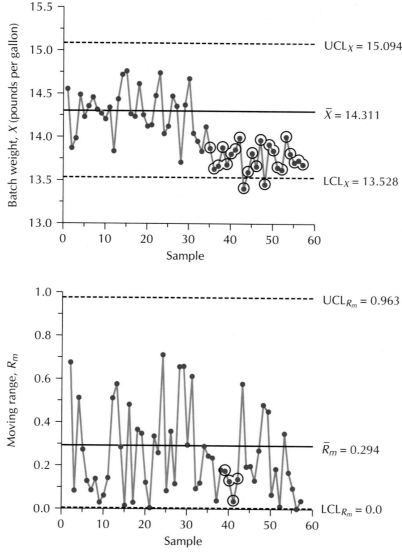

Figure 11.3 X and R_m Control Charts for the Millbase Process, Batches 1 to 57

The statistical signals indicate that the batch adjustment chart has not only reduced the average weight, but has also improved the batch-to-batch consistency. The use of the batch adjustment chart has in fact changed the system. Since the R_m chart does show out-of-control conditions past sample 31, the limits on the X chart are clearly wrong from that point. Actually, it was the X chart that showed out-of-control conditions first, so the resulting interpretation here is quite sound.

To develop control charts for the new millbase processing system, we may use the data for samples/batches 31 to 57 in Table 11.2.

1. For the R_m chart: The centerline is

$$\overline{R}_m = \frac{\sum\limits_{i=31}^{56} R_{mi}}{26} = \frac{4.99}{26}$$
$$= 0.192.$$

It may be noted that the first R_m value in Table 11.2 (R_{m30}) is not used in calculating the average moving range since it was calculated from two individuals that are now believed to come from different processes (X_{30} and X_{31}). The control limits for the R_m chart are ($D_3 = 0$ and $D_4 = 3.27$ for $n = 2$)

$$\text{UCL}_{R_m} = D_4\overline{R}_m = (3.27)(0.192) = 0.628$$
$$\text{LCL}_{R_m} = D_3\overline{R}_m = (0.0)(0.192) = 0.0.$$

2. For the X chart: The centerline is

$$\overline{X} = \frac{\sum\limits_{i=31}^{57} X_i}{27} = \frac{371.87}{27}$$
$$= 13.773.$$

The estimated process standard deviation is ($d_2 = 1.128$ for $n = 2$)

$$\hat{\sigma}_X = \frac{\overline{R}_m}{d_2} = \frac{0.192}{1.128}$$
$$= 0.170.$$

The control limits are therefore

$$\text{UCL}_X = \overline{X} + 3\hat{\sigma}_X = 13.773 + (3)(0.170) = 14.283$$
$$\text{LCL}_X = \overline{X} - 3\hat{\sigma}_X = 13.773 - (3)(0.170) = 13.263.$$

The X and R_m charts for the millbase process using the batch adjustment chart are shown in Fig. 11.4. In examining the X and R_m charts of Fig. 11.4 no statistical signals are evident. In addition to the analysis of trends and patterns on the X chart, one should also see whether the fluctuations of sample X values are becoming narrower or wider. These narrowing or widening trends may indicate a change in process variability. The result of examining the charts of Fig. 11.4 is to conclude that there is no reason to believe that the process is not in a state of statistical control, at least over the period of the 27 batches examined.

The charts in Figs. 11.3 and 11.4 demonstrate a definite improvement in terms of both the mean and variability level of the new millbase dispersion processing system. Some of the observable effects from this study are listed briefly below.

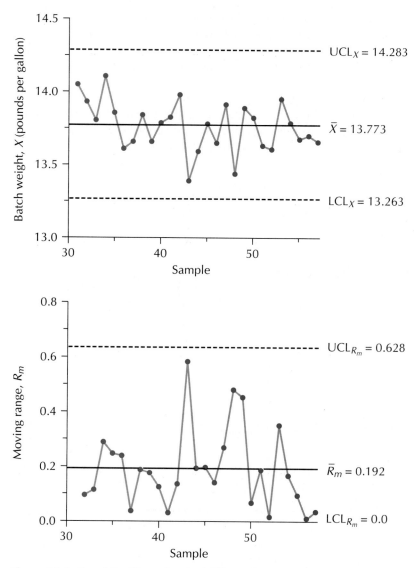

Figure 11.4 X and R_m Charts for the Millbase Data, Batches 31 to 57

- The average pigment level was lowered by over 2%, amounting to an annual savings of tens of thousands of dollars in material alone.
- The first gloss check met specifications on 18 consecutive batches, reducing the processing time by hundreds of hours. This served to prove that the theoretical batch weight was correct and it was important to maintain the process average at this level.
- In terms of average sample range, the average batch-to-batch fluctuation of pigment content was reduced by one-third.

- A solid basis was established from which to search for other improvements, including upgrading the silo weighing system and the shop floor measurement procedures.
- A similar approach could be applied to other large-volume products in the plant.

Following the implementation of the batch adjustment chart, the millbase dispersion process was continually monitored with X and R_m control charts. Figure 11.5 shows 29 new batches of data (samples 58 to 86) taken from the

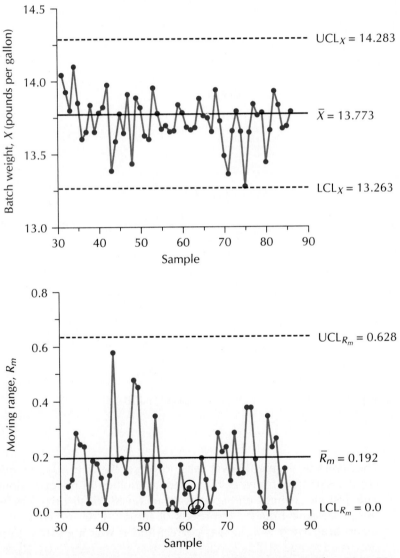

Figure 11.5 X and R_m Control Charts for Millbase Data, Batches 31 to 86

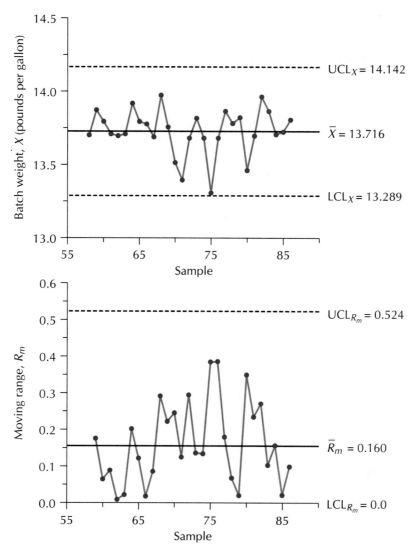

Figure 11.6 X and R_m Control Charts for the Millbase Dispersion Process, Batches 58 to 86

millbase dispersion process approximately 6 months after the study described above was performed. These data have been plotted relative to the limits and data of batches 31 to 57. An examination of Fig. 11.5 shows that in both the X and R_m charts there appears to be less variability in the newly collected data. This is most evident in the first 10 new batch weights that were obtained. In fact, there is a run of 10 points beneath the centerline on the moving range chart. Although no special cause(s) was found to explain this statistical signal, it does appear as if the process variability has been even further reduced, perhaps as a result of the added care and attention that are being given to the process.

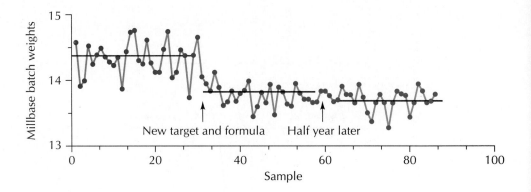

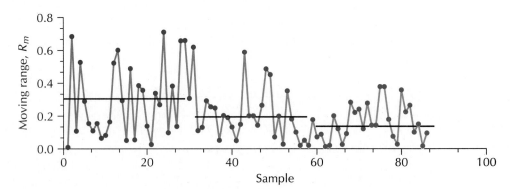

Figure 11.7 Improvement Progress of White Millbase Dispersion Process

A final set of X and R_m control charts were constructed, based only on the newly collected data. These charts are shown in Fig. 11.6. The plots of X and R_m of Fig. 11.7 represent the history of the process improvement during this study. Figure 11.8 shows a summary of the improvement in consistency and stability of the weight per gallon of the white millbase dispersion process over the period.

Summary

In summary, a control chart for individuals is less sensitive than $\overline{X}$ and R charts, but in those cases where rational subgroups are not possible, the X and R_m charts can serve to determine whether a process is statistically in control or not, and can lead to important quality and productivity improvement opportunities by proper and careful implementation. The millbase case study clearly demonstrates this fact.

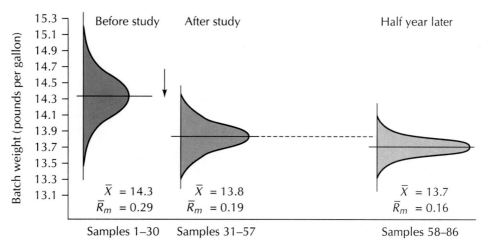

Figure 11.8 Consistency Improvement in the Millbase Dispersion Process over a Period of About Half a Year

---

11.4

Exponentially Weighted Moving Average Control Charts

As pointed out in the preceding section, X and R_m charts may not be very sensitive for detecting shifts in the process mean level or amount of process variation, particularly when the shift in the mean or in the variability is relatively small. Another weakness of X and R_m charts is the fact that these two charts are not independent of each other. In this section we introduce the use of exponentially weighted moving averages as the sample means and exponentially weighted moving deviations as the sample standard deviations for the tracking of process performance. The control charts for these moving statistics are known as *exponentially weighted moving average* (EWMA) control charts.

An exponentially weighted moving average is a moving average of past data where each data point is assigned a weight. These weights decrease in an exponentially decaying fashion from the present into the remote past. Thus the moving average tends to be a reflection of the more recent process performance because most of the weighting is allocated to the most recently collected data. The amount of decrease of the weights is an exponential function of the weighting factor, r, which can assume values between 0 and 1:

$$W_{t-i} = r(1 - r)^i,$$

where W_{t-i} is the weight associated with the observation X_{t-i}, X_t being the most recently observed data point. When a very small value of r is used, the

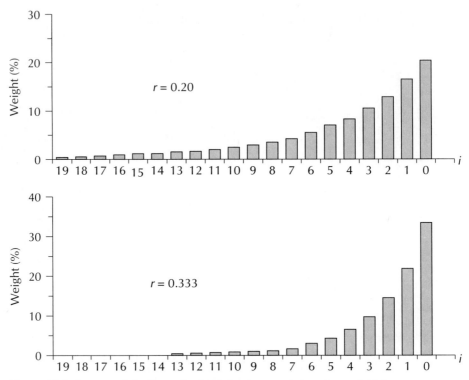

Figure 11.9 Exponentially Decaying Weights for $r = 0.2$ and $r = 0.333$

moving average at any time t carries with it a great amount of inertia from the past, hence it is relatively insensitive to short-lived changes in the process. For control chart applications where fast response to process shifts is desired, a relatively large weighting factor, say $r = 0.2$ to $r = 0.5$, may be used. To see how the weights gradually decrease as a function of r, Fig. 11.9 shows the decaying behavior for some commonly used values of r.

In selecting r, the following relationship between the weighting factor r and the sample size n for Shewhart control charts for sample means is often used;[1]

$$r = \frac{2}{n + 1}.$$

It is seen from the relationship above that if $r = 1$ is selected for the EWMA, all the weight is given to the current single observation, which is equivalent to the Shewhart control chart for individual measurements.

[1] A. L. Sweet, "Control Charts Using Coupled Exponentially Weighted Moving Averages," *Transactions of the IIE*, Vol. 18, No. 1, 1986, pp. 26–33.

Construction of EWMA Control Charts: A Step-by-Step Procedure

Let X_t represent the individual measurement of a process observed at time t, when the process mean is μ and the process standard deviation is σ_X. At each sampling time t, the value of X_t is assumed to vary according to a normal distribution with mean μ and standard deviation σ_X as long as the process is in a state of statistical control. If a shift in the mean occurs, the exponentially weighted moving averages will gradually (depending on r) move to the new mean of the process, while the exponentially weighted moving standard deviations will remain unchanged. On the other hand, if there is a shift in the process variability, the exponentially weighted moving standard deviations will gradually move to the new level, while the exponentially weighted moving averages still vary about the process mean μ. Similar to the behavior of $\overline{X}$ and R charts, when the process variability increases as a result of a special cause, there is an increasing probability of having out-of-control signals on the control chart for the mean as well.

The following step-by-step procedure leads to the construction of EWMA control charts:

1. Collect at least $k = 25$ samples of individual measurements $X_1, X_2, \ldots,$ X_k (preferably, k should be 40 to 50).
2. Calculate estimates of the process mean and standard deviation by

$$\overline{X} = \sum_{t=1}^{k} \frac{X_t}{k}, \qquad \text{sample mean}$$

$$s_X = \left[\sum_{t=1}^{k} \frac{(X_t - \overline{X})^2}{k - 1} \right]^{1/2} \qquad \text{sample standard deviation.}$$

3. Compute exponentially weighted moving averages, A_t, and exponentially weighted moving standard deviations, V_t, as follows:

$$A_t = rX_t + (1 - r)A_{t-1}, \qquad \text{where } A_0 = \overline{X}$$
$$V_t = rD_t + (1 - r)V_{t-1}, \qquad \text{where } D_t = |X_t - A_{t-1}| \quad \text{and} \quad V_0 = s_X.$$

4. Calculate control limits and centerlines for A_t's and V_t's by

$$\text{UCL}_A = \overline{X} + A^* s_X \qquad \text{Centerline} = \overline{X}$$
$$\text{LCL}_A = \overline{X} - A^* s_X$$
$$\text{UCL}_V = D_2^* s_X \qquad \text{Centerline} = s_X d_2^*,$$
$$\text{LCL}_V = D_1^* s_X$$

where the constants A^*, d_2^*, D_1^*, and D_2^* are as listed in Table 11.3.
5. Plot all k A_t values on the A chart and all k V_t values on the V chart and interpret the charts to determine if the process is in statistical control in terms of both the variability and the mean.

TABLE 11.3	Constants for EWMA Control Charts					
Weighting Factor, r	Equivalent Sample Size, n	For Means: A^*	For Standard Deviations			
			D_1^*	D_2^*	d_2^*	
0.050	39	0.480	0.514	1.102	0.808	
0.100	19	0.688	0.390	1.247	0.819	
0.200	9	1.000	0.197	1.486	0.841	
0.250	7	1.132	0.109	1.597	0.853	
0.286	6	1.225	0.048	1.676	0.862	
0.333	5	1.342	0.000	1.780	0.874	
0.400	4	1.500	0.000	1.930	0.892	
0.500	3	1.732	0.000	2.164	0.921	
0.667	2	2.121	0.000	2.596	0.977	
0.800	—	2.449	0.000	2.990	1.030	
0.900	—	2.714	0.000	3.321	1.076	
1.000	1	3.000				

Source: Adapted from A. L. Sweet, "Control Charts Using Coupled Exponentially Weighted Moving Averages," *Transactions of the IIE,* Vol. 18, No. 1, 1986, pp. 26–33.

Millbase Dispersion Process Revisited

Let us reconsider again the first 30 samples/batches collected from the millbase dispersion process to illustrate the application of EWMA control charts. First, we calculate the average and standard deviation of the data:

$$\overline{X} = 14.311 \text{ pounds/gallon,}$$

which is the previously calculated $\overline{X}$ for the first 30 batch averages, and

$$s_X = \left[\sum_{t=1}^{30} \frac{(X_t - 14.311)^2}{30 - 1} \right]^{1/2}$$

$$= 0.266 \text{ pound/gallon.}$$

After choosing a weighting factor of $r = 0.333$, we obtain the following constants from Table 11.3:

$$A^* = 1.342, \quad D_1^* = 0, \quad D_2^* = 1.780, \quad d_2^* = 0.874.$$

The centerlines and control limits are calculated below.

1. For the chart of moving averages, A_t:

$$CL_A = \overline{X}$$

$$= 14.311$$

$$UCL_A = \overline{X} + A^* s_X$$

$$= 14.311 + (1.342)(0.266)$$

$$= 14.668$$

$$\text{LCL}_A = \overline{X} - A^* s_X$$

$$= 14.311 - (1.342)(0.266)$$

$$= 13.954.$$

2. For the chart of moving deviations, V_i:

$$\text{CL}_V = d_2^* s_X$$

$$= (0.874)(0.266)$$

$$= 0.233$$

$$\text{UCL}_V = D_2^* s_X$$

$$= (1.780)(0.266)$$

$$= 0.474$$

$$\text{LCL}_V = D_1^* s_X$$

$$= (0.0)(0.266)$$

$$= 0.0.$$

To start the series of moving averages and moving deviations, we need some initial values for the A's and V's, which can be estimated by $A_0 = \overline{X} = 14.311$ and $V_0 = s_X = 0.266$, as mentioned previously. Now, using the first data point, $X_1 = 14.56$, the value of the first moving average is given by

$$A_1 = rX_1 + (1 - r)(A_0)$$

$$= (0.333)(14.56) + (1 - 0.333)(14.311)$$

$$= 14.394.$$

Since $D_1 = |X_1 - A_0| = |14.56 - 14.311| = 0.249$, the value of the first moving deviation is given by

$$V_1 = rD_1 + (1 - r)(V_0)$$

$$= (0.333)(0.249) + (1 - 0.333)(0.266)$$

$$= 0.261.$$

The moving averages and moving deviations for the first 30 samples of the millbase study are listed in Table 11.4.

Figure 11.10 presents the A chart and V chart for the data given in Table 11.4. Since the control limits of the A chart are constructed using the sample standard deviation of the original data, it is important first to check to determine if the V chart shows stability of the process variation. An examination of the V chart shows a run of 8 points beneath the centerline. It may be recalled that we did not see a run of 8 points in our study of the first 30 samples using the R_m chart. However, there is a basic difference between the Shewhart control charts and the EWMA control charts in their method of construction and interpretation. The Shewhart charts plot independent sample data points, each of which is interpreted according to the probability law of the sampling distribution of the statistic in question. Inferences about possible shifts in the process

TABLE 11.4	**Exponentially Weighted Moving Statistics for the Millbase Data (pounds per gallon)**			
Batch or Sample	Raw Data, X_t	Moving Average, A_t	Absolute Deviation, D_t	Moving Deviation, V_t
1	14.56	14.394	0.249	0.261
2	13.88	14.223	0.514	0.345
3	13.98	14.142	0.243	0.311
4	14.50	14.261	0.358	0.327
5	14.22	14.247	0.041	0.232
6	14.36	14.285	0.113	0.192
7	14.46	14.343	0.175	0.186
8	14.32	14.335	0.023	0.132
9	14.27	14.314	0.065	0.110
10	14.20	14.276	0.114	0.111
11	14.35	14.301	0.074	0.099
12	13.84	14.147	0.461	0.219
13	14.43	14.241	0.283	0.240
14	14.72	14.401	0.479	0.320
15	14.75	14.517	0.349	0.330
16	14.27	14.435	0.247	0.302
17	14.23	14.367	0.205	0.270
18	14.60	14.444	0.233	0.258
19	14.25	14.380	0.194	0.237
20	14.12	14.293	0.260	0.244
21	14.13	14.239	0.163	0.217
22	14.47	14.316	0.231	0.222
23	14.73	14.454	0.414	0.286
24	14.02	14.309	0.434	0.335
25	14.11	14.243	0.199	0.290
26	14.48	14.322	0.237	0.272
27	14.36	14.335	0.038	0.194
28	13.70	14.123	0.635	0.341
29	14.36	14.202	0.237	0.306
30	14.66	14.355	0.458	0.357

parameters are made indirectly through the distribution patterns of the data on a Shewhart chart. An exponentially weighted moving statistic is directly an estimate of the corresponding process parameter. Therefore, a series of EWMA data on the chart tends to move slowly to the new level following a shift in the process, or will vary about the centerline with small fluctuations when the process is in control. In general, it is rare to have data points fall outside the control limits on an EWMA chart, so the detection of shifts is based primarily on trend lines of the data. Additionally, it is very common to see runs above

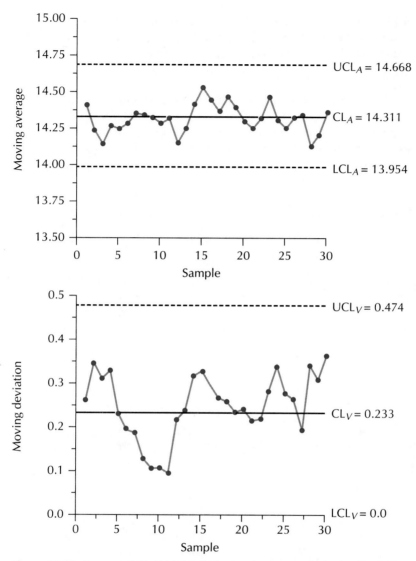

Figure 11.10 Exponentially Weighted Moving Average and Moving Deviations Control Charts for the White Millbase Data, Samples 1 to 30

and below the chart centerlines, because the use of the weighting procedure described previously actually introduces inertia/correlation into the moving averages and deviations.

A run of 8 points beneath the centerline on the V chart does not necessarily constitute a statistical signal in this case. Apart from the run, the moving deviations chart displays no obvious trendlike patterns. An examination of the

A chart shows no systematic tendencies, and thus we may conclude that the process mean is stable.

Fundamental Process Improvement

As has been described previously, at this point in the millbase dispersion process improvement case study, a batch adjustment procedure was developed and implemented. Twenty-seven additional samples were collected to judge the effectiveness of the new procedure. Table 11.5 lists the moving averages and deviations for these 27 additional batches. The moving averages and de-

TABLE 11.5	Exponentially Weighted Moving Statistics for the Millbase Data (Samples 31 to 57: $A_{30} = 14.355$, $V_{30} = 0.357$)			
Batch or Sample	Raw Data, X_t	Moving Average, A_t	Absolute Deviation, D_t	Moving Deviation, V_t
31	14.04	14.250	0.315	0.343
32	13.94	14.147	0.310	0.332
33	13.82	14.038	0.327	0.330
34	14.11	14.062	0.072	0.244
35	13.86	13.995	0.202	0.230
36	13.62	13.870	0.375	0.278
37	13.66	13.800	0.210	0.255
38	13.85	13.817	0.050	0.187
39	13.67	13.768	0.147	0.174
40	13.80	13.779	0.032	0.127
41	13.84	13.799	0.061	0.105
42	13.98	13.859	0.181	0.130
43	13.40	13.706	0.459	0.240
44	13.60	13.671	0.106	0.195
45	13.80	13.714	0.129	0.173
46	13.66	13.696	0.054	0.134
47	13.93	13.774	0.234	0.167
48	13.45	13.666	0.324	0.219
49	13.90	13.744	0.234	0.224
50	13.83	13.773	0.086	0.178
51	13.64	13.728	0.133	0.163
52	13.62	13.692	0.108	0.145
53	13.97	13.785	0.278	0.189
54	13.80	13.790	0.015	0.131
55	13.70	13.760	0.090	0.117
56	13.71	13.743	0.050	0.095
57	13.67	13.719	0.073	0.088

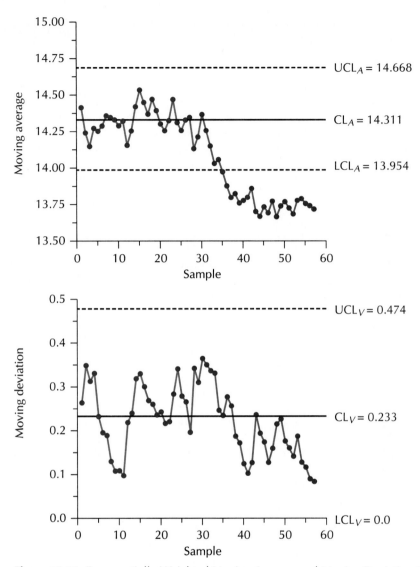

Figure 11.11 Exponentially Weighted Moving Average and Moving Deviation Control Charts for the White Millbase Data (Limits Based on Samples 1 to 30)

viations for these samples are shown in Fig. 11.11. The control limits displayed in Fig. 11.11 are based on the first 30 samples taken from the millbase process. As is evident from an examination of Fig. 11.11, the mean and variability levels are quite different for the samples generated with the new batch adjustment procedure. The effectiveness of the new procedure is clearly demonstrated on the control charts.

	Raw Data,	Moving Average,	Absolute Deviation,	Moving Deviation,
TABLE 11.6 Exponentially Weighted Moving Statistics for the Millbase Data (Samples 31 to 57)				
Batch or Sample	X_t	A_t	D_t	V_t
31	14.04	13.865	0.263	0.203
32	13.94	13.890	0.075	0.160
33	13.82	13.866	0.070	0.130
34	14.11	13.948	0.244	0.168
35	13.86	13.918	0.088	0.141
36	13.62	13.819	0.298	0.194
37	13.66	13.766	0.159	0.182
38	13.85	13.794	0.084	0.149
39	13.67	13.753	0.124	0.141
40	13.80	13.768	0.047	0.110
41	13.84	13.792	0.072	0.097
42	13.98	13.855	0.188	0.127
43	13.40	13.703	0.455	0.236
44	13.60	13.669	0.103	0.192
45	13.80	13.713	0.131	0.172
46	13.66	13.695	0.053	0.132
47	13.93	13.773	0.235	0.166
48	13.45	13.666	0.323	0.219
49	13.90	13.744	0.234	0.224
50	13.83	13.772	0.086	0.178
51	13.64	13.728	0.132	0.163
52	13.62	13.692	0.108	0.145
53	13.97	13.785	0.278	0.189
54	13.80	13.790	0.015	0.131
55	13.70	13.760	0.090	0.117
56	13.71	13.743	0.050	0.095
57	13.67	13.719	0.073	0.088

Since implementation of the new batch adjustment procedure effectively created a new processing system, the samples collected under the operation of the new system may be used to create a new set of A and V charts. Using the data from samples 31 to 57, centerlines and control limits for the new charts may be calculated. First, we calculate the average and standard deviation of the data using samples 31 to 57:

$$\overline{X} = \sum_{t=31}^{57} X_t/27 = 13.777 \text{ pounds/gallon},$$

$$s_X = \left[\sum_{t=31}^{57} \frac{(X_t - 13.777)^2}{27 - 1} \right]^{1/2} = 0.172 \text{ pound/gallon}.$$

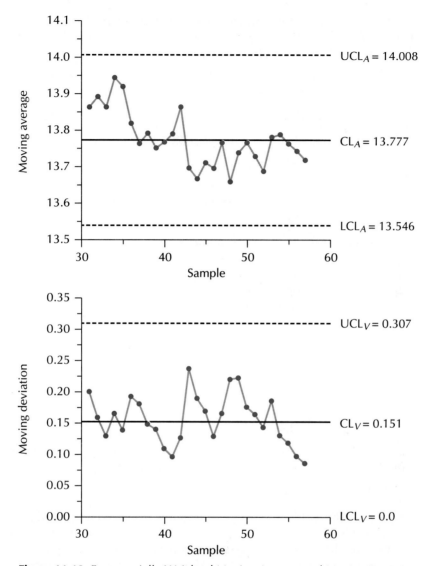

Figure 11.12 Exponentially Weighted Moving Average and Moving Deviation Control Charts for the White Millbase Data (Limits Based on Samples 31 to 57)

The centerlines and control limits are calculated below.

1. For the chart of moving averages A_t:

$$\text{CL}_A \;=\; \overline{X}$$
$$= 13.777$$
$$\text{UCL}_A = \overline{X} + A^* s_X$$
$$= 13.777 + (1.342)(0.172)$$
$$= 14.008$$
$$\text{LCL}_A = \overline{X} - A^* s_X$$
$$= 13.777 - (1.342)(0.172)$$
$$= 13.546.$$

2. For the chart of moving deviations, V_t:

$$\text{CL}_V \;=\; d_2^* s_X$$
$$= (0.874)(0.172)$$
$$= 0.151$$
$$\text{UCL}_V = D_2^* s_X$$
$$= (1.780)(0.172)$$
$$= 0.307$$
$$\text{LCL}_V = D_1^* s_X$$
$$= (0.0)(0.172)$$
$$= 0.0.$$

The moving averages and deviations for samples 31 to 57 of the millbase study are listed in Table 11.6 (see p. 370). To start the series of moving averages and moving deviations, we need some initial values for the A's and V's, which can be estimated by $A_{30} = \overline{X} = 13.777$ and $V_{30} = s_X = 0.172$, as mentioned previously. The control charts associated with the data in Table 11.6 are displayed in Fig. 11.12. The V chart appears to show no statistical signals, suggesting that the process variability is constant. The moving averages on the A chart, however, seems to indicate a downward trend in the mean from the fifth batch on, although there are no out-of-control points. This slowly developing trend was not detected from the X and R_m charts shown in Fig. 11.4.

Exercises

11.1. For the data given in the table, calculate the moving ranges, R_m, based on a sample size of $n = 2$, $n = 3$, and $n = 4$.

Sample	X_i	Sample	X_i
1	11	6	9
2	13	7	8
3	9	8	12
4	14	9	14
5	10	10	10

11.2. Describe the circumstances under which it is difficult or does not make sense to use $\overline{X}$ and R charts to study a process, and therefore an investigator must use control charts for individual measurements. Give several examples where control charts for individual measurements should be used instead of $\overline{X}$ and R charts. What concerns or cautions should be exercised in using X and R_m charts as compared with $\overline{X}$ and R charts?

11.3. Use the data from Exercise 6.16 to construct X and R_m charts by using only the first measurement (X_1) from each subgroup.
   a. Compute all the sample moving ranges using $n = 3$.
   b. Calculate $\overline{X}$ and $\overline{R}_m$.
   c. Calculate an estimate for the process standard deviation.
   d. Calculate control limits for the X and R_m charts.
   e. Construct X and R_m charts for these data and plot the data.
   f. Evaluate these charts and comment on their appearance. Identify all statistical signals that indicate out-of-control conditions.

11.4. Charting can be continued from Exercise 11.3 using data from Exercise 6.17, again using only the first measurement (X_1) in each subgroup listed there.
   a. Compute all the sample moving ranges using $n = 3$.
   b. Continue charting using the control limits and centerlines from the first 35 points. Comment on the appearance of the charts.
   c. Construct new X and R_m charts based only on the new data. Comment on the appearance of these charts and contrast them to the charts made in part (b). Show all calculations, and explain the reasoning behind the comments you make.

11.5. Consider the third measurement (X_3) from each subgroup listed in Exercise 6.16. Compute all the moving ranges, centerlines, and control limits, and construct X and R_m control charts for the following cases:
   a. Moving ranges based on $n = 2$.
   b. Moving ranges based on $n = 3$.
   c. Moving ranges based on $n = 4$.
   d. Compare the three sets of control charts constructed above. Comment on the differences between the charts. Be specific.

11.6. In this chapter it is stated that the individual observations used to make X and R_m charts need to follow a normal distribution if zone rules are to be used in analyzing the X chart. Explain why this is so.

11.7. A chemical process that has been operating for some time has produced the data listed in the table, which are pH measurements of a chemical reactant taken once every hour.

Sample	X_i	Sample	X_i
1	4.7	23	5.1
2	4.9	24	5.7
3	4.5	25	4.8
4	5.4	26	4.7
5	4.7	27	4.9
6	4.7	28	4.7
7	4.8	29	5.2
8	4.8	30	4.9
9	4.7	31	5.1
10	5.0	32	5.6
11	5.3	33	4.9
12	4.8	34	5.2
13	5.4	35	5.0
14	5.6	36	5.0
15	4.8	37	5.0
16	5.6	38	4.9
17	4.7	39	5.1
18	6.0	40	4.9
19	4.4	41	5.1
20	4.3	42	5.0
21	5.2	43	4.9
22	4.8	44	4.9

a. Construct a dot diagram for the individual measurements. Comment on its general appearance and what this might mean.
b. Compute all the sample moving ranges using $n = 2$.
c. Calculate $\overline{X}$ and $\overline{R}_m$.
d. Calculate an estimate for the process standard deviation.
e. Calculate control limits for the X and R_m charts.
f. Construct X and R_m charts for these data and plot all points.
g. Evaluate these charts and comment on their appearance. Identify all statistical signals that indicate out-of-control conditions.

11.8. Forty consecutive samples of size $n = 1$ were collected from an assembly process during an 8-hour shift. This process had not been studied previously by charting, and some problems with poor quality had been noted. The process measurements, the time (in minutes) required to assemble a product using a bench fixture, are shown in the table.

Sample	X_i	Sample	X_i
1	2.09	21	3.42
2	3.81	22	3.16
3	1.93	23	3.21
4	1.49	24	3.59

Sample	X_i	Sample	X_i
5	3.29	25	3.05
6	4.25	26	1.97
7	2.69	27	3.58
8	2.88	28	3.16
9	3.54	29	3.62
10	3.56	30	1.80
11	3.56	31	2.48
12	2.34	32	3.67
13	4.35	33	2.15
14	1.48	34	3.14
15	2.38	35	3.27
16	2.79	36	1.86
17	1.29	37	1.71
18	3.35	38	4.10
19	5.29	39	2.28
20	3.20	40	1.99

a. Is it appropriate to use X and R_m charts to study this process? Why or why not?

b. If you were studying this process, what type of sampling scheme would you develop? Be specific. Give the rationale for your choice of sample size, sampling frequency, number of samples, and so on.

11.9. X and R_m charts are being constructed for a chemical distillation process. Samples of size $n = 2$ have been used to obtain the moving range values. Based on 50 samples, $\overline{X}$ and $\overline{R}_m$ have been calculated to be 10.0 and 3.0, respectively. What are the control limits for the X and R_m charts?

11.10. The temperature (in degrees Celsius, °C) within a refrigeration unit has been recorded every 2 hours. These temperature data are shown in the table.

Sample	X_i	Sample	X_i
1	0.6	16	−1.3
2	−2.0	17	−4.7
3	−3.3	18	0.0
4	−9.6	19	−4.2
5	−5.2	20	−2.1
6	−5.3	21	−1.8
7	−0.6	22	1.6
8	−4.1	23	−4.7
9	−9.4	24	−2.3
10	−0.4	25	3.9
11	−4.7	26	0.4
12	0.4	27	0.1
13	−1.6	28	−0.2
14	−5.3	29	−0.8
15	−0.4	30	−4.3

a. Construct a normal probability plot using the individual measurements and comment on its appearance.

b. Using $n = 2$, calculate the moving ranges. Calculate $\overline{X}$, $\overline{R}_m$, and an estimate for the process standard deviation.

c. Calculate control limits for the X and R_m charts, and construct X and R_m charts for these data using all 30 observations.

d. Evaluate these charts and comment on their appearance. Identify all statistical signals that indicate out-of-control conditions.

11.11. Following the analysis of the data described in Exercise 11.10, an improvement action was made to the temperature control device of the refrigeration unit. To assess the effect of the improvement action, more data were collected and are shown in the table.

Sample	X_i	Sample	X_i
1	0.8	16	-3.1
2	-4.4	17	0.0
3	-1.8	18	-1.7
4	-1.3	19	-1.2
5	-3.4	20	-1.0
6	1.0	21	-3.3
7	0.7	22	-2.6
8	0.0	23	-2.2
9	-0.6	24	-4.5
10	-1.3	25	-1.4
11	-0.3	26	-2.1
12	-2.7	27	-1.7
13	-7.2	28	-0.9
14	-5.4	29	-3.7
15	-2.0	30	-2.6

a. Calculate moving ranges for each new measurement (use $n = 2$).

b. Continue charting using the control limits and centerlines for the first 30 measurements. Comment on the appearance of the charts.

c. Construct new X and R_m charts based on the new data only. Comment on the appearance of these charts and contrast them to the charts made in part (b). Show all calculations, and explain the reasoning behind the comments you make.

11.12. It is desired to study the chemical concentration of an acid bath via X and R_m control charts. Thirty individual concentration values have been collected and moving range values have been calculated based on a sample size of 2. An examination of the data reveals that

$$\Sigma X_i = 596.1 \quad \text{and} \quad \Sigma R_{mi} = 85.8.$$

a. Calculate the centerlines and control limits for the X and R_m charts.

b. Ten additional concentration levels have been collected as shown in the table. Calculate the missing R_m values.

Sample	31	32	33	34	35	36	37	38	39	40
X_i	21	24	23	26	24	26	18	21	19	20
R_{mi}	1.8									

The normal probability plot shown here has been prepared based on the first 30 concentration levels. X and R_m control charts were constructed for the first 30 points and showed no extreme points or runs.

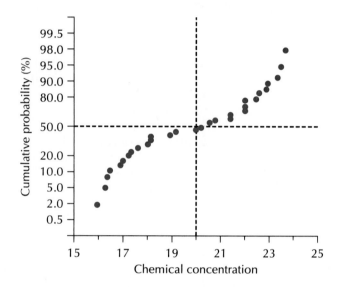

c. Comment on the appearance of the normal probability plot. Will the appearance of the plot in this case have any impact on the further interpretation of the X and R_m control charts? Why or why not?

11.13. The moisture content of corn kept in a bulk storage facility is under study. Daily measurements of the moisture content have been recorded for 40 consecutive days and are shown in the table.

Sample	X (% moisture)	Sample	X (% moisture)
1	14.3	21	17.0
2	18.7	22	15.2
3	15.6	23	14.4
4	14.9	24	15.4
5	10.0	25	17.7
6	16.2	26	14.2
7	17.5	27	17.6
8	16.4	28	13.4
9	13.8	29	14.5
10	13.3	30	14.8

(continued)

Sample	X (% moisture)	Sample	X (% moisture)
11	13.5	31	12.2
12	19.1	32	16.5
13	16.8	33	10.4
14	14.7	34	14.3
15	15.3	35	12.9
16	15.6	36	12.8
17	14.1	37	16.0
18	14.6	38	12.6
19	12.8	39	14.7
20	16.5	40	10.9

Compute all the moving ranges, centerlines, and control limits, and construct X and R_m control charts for the following cases:

a. Moving ranges based on $n = 2$.

b. Moving ranges based on $n = 3$.

c. Moving ranges based on $n = 4$.

d. Compare the three sets of control charts constructed above. Comment on the differences between the charts. Be specific.

11.14. Use the data shown in the table to construct EWMA charts for the process that generated it. Use $r = 0.200$.

Sample	X_i	Sample	X_i
1	1.5	21	1.6
2	0.5	22	1.7
3	0.8	23	1.3
4	-0.8	24	1.2
5	1.1	25	-0.1
6	0.7	26	-0.7
7	1.3	27	-0.5
8	2.0	28	-0.6
9	-0.5	29	0.1
10	1.6	30	-0.9
11	1.0	31	-2.8
12	0.4	32	0.2
13	0.7	33	-1.4
14	0.8	34	0.8
15	1.6	35	-0.4
16	0.7	36	-1.8
17	1.2	37	-0.6
18	0.1	38	-2.1
19	0.3	39	-0.6
20	1.6	40	-1.5

a. Calculate estimates of the process mean and standard deviation.

b. Calculate A_t and V_t for each observation.

c. Calculate control limits and centerlines for both charts.

d. Construct an A chart and a V chart for the process and plot all the points. Interpret the charts and make comments on any patterns you see. What conclusions can you draw about the behavior of the process? Be thorough in your explanation.

11.15. Consider the data in Exercise 11.14.

a. Construct X and R_m charts using $n = 2$. Interpret the charts.

b. Construct EWMA charts using $r = 0.2$. Interpret the charts.

c. Comment on the relative appearance of the two sets of charts constructed in parts (a) and (b). Are there significant differences? If so, how can these be accounted for or explained?

11.16. The data shown in the table were collected from an automated chemical process. The measurements represent the deviation of the concentration of a certain chemical (%) from the target value for that factor.

Sample	X_i	Sample	X_i
1	-0.01	21	0.75
2	-0.23	22	-1.07
3	-0.80	23	-0.71
4	-0.17	24	0.23
5	-1.00	25	1.06
6	-1.20	26	-2.65
7	0.33	27	-1.17
8	0.89	28	-2.61
9	-1.18	29	-1.85
10	-1.64	30	-1.68
11	2.07	31	-1.94
12	-0.77	32	-1.83
13	0.83	33	-2.50
14	0.82	34	-0.56
15	0.03	35	-2.58
16	-0.90	36	-2.46
17	1.57	37	-1.84
18	-0.16	38	-2.11
19	-0.52	39	-2.14
20	-0.23	40	-1.03

a. Calculate estimates of the process mean and standard deviation.

b. Calculate A_t and V_t for each observation using r values of 0.2, 0.333, and 0.4.

c. Construct A and V charts for r values of 0.2, 0.333, and 0.4. Interpret all three sets of charts.

d. Considering the three sets of charts developed in part (c), what effect(s) does changing r appear to have on the charts? What factors might influence the selection of a value for r?

11.17. A wire drawing process with large variability is being studied in hopes of discovering opportunities for improvement. The data shown in the table are the measured diameters of the wire (in 0.001 inch), and were collected every 5 minutes for a period of approximately 3.5 hours.

Sample	X_i	Sample	X_i
1	110.5	21	111.9
2	108.4	22	109.0
3	112.0	23	114.1
4	109.5	24	109.9
5	112.0	25	114.2
6	109.4	26	110.4
7	110.2	27	110.3
8	109.9	28	110.4
9	112.1	29	113.3
10	107.7	30	109.6
11	109.3	31	110.6
12	112.6	32	110.0
13	110.7	33	112.2
14	109.3	34	111.5
15	106.3	35	109.5
16	107.4	36	109.1
17	109.3	37	109.5
18	108.7	38	110.1
19	109.2	39	111.0
20	108.1	40	109.1

a. Calculate estimates of the process mean and standard deviation.
b. Calculate A_t and V_t for each observation using $r = 0.333$.
c. Calculate control limits and centerlines for both charts.
d. Construct an A chart and a V chart for the process and plot all points. Interpret the charts and make comments on any patterns you see. What conclusions can you draw about the behavior of the process? Be thorough in your explanation.

11.18. A change was made to the wire drawing process described in Exercise 11.17 in an attempt to improve its behavior, and the data collection activity was continued. The data are shown in the table.

Sample	X_i	Sample	X_i
1	110.6	16	108.3
2	110.0	17	109.8
3	112.1	18	110.4
4	111.5	19	110.6
5	109.6	20	109.0
6	109.2	21	109.4
7	109.5	22	109.1
8	110.0	23	111.2

Sample	X_i	Sample	X_i
9	110.9	24	110.6
10	109.2	25	108.6
11	110.0	26	110.8
12	110.8	27	108.4
13	111.4	28	108.8
14	110.8	29	110.1
15	109.6	30	111.4

a. Using the control limits and centerlines developed in Exercise 11.17, plot the newly collected data.
b. Comment on the behavior of the new data. Interpret the extended charts and make appropriate comments.
c. Construct charts based only on the new data. Comment on the appearance of these new charts and compare them to the charts you made in part (b).

11.19. The temperature of a chemical reactant has been determined to be a crucial factor in obtaining satisfactory yield from a chemical process. The nominal value for this temperature is supposed to be 150°C. The temperature of the reactant is recorded every hour for 40 consecutive hours. These temperature measurements, recorded as the deviation from the desired value of 150°C, are shown in the table.

Sample	X_i	Sample	X_i
1	−4.53	21	−4.87
2	7.52	22	5.12
3	−3.05	23	−2.73
4	3.31	24	−2.11
5	13.27	25	−2.29
6	−0.68	26	5.45
7	8.13	27	0.92
8	−0.35	28	4.26
9	5.74	29	0.03
10	3.92	30	5.14
11	2.43	31	−3.91
12	−0.91	32	3.68
13	−0.67	33	11.49
14	5.76	34	5.45
15	2.44	35	3.93
16	−0.96	36	10.80
17	8.72	37	−1.28
18	13.29	38	10.52
19	2.41	39	0.91
20	−1.82	40	−1.84

a. Construct X and R_m charts using $n = 3$. Interpret the charts.
b. Construct EWMA charts using $r = 0.5$. Interpret the charts.
c. Comment on the relative appearance of the two sets of charts constructed in parts (a) and (b). Are there significant differences? If so, how can these be accounted for or explained?

11.20. A seamless tubing extrusion process that is producing a large number of "wall breakouts" is being scrutinized in hopes of improving its performance. The tubing is extruded and cut into 10-foot lengths. One wall thickness measurement is made for each of 40 consecutive lengths, with the data shown in the table. It is felt that the process may be prone to exhibit abrupt but short-lived shifts in the mean level.

Sample	X_i	Sample	X_i
1	11.7	21	10.7
2	11.1	22	12.3
3	11.4	23	9.9
4	10.7	24	10.7
5	9.4	25	10.8
6	11.0	26	10.6
7	10.2	27	11.4
8	11.0	28	9.9
9	10.9	29	11.0
10	11.4	30	11.7
11	11.9	31	10.8
12	10.9	32	11.4
13	11.7	33	12.0
14	12.4	34	9.8
15	10.3	35	11.4
16	11.4	36	10.5
17	12.7	37	10.4
18	11.6	38	11.2
19	10.4	39	10.4
20	11.3	40	11.2

a. Calculate estimates of the process mean and standard deviation.
b. Select an appropriate value for r based on the nature of the process. Explain the rationale for your selection. Calculate A_t and V_t for each observation.
c. Calculate control limits and centerlines for both charts.
d. Construct an A chart and a V chart for the process and plot all points. Interpret the charts and make comments on any patterns you see. What conclusions can you draw about the behavior of the process? Be thorough in your explanation.

11.21. After a change is made to the extrusion process from Exercise 11.20, more data are collected, as shown in the table.

Sample	X_i	Sample	X_i
1	13.2	16	12.6
2	13.0	17	13.6
3	12.3	18	14.9
4	14.0	19	14.7
5	13.7	20	12.0
6	13.5	21	12.5
7	12.2	22	12.7
8	13.8	23	12.7
9	13.9	24	13.1
10	13.4	25	12.5
11	12.1	26	13.7
12	13.7	27	14.3
13	12.8	28	12.4
14	13.0	29	14.4
15	12.9	30	14.1

a. Using the control limits and centerlines developed in Exercise 11.20, plot the newly collected data.

b. Comment on the behavior of the new data. Interpret the extended charts and make appropriate comments.

c. Construct charts based only on the new data. Comment on the appearance of these new charts and compare them to the charts you made in part (b).

d. Does it appear that the change made to the process will affect the "wall breakout" problem? Why or why not? Be thorough in your explanation.

11.22. Discuss the similarities and differences between Shewhart-type control charts for individual measurements and EWMA charts for individual measurements in terms of:

a. The assumption about the relationship among individual points on the charts.

b. The general manner(s) in which special causes due to shifts in the mean show up on the charts.

c. The relative appearance of X and R_m and EWMA charts for the same data where frequent, short-lived shifts in the mean are occurring.

C H A P T E R

12

Cumulative-Sum Control Charts

12.1

Introduction

In Chapter 11 we saw that a control chart could be developed that made use of the recent past history of the process in projecting the process status with respect to control or stability. This chart, the exponentially weighted moving average chart, plotted an average statistic where the last few observations of the process were weighted from the present, backward in time, in exponentially decreasing importance. A similar type of control chart that can be effectively used for process control purposes is the *cusum chart*, where "cusum" stands for "cumulative sum." The statistic to be summed may be individual measurements, sample means, sample ranges, sample fraction defectives, sample number of defects, and so on.

The use of a series of cumulative sums of sample data for on-line process control was developed by E. S. Page[1] and G. A. Barnard[2]. It is based on a method using the sequential likelihood ratio test (see A. Wald[3]), where, as each new sample becomes available, a test is conducted to determine whether or not the process mean deviates by at least a specified amount from a target value. In 1962, N. L. Johnson and F. C. Leone[4] generalized the cusum charts to Poisson

---

[1] E. S. Page, "Continuous Inspection Schemes," *Biometrika*, Vol. 41, 1954, pp. 100–114.
[2] G. A. Barnard, "Control Charts and Stochastic Processes," *Journal of the Royal Statistical Society, Ser. B*, Vol. 21, 1959, p. 239.
[3] A. Wald, *Sequential Analysis*, Wiley, New York, 1947.
[4] N. L. Johnson and F. C. Leone, "Cumulative Sum Control Charts: Mathematical Principles Applied to Their Construction and Use," Parts I, II, and III, *Industrial Quality Control*, Vol. 18, No. 12, Vol. 19, No. 1, and Vol. 19, No. 2, 1962.

and binomial random variables as well as for testing for shifts in variance and range.

Common to all cusum control charts that are developed according to a sequential likelihood ratio is the idea of hypothesis testing between two alternative quality levels, one acceptable and one rejectable. These cusum charts are referred to in this chapter as traditional cusum control charts. To construct the appropriate charts, both of the average quality levels must be specified. Such a prerequisite raises a number of philosophical questions and sometimes causes difficulty in their applications. Due to differences in the basis of chart construction, these traditional cusum charts are different from the Shewhart charts in terms of their interpretation and, more important, their appropriate areas of application. We discuss them in more detail in this chapter.

Over the last decade, there has been a strong movement away from the concepts and methods of acceptance sampling in proportion to the growth in the philosophy and methods of continual, never-ending improvement. Some of the conceptual as well as practical concerns associated with the use of traditional cusum control charts are discussed in Section 12.3. In Section 12.4 we discuss the general interpretation of traditional cusum control charts.

A cumulative statistic is generally more efficient than single-sample data points in control chart applications. This is generally true for cusum control charts for a variety of statistics, including individual measurements, sample means, sample variances, ranges, and statistics using attribute data. Some cusum control charts that are more Shewhart-like, rather than based on sequential probability tests, are presented with examples in Section 12.5. Section 12.6 presents some graphical methods to aid in the use and interpretation of Shewhart-like cusum charts.

12.2

Traditional Cusum Control Charts

In this section we review the construction and interpretation of one-sided and two-sided cusum control charts traditionally used for the control of a process mean under the assumption that the data follow a normal distribution.

One-Sided and Two-Sided Cusum Control Schemes

The traditional cusum control chart is developed according to the concept of hypothesis testing using the sequential likelihood ratio.[5] The null and alternative hypotheses are stated as follows:

$$H_0: \quad \mu = \mu_0 \tag{12.1}$$
$$H_1: \quad \mu = \mu_1 \quad (\mu_1 > \mu_0),$$

---

[5] A. Wald, op. cit.

where μ is the mean of X and both type I and type II errors, with probabilities α, and β, respectively, are specified beforehand. A decision cannot be reached to accept either H_0 or H_1 as long as the following is true:

$$\frac{\beta}{1-\alpha} < \text{likelihood ratio for the data} < \frac{1-\beta}{\alpha}. \tag{12.2}$$

Equation (12.2) can be used to construct a one-sided control chart for detecting upward shifts in the mean. If we have a time-ordered sequence of independent sample observations, $X_1, X_2 \ldots , X_t$, from a normal population with a variance σ_X^2 but an uncertain mean, it can be shown that the ratio in Eq. (12.2) will take the form

$$\frac{\sigma_X^2}{\Delta_1} \ln \frac{\beta}{1-\alpha} < S_t < \frac{\sigma_X^2}{\Delta_1} \ln \frac{1-\beta}{\alpha}, \tag{12.3}$$

where $\Delta_1 = \mu_1 - \mu_0$ is the difference between the two hypothesized means or the shift in the process mean from μ_0, and

$$S_t = \sum_{i=1}^{t} \left\{ X_i - \left(\mu_0 + \frac{\Delta_1}{2} \right) \right\} \tag{12.4}$$

is the cumulative sum of sample deviations of the data from the average of the two means. A plot of a sequence of S_t's may be examined on a chart with the upper and lower acceptance limits determined from Eq. (12.3) as shown in Fig. 12.1.

According to the procedure of sequential sampling, a decision would have been reached to stop sampling and accept $\mu = \mu_0$ on the sixth sample when the cumulative sum of deviations falls below the lower acceptance limit and accept $\mu = \mu_1$ upon the eighteenth sample when the sum exceeds the upper acceptance limit as shown in Fig. 12.1. In one-sided control chart applications, the upper acceptance limit becomes the upper control limit, while the lower acceptance limit has no practical meaning. The illustration in Fig. 12.1 would be interpreted as having one out-of-control signal on the eighteenth sample, while the sampling would continue without taking any action for the sixth sample. Similarly, a one-sided control chart can be constructed for detecting only downward shifts in the mean.

When it is desirable to detect a process mean shift from the target μ_0 in either direction, one could use a pair of one-sided cusum charts to monitor the process for upward and downward shifts separately. It is possible, however, to maintain one control chart of the cusums to detect both types of the shifts. Let the two off-target mean values of interest be denoted by $\mu_1 > \mu_0$ and $\mu_2 < \mu_0$ with type II error risks β_1 and β_2 and a type I error risks of 2α. By combining two one-sided decision criteria of Eqs. (12.2) and (12.3), it can be shown that the upper and lower control limits for the centered cumulative sum, $\Sigma(X_i - \mu_0)$, are

$$\text{UCL} = \frac{\sigma_X^2}{\Delta_1} \ln \frac{1-\beta_1}{\alpha} + t\left(\frac{\Delta_1}{2}\right) \tag{12.5}$$

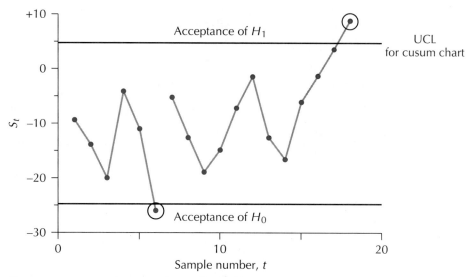

Figure 12.1 Cusum Sequential Sampling Scheme

$$LCL = \frac{\sigma_{\bar{X}}^2}{\Delta_2} \ln \frac{1 - \beta_2}{\alpha} + t\left(\frac{\Delta_2}{2}\right), \qquad (12.6)$$

where $\Delta_1 = \mu_1 - \mu_0$ and $\Delta_2 = \mu_2 - \mu_0$. Since Eqs. (12.5) and (12.6) for the control limits are functions of the sampling number t, the upper and lower control limits are linear trend lines for the cusum chart, instead of horizontal lines. The resulting cusum control chart scheme is shown in Fig. 12.2(a). In general, for any amount of upward shift and downward shift in the mean, Δ_1 and Δ_2, a cusum control chart may be used for a process that is subject to two-sided shifts. The chart in Fig. 12.2(a) shows one signal of a downward shift and one of an upward shift.

The general model for the cusum control chart above is that of a sequential sampling scheme as shown in Fig. 12.2(b), with the three decision zones for the acceptance of the corresponding hypotheses: H_0: $\mu = \mu_0$, H_1: $\mu = \mu_0 + \Delta_1$, and H_2: $\mu = \mu_0 + \Delta_2$. The shaded area is an indecision zone. Sampling continues until a sample cumulative sum falls into any one of the three decision zones. In control chart applications, only the upper and lower outer trend lines are used for the detection of shifts in the mean. It is noted that in all cases the cusums are cumulative sums of sample deviations from the target process mean μ_0.

Cusum Control Charts with a V Mask

If the upward and downward shifts are of the same magnitude (i.e., if $\Delta_1 = -\Delta_2 = \Delta$), a V mask with its two symmetrical arms representing the upper and lower control limits is often used for the analysis of a cusum chart. Since

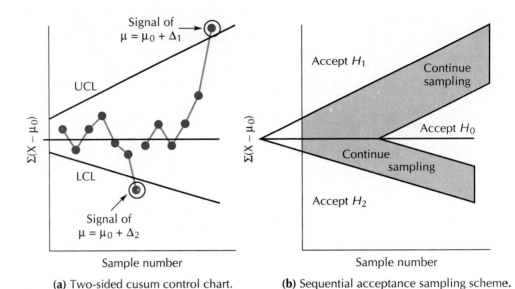

(a) Two-sided cusum control chart.   (b) Sequential acceptance sampling scheme.

Figure 12.2 Two-Sided Sequential Probability Ratios for Acceptance Sampling and for Cusum Control Charting

a nonrandom pattern begins to form in a sequence of sample cusums only after a shift has occurred, it is sufficient to examine the more recent data for signals, instead of the entire sequence. For this reason, the upper and lower control limits of the two-sided chart of Fig. 12.2(a) are flipped to form the V mask that can be moved around on a chart as shown in Fig. 12.3. The chart analysis is accomplished by positioning the V mask parallel to the horizontal axis with its center, O, on the most recent sample cusum, S_t, to observe if any previous data are covered by either arm. When one or more of the previous sample cusums are covered by one of the arms, a signal is therefore evident that a process shift has occurred. An upward shift is signified if one or more data points are covered by the lower arm, while the shift detected is downward if the upper arm covers any of the cusum data.

Figure 12.3 illustrates a typical cusum chart application with a V mask where an out-of-control condition has been detected at the tth sample. The upward shift in the mean may have actually occurred three to seven periods earlier, as indicated on the chart. Information about the posssible time of the actual shift in the process is one of the advantages of cusum charts. Additionally, if the actual time of the shift is judged to be at the $(t - m)$th sample, an estimate of the magnitude of the actual shift could be calculated by

$$\text{estimate of } \Delta = \frac{S_t - S_{t-m}}{m},$$

where

$$S_t = \sum_{i=1}^{t} (X_i - \mu_0).$$

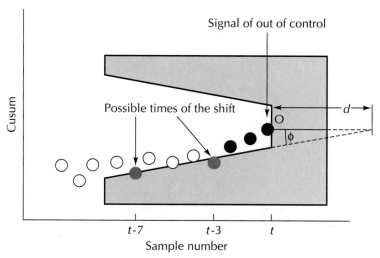

Figure 12.3 Cusum Chart Using a V Mask

The V mask can be constructed by determining the lead distance, d, and angle, ϕ, as shown in Fig. 12.3, for $\beta_1 = \beta_2 = \beta$ from the formulas

$$d = \frac{2}{\Delta}\left[\frac{\sigma_{\bar{X}}^2}{\Delta}\ln\frac{1-\beta}{\alpha}\right] \quad \text{and} \quad \tan\phi = \frac{h\Delta}{2},$$

where h is a factor that scales one sampling interval as h cusum units.

Traditional Cusum Control Charts for Sample Means

The foregoing presentation of the cusum control chart has been based on individual sample measurements. If independent samples of size n are taken, and sample means are calculated,

$$(\bar{X}_1, \bar{X}_2, \ldots, \bar{X}_t),$$

all the cusum chart construction procedures will remain the same with slight modifications in the formulas for control limits. The control limits are now given by

$$\text{UCL} = \frac{\sigma_{\bar{X}}^2}{n\Delta_1}\ln\frac{1-\beta_1}{\alpha} + t\left(\frac{\Delta_1}{2}\right) \tag{12.7}$$

$$\text{LCL} = \frac{\sigma_{\bar{X}}^2}{n\Delta_2}\ln\frac{1-\beta_2}{\alpha} + t\left(\frac{\Delta_2}{2}\right), \tag{12.8}$$

where $\Delta_1 = \mu_1 - \mu_0$ and $\Delta_2 = \mu_2 - \mu_0$. The cusum statistic is given by

$$S_t = \sum_{i=1}^{t}(\bar{X}_i - \mu_0).$$

The formulas for the lead distance and the slope of the control limits for a V mask with $\Delta_1 = -\Delta_2 = \Delta$ and $\beta_1 = \beta_2 = \beta$ are

$$d = \frac{2}{\Delta}\left[\frac{\sigma_{\bar{X}}^2}{n\Delta}\ln\left(\frac{1-\beta}{\alpha}\right)\right]$$

$$\tan\phi = \frac{h\Delta}{2}.$$

Traditional Cusum Control Charts for Sample Ranges

The use of sample cumulative statistics for process variability control can be developed similarly on the basis of sequential probability ratio tests. For the case where sample ranges are used in detecting upward shifts in σ_X from σ_0 to σ_1 with type I and type II error risk α and β, respectively, a one-sided upper control limit may be determined by the following equations:

$$d_u = \frac{\ln[(1-\beta)/\alpha]}{\ln[1-\sigma_0/\sigma_1]} \tag{12.9}$$

$$\phi = \tan^{-1}\left[\frac{2\ln(\sigma_0/\sigma_1)}{1-\sigma_0/\sigma_1}\right].$$

The sample cusum statistic is

$$S_t = \sum_{i=1}^{t}\frac{R_i}{w\sigma_0},$$

where

$R_i = i$th sample range

$\sigma_0 = \sigma_X$ under null hypothesis

w = a constant varying according to sample size n (see Table 12.1).

TABLE 12.1 Constant for Determining Cusum for Sample Ranges	
Sample Size, n	w
3	0.233
4	0.188
5	0.160
6	0.142
7	0.128
8	0.118
9	0.110
10	0.103

Source: Footnote 4, but based on the work of D. R. Cox, "The Use of Range in Sequential Analysis," *Jour. of Royal Stat. Soc.,* B, Vol. II, 1949.

12.3

Discussion of Traditional Cusum Control Charts

Traditional Cusum Charts in Process Control

In control chart applications, action is taken for special-cause verification only if the chart shows an out-of-control signal. Consequently, no actions are taken if the sample cumulative sum S_t is inside the area between the upper and lower control limits or within the two arms of a V mask. These are indecision regions. Since these indecision regions, as well as the control limits, are functions of the shifts, Δ_1 and Δ_2, it is extremely important to define the magnitude of these shifts precisely for an effective design of such control charts. When a sample cusum falls in one of the indecision regions, we cannot be certain that the process is in control at mean μ_0 even though the chart is telling us that the process is probably not performing at the rejectable level, μ_1 or μ_2. Obviously, the larger the size of the indecision regions, the more uncertain will be the control chart in terms of the process operation at μ_0.

Cusum charts of the type described above have been in use for more than 30 years, particularly in the study of chemical processes. Both practitioners and researchers have shown that cusum charts are more effective in detecting small shifts than the Shewhart control charts. Shifts of 1 standard deviation or less are often referred to as small shifts. Even with a shift of one standard deviation, the uncertainty still ranges from $-2.25 \sigma_X$ to $2.89 \sigma_X$. Such uncertainty will increase almost exponentially for very small shifts. This means that the control limits become far away from the centerline when the two hypothesized means are very close.

Practical Concerns of Traditional Cusum Control Charts

The necessity of having to specify the shift(s) in advance for the traditional method of cusum chart construction presents a serious concern in the applications of such charts. With some exceptions where the process engineers can predetermine the shifts from physical considerations, most applications end up with somewhat arbitrary selection of the shifts just to start the charting. The consequence of using arbitrary shifts for cusum charting would be analogous to that of the engineering specifications selected for convenience. Except in the case of process control, the process might be undercontrolled if the shift selected is too large and overcontrolled if the shift selected is too small.

The traditional cusum control charts should not be applied in general for process control, except in some special situations where the process mean is known to vary and not to be controlled by physical means. Two fundamental concepts of statistical thinking that we have discussed earlier would essentially be ignored if these cusum charts are used indiscriminately for general process control. These concepts are

1. **Common-cause versus special-cause variability.** It is true that some processes do not possess a constant mean due to their inherent physical

characteristics, such as continuously changing process behavior within a cycle between tool changes or between chemical replenishments. But such process behavior is likely to vary according to certain systematic patterns, hence be more predictable. These systematic changes are chronic phenomena of the process, not special or sporadic causes. Control charts are designed for the detection of unpredictable occurrences of the special causes, not the common causes. In general, chronic problems are problems of system design; they should not be solved at the production process by control charts. Making the distinction between the factors that generate a "routine" and those that cause the routine to break down, as pointed out by Shewhart, is fundamentally important and therefore must be built into any statistical methods for on-line process control. Traditional cusum charts actually mix the common and special causes together as can be seen from Eqs. (12.4), (12.5), and (12.6), where σ and Δ are used simultaneously to set the control limits. To view it still another way, the traditional method of cusum charts explicitly allows the process mean to vary, thereby failing to establish a constancy of purpose in process control.

2. **Analytic versus enumerative studies.** The derivation of the decision limits for cusum charts was originally developed for applications in acceptance sampling. The objective of acceptance sampling is to arrive at a decision of either accepting or rejecting a fixed and generally large lot of products on the basis of a small sample of observations of that lot. It is, therefore, a problem of enumerative studies. The performance behavior of a process, on the other hand, is subject to unpredictable changes at all times, hence a problem of analytic studies, as discussed in Chapter 3. In general, statistical methods that are developed for one type of study are not directly applicable to the other without proper modifications. Recall that cusum charts were developed in the 1950s, when the predominant thinking in statistical quality control was product control relying on acceptance sampling methods. One might speculate that few practitioners and researchers were concerned with the adaptation of acceptance sampling to process control. Indeed, research papers on the refinement of and case study reports on the use of such cusum control charts have flourished in the literature right to the time of this writing. Ignoring the important distinction between analytic and enumerative studies may well have contributed to the development of such cusum charts for over a period of 30 years, with possibly questionable results.

Having mentioned some of the problems associated with the traditional methods of cusum control charts, we should not overlook the fact that they are sensitive in detecting small shifts. Such sensitivity, however, is the result of using a cumulative sum as a sample statistic, not because of the use of the likelihood ratios in chart construction. In Section 12.5 we present control charts of cumulative sums of individual measurements and sample means that are Shewhart-like in their construction and interpretation.

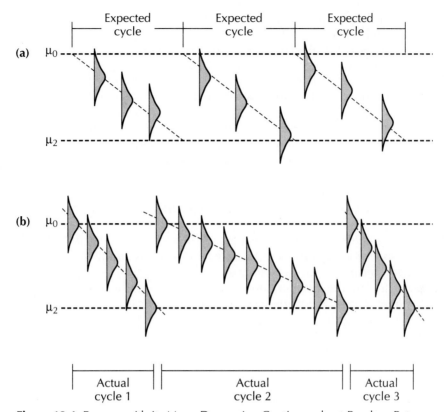

Figure 12.4 Process with Its Mean Decreasing Continuously at Random Rates

Special Situations for Traditional Cusum Control Charts

Shewhart control charts are developed on the principle that a process is said to be in statistical control if it is subject only to a fixed amount of common-cause variability. Most real-world production processes and service operations do not actually have a constant mean but may be so conceptualized for the purpose of statistical analysis if the common-cause variability is relatively small and random. Such conceptualization, however, may not be meaningful or feasible in situations where the process mean is known to change continuously in a somewhat predictable manner. Any process that has certain inherent growth or decaying effects does not have a constant mean and should not be treated as one that does if the amount of change in the mean is large relative to its variability. One such process is illustrated in Fig. 12.4, where the process mean continuously decreases but the rate of decrease is a random variable. Although an expected cycle time may be determined from theoretical/physical or empirical reasoning, for a process with changing mean such as the one shown in Fig. 12.4(a), the actual time each cycle decreases to the rejectable level μ_2 will vary due to the random nature of some common-cause effects on the mean, as Fig. 12.4(b) illustrates.

To know the dynamics of the changes in the mean is essential in system design of a process for physical control. Where physical control of such process dynamics is neither technologically nor economically feasible, one may have to leave the control at the process while it is in operation. It is only in such a situation that one relies on sample observations of the process to provide the necessary information for control. In fact, such control actions usually involve rather simple adjustments of a control device or replenishments of certain materials. In other words, the cause of the out-of-control condition is typically known for such applications. The traditional cusum control charts may prove to be an efficient method for these situations.

It is of prime importance to understand the fundamental difference between the Shewhart control charts and cusum control charts in their basic objectives. Shewhart control charts are designed for the purpose of detecting the occurrence of special causes relative to an established common-cause system with a constant mean (or any other process parameter). The traditional cusum control charts, on the other hand, are designed for the purpose of monitoring the movement of the dynamic process mean (or any other parameter). Since the dynamic nature of a changing mean is an integral part of the process characteristics, it is a common cause of such processes. Hence the traditional cusum charts could be viewed as a device to help control the common causes rather than the special causes.

It is also important to note that such processes as discussed above are still subject to the disturbance from special, sporadic causes, in addition to variable common causes. Since special causes produce a shift in the parameter with generally unknown magnitudes, one cusum control chart based on sequential hypothesis testing cannot be relied on for the control of both special and common causes. Therefore, either an entirely different control measure such as a Shewhart control chart or a separate analysis of the available sample data must be instituted to detect the occurrence of special causes when a traditional cusum chart is used. This topic will not be discussed further in this book.

12.4

Interpretation of Cusum Control Charts

Due to the fact that the sample data on a cusum chart are not independent of each other, it is more difficult to detect nonrandom patterns of the data when they are all within control limits. All the zone rules (except rule 1) that were discussed for analyzing the Shewhart $\overline{X}$ charts are not applicable for cusum charts. Since each sample point on the chart is a cumulative statistic that carries with it a memory of the past data, a long sequence of data points could wander far away from the centerline before returning to the centerline. Therefore, the actual location of the cusums is not a critical pattern for chart interpretation.

If a long sequence of cusums randomly varies about a line that is a distance away from but parallel to the centerline, it is possible that the initial estimate of the centerline (μ_0) is in error. In general, a trend line with a sharp increase or decrease in the slope that persists for several samples is an indication of a possible shift in the process parameter. Another possible signal of a sustained shift is indicated by an almost linear drift of a long sequence of data. Some of these drifts are illustrated in Fig. 12.5, which is a plot of cumulative differences of daily mold changes from a target average of nine changes per day. It appears that during the sampling period of 72 days, there were at least three upward shifts and one downward shift in the mean number of mold changes each day. When these 72 daily mold changes are individually plotted on a usual Shewhart chart as shown in Fig. 12.5(a), it is not possible to see any signals of an out-of-control condition. In comparison, therefore, a cusum chart is more sensitive to small but sustained shifts in the mean.

Sometimes, the high degree of sensitivity of cusum charts might produce false signals, due simply to the way the cumulative differences are calculated. In other words, if the reference value, m, or a selected target value from which each sample data is differenced, is different from the true process average, false signals might be formed from time to time without any actual shift. Therefore, the interpretation of a cusum chart requires extreme care. Figure 12.5(c) and (d) are cusum plots of the same sample data using two different reference values. Although the four signals that were indicated in Fig. 12.5(b) are also easily seen in these two plots, more signals are seen in both plots (c) and (d). These plots demonstrate the importance of accurate estimation of the process mean. It is also noted that cusum charts are relatively ineffective in detecting sudden shifts that are short-lived.

12.5

Shewhart-Like Cusum Control Charts

The success of traditional cusum control charts depends to a large extent on the accurate estimation of both the target process parameter value and the size of the shift the control chart is designed to detect. Often such accurate estimation or specification of the parameter values is difficult to obtain. In these situations, it is best not to use the traditional cusum charts with arbitrary selection of the required parameter values. Instead, some other control charts should be considered.

To take advantage of the increased sensitivity of a cumulative statistic, a control chart can be constructed for sample cusums, following the principle of Shewhart control charts. As will be seen, such charts do not require an alternative hypothesis. We will first discuss the use of such a cusum chart for the control of a process mean based on individual measurements. The Shewhart-like cusum control charts for sample means are also presented in this section.

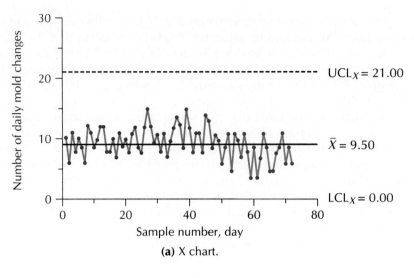

(a) X chart.

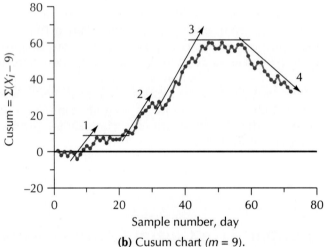

(b) Cusum chart *(m = 9)*.

Figure 12.5 Charts of Data on Mold Changes

Shewhart-Like Cusum Control Charts for Individual Measurements

A cusum control chart can be used effectively for plotting individual measurement data. In some regards, this cusum is similar to an EWMA, as each summarizes all past sample data in one current sum or average. Let X_1, X_2, . . . , X_t be individual measurements observed up to the current sample t; then their sum can be expressed by

$$S_t = \sum_{i=1}^{t} (X_i - m) = S_{t-1} + (X_t - m),$$

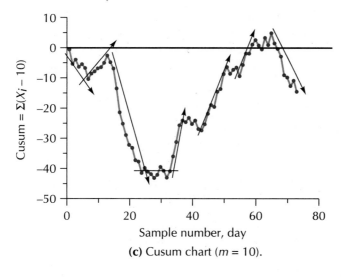

(c) Cusum chart (*m* = 10).

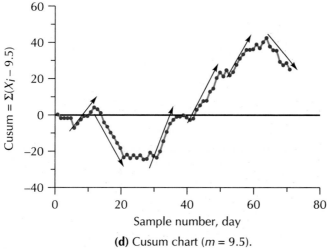

(d) Cusum chart (*m* = 9.5).

Figure 12.5 *(continued)*

where m is any reference value, usually set as the process mean value. As such, S_t is actually a cumulative sum of sample deviations from the mean. If a process is in control and the true process average is equal to m, then the cusums, S_i, will fluctuate about m without exhibiting any trends. However, when the process is shifted up or down from m, the cusums will gradually form a trend upward or downward, respectively. For a sustained shift in a process, the trend is linear. Therefore, in examining a cusum chart, it is important to look for trend lines of certain consecutive sums.

The process is assumed to be in control if no trends, linear or otherwise, can be detected from the data on the chart. It is to be noted that the actual level of the cusums is usually immaterial. As long as a sequence of cusums vary about a horizontal line, it indicates that the process mean is constant without any shifts. Similar to the exponentially weighted moving average charts, the cusum charts are more effective in detecting small shifts because of the cumulative effect.

Conventional use of cusum charts employs the accept/reject idea of statistical hypothesis tests, where both a rejectable mean and an acceptable mean must be specified together with their corresponding risks. What follows now is an alternative approach of the cusum charts for individual measurements, which are assumed to have a common normal distribution with mean μ_X and standard deviation σ_X.

Given sample individual measurements, X_i, $i = 1, 2, \ldots, t$, we can standardize them by

$$Z_i = \frac{X_i - \mu_X}{\sigma_X}.$$

Then, instead of summing the X's, we sum the Z's. The sum of Z's will have a normal distribution with mean $= 0$ and variance $= t$ [since the variance of each Z_i is 1.0, the variance of t independent Z_i's will be $(t)(1.0) = t$]. Now we can write the cusum as follows:

$$S_t^* = \frac{\sum_{i=1}^{t} Z_i}{\sqrt{t}},$$

which follows a standard normal distribution. This will allow us to plot the S_t^* on a control chart with constant control limits of ± 3 and a centerline of zero. Hence a relatively simple control chart with the same appearance of Shewhart charts can be constructed for the cusum as defined above.

Construction of the Cusum Chart: A Step-by-Step Procedure

1. Collect at least $k = 25$ individual measurements in time order, X_1, X_2, $\ldots$, X_k.
2. Compute $\overline{X}$ and s_X from the data. These are used as estimates of the true mean and true standard deviation of the X's, respectively.

$$\overline{X} = \frac{\sum_{i=1}^{k} X_i}{k}, \qquad s_X = \left[\sum_{i=1}^{k} \frac{(X_i - \overline{X})^2}{k - 1} \right]^{1/2}.$$

3. Standardize all the X's into Z's for $i = 1, 2, 3, \ldots, k$:

$$Z_i = \frac{X_i - \overline{X}}{s_X}.$$

4. Sum the Z's cumulatively for each t, $t = 1, 2, 3, \ldots, k$:

$$\text{sum}_t = \sum_{i=1}^{t} Z_i.$$

5. Obtain the standardized cusum for each t, $t = 1, 2, 3, \ldots, k$:

$$S_t^* = \frac{\text{sum}_t}{\sqrt{t}}.$$

6. Plot the S_t^* on the standardized cusum chart, where

$$\begin{aligned}
\text{centerline} &= \quad 0 \\
\text{UCL} &= \quad 3 \\
\text{LCL} &= -3.
\end{aligned}$$

7. Interpret the chart, looking especially for possible trends in the sums. If special causes exist to explain the out-of-control trends, the chart should be revised by excluding the data groups in the trends.

Example

This example uses the data of the white millbase case study of Chapter 11 (batches 31–57) to illustrate the construction and interpretation of the standardized cusum control charts. It was shown earlier that the sample mean and sample standard deviation of this set of individual measurements are

$$\begin{aligned}
\overline{X} &= 13.777 \\
s_X &= \quad 0.1720.
\end{aligned}$$

The calculations, following the step-by-step procedure above, are shown in Table 12.2. With the standardization of the raw data, the control chart for the cusums to monitor the process mean is always the same, as shown in Fig. 12.6; that is, the control limits and centerline are always 3, -3, and 0, respectively.

It is seen from the cusum chart of Fig. 12.6 that all the cusums are above the centerline and exhibit a downward sloping trend. This could be due to a number of different reasons, some of which are listed below.

1. If a downward shift in the mean occurred during the collection of the k samples, the cumulative sums would slope downward. Similarly, an upward trend would correspond to a shift up in the mean.
2. A slow but continuous drift in the process mean could cause the cusums to exhibit a nonlinear trend up or down.
3. Since both the process mean and the process standard deviation were estimated from the sample data, any estimation error in either or both parameters would create a bias in standardizing the raw cusums, which could then behave as nonrandom patterns on the chart.
4. A few data points reflecting short-lived extreme impulses in the early

| | TABLE 12.2 | Standardized Cumulative Sums of the Millbase Batch Weights (Batches 31 to 57) | | |

Sample	X_i	Z_i	sum_t	S_t^*
1	14.04	1.567	1.567	1.567
2	13.94	0.980	2.547	1.801
3	13.82	0.276	2.823	1.630
4	14.11	1.978	4.801	2.401
5	13.86	0.511	5.312	2.376
6	13.62	−0.898	4.414	1.802
7	13.66	−0.663	3.751	1.418
8	13.85	0.452	4.203	1.486
9	13.67	−0.605	3.598	1.199
10	13.80	0.158	3.756	1.188
11	13.84	0.393	4.149	1.251
12	13.98	1.215	5.364	1.548
13	13.40	−2.189	3.175	0.881
14	13.60	−1.015	2.160	0.577
15	13.80	0.158	2.318	0.599
16	13.66	−0.663	1.655	0.414
17	13.93	0.921	2.565	0.622
18	13.45	−1.896	0.669	0.158
19	13.90	0.745	1.414	0.324
20	13.83	0.335	1.749	0.391
21	13.64	−0.781	0.968	0.211
22	13.62	−0.898	0.070	0.015
23	13.97	1.156	1.226	0.256
24	13.80	0.158	1.384	0.283
25	13.70	0.428	0.956	0.191
26	13.71	−0.370	0.586	0.115
27	13.67	−0.605	−0.019	−0.004

portion of the data set would require a long time to be smoothed or settled down, thereby forming a trend on the chart.

5. A combination of two or more of the foregoing causes.

The downward trend of the 27 millbase data is quite obvious, as shown in Fig. 12.6. In the actual case study of this process, discussed initially in Chapter 11, the X and R_m charts did not show any nonrandom patterns, although data early in the sequence were large. Consequently, no search was attempted to find special causes. From the cusum plot here (Fig. 12.6), there seems to be some special causes at work.

For our purposes of illustration of cusum charts, suppose that we remove the first five data points from the set and revise the chart. The sample mean and the sample standard deviation, calculated from the remaining 22 data points, are different. A standardized cusum chart based on the new estimates

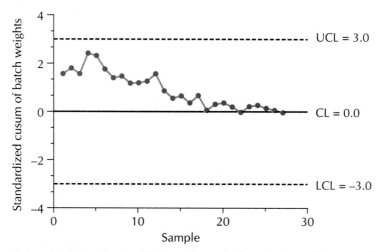

Figure 12.6 Standardized Cusum Control Chart for the Millbase Batches 31 to 57

is shown in Fig. 12.7. One may now conclude that the pigment dispersion process was in a state of statistical control when it produced the last 22 batches of the white millbase. It is to be noted that the cusums are randomly distributed not only about a horizontal line but also about the estimated mean. The last cusum, S_k^*, will always be on or very near the centerline because we used $\overline{X}$ for the centerline as well as the reference value in calculating the sums.

To continue our illustration of the cusum charts, let us suppose that the mean weight of the white millbase was increased by 0.5 pound per gallon from samples 13 to 17 (Table 12.2), due to some special causes during that period of time. In other words, if there was a short-lived special cause that shifted the

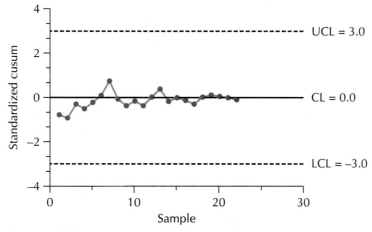

Figure 12.7 Revised Cusum Chart for the Millbase Batches 36 to 57

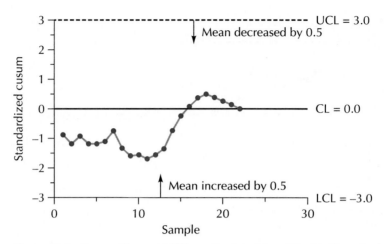

Figure 12.8 Cusum Chart of Millbase Batches 36 to 57 with Hypothetical Shifts

process mean up by 0.5 pound per gallon for five sampling periods, we would like to see how a cusum chart using the data with the increase of 0.5 pound to each of the five samples would behave. Such a chart is shown in Fig. 12.8. It is interesting to observe in Fig. 12.8 the obvious nonrandom trend formed by the five points that reflect the hypothetical shift in the mean. This illustration demonstrates the general capability of cusum charts to quickly identify shifts in the mean. Another advantage of the cusum chart is its ability to help determine the time of the shift more precisely by locating the starting point of a trend on the chart. Also, the slope of a trend line can be used to estimate the magnitude of the shift or the impact of a known change that occurred in the system. To complement the cusum chart when used primarily for the control of process means based on individual measurement data, an independent R_m chart for the process variability may be used. This combination would allow us to monitor the process by analyzing the data both individually and collectively.

Since the interpretation of a cusum chart is basically a search for possible trends in the data and is not so much concerned with the actual level of the data on the chart, one may even use just a plot of the cusums without control limits. A cusum plot of individual measurement data can be very revealing to possible changes in a process by identifying trend lines in the data. As such, the charting of cusums can be applied to many different types of data including operational and administrative records.

Shewhart-Like Cusum $\overline{X}$ Control Charts

A Shewhart control chart may be constructed for a cusum statistic calculated from a time-ordered sequence of independent sample means. If the process quality variable X has a mean μ_X and a standard deviation σ_X, a cumulative sum of the differences between the sample means and μ_X,

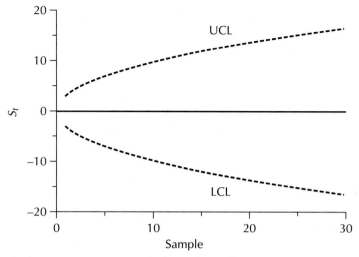

Figure 12.9 Cusum Control Chart of Sample Means

$$S_t = \sum_{i=1}^{t} (\overline{X}_i - \mu_X),$$

(12.10)

has a mean equal to zero and a variance equal to $(t/n)(\sigma_X^2)$, where n is the sample size. A control chart with 3 sigma control limits for such a cusum would look like the one shown in Fig. 12.9. Alternatively, if the cumulative statistic is defined in the units of its standard deviation as

$$S_t^* = \frac{S_t}{\sigma_X \sqrt{t/n}},$$

(12.11)

this cusum is distributed approximately normal with mean zero and standard deviation 1 for a sample size n. A much simpler control chart for the S_t^* is shown in Fig. 12.10.

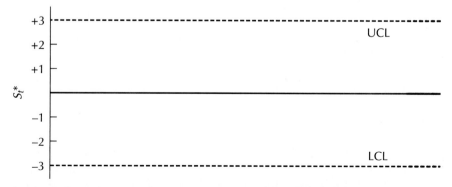

Figure 12.10 Standardized Cusum Control Chart for Sample Means

Example. We will now use the second set of 25 simulated samples (samples 26 to 50) of the sampling experiment of Chapter 8 to illustrate the construction and interpretation of the cusum chart described above. Based on the first set of 25 samples, it was shown that

$$\overline{\overline{X}} = -0.072 \quad \text{and} \quad \overline{R} = 3.52,$$

which are our estimates of the true process mean and range. Recall that the sample size is $n = 5$. Assuming a normal distribution for the individual measurements, we estimate the standard deviation of the sample means as

$$\frac{\sigma_X}{\sqrt{n}} = \frac{3.52}{2.326\sqrt{5}} = 0.675.$$

Two alternative ways to construct the Shewhart cusum chart with 3 sigma limits are described as follows.

Method 1: Variable Control Limits

$$\text{UCL} = -0.072 + (3)(0.675)(\sqrt{t}) = -0.072 + 2.025\sqrt{t}$$
$$\text{LCL} = -0.072 - (3)(0.675)(\sqrt{t}) = -0.072 - 2.025\sqrt{t}.$$

The sample cusum statistic for this chart is

$$S_t = \sum_{i=26}^{t} [\overline{X}_i - (-0.072)].$$

The 25 sample cusums, $t = 26, 27, \ldots, 50$, are plotted in Fig. 12.11(a). It is seen from the chart that a very strong trend line appears to have started from sample 31 and persists until the thirty-ninth cusum value. The actual out-of-control signal shows up at the thirty-fifth sample on this chart. Although the cusums begin to settle down at a new horizontal level from sample 41, they are all outside the lower control limit. Note that the simulated shift in process variability for samples 41 to 45 could not be detected from such a cusum chart.

Method 2: Constant Control Limits at ±3.
The sample cusum statistic for this chart is

$$S_t^* = \sum_{i=26}^{t} \frac{\overline{X}_i - (-0.072)}{0.675\sqrt{t}}.$$

This cusum chart is shown in Fig. 12.11(b). The appearance of the standardized cusums in Fig. 12.11(b) is very similar to the ones in Fig. 12.11(a). For this set of simulated experimental data, the approach using cumulative statistics does not show any advantage over the $\overline{X}$ and R charts. In fact, the $\overline{X}$ charts presented in Chapter 8 even revealed some information about possible shifts in the ranges for samples 41 to 45. When compared with the $\overline{X}$ and R charts for the same data, the Shewhart cusum chart presented here detects the shift in the mean much more slowly.

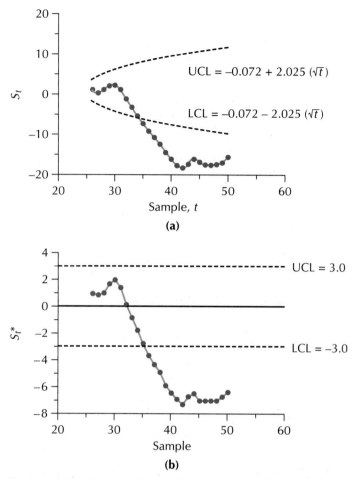

UCL $= -0.072 + 2.025\,(\sqrt{t}\,)$

LCL $= -0.072 - 2.025\,(\sqrt{t}\,)$

(a)

UCL $= 3.0$

LCL $= -3.0$

(b)

Figure 12.11 Shewhart Cusum Control Chart of Sample Means: Simulated Data

There are two important differences between the two types of charts that will help explain the relative sensitivities. First, the Shewhart cusum chart is constructed without a given mean value to detect. The second difference is in the size of the type I errors. A 3 sigma distance is set from the centerline to locate the control limits on the Shewhart cusum charts, while the traditional cusum charts generally use a type I error of 0.005, which is equivalent to the use of about 2.5 sigmas.

12.6

Cusum Plots for Data Analysis

The technique of cusum charting requires a specified reference or target value, m. Often, it is obtained by determining a midvalue between an acceptable mean and a rejectable mean, or the midvalue between the upper and lower specification limits, or simply a desirable target. Some of the drawbacks and concerns of such practice were discussed earlier. Nevertheless, for individual measurements or observations the cumulative sum is a very useful statistic for process control as well as many other applications where the strict notion of statistical control might not even exist. Examples of the latter include monthly sales figures, foreign trade balances, and holiday weekend traffic fatalities. These are individual statistics of vital importance, and it is often desirable to monitor them and, hopefully, control their destiny. It may be desirable to analyze such data using a cusum statistic but without the use of control limits. We will refer to such graphs as *cusum plots*. We use the following example to illustrate the analysis of data using cusum plots.

Analysis of Cusum Data for Polyol Moisture Content

A chemical plant was routinely analyzing the moisture content of a certain material, polyol. Moisture variation in the polyol causes some difficulty in controlling the reactivity when mixed with isocynate that is used in a foam molding process. A record of 27 results of the polyol moisture analysis over a period of 4 months was studied to assess the consistency of the moisture in the polyol. These measurements are listed in the second column of Table 12.3 and plotted on an X chart as shown in Fig. 12.12(a). It is easy to see from the X chart that signals of out-of-control conditions show up on the seventeenth and eighteenth observations. The cause of the problem was identified and, subsequently, results 17 and 18 were eliminated for the purpose of constructing a revised X chart, shown in Fig. 12.12(b). Now the X chart appears to have no statistical signals of out-of-control conditions. One might conclude that the moisture content was in statistical control, varying about an average of 2.074%. Judging from the specification limits of 2.00% and 2.25%, the average moisture content was on the low side. Five of the remaining 25 results fell below the lower specification limit. At the same time, the plant was experiencing problems with the reactivity downstream due to the variable moisture content in the polyol. Since the moisture content was not meeting the specification, someone suggested conducting a thorough investigation of the silo in which the polyol was stored and its sealing method for possible improvements.

An engineer from R&D thought the complete operations record of the material storage system should be studied together with the sample data of the moisture analysis. In particular, the shipment dates of polyol tankers should be checked for possible correlations with the changes in moisture content. He further suggested plotting the cusums of the moisture data, to detect possible small shifts in the mean moisture level. Accordingly, using $m = 2.125$, which was the average of the upper and lower specification limits, the cusums were computed as listed in the last column of Table 12.3 and plotted as shown in Fig. 12.13(a). Figure 12.13(b) shows another cusum plot of the moisture data. These cusums are computed using the reference value of 2.075, which was the revised mean moisture after samples 17 and 18 were excluded. Note that the

	Moisture,	Difference,		
Sample	X_i (%)	$X_i - m$	$\sum_{i=1}^{t} (X_i - m)$	

TABLE 12.3   Data of Polyol Moisture Content

Sample	X_i (%)	$X_i - m$	$\sum_{i=1}^{t} (X_i - m)$	
1	2.29	0.165	0.165	
2	2.22	0.095	0.260	
3	1.94	−0.185	0.075	
4	1.90	−0.225	−0.150	
5	2.15	0.025	−0.125	New tanker
6	2.02	−0.105	−0.230	
7	2.15	0.025	−0.205	New tanker
8	2.09	−0.035	−0.240	
9	2.18	0.055	−0.185	New tanker
10	2.00	−0.125	−0.310	
11	2.06	−0.065	−0.375	
12	2.02	−0.105	−0.480	
13	2.15	0.025	−0.455	
14	2.17	0.045	−0.410	New tanker
15	2.17	0.045	−0.365	
16	1.90	−0.225	−0.590	
17	1.72	−0.405	−0.995	
18	1.75	−0.475	−1.470	
19	2.12	−0.005	−1.475	
20	2.06	−0.065	−1.540	
21	2.00	−0.125	−1.665	
22	1.98	−0.145	−1.810	
23	1.98	−0.145	−1.955	
24	2.02	−0.105	−2.060	
25	2.14	0.015	−2.045	New tanker
26	2.10	−0.025	−2.070	
27	2.05	−0.075	−2.145	

plots of Fig. 12.13 are not the conventional control charts, as they have neither a centerline nor control limits. Each is just a plot of the cusums.

In examining these cusum plots, we see from Fig. 12.13(a) that the average moisture level was less than 2.125, the specified reference value, since the cusums continuously were drifting downward from the 0 line. We also see from both plots of Fig. 12.13 a number of trend lines, each of which could be a reflection of a certain shift in the process mean. The question now to be asked is whether or not each trend is an indication of some actual shift in the mean moisture. Trend line a_2 of Fig. 12.13(a) is the same signal that was found from the X chart of Fig. 12.12(a). For the five trend lines shown in Fig. 12.13(b), however, we must examine each of them carefully because the X chart of Fig. 12.12(b), which is based on the same 25 data points, shows no out-of-control signals.

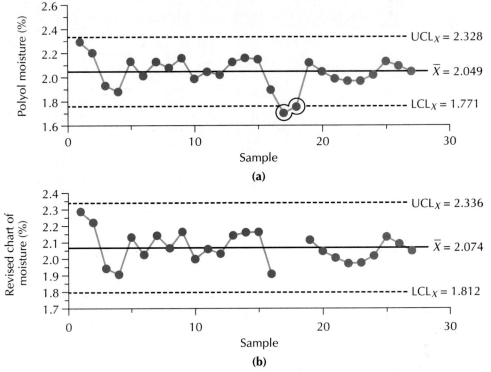

Figure 12.12 X Charts of the Polyol Moisture Data

Whether an observable trend on a cusum plot is a signal of a shift or not depends on the slope of the line and the number of successive points on the line. In general, the larger the slope of the line and/or the more points on the line, the more likely it is that the apparent signal is a true signal. This can be reasoned from the nature of an actual shift of the process mean. If the shift in the mean is large and sustained, the sample cusum data will exhibit a long, very steep trend line. If the shift is small but sustained, the trend line will be drifting slowly away from a horizontal line. Therefore, in reading a cusum plot, we should search for long trend lines. The size of the slope is more difficult to judge because it is relative to the scales of the plot. Cusum plots are less effective in identifying small to medium shifts that are short-lived.

With the exception of the first trend line b_1, the other four lines, b_2 to b_5 in Fig. 12.13(b), appear to correspond closely with the times of new shipments (marked "new tanker" in Table 12.3). The slopes of these lines are comparable and the lengths of these lines are at least four points. This observation led to further study of the entire operation of the polyol replenishment method.

Graphical Method of Interpreting Cusum Plots

The graphical interpretation of a cusum plot just described can be quite useful for analyzing many kinds of data that are available or taken routinely as a

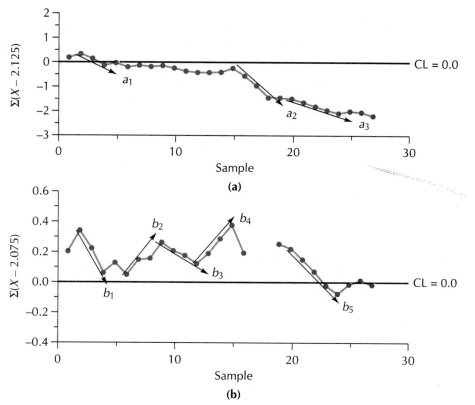

Figure 12.13 Cusum Plots of Polyol Moisture Content

matter of record. As an aid in such applications, we present the following graphical method of cusum plotting and interpretation.

Let a series of sample data, $X_1, X_2, \ldots, X_k$, be the individual observations of a normal variable with an unknown mean μ_X and a standard deviation σ_X. Suppose that the desirable or target value of μ_X is m, which may differ from μ_X by δm. Furthermore, μ_X itself is subject to change to a level $\mu_X + \delta\mu_X$ and will remain at the new level for at least one sampling interval. Thus at each sampling, X_t is an observation of a process with the mean that is different from the reference value m by $(\delta m + \delta\mu_X)$. Since a cusum S_t is calculated by

$$S_t = \sum_{i=1}^{t} (X_i - m),$$

its expectation depends on $(\delta m + \delta\mu_X)$.

If $(\delta m + \delta\mu_X) = 0$ for all samples through t, the expected value of S_t will be zero. Under this condition, the series of t cusums will fluctuate randomly about a zero line on a cusum plot without any significant trend lines. When such a random distributional behavior is shown in a cusum plot, usually no action is taken, although the mean of the underlying process is not necessarily equal to m. For instance, a cusum plot may exhibit the random distribution if $\delta m = -\delta\mu_X$ at every sampling.

If $(\delta m + \delta\mu_X) \neq 0$ at least at some of the samplings, the cusum data will not vary about a zero line and/or may show some trend lines. It is this nonrandom data behavior we are trying to detect when we read a cusum plot. It is noted that when the nonrandom pattern is detected from a cusum plot, there is still no way to determine whether it is due to $\delta m \neq 0$, $\delta\mu_X \neq 0$, or both. To simplify the application and the development of the graphical cusum analysis that follows, we standardize S_t (sum_t, see page 399) and the quantity $(\delta m + \delta\mu_X)$, which is to be denoted by

$$\partial = \frac{\delta m + \delta\mu_X}{\sigma_X}.$$

As discussed earlier, a trend line is formed by a sequence of cusum data according to the magnitude of the "shift" in the mean and the length that such a shift has been sustained. Since we like to identify a trend line with a small α-risk or a very low probability of being wrong, small shifts require a long sequence of cusum data points to form a noticeable trend line. Table 12.4 is a suggested list of the rules for trend line identification. As an example, if a trend line is shown in a cusum plot with six points (five increments or decrements), Table 12.4 indicates that such a trend line signals a shift in the mean by at least 0.60 standard deviation from m, the reference value, with a risk of 0.16%. Such a line is more likely to show with a small slope. The trend line for a given size of shift may show up with varying slopes on a plot, according to the relative scales between the cusum axis and the sample number axis.

Cusum Paper for Graphical Interpretation

To facilitate the trend identification of cusum plots, Fig. 12.14 provides a reference chart for the different slopes for plots where the interval between two sample numbers is equal to 1 standard deviation, which is also the unit for the

TABLE 12.4 Cusum Trend Line Identification Rules

Number of Points on the Line, n	Shift Is Greater Than $\partial \geq$	Size of Type I Error, α
8	0.25	<0.0017
7	0.40	<0.0017
6	0.60	<0.0016
5	0.90	<0.0012
4	1.25	<0.0012
3	1.75	<0.0016
2	3.00	<0.0014

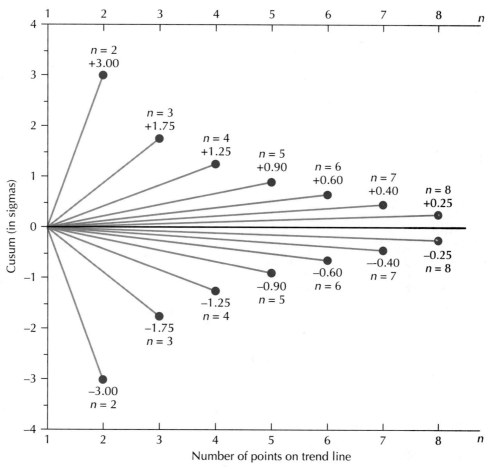

Figure 12.14 Reference Chart for Cusum Trend Line Identification (Horizontal Scale Is in Unit Standard Deviations)

cusums. With such a graph paper for plotting the cusums, if a trend line running through n data points and having a slope that is at least as large as the one shown in the reference chart for the same n points, the trend line is to be identified as a signal of a shift as big or bigger than the ∂ value indicated in the chart. For example, if a five-point line shows up in a cusum plot with a slope falling between the third and fourth lines from the top of the chart, we would identify it as a signal with a possible shift of at least 0.90 standard deviation above the reference value m. Note that the reference chart only goes up to $n =$ 8. Any trend line with nine or more data points in a row could be considered a signal even when the slope may be extremely small.

An appropriately designed graph paper together with the corresponding reference chart is needed for a given application, depending on the number of data points that can be plotted on one sheet of paper. For a large data set, each sample interval may be chosen as a convenient fraction of 1 standard deviation.

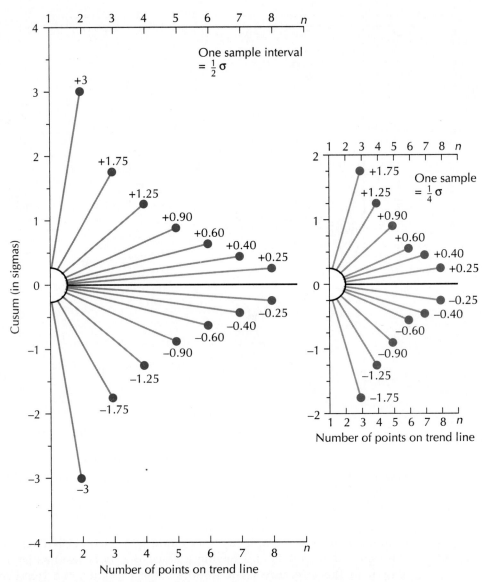

Figure 12.15 Cusum Reference Charts for One-Half and One-Fourth Scales

Figure 12.15 shows two such reference charts for graph papers with sample intervals scaled as one-half and one-fourth of the standard deviation of the data.

Figure 12.16 shows a cusum paper with a sample interval scale of $\frac{1}{4}\sigma_X$ together with the corresponding reference chart for checking trend lines. To use the cusum paper with a reference chart for graphical interpretation of the plot, an estimate of the standard deviation of the data is needed. This is usually

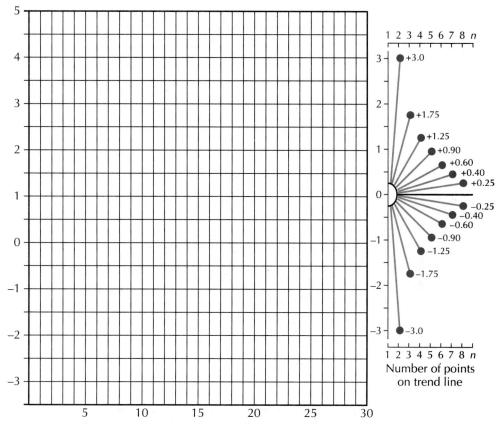

Figure 12.16 Cusum Paper for Graphical Identification of Signals

obtained from the sample standard deviation of the same data set that is to be plotted. Initially, such estimates might not be very good. Therefore, revised estimates may be used for subsequent applications.

Example. The polyol moisture data (excluding samples 17 and 18), described previously, will be used to illustrate the use of cusum paper with a reference chart. The sample standard deviation of the 25 observations is 0.087%, as seen from the X chart of Fig. 12.12(b). The cusums, with a reference value $m = 2.075$, and their standardized values are shown in Table 12.5.

Now we plot these standardized cusums on cusum paper with one-fourth scale, as shown in Fig. 12.17. It is seen from this plot that there are four significant trend lines with about the same slopes, two of which are upward shifts ($\partial > 0$) while the other two are downward ($\partial < 0$) shifts. When these are compared with the reference chart on the right-hand side of Fig. 12.17, it appears that these four signals may have been caused by a shift of about 1.25 to 1.75 standard deviations from the estimated mean of 2.075%. Using the

TABLE 12.5 Data for the Polyol Moisture Example		
Sample	$\sum_{i=1}^{t} (X_i - 2.075)$	$\sum_{i=1}^{t} (X_i - 2.075)/\sigma_X$
1	0.215	2.471
2	0.360	4.138
3	0.225	2.586
4	0.050	0.575
5	0.125	1.437
6	0.070	0.805
7	0.145	1.667
8	0.160	1.839
9	0.265	3.046
10	0.190	2.184
11	0.175	2.011
12	0.120	1.379
13	0.195	2.241
14	0.290	3.333
15	0.385	4.425
16	0.210	2.414
17	0.255	2.931
18	0.240	2.759
19	0.165	1.897
20	0.070	0.805
21	−0.025	−0.287
22	−0.080	−0.920
23	−0.015	−0.172
24	0.010	0.115
25	−0.015	−0.172

current estimate of the standard deviation of 0.087%, these shifts amount to about 0.11 to 0.15% in actual moisture content. These are very large shifts relative to the tolerance interval of (2.25% − 2.00%) = 0.25%.

Another concern in this example is the current estimate of the mean moisture content because ∂ consists of δm and $\delta \mu_X$. In this case, however, the two upward shifts and the two downward shifts are about the same in magnitude, and hence the actual average moisture content may be very close to 2.075% as estimated. It seems, therefore, effort should be focused on the study of factors that changed the mean. If actions could be taken to reduce the moisture variation caused by new shipments of the polyol, its moisture content would be significantly more consistent.

This example demonstrates the utility of cusum plots for further improvements of a process that may already have been in statistical control, as evidenced by some other control charts. It is to be noted, however, that the use of cusum plots does not require other control charts as a prerequisite.

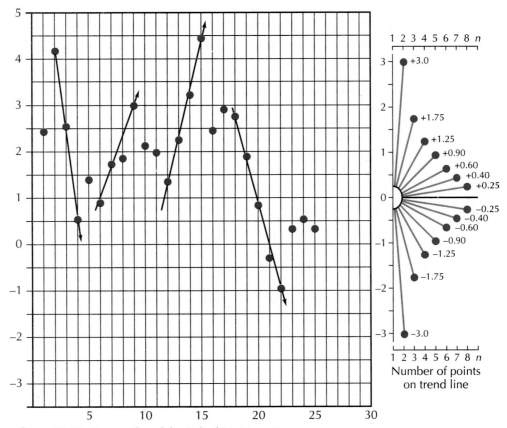

Figure 12.17 Cusum Plot of the Polyol Moisture Data

The application of cusum plots is not limited to individual measurements. Cusum plots could be used for any type of sample data including attributes and recorded statistics. Use of cusum paper with a reference chart is a very good way to monitor the progress of a certain event for the purpose of evaluating the trend or impact of certain modifications or initiatives that were introduced in the system. For example, the results of a sales promotion could be monitored using a cusum plot to evaluate the effectiveness of the advertising program in terms of its capability in reaching the goal as well as the time it takes to reach the goal.

Another example would be to keep a plot of accident occurrences in a plant so as to identify their major causes. Accident data, however, often follow a distribution other than normal because of the physical nature/occurrence of accidents. The number of accidents in a given period of time is not a continuous variable; it is a discrete variable that is either "yes" (occurs) or "no" (does not occur) at each particular instant. Many other data are of this discrete or attribute nature, each of which may be described by its apppropriate probability distribution for proper treatment. All these data have one thing in common. They are either collected and/or recorded as individuals (i.e., $n = 1$ at each sampling).

Understanding the inherent nature of the data at hand is necessary for adequate choice of the probability distribution for its analysis. For instance, a list of percentage data of the Consumer Price Index may be treated as continuous measurements, while the percentage of defective data of a lot of products must be analyzed by a discrete probability distribution. This is because the underlying variable is the number of defective items in the lot, which is discrete, and each item in the lot can only be defective or nondefective, which is an attribute. Shewhart control charts are available to handle these attribute data. The following chapter is devoted to a discussion of such control charts.

Exercises

12.1. A manufacturing plant is considering the use of a traditional cusum chart for monitoring the inventory of a certain raw stock for a product line. It intends to control the worth of this inventory between 12 and 18% of the total dollar value of the production. Weekly records of these percentage data are to be charted for management control. The inventory policy is designed to maintain an average stock level that amounts to 15% of the production volume. Experience has shown that the actual stock level varies with a standard deviation of 1.25%. Using $\alpha = 0.05$ and $\beta_1 = \beta_2 = 0.10$, construct a traditional cusum chart by determining the lead distance d and the angle ϕ for the V mask and choosing the appropriate scales for the axes.

12.2. a. Plot the cusums of the past 18 weekly inventory data listed in the table on the cusum chart constructed in Exercise 12.1.

Week	1	2	3	4	5	6	7	8	9
Stock (%)	12.5	12.1	13.3	14.8	15.2	16.4	17.1	15.9	17.6
Week	10	11	12	13	14	15	16	17	18
Stock (%)	12.1	13.9	15.6	16.9	16.7	16.1	12.8	14.9	15.4

b. Interpret the chart to help assess how well management's inventory policy was carried out during the last 18 weeks. In particular, state any concerns that you might have from reading the chart and recommend any action for improvement.

12.3. Since the implementation of an SPC program in the foundry department in January 1990, the engineers made two changes in the foundry operations, which appear to have resulted in a 4% reduction in the scrap rate from the 1989 average rate of 28% (the standard deviation was 2.5%). The table summarizes the weekly scrap rate data for the first half of 1990.

Construct a traditional cusum chart to analyze the data for the purpose of verifying the engineer's claim regarding the scrap reduction. Specifically, identify the time and estimate the magnitude of the average reduction for each of the two changes in foundry operation on the basis of your cusum chart analysis. Use values of $\alpha = 0.05$ and $\beta_2 = 0.10$.

Week	1	2	3	4	5	6	7	8	9	10
Scrap Rate (%)	26.4	25.3	31.8	27.1	28.5	20.2	19.3	22.7	20.4	19.5

Week	11	12	13	14	15	16	17	18	19	20
Scrap Rate (%)	21.7	24.0	22.6	23.1	21.9	23.0	21.8	16.4	17.9	16.8

Week	21	22	23	24	25	26
Scrap Rate (%)	18.1	17.2	16.9	18.1	17.6	18.3

12.4. As a part of an overall quality improvement program, a textile manufacturer decides to initiate a cusum chart to monitor the number of imperfections found in each bolt of cloth inspected. The data from the last 34 inspections are recorded in the table.

Bolt	1	2	3	4	5	6	7	8	9	10	11	12	13	14	15	16	17
Imperfections	34	25	30	29	20	26	22	29	28	27	23	32	21	42	21	26	32

Bolt	18	19	20	21	22	23	24	25	26	27	28	29	30	31	32	33	34
Imperfections	24	31	28	26	29	27	21	25	32	24	26	22	27	20	25	24	22

a. Construct a traditional cusum chart with $\mu_1 = 30$, $\alpha = 0.05$, and $\beta_1 = 0.10$. Revise the chart as necessary until the chart exhibits statistical control.

b. What would you do if you were told that bolts 1 and 14 were actually the results of a different cloth?

c. What would you do if bolts 5, 15, and 31 were only half the size of the other bolts?

d. What would you do to defend your cusum chart if someone tells you that the number of defects are attribute data and therefore a control chart for attributes must be used? Modify the chart if you deem it appropriate.

12.5. The filling process of Exercise 6.19 finally was brought into statistical control with the followng results:

$\overline{\overline{X}} = 1.0029$ pounds (over the 20-pound stated on the bags)

$\overline{R} = 0.1805$ pound

Sample size = 5

Use the results above as the basis to estimate the true process mean μ_0 and standard deviation σ_0. Set up a traditional cusum control chart for the sample means with the following information:

$$\mu_1 = 0.500 \text{ pound}$$
$$\alpha = 0.05 \quad \text{and} \quad \beta_2 = 0.05$$

Suppose that the cusum chart is employed to monitor the process. Plot on the chart the 15 samples shown in the table, detect any signals of out-of-control conditions, and make any comments about the process behavior over this period of time.

Sample	$\overline{X}_i$	Sample	$\overline{X}_i$	Sample	$\overline{X}_i$
1	0.990	6	0.992	11	1.014
2	1.002	7	1.010	12	0.994
3	0.992	8	0.982	13	0.998
4	1.004	9	1.004	14	0.970
5	0.974	10	0.998	15	1.002

12.6. Suppose that we review Exercise 6.19 with the help of the cusum control chart of Exercise 12.5. For the 20-pound dog food bags, the filling process was apparently set at a nominal average weight of 21 pounds per bag. Why do you think the company was doing it this way? Now the filling process is in control with reduced variability. Do you think the company should keep the nominal at the current level? Why or why not? Do you think the nominal weight could be lowered to about 20.5 pounds per bag? If so, what is the risk of having underweight bags, and what are some measures the company should take to minimize that risk?

12.7. The data in the table are hourly measurements of the width of the slot on a terminal block.

Sample	X_i	Sample	X_i	Sample	X_i
1	0.46	11	0.52	21	0.52
2	0.47	12	0.50	22	0.49
3	0.43	13	0.56	23	0.49
4	0.47	14	0.55	24	0.53
5	0.47	15	0.52	25	0.50
6	0.46	16	0.57	26	0.43
7	0.48	17	0.50	27	0.45
8	0.48	18	0.46	28	0.44
9	0.52	19	0.55	29	0.48
10	0.50	20	0.58	30	0.46

a. Construct a Shewhart-like cusum chart with variable control limits for individuals using the sample mean and sample standard deviation from the 30 data points as the normal distribution parameter estimates. Plot the cusum data and interpret the chart.

b. Repeat part (a) with a standardized Shewhart-like cusum chart.

c. Compare your observations from the charts in parts (a) and (b).

12.8. By referring to Exercise 12.7, it is learned now that the slot width may vary after it is cooled down to room temperature, and the operator might not have been consistent as to the timing of taking each measurement. Since this "time" effect could have been a special cause for any out-of-control conditions, it is suggested that any out-of-control data points should be eliminated for chart revision.

a. Carry out the necessary revisions of the cusum chart according to the above suggestion.

b. Discuss the pros and cons of the suggested method of revision and recommend a method to handle the data under such situations.

12.9. Repeat Exercise 12.5 using a standardized Shewhart-like cusum control chart and compare your observations from this chart with the results of Exercise 12.5.

12.10. Make a cusum plot of the data of Exercise 12.3 and identify possible shifts in the mean scrap rate during the first half of 1990. Also estimate the reduction in average scrap rate for each shift that you identify.

12.11. Plot the percentage of inventory data given in Exercise 12.2 on a cusum plot, assuming that the data follow a normal distribution with a standard deviation of 1.25%. Interpret the plot and summarize your observations of the plot to compare with the results of Exercise 12.2.

12.12. Referring to the data on defects per bolt of cloth given in Exercise 12.4, suppose that we eliminate data points 1, 5, 14, 15, and 31. Use a cusum plot of the remaining 29 data points to analyze the defect occurrences based on an assumed normal distribution with a standard deviation estimated from the data.

12.13. The data on weld joint strength of Exercise 3.3 were said to have been taken from a process known to be in statistical control. Use the following techniques to check whether in fact the data support the hypothesis that the process is in control, at least with respect to the process mean.

a. A Shewhart-like cusum control chart for standardized sample means.

b. A cusum plot.

12.14. An automated process has generated the data shown in the table. Construct a standardized cusum chart for the process.

Sample	X_i	Sample	X_i
1	0.72	19	−1.25
2	−3.60	20	2.38
3	0.51	21	−0.98
4	−3.56	22	−3.19
5	−3.08	23	1.70

(continued)

Sample	X_i	Sample	X_i
6	0.69	24	-3.25
7	-2.26	25	0.00
8	-2.36	26	1.26
9	1.38	27	3.12
10	5.90	28	2.49
11	1.46	29	0.03
12	-0.97	30	-2.39
13	-0.36	31	-0.51
14	2.45	32	4.07
15	-1.82	33	0.51
16	0.28	34	1.85
17	-2.50	35	4.20
18	1.51		

a. Calculate $\overline{X}$ and s_X from the data.

b. Calculate standardized values (Z's) for all the X's.

c. Compute the standardized cusum statistic (S_t^*) for each t based on the Z values. Construct a standardized cusum chart for the process, and plot the S_t^*.

d. Interpret the chart. Do you notice any trends? If so, what do they imply is taking place in the process? At what points in time do the observed changes occur? Be thorough in your discussion and explain your reasoning.

12.15. It is desired to determine the time of occurrence of small mean shifts that are known to take place in a certain process. Data from this process are collected for the purpose of constructing a standardized cusum chart and are shown in the table.

Sample	X_i	Sample	X_i
1	10.7	21	13.0
2	10.3	22	12.1
3	12.8	23	11.9
4	9.2	24	11.5
5	8.3	25	11.5
6	9.0	26	10.9
7	10.2	27	12.4
8	8.9	28	10.7
9	10.2	29	11.9
10	10.7	30	10.8
11	8.5	31	9.2
12	9.7	32	11.2
13	9.0	33	8.9
14	9.7	34	9.0
15	8.9	35	9.7

Sample	X_i	Sample	X_i
16	9.1	36	10.3
17	11.8	37	9.2
18	11.0	38	10.2
19	9.4	39	10.0
20	8.0	40	9.9

a. Calculate $\overline{X}$ and s_X from the data.
b. Calculate standardized values (Z's) for all the X's.
c. Compute the standardized cusum statistic (S_t^*) for each t based on the Z values. Construct a standardized cusum chart for the process, and plot the S_t^*.
d. Interpret the chart. Do you notice any trends? If so, what do they imply is taking place in the process? At what points in time do the observed changes occur? Be thorough in your discussion and explain your reasoning.

12.16. A process is being studied and the chosen charting method is the standardized cusum chart. The data collected from the process are shown in the table.

Sample	X_i	Sample	X_i
1	6.43	19	3.90
2	6.70	20	4.05
3	8.22	21	4.70
4	7.48	22	5.26
5	8.33	23	4.19
6	8.06	24	5.21
7	7.49	25	5.05
8	7.66	26	4.93
9	7.79	27	4.60
10	6.24	28	5.76
11	3.90	29	5.03
12	4.12	30	5.45
13	6.80	31	4.63
14	5.99	32	5.71
15	4.40	33	5.68
16	3.05	34	3.79
17	4.24	35	5.03
18	6.23		

a. Calculate $\overline{X}$ and s_X from the data.
b. Calculate standardized values (Z's) for all the X's.
c. Compute the standardized cusum statistic (S_t^*) for each t based on the

Z values. Construct a standardized cusum chart for the process, and plot the S_t^*.

    d. Interpret the chart. Do you notice any trends? If so, what do they imply is taking place in the process? At what points in time do the observed changes occur? Be thorough in your discussion and explain your reasoning.

12.17. An engineer studying a process on one of his automated production lines wishes to construct a standardized cusum control chart with the data shown in the table.

Sample	X_i	Sample	X_i
1	4.70	21	2.96
2	5.26	22	3.34
3	4.19	23	3.40
4	5.21	24	3.20
5	5.05	25	2.74
6	4.13	26	5.67
7	4.60	27	3.46
8	5.76	28	4.70
9	5.03	29	3.94
10	5.45	30	3.78
11	4.63	31	4.70
12	5.71	32	3.90
13	5.68	33	4.12
14	3.79	34	5.77
15	5.03	35	5.99
16	3.37	36	4.41
17	4.19	37	3.05
18	4.95	38	4.27
19	2.84	39	5.23
20	3.62	40	3.91

    a. Calculate $\overline{X}$ and s_X from the data.

    b. Calculate S_t^* for each t.

    c. Construct a standardized Shewhart cusum control chart for the process, and plot the S_t^*.

    d. Interpret the chart. Comment on any trends you see, and explain what they mean in terms of the process. At what points in time do any observed changes occur? Be thorough in your discussion and explain your reasoning.

12.18. Use the process data that were collected and are shown in the table to make a standardized cusum control chart for the process. The data are averages of samples of size $n = 4$, collected hourly on a ring broaching process. Assume that σ_X is known to be 2.0.

Sample	$\overline{X}_i$	Sample	$\overline{X}_i$
1	8.6	19	7.0
2	6.5	20	8.2
3	6.3	21	6.9
4	4.7	22	8.2
5	5.3	23	8.7
6	6.5	24	6.5
7	6.4	25	7.7
8	6.9	26	7.0
9	5.1	27	6.8
10	5.8	28	7.7
11	5.6	29	6.9
12	6.9	30	7.1
13	5.4	31	8.8
14	8.7	32	9.0
15	8.3	33	7.4
16	10.8	34	6.0
17	7.2	35	7.2
18	6.3		

a. Calculate $\overline{\overline{X}}$ from the data.
b. Calculate S_t^* for each t.
c. Construct a standardized cusum control chart for the process and plot the S_t^*.
d. Interpret the chart. Comment on any trends you see, and explain what they mean in terms of the process. At what points in time do any observed changes occur? Be thorough in your discussion and explain your reasoning.

12.19. A process is subjected to study for the purpose of detecting the time at which certain mean shifts occur. Assume that the intended process mean is 10.00 units and that a mean shift to 12.00 units is know to occur occasionally. Use the technique referred to in the text as traditional cusum charting to analyze the data collected, shown in the table. Use $\mu_0 = 12.00$, $\mu_1 = 14.00$, $\sigma_{\overline{X}}^2 = 4.00$, and let $\alpha = 0.05$ and $\beta = 0.10$.

Sample	X_i	Sample	X_i
1	12.7	19	13.0
2	12.3	20	11.4
3	14.8	21	10.0
4	11.2	22	11.2
5	10.3	23	13.2
6	11.0	24	10.9
7	12.2	25	11.0

(continued)

Sample	X_i	Sample	X_i
8	10.9	26	11.7
9	12.2	27	12.3
10	12.7	28	11.2
11	10.5	29	12.2
12	11.7	30	12.0
13	11.0	31	15.0
14	10.8	32	14.1
15	11.7	33	13.9
16	10.9	34	13.5
17	11.1	35	15.5
18	13.8		

      a. Calculate the UCL for the chart.

      b. Calculate the cumulative sum statistic, S_t, for each t.

      c. Construct the one-sided traditional cusum control chart for the process, and plot all the S_t's.

      d. Interpret the chart and comment on the patterns and/or trends you see. What behavior in the process may have produced them? At what times do any observed changes occur? Be thorough in your discussion and explain your reasoning.

12.20. Use the same data that were presented in Exercise 12.19, but this time construct a traditional two-sided cusum control chart for the process. Use $\mu_0 = 12.00$, $\mu_1 = 14.00$, $\mu_2 = 10.00$, $\sigma_{\bar{X}}^2 = 4.00$, and let $\alpha = 0.05$ and $\beta_1 = \beta_2 = \beta = 0.10$.

      a. Calculate the UCL and the LCL for the chart.

      b. Calculate the cumulative sum statistic, S_t, for each t.

      c. Construct the two-sided traditional cusum control chart for the process, and plot all the S_t's.

      d. Interpret the chart and comment on the patterns and/or trends you see. What behavior in the process may have produced them? At what times do any observed changes occur? Be thorough in your discussion and explain your reasoning.

12.21. Use the data shown in the table to construct a traditional cusum chart for sample means. Use $\mu_0 = 1.00$, $\mu_1 = 2.00$, $\mu_2 = 0.00$, $\sigma_{\bar{X}}^2 = 1.00$, $n = 5$, and let $\alpha = 0.025$ and $\beta = 0.10$.

Sample	$\overline{X}_i$	Sample	$\overline{X}_i$
1	1.3	16	1.1
2	0.7	17	0.7
3	0.7	18	0.9
4	1.6	19	0.7
5	1.1	20	0.5

Sample	$\bar{X}_i$	Sample	$\bar{X}_i$
6	1.0	21	1.5
7	0.8	22	1.0
8	0.3	23	0.9
9	0.2	24	0.3
10	0.7	25	1.1
11	1.6	26	0.3
12	1.0	27	0.4
13	1.3	28	0.5
14	1.4	29	0.5
15	0.7	30	-0.7

a. Calculate the UCL and the LCL for the chart.
b. Calculate the sample cusum statistic, S_t, for each t.
c. Construct the traditional cusum chart for the sample means, and plot all the S_t's.
d. Interpret the chart and comment on the patterns and/or trends you see. What behavior in the process may have produced them? At what times do any observed changes occur? Be thorough in your discussion and explain your reasoning.

C H A P T E R

13

Shewhart Control Charts for Attribute Data

13.1

Data Characterization by Attributes

Observation by Attributes

To this point, the quality assessment criteria for manufactured goods have been of the variable measurement type (e.g., diameter in inches, filled weight in ounces). However, some quality characteristics are more logically defined in a "presence of or absence of" sense. Such situations might include surface flaws on a sheet metal panel, cracks in drawn wire, color inconsistencies on a painted surface, voids, flash or spray on an injection molded part, or wrinkles on a sheet of vinyl. Such nonconformities or defects are often observed visually or according to other sensory criteria and cause a part to simply be defined as a defective part. In these cases we refer to quality assessment as being made by attributes.

Many quality characteristics that could be made by measurements (variables) are often not done as such in the interest of economy. During an inspection procedure a go/no-go gage may be employed to determine whether or not a variable characteristic falls within the part specification or not. Parts that fail such a test are simply labeled "defective." The opportunity for making variable measurements may be great in a given plant, but the construction and continuing use of $\overline{X}$ and R charts is more time consuming and expensive. Attribute measurements may be usefully employed to identify the presence of problems that can then be attacked by the use of the more sensitive $\overline{X}$ and R charts.

Some Important Definitions

Before developing control charts for attribute data it is necessary to define clearly some important terms. The following definitions are required in working with attribute data.

Defect.   A fault or nonconformity that causes an article or an item to fail to meet specification requirements is called a defect. Each instance of an article's lack of conformity to specification is a defect.

Defective.   An item or article with one or more defects is a defective item.

Number of Defectives.   In a sample of n items, d is the number of defective items in the sample.

Number of Defects.   In a sample of n items, c is the number of defects in the sample. An item may be subject to many different types of defects. As an item is inspected, every instance of each type of defect is noted. The sum of all defects across the sample is c.

Fraction Defective.   The ratio of the number of defective items in the sample, d, to the total number of items in the sample, n. The sample fraction defective p is therefore equal to d/n.

Operational Definitions

The most difficult aspect of quality characterization by attributes is the precise determination of what constitutes the presence of a particular defect. This is so since many attribute defects are visual in nature and therefore require a certain degree of judgment. This is also so because of the inability to discard the product control mentality. For example, a scratch that is barely observable by the naked eye may not be considered a defect in a product control sense. However, the physical mechanism within the process causing scratches may well be the same for both such a small scratch and much larger ones. Remember, it is the purpose of the control chart to signal the presence of faults in the process so that they may be corrected. Furthermore, human variation is generally considerably larger in attribute characterization, as is pointed out in Fig. 13.1. It is therefore important that precise and quantitative operational definitions be set down for all to observe uniformly when attribute quality characterization is being used. The length or depth of a scratch may be specified, or the diameter of a surface blemish, or the length of a flow line on a molded part.

The issue of the product control versus process control way of thinking about defects is a crucial one. From a product control point of view, scratches on an auto grille should be counted as defects only if they appear on visual surfaces that would directly influence part function. From a process control

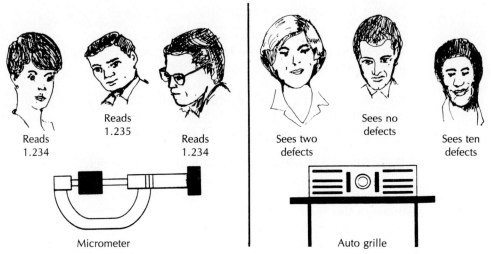

Figure 13.1 Human Variability in Attribute Quality Characterization

point of view, however, scratches on an auto grille should be counted as defects regardless of where they appear since the mechanism creating scratches does not know if it is a visual surface or not. Furthermore, by counting all scratches, the sensitivity of the statistical charting instrument used to identify the presence of defects and lead to their diagnosis will be increased considerably. From a process control perspective we may actually wish to be "harder" on the part than if we are only worried about whether to ship or not! The issue is one of chart sensitivity and fault identification and diagnosis.

A major problem with the product control way of thinking about part inspection is that when attribute quality characterization is being used, not all defects are observed and noted. Often, the first occurrence of a defect that is detected immediately causes the part to be scrapped. Such data are recorded

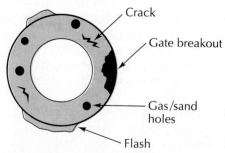

Figure 13.2 Example of Defect Identification in an Attribute Quality Characterization Situation (Engine Valve Seat Blank)

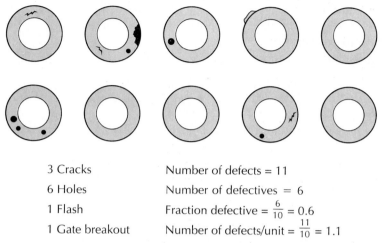

3 Cracks	Number of defects = 11
6 Holes	Number of defectives = 6
1 Flash	Fraction defective = $\frac{6}{10}$ = 0.6
1 Gate breakout	Number of defects/unit = $\frac{11}{10}$ = 1.1

Figure 13.3 Example of Sample Result for Attribute Quality Characterization

in scrap logs, which then present a biased view of what the problem may really be. One inspector may be keen on scratch defects on a molded part and therefore will seem to see these first. Another may think splay is more critical, so his "data" tend to reflect this type of defect more frequently. The net result is that such data may then mislead those who may be trying to use it for process control and improvement purposes.

Figure 13.2 provides an example of the occurrence of multiple defects on a part. It is essential from a process control standpoint to observe and note carefully each occurrence of each type of defect. Figure 13.3 shows a typical sample result and the careful observation of each occurrence of each type of defect. In Fig. 13.3 the four basic measures used in attribute quality characterization are defined for the sample in question. Some other examples of quality characterization by attributes might be

1. *Final inspection of an assembled engine.* Defects:
   - Missing oil dipstick
   - Paint runs
   - Proper ID tag
   - Dyno test (pass/fail)
   - Crankcase bolts missing
2. *Automobile radiator grille after injection molding process.* Defects:
   - Splay
   - Oil/grease
   - Flash
   - Scratches/scuffs
   - Black spots
   - Voids
   - Sinks
   - Flow lines
   - Burns
   - Short shots
3. *Accounts payable process.* Defects:
   - No purchase order
   - Units conversion issue

- Vendor number missing
- Missing approval signature
- No packing slip
- Material not received

13.2

Control Chart for Fraction Defective

Consider an injection molding machine producing a molded part, an instrument panel for an automobile, at a steady pace. Suppose that the measure of quality conformance of interest is the occurrence of four types of defects: flash, splay, voids, and short shots. If a part has even one occurrence of any of these defects, it is considered to be nonconforming (i.e., a defective part). Recently, the scrap rate of this process has been quite high and it is decided to employ the statistical control chart idea to attempt to determine the source(s) of the problem.

Samples of $n = 10$ consecutive panels were examined every 45 minutes. Each incidence of each type of defect was noted. Also, the number of panels in each sample with at least one defect present was noted (i.e., the number of defectives, d). For the purpose of getting an overall impression of the performance of the process, the data for each shift, 10 samples of $n = 10$ panels each, were collected into a "shift sample" of $n_s = 100$ panels. A fraction defective control chart, a p chart, is to be used to study the process. The data in Table 13.1 show the number of panels in each shift sample of $n_s = 100$ that contain at least one defect and the associated fraction defective data for 30 consecutive shifts.

The process characteristic of interest here is the true process fraction defective μ_p, often denoted as p', so that each sample result is converted to a fraction defective p as shown in Table 13.1,

$$p = \frac{d}{n}.$$

The data (fraction defective p) are plotted for 30 successive samples of the size $n_s = 100$. This plot is shown in Fig. 13.4, where a centerline has been established on the chart by averaging all 30 p values.

The individual values for the sample fraction defective, p, vary considerably and it is difficult at this point to determine from the plot if the variation about the average fraction defective $\bar{p}$ is due solely to the forces of common causes. Several of the p values are quite low, perhaps indicating that some sporadic behavior is evident in the chart. To help us resolve this problem, the Shewhart control chart idea previously used for variable observations ($\overline{X}$ and R) can be applied. But to develop the Shewhart control chart for the attribute quality characteristic observed as the fraction defective in a sample or subgroup, we will need to know more about how defective items rise, in a frequency sense, when subgroups of several such items are collected.

TABLE 13.1 Fraction Defective Data for 30 Shift Samples for Instrument Panel Assembly Example		
Sample	Number Defective	Fraction Defective
1	7	0.07
2	8	0.08
3	6	0.06
4	8	0.08
5	6	0.06
6	8	0.08
7	3	0.03
8	5	0.05
9	9	0.09
10	7	0.07
11	7	0.07
12	9	0.09
13	8	0.08
14	7	0.07
15	8	0.08
16	10	0.10
17	10	0.10
18	5	0.05
19	12	0.12
20	11	0.11
21	8	0.08
22	10	0.10
23	4	0.04
24	10	0.10
25	7	0.07
26	7	0.07
27	9	0.09
28	8	0.08
29	10	0.10
30	10	0.10
	237	

Binomial Sampling Situation

Suppose that we have an hour's production of 1000 parts, of which 50 are defective and 950 are nondefective. If we randomly pick one part from these 1000 parts, there will be two possible outcomes: we get either a defective or a nondefective. Suppose that we record the outcome as "N" for nondefective and "D" for defective. We mix the 1000 parts thoroughly after replacing the first part and randomly pick another part from the lot, recording the outcome as "D" or "N." Remember, we are putting back the part after we have sampled. If we repeat the sampling five times, at the end we will have recorded the

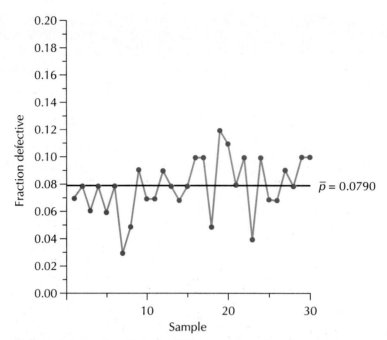

Figure 13.4 Fraction Defective Values for Samples of $n = 100$ Instrument Panels Collected over 30 Consecutive Shifts

result of five trials or outcomes. We could think of this as a sample of $n = 5$ parts, some of them being defective, some of them not.

Now, the problem we are interested in is determining the respective probabilities of getting either 0, 1, 2, 3, 4, or 5 defectives out of five trials or outcomes. In other words, out of five outcomes, we may get no defectives, 1 defective, 2 defectives, 3 defectives, 4 defectives, or even all 5 defectives, and we are interested in knowing the probabilities associated with each of these possible outcomes. In this case we know that there are 50 defective and 950 nondefective parts in the 1000 parts. Therefore, the true fraction defective for these parts is

$$p' = \frac{50}{1000} = 0.05,$$

which may be interpreted as the probability of getting a defective from a single trial. Then, the probability of getting a nondefective on each independent trial is

$$1 - p' = \frac{950}{1000} = 0.95.$$

Note that total probability $[p' + (1 - p') = 0.05 + 0.95]$ is 1.0.

Let us denote the sample size by n. Since we took five random trials, $n = 5$. Now let us try to compute the probability of getting a specific sample result, namely, the combination of 1 defective and 4 nondefectives, that is, $P(1D, 4N)$, in a sample of $n = 5$. If the five draws are independent, that is, the result for one draw does not depend on the specific result of another draw, the probability of obtaining 1 defective and 4 nondefectives is simply the product of the individual trial outcome probabilities,

$$P(1D, 4N) = (p')(1 - p')(1 - p')(1 - p')(1 - p')$$
$$= (0.05)(0.95)(0.95)(0.95)(0.95) = (0.05)^1(0.95)^4 = 0.041.$$

The probability above relates to one particular occurrence of 1 defective and 4 nondefective parts. But we can obtain 1 defective and 4 nondefectives in the following five possible ways, taking the order of the trials into account:

NNNND,   NNNDN,   NNDNN,   NDNNN,   DNNNN.

In other words, we could have gotten the defective on either the first draw, the second, or the third, fourth, or fifth draws. For this situation it is easy to enumerate the number of ways we can get 1 defective and 4 nondefectives in a sample of $n = 5$. But if the sample size is $n = 20$, how many ways can you obtain, for example, 4 defectives and 16 nondefectives? The answer is 4845. Clearly, we need a general formula to determine the number of *combinations* we could obtain. This is given by

$$\binom{n}{d} = \frac{n!}{d!\,(n - d)!},$$

where $n!$ is read as "n factorial" (e.g., $5! = 5 \cdot 4 \cdot 3 \cdot 2 \cdot 1 = 120$). In our case, $n = 5$ and $d = 1$ (defective). Thus

$$\binom{5}{1} = \frac{5!}{1!\,(5 - 1)!} = \frac{5!}{1!\,4!} = 5.$$

We saw above that $P(1D, 4N) = (0.05)^1(0.95)^4$ is the probability of only one combination. Since we can get 1 defective and 4 nondefectives in $\binom{5}{1} = 5$ ways, the total probability of getting 1 defective and 4 nondefectives, which we now denote simply as $P(d = 1)$, will be

$$P(d = 1) = \binom{5}{1}(0.05)^1(0.95)^4 = (5)(0.041) = 0.205. \tag{13.1}$$

We note here that in the sample above the true fraction defective, p', in the lot of 1000 parts is $50/1000 = 0.05$, but the sample result we obtained (1 defective and 4 nondefectives) constitutes a sample fraction defective of $p = 1/5 = 0.20$. We note further that this result will occur about 20% [i.e., $P(d = 1) = 0.205$] of the time! That is, the sample will produce a fraction defective four times the true lot (population) fraction defective for 20% of the samples so chosen. Clearly the forces of chance (common-cause) variation are at work and produce uncertainty in our sample result as it reflects the true situation at hand.

A number of other results could have occurred when we drew the sample (five successive trials). We could have gotten

d	$\binom{n}{d}$	$P(d)$
0D + 5N (0 defective, 5 nondefective)	0	1	0.7734809
1D + 4N (1 defective, 4 nondefective)	1	5	0.2036266
2D + 3N (2 defective, 3 nondefective)	2	10	0.0214344
3D + 2N (3 defective, 2 nondefective)	3	10	0.0011281
4D + 1N (4 defective, 1 nondefective)	4	5	0.0000297
5D + 0N (5 defective, 0 nondefective)	5	1	0.0000003

Total = 1.0000000

It is noted that the total probability of all the possible results adds up to 1.0.

In general, the binomial probability function is given by

$$P(d) = \binom{n}{d}(p')^d(1 - p')^{n-d} \tag{13.2}$$

for $d = 0, 1, 2, \ldots, n$, and where d corresponds to the number of defectives, n corresponds to the sample size, $(n - d)$ to the number of nondefectives, and p' and $(1 - p')$ are, respectively, the probabilities associated with the occurrence of a defective and a nondefective on a single trial. The sum of the probabilities of all the possible results from $d = 0$ to $d = n$ will then be given by

$$\sum_{d=0}^{n} P(d) = \sum_{d=0}^{n} \binom{n}{d}(p')^d(1 - p')^{n-d} = 1. \tag{13.3}$$

The left-hand side terms of Eqs. (13.2) and (13.3) are commonly written as

$$b(d; n, p') = \binom{n}{d}(p')^d(1 - p')^{n-d} \tag{13.4}$$

$$B(x; n, p') = \sum_{d=0}^{x} \binom{n}{d}(p')^d(1 - p')^{n-d}, \tag{13.5}$$

where $x; n, p'$ in parentheses means the probability of x or less for a given n and p'. The lowercase letter b outside the parentheses on the left-hand side of Eq. (13.4) means the probability of getting exactly d defectives, whereas the capital letter B outside the parentheses on the left-hand side of Eq. (13.5) means the cumulative probability of getting up to and including x or at the most x defectives.

Equations (13.4) and (13.5) are called the *probability function* and *cumulative probability function*, respectively, of the binomial distribution. The binomial distribution is a discrete distribution and serves as a very useful tool in dealing with statistical problems involving repeated trials where the characteristic observed will have either one of two outcomes. We present its applications after we discuss its important properties.

Assumptions of the Binomial Distribution.   In the development of the binomial distribution (probability law) above, we have made several important assumptions about the sampling situation. These assumptions have, in fact, led us to the mathematical formulation that we refer to as the binomial probability model. These assumptions are summarized below:

1. *Only two outcomes are possible for each trial.* Parts can be classified as either defective or nondefective. Each item sampled must be either good or bad; it can be nothing else. The probabilities associated with a defective and a nondefective must therefore add to 1.0.
2. *Sampling with replacement or from a stable process.* By replacing the sampled part after each trial we have ensured that the true fraction defective p' will remain constant from trial to trial. In control chart applications, the true fraction defective p' will remain constant if the sampled process is stable, or in statistical control.
3. *Independence of successive trials.* We have assumed that the outcome of a particular trial is in no way influenced by the outcomes of other trials. Hence we have been able to determine the joint probability of the outcome of the group of trials (total sample) as the product of the probabilities associated with the occurrence of each individual trial.

Properties of Binomial Distribution.   Several important properties of the binomial distribution are now discussed.

1. The binomial distribution has two parameters, n and p'. If we look at Eqs. (13.4) and (13.5), we find that we need only specify n and p' to find the various probabilities.
2. An interesting property of the binomial distribution is illustrated in Figs. 13.5, 13.6, and 13.7 for the case $n = 5$ and for $p' = 0.5$, $p' = 0.20$, $p' = 0.80$, respectively. When $p' = 0.50$, Eq. (13.4) becomes

$$b(d; n, p') = \binom{n}{d}(0.5)^d(1 - 0.5)^{n-d} = \binom{n}{d}(0.50)^n$$

   and the shape of the binomial distribution, with $p' = 0.50$, is symmetrical, as shown in Fig. 13.5. Note that if p' is less than 0.50, it is more likely that d will be small rather than large compared to n. This is shown in Fig. 13.6 for $n = 5$ and $p' = 0.20$. A probability distribution which has a shape like that of Fig. 13.6 is called *positively skewed,* where the long tail is on the right side. Similarly, if p' is more than 0.50, it is more likely that d will be large rather than small compared to n. This is shown in Fig. 13.7 for $n = 5$ and $p' = 0.80$. A probability distribution that has a shape like that of Fig. 13.7 is called *negatively skewed,* where the long tail is on the left side.
3. Mean and variance:
   a. Mean of d, $\mu_d = np'$.
   b. Variance of d, $\sigma_d^2 = np'(1 - p')$.

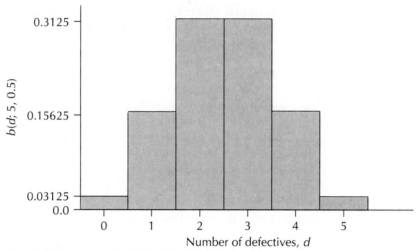

Figure 13.5 Symmetrical Binomial Distribution

Examples of the Use of the Binomial Probability Model.   The binomial distribution can be applied in cases where there are only two possible outcomes for each trial: conforming and nonconforming, success and failure, defective and nondefective, and so on. Several examples are given as follows.

Example.   Suppose we want to find the probability that exactly 9 out of 10 gears are bad, given the probability that any one of these gears will be bad is $p' = 0.80$.

$$b(9; 10, 0.80) = \binom{10}{9}(0.80)^9(0.20)^1 = 0.268$$

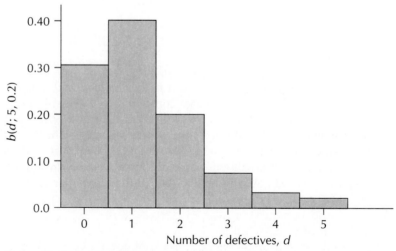

Figure 13.6 Positively Skewed Binomial Distribution

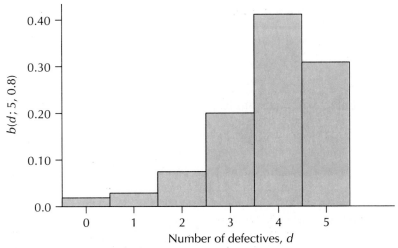

Figure 13.7 Negatively Skewed Binomial Distribution

Example. What is the probability that at least 9 out of 10 gears are bad if $p' = 0.80$?

Remember that "at least 9" means 9 or above, which means that either 9 or 10 are bad (since $n = 10$). In other words, we are asking for the sum of the probabilities of exactly 9 and exactly 10 bad out of 10:

$$b(9; 10, 0.80) + b(10; 10, 0.80) = 0.268 + 0.107 = 0.375.$$

Example. What is the probability that at most 8 of these gears are bad if $p' = 0.80$?

$$B(8; 10, 0.8) = \sum_{d=0}^{8} b(d; 10, 0.80)$$

since "at most 8" means anything from $d = 0$ to 8. But rather than evaluate each of the 9 (0 to 8) probabilities required for this sum, it is easier to find the probability that at least 9 gears are bad and subtract the result from 1, the total probability. We thus obtain $(1 - 0.375) = 0.625$.

If n is large, the calculation of the binomial distribution can become quite tedious. Fortunately, good approximations are available for binomial probability calculations when the sample size is very large. One such approximation for small p' values is given by the Poisson probability distribution discussed later in this chapter.

Control Limits for the p Chart

In the preceding section it was shown that for random sampling under the assumptions of:

1. Part is either defective or nondefective,
2. Sampling with replacement, and
3. Independent trials,

the occurrence of the number of defectives d in a sample of size n is explained probabilistically by the binomial law:

$$P(d \text{ defectives in a sample of } n \text{ items}) = P(d)$$

$$P(d) = \frac{n!}{d!\,(n-d)!}(p')^d(1-p')^{n-d}. \tag{13.6}$$

In quality control work, it has been found useful to use the fraction defective p for charting purposes. Since the sample fraction defective p is simply the number of defectives d divided by the sample size n, the mean and standard deviation for p may be obtained using theorems 1 and 3 on page 76 in Chapter 3. The mean for p is

$$\begin{aligned}
\mu_p &= E\left(\frac{d}{n}\right) \\
&= \frac{1}{n}E(d) \\
&= \frac{1}{n}np' \\
&= p'.
\end{aligned} \tag{13.7}$$

The variance of p is

$$\begin{aligned}
\sigma_p^2 &= \text{Var}\left(\frac{d}{n}\right) \\
&= \frac{1}{n^2}\text{Var}(d) \\
&= \frac{1}{n^2}np'(1-p') \\
&= \frac{p'(1-p')}{n}.
\end{aligned}$$

The standard deviation of p is therefore

$$\sigma_p = \sqrt{\frac{p'(1-p')}{n}}. \tag{13.8}$$

Hence, if p' were known, the centerline of the p chart would be at $\mu_p = p'$ and the control limits would be set at

$$p' \pm 3\sigma_p \tag{13.9}$$

or

$$p' \pm 3\sqrt{\frac{p'(1-p')}{n}}. \tag{13.10}$$

Generally, p' is not known, so we must estimate p' from the sample data obtained. Given k rational samples of size n, the true fraction defective p' may be estimated by

$$\bar{p} = \sum_{i=1}^{k} \frac{p_i}{k},$$ (13.11)

where $p_i = d_i/n_i$ for the ith sample. A more general form is given by

$$\bar{p} = \frac{\sum_{i=1}^{k} d_i}{\sum_{i=1}^{k} n_i}.$$ (13.12)

The second formula is more general, since it is valid whether or not the sample size is the same for all samples. The first formula should only be used if the sample size n is the same for all k samples.

Therefore, given $\bar{p}$, the control limits for the p chart are then given by

$$\text{UCL}_p = \bar{p} + 3 \sqrt{\frac{\bar{p}(1 - \bar{p})}{n}}$$

$$\text{LCL}_p = \bar{p} - 3 \sqrt{\frac{\bar{p}(1 - \bar{p})}{n}}.$$ (13.13)

Thus only $\bar{p}$ has to be calculated for at least 25 samples of size n to set up a p chart. Sometimes the calculation for the lower control limit may yield a value of less than zero. The binomial distribution is not symmetric and has a lower bound of $d = 0$. In this case a lower control limit of zero is used.

Return to the Injection Molding Example of p Chart Application

In this example we have been examining a process for injection molding of a portion of an automobile instrument panel. At one point in the process the partially completed panel is examined for the presence of four different types of molding defects. Recall that the data for 30 samples of size $n = 100$ were given in Table 13.1 and the fraction defective values were plotted in Fig. 13.4. Given the data in Table 13.1 and the results above based on the binomial distribution, the centerline and the control limits for the p chart for these 30 samples can be determined as shown below. Figure 13.8 shows the p chart for the fraction defective data of Table 13.1.

$$\bar{p} = \frac{\text{sum of all defectives}}{\text{total number of units}}$$

$$= \frac{237}{3000}$$

$$= 0.0790$$

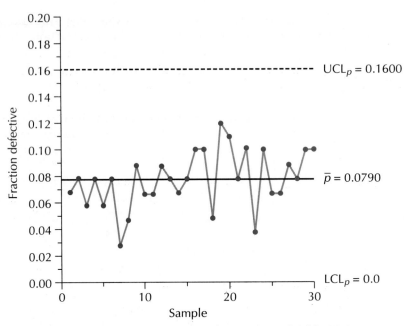

Figure 13.8 p Chart for the Fraction Defective Data of Table 13.1

For $\bar{p} = 0.0790$ and $n = 100$,

$$\hat{\sigma}_p = \sqrt{\frac{\bar{p}(1 - \bar{p})}{n}}$$

$$= \sqrt{\frac{(0.0790)(1 - 0.0790)}{100}}$$

$$= 0.0270$$

$$\text{UCL} = \bar{p} + 3\hat{\sigma}_p$$

$$= 0.0790 + (3)(0.0270)$$

$$= 0.1600$$

$$\text{LCL} = \bar{p} - 3\hat{\sigma}_p$$

$$= 0.0790 - (3)(0.0270)$$

$$= -0.002$$

$$= 0.0.$$

The interpretation of the p chart follows the same guidelines as other Shewhart charts in a general sense. That is, we look for the appearance of four basic types of behaviors in the data:

1. No points should exceed the control limits.
2. The points should be distributed randomly within the control limits (i.e.,

more points should be in the middle of the chart and fewer toward the control limits).
3. The points should show no evidence of trends or recurring cycles.
4. The points should look quite random with time (i.e., no patterns such as runs above or below the centerline are evident).

Since the p values do not follow the normal distribution, the rules based specifically on the three zones between the centerline and either of the control limits cannot be used (discussed in Chapter 6). However, rules for runs and trends can be employed.

Examination of the p chart in Fig. 13.8 shows that no signals are present that would indicate the process is out of control. It would appear that a process only subject to common-cause variation is present. But while the process seems to be stable over the period studied, the fraction defective rate is quite high, on average at 8%; clearly this is unacceptable. And the chart indicates that the problem is a system problem.

An investigation was launched to determine the cause(s) of the high defective rate on the panels. The defect data were examined and it was determined that one of the defects, flash, was accounting for over 50% of all the defects. The process in question was examined and it was determined that certain molding process parameters—the pressure settings—needed some adjustment. The mold pressure settings were changed and the data collection process was continued.

Table 13.2 provides the fraction defective data for the 30 shifts that followed the adjustments on the mold pressure. These data are plotted as a continuation of the original chart (Fig. 13.8) using the same centerline and control limits and are shown in Fig. 13.9. It is clear from the general appearance of the chart that a significant change has occurred in the process. In examining Fig. 13.9 more carefully, we see that a run of 12 consecutive points below the centerline begins at sample 31, a run of 8 in a row below the centerline begins at sample 44, and a run of 8 in a row below the centerline begins at sample 53. In fact, only two of the second 30 sample fraction defective values fall at or above the centerline. It is clear that the process has changed—in particular, that the fraction defective rate has decreased considerably, perhaps as much as halved.

The centerline and control limits for the p chart for samples 31 to 60 were recalculated using only the data from those samples. The calculations are given below and the resulting p chart for these samples is shown in Fig. 13.10.

$$\bar{p} = \frac{\text{sum of all defectives}}{\text{total units}}$$

$$= \frac{114}{3000}$$

$$= 0.0380$$

$$\hat{\sigma}_p = \left[\frac{0.038(1 - 0.038)}{100} \right]^{1/2}$$

$$= 0.0191$$

TABLE 13.2	Fraction Defective Data for 30 Shift Samples After the Pressure Adjustments Were Made for the Instrument Panel Molding Example	

Sample	Number of Defectives	Fraction Defective
31	5	0.05
32	6	0.06
33	2	0.02
34	5	0.05
35	6	0.06
36	3	0.03
37	2	0.02
38	4	0.04
39	3	0.03
40	6	0.06
41	4	0.04
42	3	0.03
43	8	0.08
44	3	0.03
45	2	0.02
46	4	0.04
47	2	0.02
48	2	0.02
49	5	0.05
50	3	0.03
51	5	0.05
52	8	0.08
53	2	0.02
54	1	0.01
55	2	0.02
56	1	0.01
57	7	0.07
58	4	0.04
59	4	0.04
60	2	0.02
	114	

$$\text{UCL}_p = 0.0380 + (3)(0.0191)$$
$$= 0.0953$$
$$\text{LCL}_p = 0.0380 - (3)(0.0191)$$
$$= -0.0193$$
$$= 0.0.$$

From Fig. 13.10 two things are clear. First, the process is still running in a stable fashion after the machine adjustment. A careful examination of the data indicates that none of the rules that apply here for out-of-control conditions are violated. Second, the fraction defective rate has been reduced dramatically, to

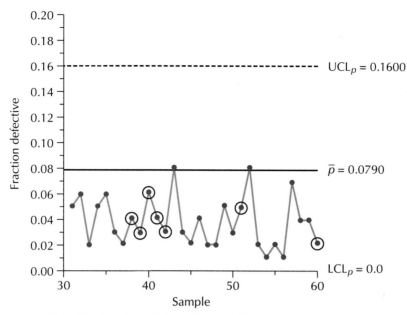

Figure 13.9 Continuation of the p Chart of Figure 13.4

about one-half of its original level. A common-cause/system problem has been identified, adjustments have been made to the process, and the continued charting has demonstrated through the data that a significant improvement in the process has taken place. Figure 13.11 summarizes the charting of this process through the 60 samples and clearly shows the nature and magnitude of the change in the system.

13.3
Variable-Sample-Size Considerations for the p Chart

It is often the case that the sample size may vary from one time to another as data for the construction of a p chart are obtained. This may be the case if the data have been collected for other reasons or if a sample constitutes a day's production and production rates vary from day to day. Since the limits on a p chart depend on the sample size n, some adjustments must be made to ensure that proper interpretation of the chart is made.

There are several ways that the variable-sample-size problem can be handled. Some of the more common approaches are

1. Compute separate limits for each subgroup. This approach certainly leads to a correct set of limits for each sample but requires continual calculation of the control limits and a somewhat messy-looking control chart.

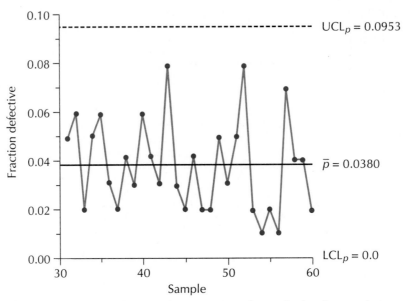

Figure 13.10 p Chart for Samples 31 to 60 with Recalculated Control Limits

2. Determine an average sample size $\bar{n}$ and a single set of control limits based on $\bar{n}$. This method may be appropriate if the sample sizes do not vary greatly, perhaps no more than about 20%. However, if the actual n is less than $\bar{n}$, a point above the control limit based on $\bar{n}$ may not be above its own true upper control limit. Conversely, if the actual n is greater than $\bar{n}$, a point may not show out-of-control, when in fact it really is.

3. A third procedure for handling the variable-sample-size problem for the p chart is to express the fraction defective p in standard deviation units; that is, $Z = (p - \bar{p})/\hat{\sigma}_p$ becomes the value plotted on a control chart where the centerline is zero and the control limits are set at ± 3.0. This standardizes the plotted value even though n may be varying. Note that $Z = (p - \bar{p})/\hat{\sigma}_p$ is a familiar form; recall the standard normal distribution, $N(0, 1^2)$. For this method the continued calculation of the standarized variable is somewhat tedious, but the chart has a clean appearance, with constant limits always at ± 3.0 and a constant centerline at 0.0.

Example: Variable-Sample-Size p Chart— Manufacture of a Package Tray

The following example deals with a process that manufactures the package tray that is used in a line of hatchback economy automobiles. The tray starts with a wood fiber/polymer composite material that is extruded into thin sheets. This material is then heated in a furnace and molded in a press that forms the shape of the tray. As part of the molding process the tray has a carpetlike (cloth)

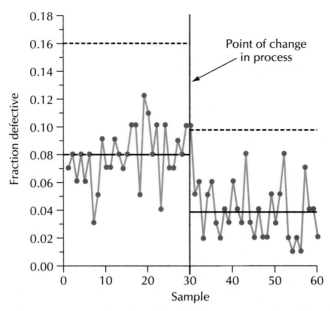

Figure 13.11 p Chart for All of the Samples Analyzed for the Instrument Panel Assembly Example

material applied by an adhesive to the surface, after which the formed and carpeted tray is trimmed to size. The process we are describing here was developed several years ago to replace the existing tray, which was made of cardboard and lacked sufficient strength for some extreme loadings that had been experienced in field use. The process was launched on a pilot basis in the plant that ultimately was to manufacture this package tray. A semimanual operation was developed to study the processing parameters for the furnace heating and the molding process to develop the operating conditions for the scaled-up process.

In an effort to evaluate the performance under operating conditions, the process was run semimanually one shift per day. Depending on machine adjustments and other factors, somewhere between 200 and 300 trays could be made per shift. After some initial experiments were run to determine a first set of operating conditions, the process was run for several weeks, inspecting each tray for a series of defects, making minor process adjustments along the way, and carefully noting what the adjustments were and when they were made. After 30 shifts of data were available, a p chart was initiated and used to evaluate the influence of changes made in the process.

There were seven different defects noted in inspecting the trays. These were

- Cloth coverage on the hinge.
- Carpet coverage.

- Bleed-through (substrate showing through the carpet).
- Soiled carpet surface.
- Improper flex on hinge.
- Chips, scratches, abrasions on back surface.
- Wrinkles in the carpet.

Table 13.3 provides defective and fraction defective data for the first 30 samples collected. Remember, a sample is really an entire shift's production. For this example we construct and examine three different p charts, one with individual limits for each sample, one with constant limits based on an average sample size, and one based on the standardization of the observed fraction defective value.

TABLE 13.3 Data for Package Tray Example						
Sample	n	d	p	LCL_p	UCL_p	z
1	238	11	0.046	0.026	0.131	−1.84
2	245	18	0.073	0.027	0.130	−0.28
3	270	17	0.063	0.029	0.127	−0.94
4	207	15	0.072	0.022	0.134	−0.31
5	251	11	0.044	0.027	0.129	−2.03
6	254	15	0.059	0.028	0.129	−1.14
7	236	19	0.081	0.026	0.131	0.13
8	245	20	0.082	0.027	0.130	0.19
9	246	35	0.142	0.027	0.130	3.74
10	269	14	0.052	0.029	0.127	−1.60
11	223	7	0.031	0.024	0.132	−2.61
12	246	42	0.171	0.027	0.130	5.40
13	262	14	0.053	0.029	0.128	−1.50
14	258	15	0.058	0.028	0.128	−1.21
15	232	20	0.086	0.025	0.131	0.45
16	219	9	0.041	0.024	0.133	−2.05
17	263	23	0.087	0.029	0.128	0.55
18	244	11	0.045	0.027	0.130	−1.93
19	274	21	0.077	0.030	0.127	−0.10
20	245	37	0.151	0.027	0.130	4.24
21	233	16	0.069	0.026	0.131	−0.55
22	267	18	0.067	0.029	0.128	−0.66
23	254	20	0.079	0.028	0.129	0.03
24	264	16	0.061	0.029	0.128	−1.07
25	253	34	0.134	0.028	0.129	3.32
26	290	22	0.076	0.031	0.126	−0.15
27	231	9	0.039	0.025	0.131	−2.23
28	227	40	0.176	0.025	0.132	5.49
29	234	18	0.077	0.026	0.131	−0.08
30	253	15	0.059	0.028	0.129	−1.13

***p* Chart Based on Individual Limits.**   The most obvious approach to the variable-sample-size problem for fraction defective data is simply to calculate separate limits for each sample based on its own sample size. The centerline of the chart will be the average fraction defective, $\bar{p}$. Calculations for the centerline and control limits for this chart are given below.

Centerline for all samples:

$$\Sigma\, n_i = 7433, \qquad i = 1, 30$$
$$\Sigma\, d_i = 582, \qquad i = 1, 30$$
$$\bar{p} = \frac{582}{7433}$$
$$= 0.0783.$$

Control limits for sample 1:

$$\text{UCL}_{p_1} \text{ and LCL}_{p_1}: \quad \bar{p} \pm 3\sqrt{\frac{\bar{p}(1 - \bar{p})}{n_1}}.$$

For sample 1, $n_1 = 238$,

$$0.0783 \pm 3\left[\frac{0.0783(1 - 0.0783)}{238}\right]^{1/2}$$
$$0.0783 \pm 0.0527$$
$$\text{LCL}_{p_1} = 0.026, \qquad \text{UCL}_{p_1} = 0.131.$$

The control limits for each of the remaining 29 samples were calculated in a similar fashion and are given in Table 13.3. The associated p chart is shown in Fig. 13.12. From the figure we see that five of the sample p values fall above their respective upper control limits, indicating that at least for those shifts the true process fraction defective was higher than the centerline.

***p* Chart Based on Average-Sample-Size Control Limits.**   When the sample size variation is not too great, average-sample-size control limits can be determined for the p chart. The advantage is that the chart looks "clean" because only a single set of control limits is present. The disadvantage is that we cannot be sure about the interpretation of the individual p values without checking them against their actual individual limits. This need only be done for those p values near the control limits, on either side.

   Calculations for the average sample size and the associated control limits are given below. The centerline of the chart remains the average fraction defective value previously calculated:

$$\Sigma\, n_i = 7433, \qquad i = 1, 30$$
$$\bar{n} = \frac{7433}{30} = 247.78.$$
$$\bar{n} \approx 248.$$

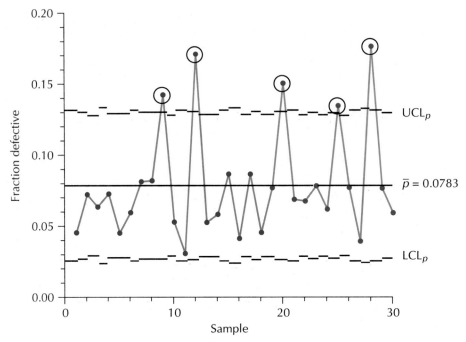

Figure 13.12 Variable-Sample-Size p Chart with Individual Limits for Each Sample: Package Tray Example

Upper and lower control limits:

$$0.0783 \pm 3 \left[\frac{0.0783(1 - 0.0783)}{248} \right]^{1/2}$$

$$\text{UCL}_p = 0.1295, \ \text{LCL}_p = 0.0271.$$

Figure 13.13 shows the associated p chart with a single set of control limits based on the average sample size. In this example, the individual sample sizes vary about the average by less than 20%, so it is likely that we would be safe in checking only those p values that are close to the limits. Samples 9 and 25 might qualify for further scrutiny. From Table 13.3 we see that sample 9 has a sample size of 246 and sample 25 has a sample size of 253. For sample 9, the sample size is very close to the average, so the interpretation of this point as being out of control, based on the average limit, is valid. In the case of sample 25, since its sample size is greater than the average, its own upper control limit would be lower than the average limit, so again, the interpretation of this point as being out of control, based on the average limit, is valid.

p **Chart Based on Standardized *p* Values.**   A p chart for varying sample sizes that is very simple to interpret can be made by standardizing the individual p

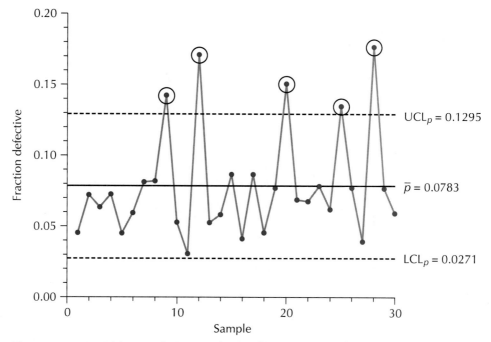

Figure 13.13 Variable-Sample-Size p Chart with Average-Sample-Size Limits: Package Tray Example

values. In this case the centerline of the chart is zero and the control limits are set at 3.0 and -3.0.

The procedure here for variable sample size is to express the fraction defective p in standard deviation units [i.e., $Z = (p - \bar{p})/\hat{\sigma}_p$]. This means that the standard deviation must be calculated for each sample because σ_p depends on the sample size. Once this is done, the standardized value for each sample can be determined. These values are given in the last column of Table 13.3. The associated p chart is shown in Fig. 13.14. In the figure we see that the interpretation—that five samples have standardized p values above the upper control limit—is the same as that for Figs. 13.12 and 13.13. Although the calculations here are a bit more tedious, the common use of computers today for chart construction negates this apparent disadvantage of this chart.

13.4

Control Chart for Number of Defectives

Sometimes it is convenient to make a control chart simply by plotting the number of defectives d in the sample instead of the fraction defective p. This is particularly true if the sample size n is constant and the fraction defective p is

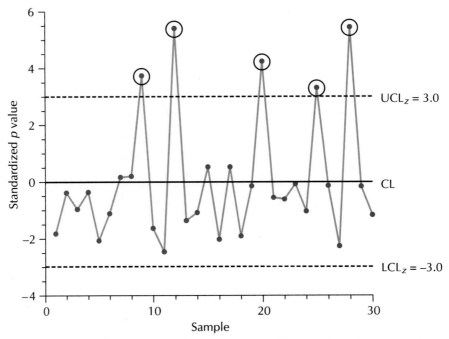

Figure 13.14 Standardized Variable-Sample-Size p Chart: Package Tray Example

quite small. Such a chart, commonly referred to as an np *chart*, provides essentially the same information as the p chart. The basic difference between the p and np charts is a scaling of the vertical axis by a constant n. The np chart requires one less calculation since $p = d/n$ for each sample need not be determined.

The centerline and control limits for the np chart are of course based on the binomial distribution. The average number of defectives expected in a sample of n is np' and the standard deviation for d is

$$\sigma_d = [np'(1 - p')]^{1/2}. \tag{13.14}$$

Hence the control limits for the np chart based on actual data are given by

$$n\bar{p} \pm 3\sqrt{n\bar{p}(1 - \bar{p})}. \tag{13.15}$$

Example: Control Chart for Number of Defectives

To illustrate the relationship between the p and the np charts, the data of Table 13.1 will again be used here in the construction of an np chart. The p chart and the np chart for the same data should look the same, the only difference being the vertical scale.

Using the first 30 samples for the instrument panel assembly example, we find that the following calculations for the centerline and the upper and lower control limits lead to the np chart shown in Fig. 13.15.

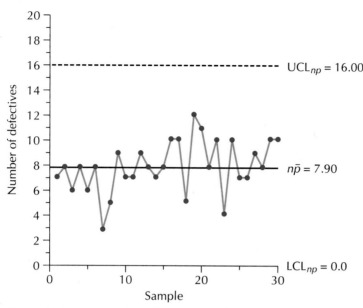

Figure 13.15 Control Chart for Number of Defectives Using Data of Table 13.1

$$\bar{p} = \frac{237}{3000}$$
$$= 0.079$$
$$n\bar{p} = 100(0.079)$$
$$= 7.9$$

Control limits:

$$n\bar{p} \pm 3\sqrt{n\bar{p}(1 - \bar{p})}$$
$$7.9 \pm 3\sqrt{7.9(1 - 0.079)}$$

$$UCL_{np} = 16.00$$

$$LCL_{np} = -0.20, \quad \text{so a lower limit of } 0.00 \text{ is used.}$$

Comparison of Figs. 13.8 and 13.15 shows the same pattern in the data, and the resulting analysis of this pattern for Fig. 13.15 is the same as for Fig. 13.8. Again, this should not be surprising since the only difference in the two charts is the constant scaling of the vertical axis.

Use of the np chart has been very limited. One reason for not using np is that if the sample size varies, not only will the control limits vary but the centerline will vary from sample to sample as well. Such a chart is more difficult to interpret. On the other hand, the np chart is really plotting d, the number of defectives in the sample, and therefore one less calculation would be required. The problem, though, is that the number of defectives is a hard number to grasp and interpret without consideration of the sample size, hence the acceptance of the use of the fraction defective, p.

13.5

Control Chart for Number of Defects

Importance of Number of Defects per Sample

The p chart deals with the notion of a defective part or item, where "defective" means that the part has at least one nonconformity or disqualifying defect. It must be recognized, however, that the incidence of any one of several possible nonconformities would qualify a part for defective status. A part with 10 defects, any one of which makes it a defective, is on equal footing with a part with only one defect, in terms of being a defective.

Often, it is of interest to note every occurrence of every type of defect on a part and chart the number of defects per sample. A sample may only be one part, particularly if interest is focusing on final inspection of an assembled product, such as an automobile, a lift truck, or a washing machine. Inspection may focus on one type of defect, such as nonconforming rivets on an aircraft wing, or multiple defects, such as flash, splay, voids, and knit lines on an injection-molded truck grille.

Considering an assembled product such as a lift truck, we find that the opportunity for the occurrence of a defect is quite large, and perhaps should be considered infinite. However, the probability of occurrence of a defect in any one spot arbitrarily chosen is probably very, very small. In this case the probability law that governs the incidence of defects is known as the *Poisson law* or *Poisson probability distribution*, where we denote c to be the number of defects per sample. It is important that the opportunity space for defects to occur be constant from sample to sample.

The Poisson distribution defines the probability of observing c defects in a sample, where c' is the mean rate of occurrence of defects per sample. To see more clearly how the Shewhart control chart model for the number of defects in a sample emerges, we now briefly examine this type of sampling situation probabilistically.

Poisson Probability Distribution

If the probability of an outcome, p', of a binomial random variable X is very small (or very large), n is large, and np' is moderately large but, say, less than 20, the binomial probability function of Eq. (13.6) can be approximated by a distribution called the Poisson distribution, with one parameter $\lambda = np'$, defined as follows:

$$P(X = x) = \frac{\lambda^x e^{-\lambda}}{x!}, \qquad x = 0, 1, 2, \ldots, \tag{13.16}$$

where λ is the mean of X.

To illustrate the use of Poisson distribution as an approximation of a binomial distribution, we will consider an example. Suppose that some patients may suffer a bad reaction from taking a certain drug. If the probability of any such patient suffering a bad reaction is estimated to be 0.025, what is the probability that exactly 3 patients out of 200 who have taken the drug will have the bad reaction? If we assume that each of the 200 patients has the same probability $p' = 0.025$ of experiencing a bad reaction, the problem is basically a binomial random variable problem. Accordingly, $P(X = 3) = [200!/(3! \ 197!)]$ $[(0.025)^3(0.975)^{197}] = 0.1399995$. By Poisson approximation, we have $\lambda = np' = (200)(0.025) = 5$, and hence $P(X = 3) = 5^3e^{-5}/3! = 0.1403739$, which is very close to the result from binomial calculation. Obviously, Eq. (13.16) is easier to work with. An acceptable rule of thumb is to use the Poisson distribution to approximate binomial probabilities if $n \geq 20$ and $p' \leq 0.05$. If $n \geq 100$, the approximation is excellent if $np' \leq 10$.

The Poisson probability distribution is a very important distribution in its own right. It is widely used in modeling various process and operational phenomena and many natural events, such as the weekly number of burned-out lights in a factory, the number of defects per bolt of a fabric, or the number of sunny days in a monsoon season. It is noted that a Poisson random variable is the number of occurrences of a certain event, which is often a rare event, in a clearly specified sample space such as a fixed time period or a fixed particular dimensional size (e.g., area, volume). The shape of a Poisson distribution is in general skewed to the right, as shown in Fig. 13.16.

Properties of Poisson Distribution.   The following are important properties of the Poisson distribution:

1. The Poisson distribution has only one parameter, λ. If we look at Eq. (13.16), we find that we only need to know the mean, λ, to find the probability for any integer value of X from zero to infinity.

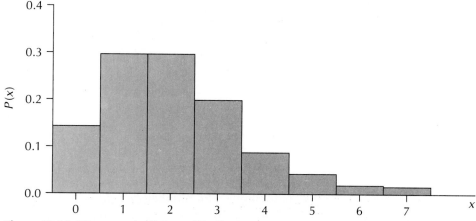

Figure 13.16 Histogram for Poisson Distribution

    2. An interesting property of the Poisson distribution is that its mean is equal to its variance;
       a. Mean of $X = \lambda$.
       b. Variance of $X = \lambda$.
    3. The mean of a Poisson variable is proportional to the size of the sampling unit. For example, if it is known that the average number of defects per square inch is 5, there are on an average 20 defects per 4 square inches and an average of 2.5 defects per $\frac{1}{2}$ square inch.

Example. Suppose that a telephone switchboard is designed to take 35 calls per minute. If it is known that the average number of calls is 20 per minute, what is the probability that in any given minute the switchboard will be overloaded?

Solution. Here the expected number of calls per minute is 20 (i.e., $\lambda = 20$). What is the probability that X will exceed 35?

$$P(X > 35) = \sum_{x=0}^{\infty} P(X = x) - \sum_{x=0}^{35} P(X = x)$$

$$= 1 - \sum_{x=0}^{35} \frac{\lambda^x e^{-\lambda}}{x!}$$

$$= 1 - \sum_{x=0}^{35} \frac{20^x e^{-20}}{x!}$$

$$= 1 - 0.999 = 0.001.$$

Example. In another application of the Poisson distribution suppose it is known that in a certain enameling process the mean number of insulation breaks per yard is 0.07. What is the probability of finding exactly two such breaks in a piece of wire 16 yards long?

Solution. Here λ is the expected number of defects in a piece of wire 16 yards long, $\lambda = 16 \times 0.07 = 1.12$. Thus

$$P(X = x) = \frac{(1.12)^x}{x!} e^{-1.12},$$

and for $x = 2$,

$$P(2) = \frac{(1.12)^2}{2!} e^{-1.12} = 0.201.$$

Construction of c Charts from Sample Data

As we have just seen above, the number of defects, c, arises probabilistically according to the Poisson distribution. One important property of the Poisson distribution is that the mean and variance are the same value. Then, given c' (in general, we have used λ), the mean number of defects per sample, the 3 sigma limits for the c chart are given by

$$c' \pm 3\sqrt{c'}. \tag{13.17}$$

When c' is unknown, we must estimate it from the data. For a collection of k samples, each with an observed number of defects, c, the estimate of c' is

$$\bar{c} = \sum_{i=1}^{k} \frac{c_i}{k}, \tag{13.18}$$

where c_i is the number of defects in the ith sample. Hence control limits for the c chart may be established from the formula

$$\text{UCL}_c = \bar{c} + 3\sqrt{\bar{c}} \tag{13.19}$$
$$\text{LCL}_c = \bar{c} - 3\sqrt{\bar{c}}.$$

If the LCL_c is calculated to be less than zero, a value of zero is used.

Example of c Chart Construction

We consider next the process of the installation of front and rear bumpers on automobiles. The sample size here is the number of cars built on a given shift, since all autos were inspected for the bumper installation process. This number varied slightly from shift to shift, but only by about 2%, so a constant average sample size of $n = 560$ was used for the construction of the c chart. A total of 11 different types of defects were possible. An explanation of the defects follows.

1. PGM 180 degrees out: one of the mounting plates was placed with the wrong orientation, 180 degrees off.
2. Rear bumper boss missing: one of the screws on the bumper was not installed.
3. Rear bumper fit.
4. PGM missing: one of the mounting plates is not present.
5. Rear bumper damaged.
6. Wrong rear bumper.
7. Front bumper damaged.
8. Bumper no stocks: the bumper was not on the line at the point that it would typically be installed, and therefore was not placed on the car.
9. Front bumper loose.
10. Front bumper boss missing: one of the screws on the bumper was not installed.
11. Rear bumper loose.

Charting was initiated after the defect data for 25 consecutive shifts were available. Table 13.4 provides the defect data for these 25 samples. A c chart

TABLE 13.4	Defect Data for First 25 Samples for the Bumper Assembly Example

Sample	Number of Defects
1	16
2	14
3	28
4	16
5	12
6	20
7	10
8	12
9	30
10	17
11	9
12	17
13	14
14	16
15	15
16	13
17	14
18	16
19	11
20	20
21	11
22	9
23	16
24	31
25	13

for all defects was constructed. The calculations for the c chart centerline and upper and lower control limits are given below.

Centerline:

$$\bar{c} = \frac{\text{total number of defects}}{\text{total number of samples}}$$

$$= \frac{400}{25}$$

$$= 16.$$

Control limits:

$$\text{UCL}_c, \text{LCL}_c = \bar{c} \pm 3\sqrt{\bar{c}}$$

$$= 16 \pm 3\sqrt{16}$$

$$\text{UCL}_c = 28$$

$$\text{LCL}_c = 4.$$

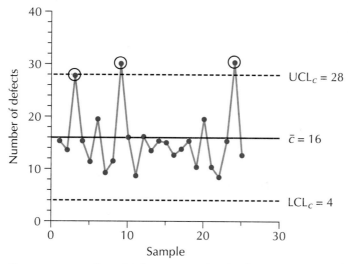

Figure 13.17 c Chart for first 25 Samples for the Bumper Assembly Process

Figure 13.17 shows the c chart for all defects for the 25 samples collected. By applying the same criteria for reading the statistical signals as were used for the p chart, it was determined that three separate instances indicating out-of-control conditions were evident in the data.

Sample	Signal
3	Point above upper control limit
9	Point above upper control limit
24	Point above upper control limit

In an effort to determine the possible reasons for the large numbers of defects associated with the three samples that have c values above the upper control limit, a Pareto diagram was constructed for all 11 defect types. This diagram is shown in Fig. 13.18.

Two of the more frequently occurring defects, front and rear bumper loose, were determined to be caused by essentially the same problem: a difficulty the assembly line workers were having in aligning the bumper with the mounting bracket under the car frame. A cartlike fixture was designed that would hold a bumper at the proper elevation and the worker would then guide the bumper into place so that it could be tightened properly. After implementing this procedural change, the process charting was continued.

Table 13.5 gives the data for the next 12 shifts after the new bumper mounting fixture was put into use. Figure 13.19 shows the c chart continuation using these data and extending the centerline and control limits for the first 25 samples. It is clear from the chart that a change in process performance has occurred. At sample 32 a run of eight consecutive points below the centerline is detected. This run continues through the end of the chart—sample 37. This

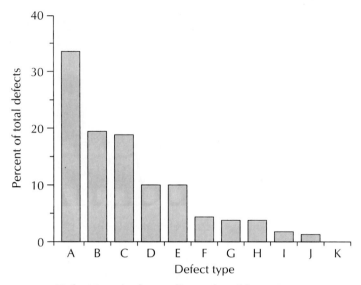

Defect type in descending order of importance are

A. Bumper no stocks: This means that the bumper was not on the line at the point that it would typically be installed, and therefore was not placed on the car.

B. Front bumper loose.

C. Rear bumper fit.

D. Rear bumper loose.

E. PGM 180 deg out: One of the mounting plates was placed with the wrong orientation—180 degrees off.

F. Front bumper damaged.

G. Wrong rear bumper

H. Rear bumper boss missing: One of the screws on the bumper was not placed.

I. Front bumper boss missing.

J. Rear bumper damaged.

K. PGM missing: One of the mounting plates is not present.

Figure 13.18 Pareto Diagram for All Defects for the Data from Samples 1 to 25 of the Bumper Assembly Process

signal is indicating that the defect rate has been reduced. It would appear that the action taken at the process has led to a significant process improvement.

13.6

Control Chart for Number of Defects per Unit

Although in c chart applications it is common to comprise a sample of only a single unit or item, the sample or subgroup may be comprised of several units. Further, the number of units per subgroup may vary, particularly if a subgroup

TABLE 13.5 Data for the Bumper Assembly Process After the Loose-Fits Problem Was Addressed	
Sample	Number of Defects
26	12
27	9
28	11
29	12
30	13
31	12
32	15
33	8
34	7
35	8
36	11
37	9

is an amount of production for the shift or day, for example. In such cases the opportunity space for the occurrence of defects per subgroup changes from subgroup to subgroup, violating the "equal opportunity space" assumption on which the c chart is based. In this case it is necessary to create another statistic to overcome this problem. Such a statistic may be the average number of defects per unit or item. The symbol u is often used to denote the average number of defects per unit:

$$u = \frac{c}{n},$$
(13.20)

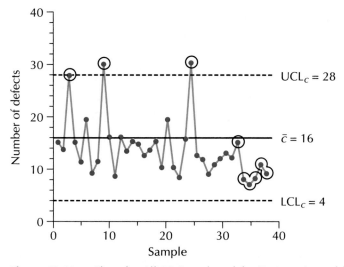

Figure 13.19 c Chart for All 37 Samples of the Bumper Assembly Process

where c is the total number of defects per subgroup of n units. For k such subgroups gathered, the centerline on the u chart is

$$\bar{u} = \frac{\text{total defects in } k \text{ subgroups}}{\text{total number of items in } k \text{ subgroups}} \tag{13.21}$$

$$= \frac{\sum\limits_{i=1}^{k} c_i}{\sum\limits_{i=1}^{k} n_i}.$$

The control limits for the u chart are then given by

$$\bar{u} \pm 3 \sqrt{\frac{\bar{u}}{n}}. \tag{13.22}$$

Example of the Use of the u Chart

The process by which moonroofs are installed in automobiles is subject to several types of critical defects:

- Wind noise.
- Water leaks.
- Binding during retraction.
- Squeaks and rattles.

Because of numerous customer complaints, the assembly plant was asked to collect some data on the occurrence of these concerns so that statistical charting and associated process study could take place. As a result of a shortage of labor time from one day to another and the time required to make the water test, the data collected varied in sample size, one sample per shift being collected.

Table 13.6 provides the defect data for 27 samples collected over a period of 9 days. The table includes the sample size for each shift, the total number of defects observed, and u, the average number of defects per unit. Upon using the data in Table 13.6, the centerline and control limits for the u chart to be constructed were determined as shown below. Since the sample size varies, the control limits will also vary from one sample to another.

$$\bar{u} = \frac{\sum\limits_{i=1}^{k} c_i}{\sum\limits_{i=1}^{k} n_i}$$

$$= \frac{706}{516}$$

$$= 1.37.$$

TABLE 13.6 Defect Data for Moonroof Installation Example					
Sample	Sample Size, n	Number of Defects per Sample, c	Average Number of Defects per Unit, u	LCL_u	UCL_u
1	16	23	1.44	0.49	2.25
2	20	30	1.50	0.59	2.16
3	26	35	1.35	0.68	2.06
4	8	12	1.50	0.13	2.61
5	22	29	1.32	0.62	2.12
6	29	35	1.21	0.72	2.02
7	31	50	1.61	0.74	2.00
8	13	15	1.15	0.40	2.35
9	28	36	1.29	0.71	2.04
10	23	38	1.65	0.64	2.10
11	19	24	1.26	0.57	2.18
12	23	32	1.39	0.64	2.10
13	14	24	1.71	0.43	2.31
14	29	34	1.17	0.72	0.72
15	27	38	1.41	0.70	2.05
16	15	25	1.67	0.46	2.28
17	22	26	1.18	0.62	2.12
18	22	24	1.09	0.62	2.12
19	14	22	1.57	0.43	2.31
20	16	17	1.06	0.49	2.25
21	22	33	1.50	0.62	2.12
22	16	21	1.31	0.49	2.25
23	14	18	1.29	0.43	2.31
24	5	9	1.80	0.00	2.94
25	13	18	1.38	0.40	2.35
26	19	26	1.37	0.57	2.18
27	10	12	1.20	0.26	2.48

An example calculation for control limits follows.

$$\text{control limits} = \bar{u} \pm 3 \sqrt{\frac{\bar{u}}{n}}$$
$$= 1.37 \pm 3 \sqrt{\frac{1.37}{n}}.$$

For sample 1, $n_1 = 16$, and therefore,

$$LCL_{u_1} = 0.49$$

$$UCL_{u_1} = 2.25.$$

Figure 13.20 shows the u chart for the first 27 samples for the moonroof installation example. Note that the control limits vary from sample to sample.

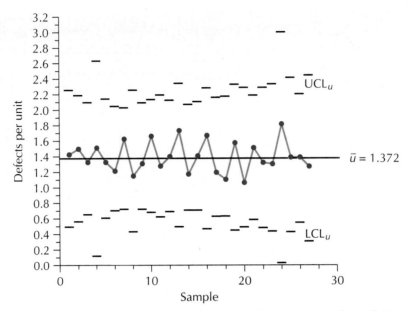

Figure 13.20 u Chart for the First 25 Samples for the Moonroof Installation Example

Applying the same criteria for reading statistical signals that were used for the p and c charts, we find no reason to believe that the process is not operating in a state of statistical control. However, the defect rate is high, almost 1.4 defects per unit. It was therefore decided to investigate the cause(s) of the large number of defects.

An investigation of the specific nature of the moonroof defects revealed that the most prevalent defect was water leaks. A check of the warranty data confirmed that this seemed to be a serious problem. After some further study it was decided to try a new type of seal. Sampling was continued after use of the new seal was begun. The data in Table 13.7 show the results of the next seven shifts. These data were plotted as a continuation of the original u chart with the centerline extended and the control limits calculated using the original $\bar{u}$ value. The results are shown in Fig. 13.21.

An evaluation of the u chart in Fig. 13.21 shows the following out-of-control signals to be evident:

Sample	Signal
31	Point below lower control limit
32	Point below lower control limit
33	Run below centerline
34	Point below lower control limit
34	Run below centerline

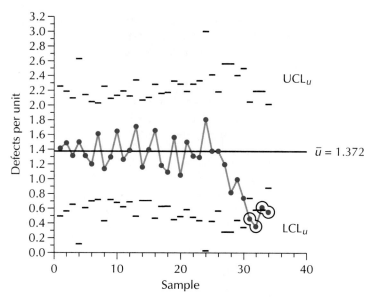

Figure 13.21 Continued u Chart for the Moonroof Example After Implementing the New Seal

		Number of	Average Number		
	Sample	Defects per	of Defects per		
Sample	Size, n	Sample, c	Unit, u	LCL_u	UCL_u
28	10	8	0.80	0.26	2.48
29	14	14	1.00	0.43	2.31
30	11	8	0.73	0.31	2.43
31	29	14	0.48	0.72	2.02
32	19	7	0.37	0.57	2.18
33	19	12	0.63	0.57	2.18
34	45	25	0.56	0.85	1.90

TABLE 13.7   Data for the Moonroof Example After the New Seal Was Implemented

It is clear that the process has changed—in particular, that the average number of defects per unit has been significantly reduced. The u values are now in the range of 0.4 to 0.7, while the original centerline for the chart was almost 1.4.

| Exercises |

13.1. What types of situations call for quality characterization by attribute data instead of variable data? Give several examples.

13.2. Is it possible to use attribute data to describe quality characteristics that are inherently of a variable measurements nature? Give an example of this type of situation and explain the possible advantages and disadvantages of doing so.

13.3. When using attribute data, why are operational definitions for defects important for proper sampling and charting?

13.4. What is meant by "using a product control mentality" in the development of operational definitions for defects for attribute control charting? What are the problems and shortcomings of this approach to the development of operational definitions?

13.5. Why might it be useful to record all occurrences of all defect types as opposed to simply denoting a part as defective as soon as the first instance of any one of many possible defects is found?

13.6. Given the sample of parts below having defect types A, B, C, and D, find:
a. The number of defects.
b. The number of defectives.
c. The fraction defective.
d. The average number of defects per unit.

Defect types: A, B, C, and D

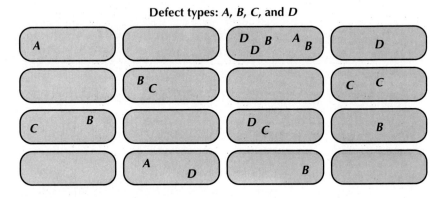

13.7. After 100% inspection of a batch of 2000 parts, it was found that there were 73 defectives.
a. How many different ways could a sample of $n = 10$ parts be selected from this batch to get 2 defectives and 8 nondefectives?
b. Find the probability of the sum of all the outcomes in part (a).
c. What is the probability of getting at most 3 defectives in a sample of size $n = 10$?

13.8. Over the last year, a manufacturer produced approximately 100,000,000 parts that were 100% inspected, and 300,000 defects on those parts were identified.
a. Find the probability of getting 3 defects in a batch of $n = 1000$ parts.
b. What is the probability of getting more than 3 defects in a sample of size $n = 1000$ parts.

13.9. Machined parts made from extruded aluminum bars are tumbled in a finishing operation to remove burrs. Samples of $n = 20$ parts are taken from each batch after it finishes the tumbling cycle and are checked for burrs. The fraction of parts with burrs is then plotted on a p chart. For some time, the process has been stable at an average fraction defective of 0.08. Occasionally, the person making the chart notices that a point falls on the lower control limit. These occurrences have been ignored because it is felt that such low p values are good and need not be worried about. His boss tells him that points on the lower control limit means the process is out of control, and therefore he needs to be looking for special causes. Comment.

13.10. Discuss the pros and cons of using a u chart instead of c chart when:
   a. The sample sizes vary from one sample to the next.
   b. The sample size remains constant.
   c. The sample size is $n = 1$ item.
   Give specific reasons in each case.

13.11. The metal body for a spark plug is made by a combination of cold extrusion and machining. The occurrence of surface cracking following the extrusion process has been shown by Pareto diagrams to be responsible for producing virtually all of the defective parts. To identify opportunities for process improvement, it was decided to construct p charts for the process. During one shift, 25 samples of size $n = 100$ were collected and the results shown in the table were obtained.

Sample	Number of Defectives	Sample	Number of Defectives
1	1	14	4
2	6	15	3
3	3	16	3
4	1	17	2
5	3	18	4
6	2	19	4
7	0	20	4
8	2	21	2
9	4	22	5
10	1	23	2
11	5	24	6
12	4	25	3
13	5		

   a. Use these data to establish the centerline and control limits for a p chart.
   b. Construct the p chart and comment on the state of the process. Sampling and charting was continued on the next shift, and the data for the next 15 samples are as shown in the table.

Sample	Number of Defectives	Sample	Number of Defectives
26	9	34	9
27	9	35	3
28	8	36	13
29	1	37	8
30	7	38	6
31	5	39	5
32	7	40	8
33	8		

c. Plot the newly collected data on the chart constructed in part (b). What can be said at this point regarding the state of the process?

13.12. Past examination of the process that produces automobile quarterpanels has demonstrated that scratches are producing the majority of defective panels. It is proposed to study the process with a c chart. The engineers involved with the process have kept track of the number of defects for samples of size $n = 100$ for the last 60 shifts, as shown in the table.

Sample	Number of Defects	Sample	Number of Defects	Sample	Number of Defects
1	3	21	1	41	9
2	12	22	2	42	11
3	4	23	0	43	9
4	9	24	4	44	7
5	6	25	2	45	5
6	12	26	0	46	5
7	5	27	5	47	9
8	8	28	3	48	7
9	5	29	1	49	5
10	9	30	0	50	10
11	8	31	3	51	7
12	9	32	5	52	3
13	4	33	6	53	8
14	3	34	4	54	7
15	10	35	0	55	10
16	3	36	2	56	8
17	5	37	4	57	5
18	8	38	0	58	6
19	8	39	2	59	10
20	6	40	1	60	6

a. Calculate the centerline and control limits for the c chart.
b. Construct the c chart. Interpret the chart and comment on what may be occurring in this process.

13.13. In casting a certain type of flywheel, it was observed that many parts were defective and being scrapped because of hot tears. Hot tears are voids in the cast metal that are likely to occur at weak points in the shape, as shown. Sampling has indicated that 3.70% of the parts cast will be defective because of one or more hot tear defects.

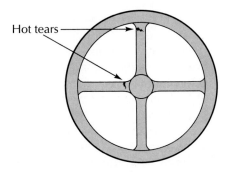

Hot tears

a. If a sample of size $n = 7$ is drawn, how many ways could two defectives occur? What is the probability of getting exactly two defectives?
b. If a shipment of 50 parts is to be sent, what is the probability that none will be defective?

13.14. It is desired to study an injection molding process with a p chart. However, because the number of parts produced per hour varies, the individuals studying the process are uncertain as to how the data should be analyzed. The data shown in the table summarize the number of parts produced and the number of defectives observed for 20 consecutive hours of operation for the process. Construct the three different variable-sample-size p charts and compare and contrast the information that each shows.

Sample	Sample Size	Number of Defectives	Sample	Sample Size	Number of Defectives
1	50	2	11	47	2
2	44	2	12	48	0
3	61	3	13	57	3
4	53	0	14	61	1
5	55	3	15	64	4
6	62	4	16	58	0
7	66	2	17	58	2
8	50	2	18	63	1
9	64	0	19	60	4
10	52	2	20	67	3

13.15. Suppose that a p chart is being used to monitor an electrostatic painting process. Parts are being painted in large batches. About 125 parts are hung by an operator on each rack, and 20 racks are painted per batch. Following the completion of the painting operation, a rack is sampled at random from the batch, and each of the parts on the rack is examined. Samples from 30 consecutive batches were collected and the number of defective parts for each was noted; $\bar{p}$ was calculated to be 0.045. Since the sample size, n, varied between 115 and 129 with an average of $\bar{n} = 124$, a p chart with limits based on $\bar{n} = 124$ was constructed. One of the p values was 0.12 for a sample of size $n = 117$. Does this point initially show up on the p chart as an indication of an out-of-control condition? Is this actually an indication of a shift in the mean fraction defective?

13.16. Why should defects such as scratches on an injection-molded part be classified as defects for the purpose of developing c charts for the molding process even if they are not on a surface that ultimately will be visible to the consumer? What is the major consequence of excluding such scratches from the data collection and charting process?

13.17. For each of the seven types of control charts presented in this chapter (p charts, variable-sample-size p charts, np charts, c charts, and u charts):
a. Explain the situation for which each is most suited.
b. Summarize the advantages and disadvantages associated with each.

13.18. In the production of a certain chemical liquid, there are a number of impurities that can be present to degrade the quality of the resulting product. Four specific impurities are of interest. Fifty samples (1 liter each) were collected, and the number of suspended particles of each type of impurity were counted. The results of this data collection activity are shown in the table.

Sample	Impurity 1	Impurity 2	Impurity 3	Impurity 4
1	6	0	0	9
2	0	1	5	9
3	4	2	2	8
4	1	2	0	1
5	3	0	2	7
6	1	5	1	5
7	2	0	0	7
8	3	1	1	8
9	4	2	3	9
10	0	0	0	3
11	3	0	4	13
12	2	2	3	8
13	3	3	4	6
14	1	3	3	5
15	3	2	2	8
16	2	0	2	7
17	3	1	1	1

(continued)

Sample	Impurity 1	Impurity 2	Impurity 3	Impurity 4
18	4	0	3	10
19	1	0	3	6
20	5	0	3	4
21	1	3	1	9
22	1	5	4	4
23	3	4	1	5
24	3	4	8	12
25	2	3	2	7
26	3	0	3	7
27	9	8	3	8
28	5	3	6	8
29	5	2	0	8
30	5	1	1	4
31	3	1	1	11
32	6	2	1	5
33	2	2	1	5
34	2	0	3	8
35	0	0	3	4
36	4	3	3	9
37	4	0	3	6
38	1	1	4	4
39	2	1	1	6
40	4	5	2	5
41	2	4	4	9
42	6	0	1	6
43	2	1	0	5
44	8	0	0	4
45	2	1	1	8
46	1	1	0	5
47	1	4	2	10
48	5	0	5	8
49	1	1	1	5
50	3	0	1	5

    a. Construct a c chart for this process from the data based on all defects, and a separate c chart for each defect. Comment on the appearance of each of the charts.

    b. Construct a Pareto chart for the defects. Which defect(s) should problem-solving efforts first focus on? Why? Consider in your answer the results from both the control charts and the Pareto diagram.

13.19. Five identical assembly lines are producing power supplies. The defective rate on one of the five lines is considerably higher than that on the other four, so an investigation into the reasons for this is begun. Samples of size $n = 50$ power supplies are gathered randomly throughout the day for each of two shifts for a total of 15 days. The results are shown in the first table.

Sample	Number of Defectives	Sample	Number of Defectives
1	6	16	4
2	4	17	2
3	5	18	5
4	3	19	4
5	1	20	10
6	8	21	6
7	2	22	2
8	4	23	4
9	3	24	4
10	3	25	5
11	7	26	4
12	2	27	6
13	6	28	2
14	2	29	3
15	4	30	1

a. Construct a p chart from these data. Show all calculations.
b. Comment on the appearance of the p chart and note any statistical signals that appear.
c. Construct a scatter diagram of shift versus the number of defectives. Assume that all odd-numbered samples were collected on the first shift, and all even-numbered samples were collected on the second shift. Comment.

One of the operators on the line suggests repairing the protective coverings on the sharp edges of the conveyor, thus eliminating the potential for scratches to occur at several locations. Sampling is continued after this improvement, with data as shown in the second table.

Sample	Number of Defectives	Sample	Number of Defectives
31	2	41	3
32	0	42	0
33	4	43	0
34	2	44	0
35	1	45	1
36	2	46	0
37	2	47	2
38	3	48	2
39	0	49	2
40	3	50	2

d. Continue the p chart with the new data, using the same limits as before.

   e. Comment on the appearance of the chart now. What can be said
      about the change that was made?
13.20. The data shown in the table were collected for a process that involves
       the assembly of odometers that are then assembled into an instrument
       panel. The sample is collected during a 15-minute period of operation
       for a robotic assembly cell. Four such periods are monitored each shift.

Sample	Size	Number of Defectives	Sample	Size	Number of Defectives
1	93	3	16	78	5
2	87	3	17	84	6
3	61	4	18	82	4
4	94	4	19	90	1
5	79	5	20	87	2
6	87	3	21	95	4
7	90	5	22	86	2
8	88	6	23	77	1
9	95	4	24	83	4
10	79	4	25	83	4
11	83	5	26	91	2
12	86	5	27	88	4
13	92	1	28	78	6
14	80	2	29	93	5
15	89	3	30	85	2

   a. Make a p chart for these data using average limits. Show all calcula-
      tions. Carefully interpret the chart.
   b. Make a p chart using individual limits for each sample. Show all
      calculations. Carefully interpret the chart.
   c. Make a standardized p chart. Again, show all calculations. Carefully
      interpret the chart.
   d. Compare the three charts and comment on any similarities or differ-
      ences you see among the charts.
13.21. The data shown in the table were collected from a process making glass
       bottles for a carbonated beverage bottling company. The sample size was
       500.

Sample	Number of Defectives	Sample	Number of Defectives
1	7	16	2
2	6	17	3
3	3	18	7
4	4	19	2
5	2	20	5

(continued)

Sample	Number of Defectives	Sample	Number of Defectives
6	4	21	3
7	3	22	3
8	5	23	5
9	3	24	4
10	8	25	6
11	10	26	0
12	1	27	5
13	5	28	5
14	8	29	4
15	3	30	4

a. Construct a p chart from these data. Show all calculations. Interpret the chart.

b. Construct an np chart from these data. Again, show all calculations and interpret the chart.

c. Compare the appearance of the two charts and comment on any similarities or differences between the charts.

13.22. An injection-molding process is being monitored for the number of defective parts it is producing. Samples of 75 parts are collected on each shift, and the average number of defectives per sample is found to be rather high. See the data in the first table.

Sample	Number of Defectives	Sample	Number of Defectives
1	10	14	14
2	4	15	8
3	2	16	4
4	10	17	6
5	2	18	8
6	4	19	6
7	8	20	8
8	12	21	16
9	6	22	22
10	10	23	2
11	10	24	8
12	14	25	16
13	4		

a. Construct a p chart from the data. Show all calculations.

b. Analyze this chart and comment on any statistical signals.

It was suggested that perhaps the p chart was not the most appropriate chart to use for this process, that instead, a c chart should be used. Again, data were collected from each shift for 25 shifts (sample size = 75) and are summarized in the second table.

Sample	Number of Defects	Sample	Number of Defects
1	10	14	9
2	8	15	13
3	12	16	9
4	10	17	12
5	12	18	11
6	13	19	16
7	14	20	17
8	25	21	25
9	10	22	19
10	7	23	14
11	12	24	27
12	14	25	15
13	10		

c. Construct a c chart from these data. Show all calculations.

d. Analyze this chart and comment on any statistical signals.

e. What, if anything, does this chart tell you that the p chart did not? What is the next step that should be taken in analyzing this process? Explain your reasoning.

13.23. An injection-molding process is making plastic gamepieces for games of checkers. There are two lines: one line makes the red pieces, and the other line makes the black pieces. Currently, the red line is producing a lot of defective parts for unknown reasons. Data collected from this line are shown in the table.

Sample	Size	Number of Defectives	Sample	Size	Number of Defectives
1	31	1	14	45	1
2	40	9	15	30	2
3	36	3	16	34	2
4	29	3	17	34	3
5	33	4	18	39	3
6	41	3	19	35	4
7	36	5	20	28	1
8	32	1	21	36	3
9	31	3	22	32	1
10	40	4	23	40	2
11	34	3	24	34	3
12	29	1	25	42	4
13	38	3			

a. Construct a p chart from this data using individual limits. Show all calculations. Carefully interpret the chart.

    b. Construct an *np* chart from this data using individual limits. Show all calculations. Carefully interpret the chart.

    c. Compare the appearance of the two charts. Which is easier to interpret?

13.24. A process takes cut lengths of wire and bends them to form paper clips. Recently, a large number of paper clips have been malformed, resulting in a large quantity of scrap for this process. Data are collected on the number of defectives produced, and the results are shown in the table.

Sample	Size	Number of Defectives	Sample	Size	Number of Defectives
1	80	4	16	70	7
2	72	2	17	71	4
3	68	3	18	82	7
4	81	3	19	73	4
5	75	4	20	67	6
6	70	4	21	80	2
7	82	5	22	79	7
8	64	1	23	81	2
9	73	1	24	81	4
10	78	5	25	78	4
11	70	3	26	79	1
12	74	4	27	84	4
13	83	1	28	66	2
14	76	12	29	74	4
15	66	4	30	80	4

    a. Construct a *p* chart from these data using average limits. Show all calculations.

    b. Analyze this chart and comment on any statistical signals.

13.25. A process is cutting lengths of plastic tubing that will be used for fish tanks and aquariums. Data are collected on the number of defective pieces of tubing that are being produced to either verify or disprove the assumption that the process is in control with respect to the number of defectives being produced. A summary of the data collected is shown in the table.

Sample	Size	Number of Defectives	Sample	Size	Number of Defectives
1	98	7	16	98	4
2	83	3	17	91	0
3	79	1	18	80	3
4	107	4	19	85	2
5	92	0	20	100	1
6	84	5	21	97	3
7	86	4	22	92	5

(continued)

Sample	Size	Number of Defectives	Sample	Size	Number of Defectives
8	91	4	23	85	8
9	108	5	24	104	4
10	99	2	25	93	3
11	89	3	26	74	6
12	102	6	27	83	2
13	78	3	28	101	5
14	84	4	29	87	5
15	101	7	30	90	1

a. Construct a standardized p chart from these data. Show all calculations.

b. Analyze this chart and comment on any statistical signals.

13.26. A process is making plastic water bottles for bicycles, and four defects can occur in the course of production: missing cap, scratches, flow lines, and excess flash. Data are collected on these defects in hopes that it will reveal where there are opportunities for improvement. The results from the first 25 samples of size $n = 100$ are shown in the table.

Sample	Missing Cap	Scratches	Flow Lines	Excess Flash
1	6	6	4	7
2	3	4	1	8
3	8	5	0	8
4	3	3	1	5
5	9	3	3	6
6	3	1	3	5
7	4	3	2	3
8	5	2	6	10
8	4	4	7	6
10	6	5	3	6
11	4	1	4	7
12	3	6	3	8
13	2	3	2	11
14	3	0	3	5
15	5	3	3	7
16	5	0	1	6
17	5	1	3	7
18	4	3	4	5
19	6	3	3	9
20	2	2	3	5
21	4	1	2	10
22	6	2	4	12
23	5	1	6	12
24	2	1	2	8
25	2	1	7	5

     a. Construct a c chart from these data. Show all calculations.
     b. Analyze this chart and comment on any statistical signals.
     c. Construct individual c charts for each of the four defects. Comment on the similarities and differences among the four charts.
     d. Make a Pareto diagram of the four defects and comment on where problem-solving efforts should be concentrated, taking into consideration both the charting results and the Pareto diagram. Explain your reasoning.

13.27. From a previous study it was found that 99% of the defects that occur in an accounts payable process at a large manufacturer are missing purchase order numbers. Data are now being collected on the number of times this occurs each day in hopes that control charts made for this process will reveal opportunities for improvement. A summary of the data collected from 30 consecutive workdays is shown in the table.

Day	Missing POs	Day	Missing POs
1	12	16	4
2	6	17	6
3	8	18	4
4	16	19	14
5	14	20	8
6	6	21	8
7	12	22	10
8	8	23	10
9	16	24	2
10	4	25	12
11	12	26	12
12	10	27	8
13	12	28	4
14	6	29	4
15	10	30	8

     a. Construct a c chart from these data. Show all calculations.
     b. Analyze this chart and comment on any statistical signals.

13.28. The final inspection for a truck engine is used to check for visual defects. The three defects that occur on this product most frequently are paint runs, improper label alignment, and misstamped serial number. An investigation into the reason behind the high defect rate is begun, with samples collected every day for 25 days. The results are shown in the table.

Sample	Number of Engines Examined	Paint Runs	Label Misaligned	Misstamped Serial No.
1	105	3	2	4
2	101	5	1	5
3	102	3	2	5
4	95	4	1	7
5	100	3	3	3
6	98	2	1	3
7	102	1	2	2
8	101	3	3	5
9	97	0	2	1
10	100	3	1	2
11	106	4	1	3
12	101	6	1	3
13	103	3	1	4
14	96	5	4	5
15	101	5	1	3
16	101	8	1	2
17	98	6	2	3
18	103	7	1	4
19	104	6	1	1
20	100	5	3	5
21	100	9	6	4
22	99	7	4	2
23	102	5	6	4
24	105	8	5	2
25	102	10	6	1

a. Construct u charts from these data. Show all calculations.
b. Analyze these charts and comment on any statistical signals that appear.
c. Make a Pareto diagram of the three defects and indicate which is occurring most frequently. Where should quality improvement efforts be focused?

13.29. A process that folds cardboard boxes that will eventually hold model airplane kits is experiencing problems with two frequently occurring defects: badly folded/mangled boxes and rips/tears in the cardboard. The data collected over the last 30 shifts are shown in the table.

Sample	Size	Bad Folds	Ripped Box
1	52	3	1
2	50	6	3
3	47	0	2

(continued)

Sample	Size	Bad Folds	Ripped Box
4	50	3	3
5	50	2	2
6	52	5	3
7	51	3	1
8	54	5	0
9	48	2	1
10	57	9	2
11	55	6	1
12	50	4	1
13	50	3	4
14	52	6	6
15	50	5	1
16	47	2	3
17	58	10	2
18	49	3	1
19	47	0	3
20	52	6	2
21	50	2	2
22	50	3	4
23	53	6	4
24	51	3	5
25	52	4	3
26	51	4	5
27	53	5	7
28	50	3	6
29	46	2	5
30	49	3	3

a. Construct u charts from these data. Show all calculations.
b. Analyze these charts and comment on any statistical signals.
c. Construct a Pareto diagram of the defects and suggest where problem-solving efforts should concentrate. Explain your reasoning.

13.30. A supplier of plexiglass sheets is concerned that the sheets produced contain an excessive number of scratches. It is decided to run a statistical study of the process to determine if it is in statistical control and to look for possible improvement opportunities. A defect is a scratch over 0.1 inch long. Each sample collected was actually 25 one-foot-square pieces of plexiglass. The study produced the data shown in the first table.

Sample	Number of Defects	Sample	Number of Defects
1	2	26	9
2	5	27	2
3	3	28	1

(continued)

Sample	Number of Defects	Sample	Number of Defects
4	3	29	3
5	5	30	1
6	4	31	4
7	6	32	4
8	0	33	5
9	5	34	4
10	5	35	4
11	4	36	2
12	4	37	5
13	3	38	4
14	6	39	6
15	4	40	5
16	0	41	3
17	6	42	1
18	3	43	2
19	6	44	2
20	11	45	2
21	6	46	3
22	5	47	6
23	6	48	7
24	7	49	3
25	7	50	6

a. Calculate the centerline and control limits for a u chart for these data. Plot the u chart and interpret it. Clearly label all out-of-control points and comment on the state of the process.

b. Assume that you were able to find the source(s) of any problems identified by the signals on the chart and that appropriate corrective actions were taken at the process. Recalculate the control limits after removing these samples.

c. Following these corrective actions, several additional samples were collected and are shown in the second table. Use the control limits calculated in part (b) as a basis to plot the new data. If you deem it appropriate, construct new control limits for the data and comment on the process.

Sample	Number of Defects	Sample	Number of Defects
51	3	76	3
52	2	77	1
53	4	78	3
54	1	79	2
55	0	80	1
56	1	81	2

Sample	Number of Defects	Sample	Number of Defects
57	1	82	0
58	0	83	1
59	3	84	3
60	0	85	4
61	2	86	2
62	1	87	1
63	3	88	2
64	0	89	0
65	3	90	3
66	3	91	0
67	2	92	2
68	0	93	0
69	1	94	1
70	2	95	2
71	3	96	1
72	1	97	1
73	4	98	3
74	5	99	1
75	3	100	0

14

Attribute Control Chart Implementation for Process Improvement: Two Case Studies

14.1

Introduction

The purpose of this chapter is to present two actual case studies that document the use of statistical thinking and methods in the improvement of processes and systems. Both case studies deal with the characterization of quality on an attribute basis. The presentations of the case studies cover not only the statistical aspects of the study, but also some of the team-building and group dynamics aspects as well. The real underlying purpose here is to develop a sense of the basic elements of the implementation methodology that must be invoked to realize full and lasting benefits from the use of statistical thinking and methods for process improvement. Although both of the situations described here are quite different, it is important to recognize the elements of commonality in attacking and solving the problem.

14.2

Implementation of SPC Methods: A Case Study on "Press 120"—Visual Defects in a Molding Process[1]

The following case study discussion is grounded in fact and details the experiences of a plastics plant in applying statistical thinking and methods for quality and productivity improvement. Some liberties have been taken with the presentation because of space limitations and in the interest of emphasizing certain key points.

Background of the Problem

A plastics plant, heavily involved in the injection molding of automobile components, was experiencing high and often erratic scrap rates on some products. As a result, the shipping requirements were producing a need to run with a considerable amount of overtime and often necessitated the running of backup dies on other presses to increase production quantity. Some troubleshooting activities were ongoing, and some improvement had been realized. As production orders increased during a time when automobile sales were increasing, the problem became more and more critical. During this time period, the people in this plant were being introduced to the concepts and techniques of statistical process control, and it was decided that an excellent opportunity existed not only to attempt to solve a pressing problem using SPC, but also for those in the plant to gain some experience in implementing the techniques and gain confidence in the power and importance of their use. The product involved in this particular case study is a radiator grille.

Formulating a Strategy for Implementation

The first step taken by those in the plant was to form a group of people who were involved in the process, including operators, setup persons, supervisors, molding engineers, quality people, area managers, and statistical methods facilitators. The group began to hold meetings to try to understand the problem and get everyone's input. It was decided to begin to collect data from the process during production and to use statistical charting to study the process variations. This required that the group address such issues as: What quality characteristics should be monitored? How should these characteristics be measured/observed? How many samples should be taken? How often should they be taken? What other useful process information should be recorded? Other relevant questions were addressed as well. Because this was the first implementation experience in this plant, it was decided that the statistical methods facilitators would assume the primary data collection and analysis responsibil-

[1] R. J. Deacon, "Reduction in Visual Defects in an Injection Molding Process," *Society of Plastics Engineers, 1983 National Technical Conference Proceedings*, p. 97.

ities. Ultimately, such activities will become a natural part of the system as the role of inspectors, supervisors, and others slowly change under the adoption of the new way of doing business.

Prior to initiating the use of control charts, an orientation meeting was held for press operators and machine setup personnel to discuss the purpose of the study. At this meeting, the important role that these people would play in effecting improvement was discussed, and the management commitment to improvement was emphasized. Suggestions were sought, and as it turned out later, information obtained from press operators and setup persons was central to solving two major problems. The operators also attended an 8-hour presentation on basic chart concepts and philosophies so they could understand the importance of using the statistical approach to problem solving.

Operational Definitions

Perhaps the most difficult task in the entire process for the group was agreeing on what the defects were and how they were to be specifically defined. It was decided to focus on visual flaws such as scratches, black spots, and flow lines, but common agreement on what was important and what was not came slowly. The group realized the importance of writing down detailed operational definitions for defects so that all concerned were "playing by the same rules" and so that the most quantitative and sensitive measures of improvement opportunities could be developed and invoked. Initially, Table 14.1 was developed to serve as a basis of observation and measurement of quality. Later, the operational definitions were modified to provide for more sensitivity in finding the root causes of the problems. The first set of operational definitions, as seen in Table 14.1, were a bit too product control oriented. For example, scratches may be acceptable in a product control sense on the nonvisual areas of the grille, but the mechanism(s) putting the scratches on the grille may put them anywhere, depending on its random orientation at some point. Hence, all scratches should be observed and counted to increase the chart sensitivity.

Sampling Frequency and Data Collection

At the onset of the study, the occurrence of defects was rather high and process behavior was erratic. Because of this, it was decided to collect samples rather frequently so that the clues to the causes could be developed based on defect data and information collected on other process conditions. The group spent many hours discussing potential special and common causes of variation (defect occurrence). This was deemed important to understanding what process data (temperatures, pressures, and material conditions) ought to be collected to provide a basis for doing the detective work that would ultimately lead to the root causes of the defects. Among other things, a process condition sheet was placed at the machine each shift, and operators and setup persons were urged to jot down observations and/or thoughts they might have during the shift.

In the early stages of a study such as this, when process behavior seems

TABLE 14.1 Operational Definitions for Defects	
Defect	Criteria
Splay	Not acceptable on high surround or horizontal grille bars
Scratches/scuffs	Not acceptable on visual surround (use grease pencil test)
Flash	Not acceptable; must be repaired
Oil	Minimal amount; should be wiped clean so that residual amount will be removed by acid etch
Grease	Not acceptable if it cannot be wiped clean
Flow lines	Not acceptable in high surround (use grease pencil test); maximum $\frac{1}{4}$-inch flow line at base of surround (see sample); flow line acceptable in lower rail and between vertical ribs
Short shots	Front surface and visual ribs must be full; acceptable—only back rib behind paint step allowed short to $\frac{1}{8}$ inch maximum (see sample)
Burns	Unacceptable on visual surfaces; minimal amount acceptable on back rib surfaces
Sinks	Unacceptable on visual surface; acceptable on bottom base rail at outer vertical ribs (see sample)
Sprue on part	Part requires being scrapped
Material buildup	No residue allowed
Black spots	Unacceptable as groups on surface; acceptable in minimal number under surface
Defective trim	Gouges and chatter unacceptable; flash standing can be repaired

erratic and the problems seem to be many, it is probably useful to collect samples quite frequently to reveal clearly the short-term and rapidly occurring changes in the process. For an attribute-based study such as this, sample sizes can be small if the occurrence of defects is great (an average of 2 to 3 defects per sample may be sufficient). As problems are found and solved, the process becomes more stable and defect rates fall, so less frequent sampling is in order, although sample size requirements will increase.

Fault Diagnosis Experience

After collecting a sufficient number of samples (25 is a bare minimum), a control chart for number of defects per item was constructed (sample size = 10). Figure 14.1 shows this initial u chart and indicates that a very erratic out-of-control process is present with a very high average defect rate ($\overline{u} = 4.5$ defects per item). To begin to shed some light on the problem, a Pareto chart for defects was also constructed and is shown in Fig. 14.2. As discussed in Chapter 6, a Pareto chart is a bar graph that indicates the relative frequencies of occurrence of the various types of defects observed and is useful in identifying major

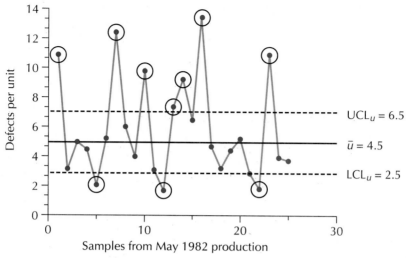

Figure 14.1 Initial Control Chart (u Chart) for All Defects from the Press 120 Molding Process

problems that may then be attacked in some precedence order. The Pareto diagram based on this initial u chart shows that black spots (degraded material at the surface of the part) constitute the vast majority of all defects at this point. For the time being, the group decided to focus its attention only on this problem.

In an effort to identify the root cause(s) of the black spot defect, the group used the "fishbone" or cause-and-effect diagram. Such a diagram, discussed in Chapter 6, is shown in Fig. 14.3; the major cause categories include the human

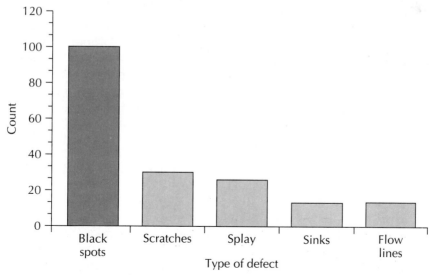

Figure 14.2 Pareto Diagram for Defects on Radiator Grille

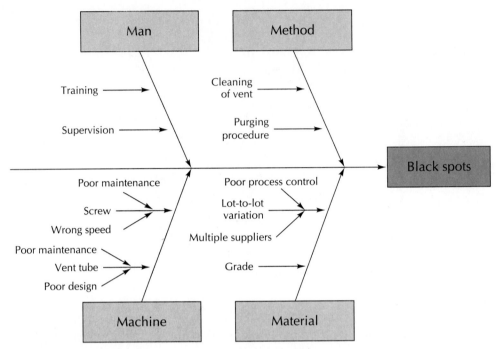

Figure 14.3 Cause-and-Effect Diagram for Black Spot Defects

element, methods, machines, and materials associated with the process. For each of these major potential cause categories, the group began to add increasing levels of detail to the diagram in an effort to reach the root cause level. For example, initial attention focused on the machine, in particular because the screw was felt to be worn out through extended processing of this raw material. It was reasoned that the screw had not been properly maintained and the material was adhering to the worn areas on the screw, which raised the temperature of the molten material to locally high levels at times, giving rise to degradation of the raw material. Particles of degraded material would then be carried along the screw and ultimately injected into the die cavity, where they would appear on the surface of the part. After considerable discussion, the group recommended that a new screw be ordered and installed.

After some delay, the new screw was installed in the machine. Statistical charting was continued all during the diagnostic period and after installation of the new screw as well. Figure 14.4 shows the u chart immediately before and after the screw was changed. Note that this chart is for black spot defects only and is based on a sample size of 5. Initially, black spots completely disappeared. However, a few days later the black spots returned with the same level of intensity as before the screw was changed. It was therefore evident that the root cause of black spots had not been properly identified. Because charting continued as it should, this fact was clearly borne out by the u chart.

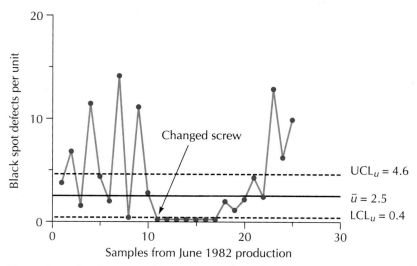

Figure 14.4 Continuation of Press 120 u Chart Showing the Effect of Changing the Screw

This points to the important role that statistical charting plays in showing clearly the short- and long-term effects (if any are present) of taking a certain remedial action. Clearly, changing the screw did have an impact on the problem of black spots, but this action did not lead to a permanent cure, so the root cause must still be active in the system. Some theorized that extensive cleaning of the inside of the barrel at the time the screw was changed may have led to the temporary reduction in black spots on the parts.

The group continued to meet to discuss possible solutions to the black spot problem. It was noted that the design of the gas vent tube on the barrel of the machine was somewhat different from that on most other machines and that operators complained of vent tube clogging and difficulty in subsequent cleaning. It was theorized that material accumulating in the vent tube port became overheated and then periodically broke free and continued down the barrel or was pushed back into the barrel during cleaning. A new vent tube port design was proposed to eliminate this problem. The maintenance department built the new vent system and installed it on the machine. Figure 14.5 shows the u chart after the new vent system was installed and indicates the immediate and lasting effect of this process change on the occurrence of black spots. Note that Fig. 14.5 is for black spot defects only.

With this major problem solved and a great deal of experience and confidence gained, the group enthusiastically continued to seek further improvement opportunities. As a result of the elimination of the black spot problem, the u chart was beginning to show signs of more stable process behavior, and the defect rate had dropped from 4.5 defects to slightly more than 1.0 defect per part on the average. The group prepared another Pareto diagram, which showed that scratches were now the most frequently occurring defect (see Fig. 14.6).

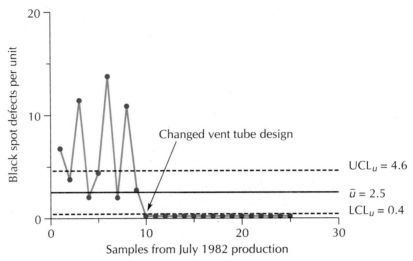

Figure 14.5 Continuation of the Press 120 *u* Chart Showing the Effect of the New Vent Tube Port Design

Several theories for the occurrence of scratches were discussed, but the most popular was that offered by a press operator, who felt that many of the molded parts were being scratched by the metal lacings of the press conveyor belt. The logical solution would be to use a continuous (vulcanized joint) belt. However, this solution was not viewed favorably because its unit cost was approximately double that of the existing belt and the installation time was

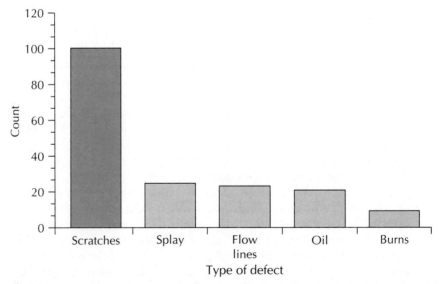

Figure 14.6 Pareto Diagram for Defects on a Radiator Grille After the New Vent Tube System Was Installed

much longer, making it more difficult to remove the belt for cleaning and maintenance purposes. A considerable amount of resistance was therefore present and a strong effort was made by some to move away from this possible root cause and look for other possible causes of scratches.

One of the statistical methods facilitators in the plant was convinced that the belt lacings were causing many of the scratches. To test the belt-lacing theory, he applied a soft latex coating to the metal lacings. The u chart with $n = 8$ for scratches only, shown in Fig. 14.7, shows that scratches disappeared completely for some time. Eventually, the soft coating wore away and broke off with belt flexing, and scratches suddenly appeared again on the u chart (see Fig. 14.7). This emphasizes the important role that continual charting plays in verifying the effect that a given remedial action may or may not have on the process. With evidence from the u chart that the belt metal lacings were causing the problem, it was now possible to convince the area supervisor that action should be taken to install a vulcanized belt on the press. As evidenced by Fig. 14.8, this led to elimination of the problem of scratches.

It is important to point out that all presses in this plant had conveyor belts with metal lacings. Once this root cause was clearly demonstrated, the logical action would be to replace the belts on all the machines. This emphasizes the fact that many problems revealed from charting on a single machine can be usefully extrapolated to other machines. Although such may not always be the case, given resources and priorities, it certainly helps in thinking through the problem of being faced with putting charts on perhaps hundreds of machines versus selectively charting a few key or representative processes.

With the scratch problem solved, the process defect rate had now fallen from 4.5 defects per part down to 0.17 defect per part. Scrap was noticeably

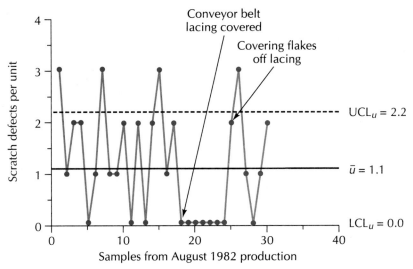

Figure 14.7 Control Chart for Scratch Defects Before and After Latex Coating Was Applied to the Belt Lacing

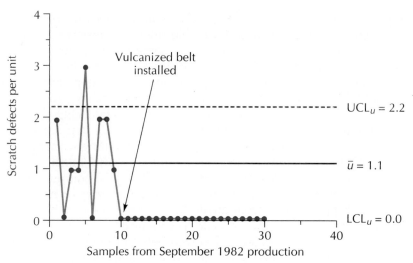

Figure 14.8 Control Chart for Scratch Defects Before and After the Bonded Belt Was Installed

reduced, overtime was eliminated, and running the backup die on another press was stopped. The economic implications here are obviously significant and need not be detailed. However, the group continued to work on improvements and began in earnest to attack the problem of splay. The splay phenomenon was found to be quite complicated, and as a result, statistical experiments were designed and conducted to develop a better understanding of the root cause of this process defect.

As an important footnote, it should be mentioned that over the period of about 3 months after the scratch problem was solved, the defect rate fell from 0.17 defect to 0.08 defect per part, for no apparent reason. The most logical explanation for this occurrence stems from the fact that as quality and productivity improvements were made, those involved in the process became more knowledgeable and concerned about the process. Press setup persons rarely needed to make adjustments, and management utilized the added free time at the press to perform more regular and thorough preventive maintenance on both the press and the mold. These factors undoubtedly would all give rise to improved process performance.

Postscript on This Case Study

The "Press 120" success story grew rapidly in fame, spreading across the company and upward to the highest levels of management. More than a year later, during a management review at this plant, the Division general manager recalled the "Press 120" story and asked to be taken to see the famous press! When he arrived in the area he immediately observed that a belt with metal lacings was installed on the press. When he asked why, the response was somewhat vague. Needless to say, the problem of holding the gains mentioned in Chapter 10 was again an issue here.

14.3

Improvement of an Accounts Payable Process

A significant inhibitor to the widespread use of statistical thinking and methods in "administrative processes" is the belief that these systems are unique and very different from manufacturing processes and that a unique and different approach and set of tools and methods is required to study them. This case study clearly demonstrates that the very same tools used in manufacturing process applications may be applied successfully to an administrative process. In this case study of an accounts payable system,[2] the following concepts and tools will be employed:

- Organization plan.
- Process flowcharts.
- Operational definitions for defects.
- p charts, c charts.
- Pareto diagram.
- Cause-and-effect diagrams.

As will become evident as this case study unfolds, the parallels between it and the injection-molding study described previously are striking.

This case study arose within a division of a major automotive manufacturer. This division was then comprised of five manufacturing plants and a central division office. Each plant has its own accounting department that handles its unique cost accounting and financial requirements, while the division accounting office handles accounting functions common to all plants in the division. One of these functions, accounts payable, is the focus of this case study. The accounts payable section in the operations accounting department processes approximately 12,000 invoices each month. The section employs 14 accountants, with each assigned to handle either the production (material that is part of the finished product) or nonproduction (spare parts) invoices for one plant (for the larger plants this task is divided among two people).

During the summer of 1985, the accounts payable section was experiencing several undesirable and costly symptoms, all of which were a concern to management. Inefficiencies had developed which slowed the payables process to the extent that a backlog of over 6000 invoices had developed. This necessitated working the accountants overtime and hiring an outside agency to open incom-

---

[2] R. J. Deacon, "Quality Improvement in an Accounts Payable System," Independent Study Project Report, Eastern Michigan University, College of Business, Ypsilanti, Mich., 1986.

ing mail and sort the invoices by plant and type (production versus nonproduction). The cost of these inefficiencies was estimated to be approximately $15,000 per month or $180,000 per year.

Organizing for Improvement

Steering Group.   Experience has shown that many quality improvement efforts have failed because management did not establish a responsive environment that was suitable for action. In an attempt to get management support and their active participation in the activities necessary to resolve the problem, a committee consisting of the area's management (manager, section supervisor, and unit supervisor) and a few statistical methods specialists was formed to guide the project. Initially, the steering group described the accounts payable system in the form of a flowchart, and operationally defined the symptoms of poor system performance as well as the various terms and acronyms used in this accounting system. Later, this committee established a goal for the project, assigned others to working-level teams, and determined the role and responsibilities of each person involved with the project, to avoid duplication of effort.

Working-Level Teams.   Two diagnostic teams were utilized to provide more detailed analysis of the problem. A group consisting of the 14 accountants and their supervisor formed one group, which concentrated on analyzing and formulating theories on potential improvements to the accounts payable system. The second team consisted of the statistical methods people in the steering group, which provided services to the accountants' team such as data analysis, meeting facilitation, and training in statistical methods.

Project Goal.   The steering committee formulated the following goal for the purpose of providing a focus for the efforts of all teams: "Identify and eliminate the faults in our system that cause inefficient processing of invoices. Eliminate the existing backlog of invoices, and institute actions as required to keep the system 'current' (invoices processed as close to, but not exceeding, 30 days) and under statistical control." The achievement of this difficult task would certainly not take place overnight and not without a lot of hard work, but the working-level teams agreed that it was a worthwhile project to undertake. These teams developed and utilized short-term action plans throughout the project to assure steady progress toward the goals.

Analysis of the Symptoms

Flowchart.   The use of a flowchart initially benefited the group through its visual illustration of the sequence of operations that should be followed by all persons involved with the accounts payable process, and subsequently was used to formulate data collection strategies. Figure 14.9 shows the flowchart for this process. A review of the flowchart for the accounts payable process indicates that the purchase notification (order), packing slip, and invoice must

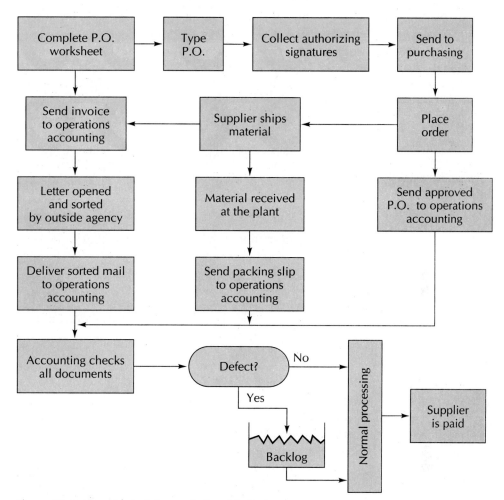

Figure 14.9 Flowchart of Accounts Payable System

exactly match each other with respect to quantity, price, and so on, before the accountant can pay the supplier. If any documentation was missing or contained errors, the invoice would go into the "backlog" until the discrepancies were resolved by the accountant (which involved time-consuming conversations with both the suppliers and the people at the plants).

Operational Definitions.   The accountants defined, in writing, the various types of symptoms encountered in the accounts payable process, as well as the acronyms commonly referred to in departmental dialect. This task served two purposes. First, the precise meaning of the words or phrases are clarified to people not directly involved with the process on a regular basis, and second, the working-level teams used them in their efforts to identify the cause of the symptom under investigation. This actually improved the efficiency of the group because theories that did not relate to the symptoms were quickly discarded, which in effect allowed more time to discuss and evaluate theories that did. Some examples of the operational definitions for the defects are

No purchase order. An instance when the packing slip and invoice are received at the accounts payable office before the purchase order.

Conversion issue. Instances where a part is specified for shipment in one unit (by the pound) but purchased in another unit (by the yard).

Vendor number differs from log. An instance where the wrong vendor number was logged in.

Logged-in nonproduction. Situations where the production parts are confused with nonproduction parts.

A review of the operational definitions reveals that

1. They are worded in very simple terms to avoid misinterpretation.
2. They are worded in a way that avoids the description of their possible cause so as not to stifle the group's creativity in future brainstorming sessions, where possible theories on the causes are formulated and discussed.

Prioritizing the Critical Problem

After a review of some historical data that only enumerated the problem, the working-level teams were assigned the tasks of collecting data in a statistical format in order to clarify whether the problem was a chronic or a sporadic issue. This information was important to the steering group because it helped them assign specific tasks to the people in the best position to carry them out. This minimized the confusion, frustration, and cost of the problem-solving effort.

As is the case with manufacturing processes, sporadic problems suggest the presence of special causes of variation which are most efficiently resolved by the people directly connected with the operation (accountant, supplier, buyer), whereas the correction of problems chronically affecting the entire system in a stable manner requires management action on the system (usually through the efforts of a larger group). The team used the Shewhart control chart for the purpose of distinguishing between these two types of process variation. It was important that everyone involved in this quality improvement effort understood the concepts of variation and various tools used in statistical process control.

The goal of the initial data collection effort for this project was to assess the stability of the accounts payable process of each division location and to identify the most common symptoms of the problem. The u chart was used for this purpose because the accountant could encounter more than one problem trying to pay the invoice, and the steering group was interested in documenting the daily failure rate for all invoices processed (which meant dealing with uneven sample sizes). The data could also be used to construct a Pareto diagram in which the symptoms are displayed in descending order in terms of their frequency. Information for a p chart on the percentage of invoices not able to be processed was also collected since the steering group was using this indicator to track the team's performance over time until the goal was achieved.

The accountants developed the data collection form that they would be using under the guidance of the statistical methods team, to ensure that the data would be rich in diagnostic content. From the statistical point of view, this involved making sure that "rational samples" were taken (separate data collection for each plant and process) to maximize the control chart's sensitivity in revealing variation due to special causes. From a team-building perspective, allowing the accountants to develop the data collection plan gave them some ownership in the project and set the stage for their support of the subsequent data analysis.

After 2 weeks of data collection, the u chart data were plotted via computer software and the p chart data were plotted by hand. The rationale for these actions is as follows:

- Even though the feedback loop with computerized charts is generally slower than that with manually maintained charts, the diagnostic capabilities of the software made its use worthwhile. Trial control limits could easily be developed (by removing samples due to special causes from the data set and/or separating the data into different time periods). Pareto analysis was performed on the symptoms automatically, and control charts for the major symptoms could easily be generated. Furthermore, once the group prioritized its effort on resolving the top concern, the original collection strategy would be discontinued in favor of another that provided more detailed information on that issue. This new data collection strategy will be discussed in more detail later.
- The steering group anticipated that in the future, each accountant would maintain a p chart at his or her desk to hold the gains. Charting the daily performance of their process by hand could be done very easily and the feedback loop was short, which would allow them to resolve the cause of any "special variation" on a timely basis.

An evaluation of the control charts revealed that the accountants were experiencing a very high (and unacceptable) failure rate in matching invoices with packing slips and purchase orders. With the exception of two plants, the control charts were stable, which indicated that the employees at the plants were doing the best they could within the constraints of the system. In other words, the system needed modification (by management) to better suit it to the division's/plant's culture and method of operation. A review of the two charts which were out of control revealed that the cause of the special variation could be tracked down and resolved immediately by the accountant (without management or group assistance). Again, one can see that the interpretation and

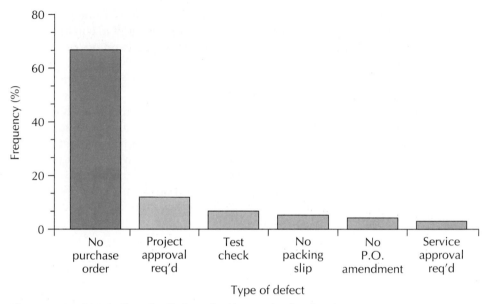

Figure 14.10 Pareto Chart for Defects for Nonproduction Invoices

reaction to the charts in the administrative area is identical to that in the manufacturing area!

The Pareto diagram for the data collection on the process for nonproduction invoices is shown in Fig. 14.10. Upon its review, the steering group immediately agreed to prioritize working on identifying and correcting the cause of the "no purchase order" symptom. A very large return on investment could be realized by the group if this larger problem were corrected, whereas much smaller gains would be made by resolving the others.

Identifying the Cause of the Problem

The first step taken to identify the cause of the "no purchase order" symptom was to list the many possible factors that could explain why the purchase orders were so slow in moving through the system. The group utilized the cause-and-effect diagram to organize the theories. Instead of using the four major categories typically found in manufacturing cause-and-effect diagrams (manpower, methods, material, and machinery), the theories were listed among the four major activities involved with the administrative system (the *plant* placing the order, the *buyer* who quotes the order, the *supplier* receiving the order, and the *accountant* who pays the bill). Figure 14.11 is an incomplete reproduction of the actual (and more detailed) cause-and-effect diagram. A review of this diagram reveals that the system is heavily dependent on the people at the plants, due to the number of possible factors that can go wrong, which slow the process.

The group utilized the information in the cause-and-effect diagram to de-

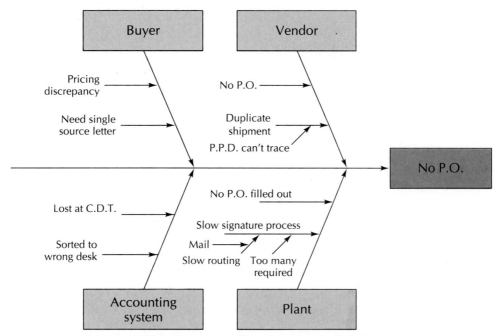

Figure 14.11 Cause-and-Effect Diagram for "No Purchase Order" Symptom

velop their second data collection strategy. The goal of this effort was to test the theories and quantify their effect on the problem. The group chose to pursue this task by randomly sampling 10 purchase orders each day where 10 critical items were recorded for each order. The selection of the 10-piece sample was based on a trade-off between the cost of collecting the data (the accountants' time constraints) and the need to guarantee the integrity of the data (to make sure that the data provide accurate information on the process). After a month of data collection, the information was entered into a computer program and analyzed several different ways. At first, the data were characterized by calculating simple statistics on such key issues of the system as

- The average time required to obtain all of the signatures/approvals required.
- The average number of signatures on the purchase order.
- The average time needed to transfer the documents from the plant to the operations accounting office.
- The average elapsed time before the supplier's invoice reached the operations accounting office.
- The percentage of the problem contributed by the type of order and the department from which it was initiated.

Afterward, the group posed a few scenarios such as "what takes place in situations where the material is ordered before the document is written and approved?" They had the computer sort and print out the data associated with such scenarios, and analyzed the data to find common elements that might explain the situation.

The analysis provided several interesting insights relative to the processing of a purchase order. It revealed that orders placed by the maintenance department (for the repair of broken equipment) quite often resulted in the "no purchase order" symptom at the accounts payable office. Analysis also showed that there was variation between the plants with respect to the number of signatures/approvals on the orders. This observation provided a major opportunity for improvement since each additional signature lengthened the approval process one day! Each manager generally reviewed these documents at the end of his or her day, so the order would not be transferred to the next person until the following day. As far as additional data collection was concerned, the group decided to discontinue the collection of the 10-piece sample (since it had accomplished its goal of revealing possible causal factors), but continued to maintain the p chart to verify the effectiveness of the remedial actions on reducing the percentage of invoices containing errors.

Developing and Implementing Remedial Actions

Upon the conclusion of the data analysis, the group developed action plans to address each issue. The accountants had identified many excellent corrective actions that had proved to be effective in similar situations in their past working experiences. Their recommendations were quite heavily dependent on the amount of support and cooperation given by the people at the plants. Therefore, the initial step in implementing remedial actions was to hold a series of orientation meetings with key people throughout the division. The operations accounting manager met with the division and plant controllers, and later the steering group met with upper management at the division's largest plant to review the project and the data analysis.

It was important to bring out the fact that the control charts continued to show process stability because a natural reaction from the managers would be to accuse the people of violating the procedures set forth by the finance manual. This may be true, but the employees were just doing the best they could with a system that was not conducive to their job objectives. This was especially true with the maintenance people, who had the task of repairing equipment that was needed for manufacturing as quickly as possible. They sometimes could not waste time filling out paperwork and getting several approvals before placing the order since they were often "under the gun" to get presses up and running.

As a result of these sessions, the plant appointed its own small team to work with the staff accountants to implement the remedial actions. The plant team consisted of the cost accounting section supervisor, two people from the maintenance department who wrote most of the orders for their department, and the plant purchasing agent (buyer). The various working-level teams implemented many remedial actions, such as:

- Revising local purchase order procedures, such as reducing the number of signatures required and increasing the dollar limit of the local buyer's authority.
- Training the salaried work force in the procedures for writing a purchase order.
- Publishing a list of account numbers not only to assure that people use the proper account, but also to advise them of the manager responsible for approving the charges.
- Using laid-off hourly employees as couriers to transport documents between staff and plant accounting offices.

The accountants used the p chart to verify the effectiveness of these actions. Breakthroughs were apparent as the failure rate dropped closer to the steering group's goal.

Holding the Gains

As of early 1987 the group had not yet achieved its goal. The working-level teams at that point were investigating additional methods to either simplify or streamline the process. The performance trends for the three main plants in the division showed considerable improvement. One plant has been able to reduce its failure rate by over 60% (monthly average from 42% in May 1985 down to 15% in March 1986) due to the aforementioned actions. A slight degradation in the performance of this plant in the last half of 1986 was attributed to the increased number of people placing orders who were inexperienced in the purchasing system. This was a signal that more training was needed at the plant. Two other plants have also experienced improvements in processing their purchase orders and invoices.

As they continued their never-ending pursuit of system improvements, the accountants continued to monitor the performance of the process via the p chart in order to hold the gains. Their task was to plot the daily reject rate, evaluate the chart's patterns, and resolve out-of-control conditions as soon as possible. Their supervisor's job was to reinforce good data collection behavior, provide any necessary coaching of the accountant, review the chart to identify long-term trends in either system improvement or degradation, and pursue further improvement actions.

To facilitate this process, the statistical methods group provided the accountants with some training on how to construct and maintain p charts. The sessions were customized to the needs of the group. In addition to teaching the charting, the quality and business philosophies of Deming and Juran were also reviewed. There was, however, a serious issue facing the statistical group. The concern was the amount of time needed to calculate the control limits associated with the variable sample sizes encountered by the accountants. Keeping in mind that the accountants are not (and probably do not care to be) statisticians, the statistical group provided each accountant with a table of upper and lower control limit values for the average ($\bar{p}$) failure rate of their process.

As a result, they were easily able to note the limit values for the number of invoices they processed, make an assessment of process stability for the day, and react accordingly.

Conclusion

The approach utilized by the group on this administrative project was essentially the same methodology that is typically used to address manufacturing problems. Even though the management of the division had not achieved its goal on the accounts payable project at the time this case study was documented, it had reduced its direct costs by approximately $7500 per month, or $90,000 per year. It does appear that the approach utilized by the group was very efficient at identifying many of the factors causing the problems within the administrative system.

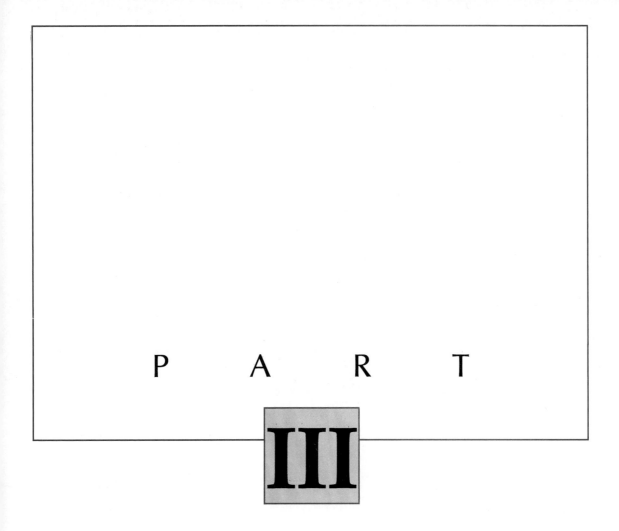

P A R T

III

Product/Process Design and Improvement

15

Conceptual Framework for Planned Experimentation

15.1

Introduction

This chapter introduces some important notions and methods for the use of designed experiments. As such, it discusses five fundamental issues central to the successful use of design of experiments for quality design and improvement.

1. The *role* of design of experiments in quality design and control.
2. Several important *guiding principles* that drive sound design of experiments practice.
3. How to mitigate the *forces of noise* in the experimental environment.
4. The *contributions of Taguchi* to planned experimentation.
5. *Contrasting philosophies* in dealing with nuisance or noise factors.

The ability to successfully design, conduct, and interpret planned experiments for product and process design and improvement depends on how well these five issues are understood and implemented.

15.2

Role of Experimentation in Quality Design and Improvement

Earlier, considerable attention was placed on two important issues:

1. The need to focus on and continually study process performance with an eye toward identifying opportunities for improvement.
2. The need to push the quality issue farther and farther upstream into the engineering design arena.

In both situations, there is a need simultaneously to study the effects that those factors perceived to be important have on product or process performance when these factors are varied or changed purposefully.

SPC/DOE Relationship

In the case of process control, we have seen that a crucial step in the "process" of process control is to diagnose or discover the fault that is causing the problem. Figure 15.1 shows the statistical process control "process" we discussed previously. Often, to get at the root cause of the problem we will need to experiment with the process, purposely changing certain factors with the hope of observing corresponding changes in the responses of the process.

Sometimes the problem is a system problem, as we saw in the gasket case study of Chapter 10. The process is in control, but the variation is too large, the defect rate is too high, and so on. This is signaling a fundamental problem that may not be revealed easily without a comprehensive study of process performance across a range of conditions, often governed by a large number of

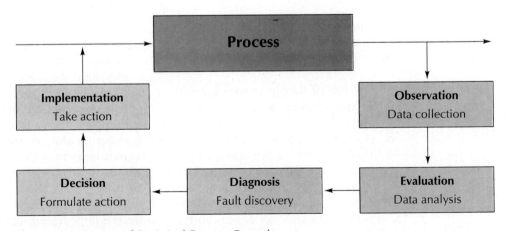

Figure 15.1 Process of Statistical Process Control

factors. Without an organized and systematic approach to experimentation, a costly and time-consuming "random walk" approach to looking for effects of change can lead to little or nothing in terms of enhanced knowledge of the process. The design of experiments methods discussed in the remainder of this book can help us to take this step in an efficient and reliable fashion.

From the results of the case study in Chapter 10 it is clear that the power of design of experiments can be greatly enhanced if the environment in which the experiments are conducted has been changed through variation reduction methods such as statistical process control. While statistical control of the process is not necessarily a prerequisite to drawing valid conclusions from the results of a designed experiment, it can greatly enhance the sensitivity of the design of the experiment in terms of its ability to detect the presence of real variable effects. A quieter process will allow the effects of small changes in the process parameters to be more readily observed.

As we will see in this chapter, there are certain fundamental design measures that we can take to ensure experimental validity. SPC, however, is an important measure that we must consider whenever possible to improve experiment sensitivity. Furthermore, in cases where statistical control of a process has been established, subsequent experimentation and the associated improvement actions taken are more likely to be realized in future operation of the process since when the process is under statistical control, the future is more predictable!

Role of DOE in the Design Process

As discussed in Chapter 1, a serious shortcoming of past approaches to quality has been the inability to deal rationally with the quality issue early in the product and process development life cycle. We have learned in the past decade, largely through the work of Taguchi and others in the United States now practicing these methods, that parameter selection at the early stages of design can be enhanced by measuring quality by functional variation during use and by using design of experiments methods. In particular, the concept of robust design, advocated by Taguchi as part of his model for the design process, shown again here in Fig. 15.2, has proved to be an effective tool for product and process design and improvement.

As we discussed previously, there is an important distinction to be made between testing and experimentation. While both have their rightful place, one should not serve as an alternative for the other. The Japanese have used experimentation—design of experiments—for parameter selection at the product and process design stage. Here the object is to experiment with various combinations of the important design parameters for the purpose of identifying the particular combination(s) that optimize certain design criteria or performance measures. In the past, the Western world has placed a great deal of emphasis on testing—life testing—by subjecting many identical units to field conditions for the purpose of determining the life expectancy of performance. When a design fails, changes are made and the product is retested. The question is:

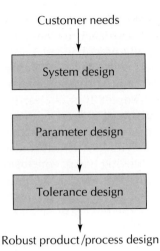

Customer needs

System design

Parameter design

Tolerance design

Robust product/process design

Figure 15.2 Taguchi's Model for the Design Process

How are those changes identified? The answer is: through a deliberate experimental approach or through more ad hoc procedures (e.g., "let's try this now"). The life testing of product performance is important but is not a substitute for experimentation to determine what ought to be tested.

15.3

Some Important Issues in Planned Experimentation

Purpose and Nature of Planned Experimentation

Scientists and engineers are commonly involved in experimentation as a means to describe, predict, and control phenomena of interest. The collection of data is a fundamental activity toward the building and verification of mathematical models, whether such models are derived from first principles or are purely empirical in nature. Comparative experiments are an important means to discern differences in the behavior of processes, products, and other physical phenomena as various factors are altered in the environment.

Too often, data analysis, modeling, and inference are stressed at the expense of the activities that embrace the planning and execution of experiments. It is assumed that valid and meaningful data are available either from passive observation of the process or from purposeful experiments and that statistical methods embrace the analysis of such data. It is, however, the planning or design stage leading toward the collection of data that is critical and needs to receive more attention, and it is here that a statistical approach to the design of experiments is so important. If experiments have been designed and conducted properly, the analysis is usually straightforward and often quite simple.

The purpose of most experimental work is to discover the direction(s) of change which may lead to improvements in both the quality and the productivity of a product or process. In the past, there has been a tendency to conduct studies farther downstream, at the process, while the use of design of experiments has been less commonly embraced by the engineering community for product design purposes. This, of course, has begun to change over the past decade, with the emphasis placed by Taguchi and others on using design of experiments for product design. As recognition of the importance of pushing the quality issue farther and farther upstream into the engineering design process grows, the role of design of experiments in the earlier stages of the product development life cycle will grow in proportion. Today, the concurrent or simultaneous engineering of products and their manufacturing processes is receiving widespread attention. Mathematical modeling, computer simulation, and the associated use of designed experiments are all playing a central role in this activity.

In this chapter we introduce some basic and important principles and methods for the design and conduct of experiments, thus laying the foundation for more detailed discussion of the methods, which follows in subsequent chapters. The recognition of these principles is crucial to the use of the tools and methods that follow.

In investigating the variation in performance of a given process, attention focuses on the identification of those factors which, when allowed to vary, cause performance to vary in some way. Some of these factors are qualitative in nature—categorical variables—while others are quantitative, possessing an inherent continuity of change. The situations we are about to examine may consider both qualitative and quantitative variables simultaneously. In fact, an important advantage of factorial designs that we will soon study is their ability to consider both types of variables within the same test plan.

But more important, there is a growing need for the precise recognition of the relative roles that factors play in governing the nature of product and process performance. Some factors are external and environmental in nature, not directly controllable but nonetheless having an important impact on performance. Some factors have a strong influence on performance on average, while others tend to influence the level of variation in performance. The specific definitions of the various roles that factors play in influencing performance are provided later, and more important, the way in which each type of factor is dealt with is discussed in detail.

Notion of the Mathematical Model

A fundamental task in design of experiments is that of selecting the appropriate arrangement of test points within the space defined by the independent variables. Although many different considerations must come into play in selecting a test plan, none can be more fundamental than the notion of the mathematical model.

Suppose that we are interested in a system involving a mean response η that depends on the input variables $x_1, x_2, \ldots, x_n$. Then we could write that

$$\eta = f(x_1, x_2, \ldots, x_n : \theta_1, \theta_2, \ldots, \theta_k).$$

That is, the mean response η could be expressed as a mathematical function f with independent variables $x_1, x_2, \ldots, x_n$ and a set of parameters $\theta_1, \theta_2, \ldots, \theta_k$.

Sometimes we know enough about the phenomenon under study so that we can use theoretical considerations to identify the form of the function f. For example, a chemical reaction may be described by a differential equation which when solved produces a theoretical relationship between the dependent and independent variables. Even if the form of f is known, the values of the parameters $\theta_1, \theta_2, \ldots, \theta_k$ will generally not be. They will need to be estimated from data collected during operation of the process using designed experiments. The observed data y are therefore represented as

$$y = f(x_1, x_2, \ldots, x_n : \theta_1, \theta_2, \ldots, \theta_k) + \epsilon,$$

where ϵ is the experimental error of observation, assumed to be NID$(0, \sigma_y^2)$.

The problem is that in most situations, little is known about the underlying mechanisms of the process. In most physical situations, at the most, we can say that f represents a relatively smooth response function. But without knowledge of the physical mechanisms, how can we proceed to explore the response surface with the ultimate goal, perhaps, of finding the values for $x_1, x_2, \ldots, x_n$ that optimize the mean response η?

When our knowledge is limited, we must rely on empirical models that serve to approximate the true but unknown model, describing relationships through the data, for example,

$$\eta = b_0 + b_1 x_1 + b_2 x_2$$
$$\eta = b_0 + b_1 x_1 + b_2 x_2 + b_{12} x_1 x_2$$
$$\eta = b_0 + b_1 x_1 + b_2 x_2 + b_{12} x_1 x_2 + b_{11} x_1^2 + b_{22} x_2^2.$$

It is important to plan experiments in such a way that several specific empirical forms can be contemplated and the best can be determined through the data once the experiment has been completed.

As an example, let us suppose that f can be represented, for a certain situation, by a simple and flexible graduating polynomial, say,

$$\eta = b_0 + b_1 x_1 + b_2 x_2 + b_{12} x_1 x_2 + b_{11} x_1^2 + b_{22} x_2^2.$$

If we can fit this function to data obtained from an experiment, we can use this equation to more clearly visualize and hence explore the complete nature of the response surface (e.g., perform optimization and sensitivity analysis). Figure 15.3 depicts this situation graphically in terms of both the shape of the true surface and its approximation based on the empirical model.

Whether explicitly recognized as such or not, most experimental studies are aimed either directly or indirectly at discovering the relationship between some performance response and a set of candidate variables which influence

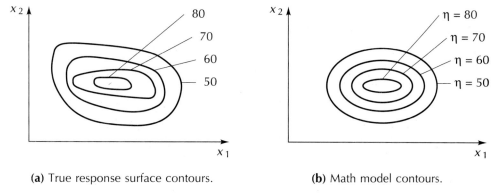

(a) True response surface contours.   **(b)** Math model contours.

Figure 15.3 Math Model Approximation to a True Response Surface

that response. In most studies the experimenter begins with a tentative hypothesis concerning the plausible model form(s) to be initially entertained. He must then select an experimental design that has the ability to produce data that will be capable of

1. Fitting the model(s) proposed (i.e., estimate the model parameters).
2. Placing the model in jeopardy in the sense that inadequacies in the model can be detected through analysis.

The second consideration above is of particular importance, to ensure that through a series of iterations the most appropriate model can be determined while others may be proved less plausible through the data. If, for example, a quadratic relationship between temperature and reaction time in a chemical process is suspected, an experiment that studies the process at only two temperatures will be inadequate to reveal this possibility. An experiment with three levels of temperature would, however, allow this possibility to be considered. On the other hand, an experiment using five levels of temperature would be unnecessary and inefficient if the true relationship were quadratic. Figure 15.4 illustrates these situations:

*   In Fig. 15.4(a), the relationship is actually curvilinear but this will never be detected by the data of the experiment because the model chosen is a straight-line model, and, only two levels of the factor are tested.
*   In Fig. 15.4(b), a poor model (straight line) has been hypothesized, but model checking will reveal this and help propose a better model because data are available to reveal the weakness of the model—the center point.
*   In Fig. 15.4(c), if the relationship is known to be a straight line, many levels of temperature in the experiment are unnecessary since only two points would be required to fit the line. The number of levels of the factors in the experiment should bear a definite relationship to the order of the model proposed to describe the process through the data.

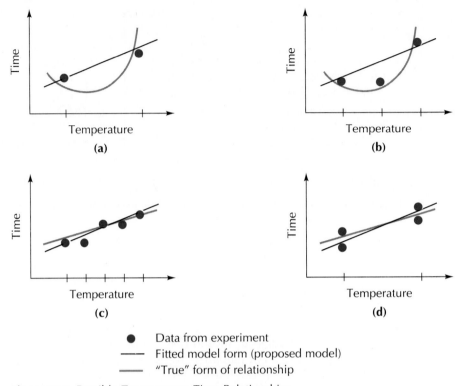

●   Data from experiment
—— Fitted model form (proposed model)
—— "True" form of relationship

Figure 15.4 Possible Temperature–Time Relationships

- In Fig. 15.4(d), if we know a priori that the relationship is a straight line, the best "test" plan would be to study only two, more extreme levels of temperature and use additional tests for replication, that is, to observe the amount of experimental error associated with the response y.

In short, the experimental design should be responsive to the mathematical model being proposed but have the ability to subject the model being considered to hazard so that if it truly is inappropriate, that can be revealed through the data.

Sequential and Iterative Experimentation

There is always the temptation to carefully design one big experiment that will consider all aspects of the problem at hand. This is dangerous, for several reasons:

1. If erroneous decisions and hypotheses about the state of affairs are made, considerable time and experimental resources may be wasted and the end

product may provide little useful information or direction in terms of what to do next.

2. If knowledge of the underlying situation is limited a priori, many factors may be suspected as being important, requiring a very large experiment in terms of number of tests. Ultimately, only a small subset of variables will be found to be of major significance.

3. In the early stages of experimentation, knowledge of the specific ranges of variables that ought to be studied is not always available. Furthermore, the metrics to be employed to define the variables, responses, or even what responses to observe may not always be clear in the early stages of experimental work.

4. One large experiment will necessarily cause the testing period to be protracted, making it more difficult to control the forces of nuisance variation.

For these reasons it is much more desirable to conduct an experimental program through a series of smaller, often interconnected experiments. Such provides for the opportunity to modify hypotheses about the state of affairs concerning the situation at hand, quickly discard variables shown to be unimportant, change the region of study of some or all of the variables, and/or define and include other measures of process performance. Experimental designs that can be combined sequentially are very useful in this regard.

Revelation of the Effects of Change

Often, the variables of importance that govern a process are not clearly known a priori. It is desirable to be able to study several variables together but independently observe the effect of a change in each of the variables. Furthermore, it may be important to know if a variable effect varies with the conditions of the process (i.e., when other variables take on varying levels). When such information is sought, the design—the arrangement of the tests—becomes very important. Consider the following example.

Suppose in a given situation that five variables seem to be important. One approach to experimentation would be to hold four constant, and run tests varying only one variable. The results might look something like Fig. 15.5. This

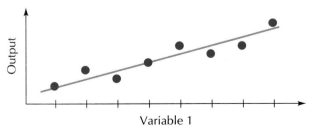

Variable 1

Figure 15.5 One-Variable-at-a-Time Experiments

process would then be repeated for each of the other variables. There are several problems with this one-variable-at-a-time approach. One major problem has to do with the fact that most variables do not influence the response of interest independent of the others. The following example illustrates this fact.

Example. Two factors, temperature and pressure, are thought to affect chemical reaction time. Two experimenters, Mr. X and Mr. Y, each run a series of tests to study the specific ways in which temperature and pressure affect reaction time. Using the one-variable-at-a-time approach, we find that the results might be entirely different, due to different selections of the fixed levels for the variables that are not currently being tested. Figure 15.6 illustrates such possible outcomes as a result of different settings of fixed variables between Mr. X and Mr. Y. It is clear from Fig. 15.6 that the effect of temperature depends on the level of pressure and vice versa. We will later refer to such behavior as signaling the presence of variable interactions.

In summary, one-variable-at-a-time experimentation suffers from significant problems.

- It generally gives rise to a large number of tests, in particular, too many levels of each factor.
- It fails to recognize variable interdependencies, which results in a highly conditional or narrow interpretation.

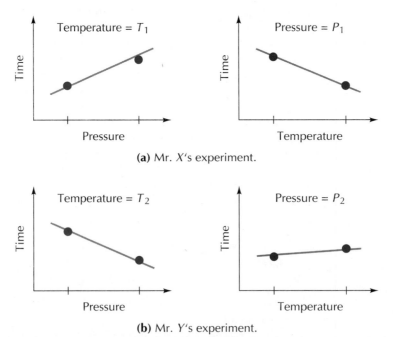

(a) Mr. X's experiment.

(b) Mr. Y's experiment.

Figure 15.6 One-Variable-at-a-Time Example Experiment

In addition, we will soon see that such experimentation creates other problems:

- It promotes a systematic test sequence where it may be easy to have unknown forces changing over the experiment—biasing the results—so validity is questionable.
- It does not lend itself well to experimental considerations such as blocking, that is, the isolation of known effects of external variables that are not being studied.

In the example that follows we discuss further the problem of test arrangement to reveal important variable effects and offer some suggestions on how to deal with this problem.

Comparative Analysis Between Two Sandmills

A vinyl plant employs vertical sandmills for the dispersion of pigments. The pigment dispersion process produces colorants for the manufacture of vinyl film. As part of a quality improvement project, the plant was considering the acquisition of a new type of sandmill, a horizontal sandmill. The manufacturer of the new machine claimed that the horizontal sandmill would improve both productivity and quality. It was decided to perform some tests on the claimed merits of the horizontal sandmill in comparison with the vertical sandmill. Arrangements were made for some trial runs on the horizontal sandmill.

The measures used to judge the relative performance of the two sandmills were fineness of the grind and color strength. The grind fineness is measured on a Hegman block, which can be read as Hegman values on a scale from 0 (coarse) to 10 (fine). In this example we are concerned only with the grind fineness in terms of Hegman values.

At first glance it seemed attractive simply to run a few trials on the horizontal sandmill and compare the results with the available production records of batch Hegman values of a representative vertical sandmill in the plant. It was quickly brought out that certain pigments were much more difficult to grind than others, depending on their hardness, or equivalently, the color of the pigment. For instance, red is soft and blue is hard. Also, it was conjectured that relative machine performance could be similar for some pigments but much different for others. Which pigment should be chosen for the trial runs on the horizontal sandmill? Or should more than one type of pigment be tested? If so, how many and what kind?

Another issue raised concerned the degree of pigment loading. The horizontal sandmill manufacturer maintained that his machine should be run with higher-percent-pigment mixes to realize the claimed merits in quality and productivity. Therefore, to run the comparative tests, at least two pigment loading levels must be chosen.

In addition to the two concerns above, there were many other factors to be considered before the trial runs could begin. One of these was to determine the settings of some key parameters, such as temperature and flow rate of the sandmill for the trial runs.

Finally, the idea of using historical Hegman values of the vertical sandmill for comparative analysis must be abandoned to avoid both large variability and any bias due to possible operational and environmental changes over time. Instead, the tests should be run side by side on vertical and horizontal sandmills to provide a more sensitive and valid comparison between the two.

It was ultimately decided for each particular machine setting that two types of pigment (red and blue) and two different pigment loadings (10% and 15%) should be tested on each sandmill. Now the questions that remain are: How many tests should be run, and how should the tests be arranged in the independent variable space? Consider the two possible arrangements of the tests as illustrated in Fig. 15.7.

Arrangement 1 is somewhat haphazard, with test conditions randomly distributed in the pigment type, pigment loading space. Therefore, it is not possible to observe the effect of changing pigment loading only, for example, since two tests do not exist for which pigment loading changes but pigment type and machine type remain fixed. When we contrast the outcomes (Hegman values) of any two trials, several things are changing simultaneously. It is

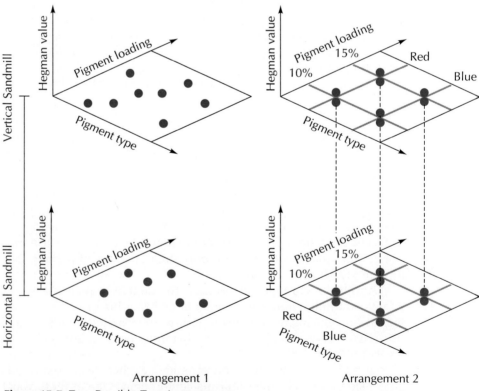

Figure 15.7 Two Possible Test Arrangements

difficult to sort out exactly what is affecting change in the outcome and what is not.

Arrangement 2 in Fig. 15.7, however, provides for the opportunity to learn much about the relationships among the three variables and the Hegman value. By selecting the same combinations of pigment type and pigment loading for each sandmill, direct contrasts in Hegman values (shown by the dashed lines) between the two sandmills can be analyzed. Some specific features of arrangement 2 follow.

1. The effect of changing any of the three variables alone can be observed.
2. The possible joint effects of two or more variables on Hegman value can be observed. For example, if the Hegman value is influenced by the type of sandmill only at a certain level of pigment loading, or is influenced differently at different levels of pigment loading, this joint effect can readily be observed with arrangement 2 but not with arrangement 1.
3. The total number of tests is generally much smaller using arrangement 2 than arrangement 1.
4. Because fewer levels are chosen in arrangement 2 for each of the variables to be tested, it is more feasible to replicate all the test conditions, as shown in Fig. 15.7. Replicated responses provide a database for the estimation of experimental error.

However, as shown, arrangement 1 is capable of estimating some curvilinear relationship of the variables on the response, if one exists. If there should be a need to fit a second- or higher-order model for a given situation, arrangement 2 may be extended or augmented to have three or more levels for some or all of the variables under study. Arrangement 2 is generally referred to as a factorial design. For the sandmill study this means to test all possible combinations of the three variables at two levels each, amounting to eight unique tests. Geometrically, the eight unique test conditions could be represented as a cube, as shown in Fig. 15.8. For example, the front lower left-hand corner represents a test for which the conditions are:

Sandmill type: vertical.
Pigment type: red.
Pigment loading: 10%.

The information obtainable from this type of test plan is quite comprehensive. In particular, we can assess independently:

- The average individual effect of each variable on the response.
- The joint, or interdependency, effect of any two variables.
- The joint, or interdependency, effect of all three variables.

Such information would constitute a much more relevant assessment of relative sandmill performance than would a simple comparative analysis. The results

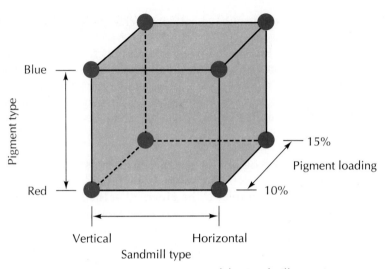

Figure 15.8 Geometric Representation of the Sandmill Experiment

of such an assessment may not only be useful for making the decision "to buy or not to buy" but could lead to the use of different loadings for different pigments on a sandmill to improve quality and productivity.

15.4

Dealing with Noise in the Experimental Environment

The experimental study of any phenomenon is made difficult by the presence of noise sometimes called experimental error. Many factors not under study are varying during the experiment. Although the variation of any one such factor may produce only a very small change in the performance measure under study, the fact that many such forces are at work simultaneously produces a noticeable and sometimes sizable level of system noise. When the system is stable, these are the forces of common-cause variation. Sometimes we refer to such variation as *white noise*, to emphasize its random nature. Such variation may cloud or mask the effect of changes in the factors under study in an experiment.

Some noise sources in the system may be of a more spurious nature. Such sporadic disturbances of often a sizable magnitude can further contribute to clouding the results of experiments and may even bias the results if such forces are occurring in a somewhat systematic fashion. Structured or systematic variation is sometimes referred to as *colored noise*. How can we better understand and perhaps mitigate the forces of noise in the experimental environment? There are several things we ought to consider to deal with noise at the planning stage. These will now be described.

General Strategies for Dealing with the Forces of Noise

Statistical Control/Stability Analysis

If the phenomenon under study is already a feasible and ongoing process, the pursuit of improvement opportunities through experimentation can be enhanced considerably by employing the techniques of statistical process control. In this way, spurious or sporadic sources of variation can be identified and through specific action removed. Achievement of a stable process will greatly contribute to our ability to observe more readily the effects of purposeful process change. Continued study in this fashion will further help us to observe the persistence of changes that might be introduced.

Once a process is stabilized, continued attack on the common-cause system will lead to a progressively "quieter and quieter" process, further enhancing our ability to observe the forces of purposeful process change through experimentation. As a process becomes "quieter" (i.e., the common-cause variation is reduced), it enhances the sensitivity of control charts in detecting the occurrences of certain special causes whose effects are relatively small.

Experimental Design Strategies

We must recognize the fact that in many situations, the notion of a stable, ongoing process has little meaning. In the early stages of product or process design, prototype or pilot-plant testing, or in comparing performance of existing and proposed new methods, the luxury of having a stable, consistent process is not present. It is perhaps for this reason (among others) that the body of knowledge we know as experimental design was cultivated in the first place. Under such situations we must strive for four things:

1. Attempt to identify major sources of variation and take action to ensure that their presence is absent from the comparisons we make within an experiment. The technique of blocking is useful for this purpose.
2. Counteract the forces of unknown systematic variation over the period of the experiment by randomization of the tests so that such variation is uniformly and randomly distributed across the trials conducted.
3. Include replication in the experimental test plan. Multiple tests at the same conditions will provide comparisons that directly measure the amount of variation/experimental error. A priori knowledge of the approximate level of variation expected will enable us to make some judgment about the amount of replication required to discern the presence of real differences in performance of a certain amount.
4. Include confirmatory testing as part of the experimental strategy. The ultimate value of designed experimentation lies with the "persistence of effects" when improvement opportunities are revealed through the experiment. Hence it will be important that additional trials are run under specific conditions determined from the analysis to verify the improvement opportunities revealed from the experiment.

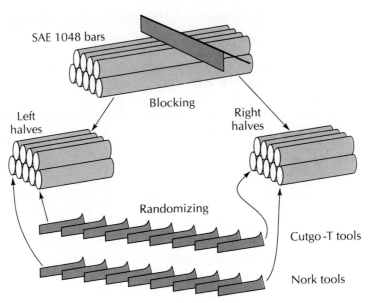

Figure 15.9 Blocking and Randomization

Two of these issues, blocking and randomization, are very fundamental issues that address the problems associated with the sensitivity and validity in making comparisons and drawing inferences on those comparisons within an experiment. We now briefly discuss these two important issues.

Blocking and Randomization in Experimental Designs

Blocking Known Sources of Variation

The example that follows illustrates the roles that the techniques of blocking and randomization play in improving the design of experiments. Nork Tool Company claims that their new cutting tools, called Nork-V tools, provide a longer life than the Cutgo-T tools for similar jobs. To check whether or not Nork's claim is valid, it is planned to machine nine 1048 steel bars using nine Nork tools and nine Cutgo-T tools, all on a single machine with the same machine setting. The tools and workpiece materials are selected as shown in Fig. 15.9.

The bar-to-bar material properties variation (mainly hardness) is known to have considerable effect on the tool life. We can see this effect in the data of Table 15.1 by looking up and down each of the columns under each tool type. The tool life values vary tremendously, even for the same tool. If we were to calculate an estimate of the experimental error using the tool life values within each column (tool), we would find these estimates inflated by the difference that the material—the bar—makes on the tool life. This large variation estimate would make it difficult to see a true difference in the two tools, if such really

TABLE 15.1 Test Results: Tool Life (Minutes)*			
Test	y_c (Cutgo-T)	y_n (Nork-V)	$d = y_n - y_c$
1 (Bar 1)	18 L	20 R	+2
2 (Bar 2)	16 R	14 L	−2
3 (Bar 3)	10 L	10 R	0
4 (Bar 4)	17 R	20 L	+3
5 (Bar 5)	17 L	19 R	+2
6 (Bar 6)	20 R	20 L	0
7 (Bar 7)	16 R	15 L	−1
8 (Bar 8)	28 L	28 R	0
9 (Bar 9)	24 L	29 R	+5

* L and R refer to the left half and right half of a bar, respectively.

exists. Therefore, each bar is cut into halves, with one machined using a Nork tool and the other a Cutgo-T tool. This arrangement will block out any bar-to-bar material differences by allowing for the observation of the difference in tool life within the same bar. In such paired comparisons a much larger portion of the difference in tool life can be attributed due to actual differences in tool performance, if such exists. Now, a measure of the difference in tool life between the two tools can be determined within a bar, where a pair of tests are more similar to each other in terms of bar material properties.

When known sources of extraneous and unwanted variation can be identified, we can design the experiment in such a way as to eliminate their influence and provide a more sensitive test of significance for the variables under study. The technique we use is referred to as blocking. Blocking is therefore an important concept in the design of experiments and is a method to eliminate the forces of extraneous variation from desensitizing the analysis when certain conditions of the experiment cannot be controlled or held constant. However, if the extraneous variation persists in the actual operating environment downstream, not just in the experimental environment, an alternative measure may be taken to include them purposefully in the designed experiment. One such measure, to be discussed later in the chapter, is robust product and process design.

Randomization for Unknown Sources of Variation

To further reduce possible bias due to material differences between the left half and the right half of each bar, the selection of either half of the bar for each test is made at random. Such random pairings would distribute evenly any possible material difference between the two halves within a bar. According to such a design the 18 test results are given in Table 15.1. The analysis can now proceed to check for a true tool life difference, estimated by the averge difference across all bars, $\bar{d}$. By comparing this average difference against the variation in the differences, the bar-to-bar variation is eliminated from the analysis.

In this case, although the average difference from the test results shows a longer life by 1 minute for the Nork-V tools, it turns out that the statistical evidence is not strong enough to conclude that a true difference exists. This could be due to the relatively large common-cause variation in the machining process that masks any real difference in the tool life. Sometimes, however, there might be other systematic or assignable factors that were not blocked or randomized. In running these tool life tests, it was learned later that two tests were performed each day with the Nork-V tool tested in the morning and the Cutgo-T tool in the afternoon. The warm-up effect of the lathe tends to give a longer tool life for the afternoon tests. Therefore, a better design of the tests should consider randomizing the tools in time in addition to the random selection between left and right halves of the bar.

Often, we tend to assume that the sample observations are independently drawn from some normal distribution(s) with a constant mean and a constant variance (i.e., a process in statistical control). Unfortunately, unexpected changes often take place in a somewhat orderly fashion over a number of runs. As a result, the observations may be correlated in time, and hence the treatment of them as independent can cause misleading conclusions. Since such systematic external sources of variation are not easily blocked out, we must employ the concept of randomization to evenly distribute them between the two variables under consideration.

In general, there are two kinds of correlations among successive runs, positive correlation and negative correlation. As an example, suppose that an inexperienced operator was assigned to run the tool life tests of Table 15.1 and he ran the first nine tests with the Nork-V tools and the remainder with Cutgo-T tools. The Cutgo-T tools may show a higher average tool life because of his learning effect, which gradually gives a longer life of any tools. Such a systematic trend would produce positively correlated results. On the other hand, the warm-up effect of the lathe produces systematically higher and lower tool lives, alternating between A.M. and P.M. This oscillating behavior would give negatively correlated results. A plan should be carefully made to determine exactly how the tests are to be randomized to deal with each kind of possible correlation.

To deal with possible positive correlations or linear trends, randomizing adjacent runs within pairs is recommended because it avoids the large differences over the long trend. An example of such a randomization is shown as arrangement 1 in Table 15.2 for the tool life experiments. For situations with possible negative correlations or oscillating trends, running each pair of two random halves both at a high and a low level of the oscillating cycle, such as both in the A.M. or both in the P.M., will avoid direct comparison between adjacent runs, such as arrangement 2 in Table 15.2.

Another Example: Need for Randomization

As another example, suppose that a very simple sandmill experiment was conducted over two successive shifts, eight tests being conducted on each shift.

TABLE 15.2 Specific Randomization Schemes for Positive and Negative Correlation Nuisance Variable Situations*				
	Arrangement 1		Arrangement 2	
Bar	Cutgo-T	Nork-V	Cutgo-T	Nork-V
1	La	Rb	LA	RA
2	Ld	Rc	RP	LP
3	Rf	Le	RP	LP
4	Lh	Rg	LA	RA
5	Lj	Ri	RA	LA
6	Rk	Ll	LP	RP
7	Rm	Ln	RP	LP
8	Rp	Lo	RA	LA
9	Lr	Rq	LP	RP

* L, left half; R, right half; A, A.M., P, P.M.; a to q are the time order of runs.

The purpose of the tests is to compare the performance of a vertical and a horizontal sandmill in the dispersion of paint pigment to a prescribed level of fineness. Suppose further that the eight tests for the vertical machine are done on the day shift, followed by the eight tests for the horizontal machine on the evening shift. The two operators may, without us realizing it, operate the process somewhat differently, and as a result influence the outcomes. The resulting data could look something like the ones shown in Fig. 15.10.

The previous results seem to favor the horizontal sandmill (i.e., the fineness values seem consistently higher). This may have nothing to do with the machine performance itself, but only be a reflection of a systematic difference in operator performance. How are we to deal with such systematic change when its presence and character are unknown to us?

Sir Ronald Fisher[1] has pointed out that in the face of systematic change of the testing environment, the fundamental precaution that we must take to guarantee validity of the test of significance is randomization. If the 16 trials were randomly assigned to the two shifts, the statistical comparison that follows would be a valid one. The variation induced by any systematically changing unknown factor would certainly adversely affect the sensitivity of the test, but since it is randomly distributed across all trials it would not bias the results. Since such systematic change could occur totally unknown to us, it will always be wise to randomize the experimental trials.

Another Viewpoint on Blocking and Randomization

Both the techniques of blocking and randomization are aimed at improving the validity and sensitivity of comparative tests. There are, however, at least two

[1] R. A. Fisher, *The Design of Experiments*, 8th ed., Hafner, New York, 1966.

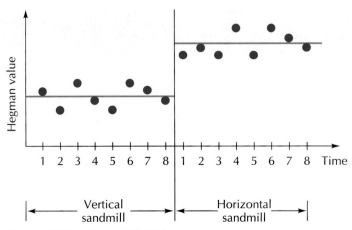

Figure 15.10 Shift-to-Shift Differences

concerns about the use of these techniques. One concerns blocking. If the factor(s) that are blocked out in our design of the tests are part of the system variation, we would not know what their effects might be in routine operation of the system. Therefore, for such cases, not only should we not block them out, we should purposely study their effects in addition to, say, the tool differences on tool life. The second concern deals primarily with randomization. If there are several systematic effects, such as within-bar material differences, machine warm-up, learning curve, and so on, to be randomized to improve the sensitivity in data analysis, their joint effects due to some dependency among them should be explicitly evaluated for better system design. In summary, when the environment surrounding the comparative tests involves too many variables that need to be blocked or randomized, one might consider them as additional factors to be experimented with using multifactor design strategies. In Chapters 16 to 20 we present some of these designs.

Summary

From the comments so far in this chapter it becomes clear to us that there are a number of characteristics or properties that are desirable in sound experimental design test plans.

1. The ability to deal simultaneously with qualitative (categorical) and quantitative factors/variables.
2. The existence of the mathematical model concept as a guide to designing the experiment, that is, explicit recognition of the relationship between the proposed mathematical model and the experimental design which responds to that model.

3. Awareness of the notion of model building and the sequential and iterative nature of experimentation. In this regard, the ability to employ experiments that serve as building blocks, one upon the other, is important.
4. The ability of an experimental design to observe separately the effects of changing factors individually (all other factors held constant) but at the same time being able to do this at several different combinations of the other factors so that factor interdependencies can be properly revealed. It is important to make comparisons that clearly show the effects of the several factors and to facilitate model identification, that is, determine just exactly which factors are important and how they are important.
5. The ability of the experimental design to observe the forces of experimental error/system variability directly, so that comparisons of performance can be made in light of the magnitude of system noise. Replication of tests provides us with the ability to do this.
6. The ability of the experimental design to mitigate the forces of nuisance variation, explicitly identifiable or not, through the techniques of blocking and randomization.

In addition to these considerations, it would of course be useful if the experimental design was

- Easy to set up and carry out.
- Simple to analyze and interpret.
- Simple to communicate or explain to others.

One important class of experimental designs that possesses many of these desirable features and is particularly useful in the formative stages of an investigation is the class of two-level factorial designs, such as the ones shown in Fig. 15.11. These experimental designs and derivatives of these designs, two-

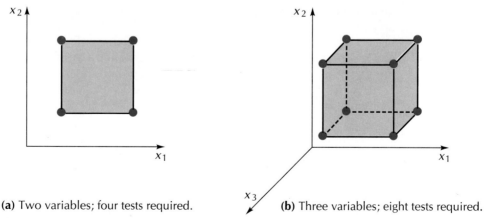

(a) Two variables; four tests required.   **(b)** Three variables; eight tests required.

Figure 15.11 Two-Level Factorial Designs

level fractional factorial designs, are discussed in considerable detail throughout the balance of the book.

15.5

Role of Planned Experimentation in Taguchi Methods

Taguchi's Contribution to Quality Design

In recent years, many U.S. companies have begun to examine new directions that they must cultivate to improve their competitive position significantly in the long term. Companies have sought to develop an overarching quality philosophy that can be implemented through policy and operating procedures to ensure that customer needs can be met competitively. One thing that has become clearly evident is the need to push the quality issue farther and farther upstream so that it becomes an integral part of every aspect of the product life cycle (e.g., marketing, product design and engineering, process engineering, production, etc.).

Presently, a great deal of attention is being given to the development of concepts and methods for product and process design and engineering. It is recognized by many that it is through engineering design that we have the greatest opportunity to influence the ultimate delivery of products that meet customer needs and expectations. In the last few years, the role of statistical methods for quality and productivity improvement has received considerable attention. In particular, the approach and techniques of Taguchi, commonly referred to as "Taguchi methods," have been widely adopted. These methods and the general issue of parameter design are the subject of many recent books and papers.[2]

The Taguchi approach to quality design has a number of significant strengths which ought to be exploited. In particular, Taguchi has placed a great deal of emphasis on the importance of minimizing variation as a means to improve quality and of bringing the mean of the process to the design-mandated target. The idea of designing products whose performance is insensitive to environmental conditions, and making this happen at the design stage through the use of design of experiments, have been cornerstones of the Taguchi methodology for quality engineering. The strengths of Taguchi's approach lie in the following areas:

1. Emphasis on quality during the engineering design process.
2. Recognition of the relative roles of factors in influencing the performance of the product/process.

---

[2] The reader is directed to entries in the References at the end of the book for Taguchi, Phadke, Kackar, Sullivan, and Clausing.

3. The emphasis on variation reduction in quality engineering: use of the loss function and signal-to-noise ratio.
4. Robust design: the parameter design concept.

Engineering Design Process

As discussed in Chapter 1, Taguchi views the design process as evolving in three distinct phases or steps: system design, parameter design, and tolerance design (refer to Fig. 15.2). It is perhaps this broad umbrella placed over his concepts and methods for quality design and improvement that makes this approach so readily embraceable by the engineering community. As we have said earlier, Taguchi considers engineering design the central issue and statistical methods as just one of several tools to accomplish his objectives in engineering.

System design is the initial phase of any engineering design problem wherein knowledge from specialized fields (chemistry, electronics, etc.) is applied to develop the basic design alternative. This design alternative is then refined in the phases of parameter design and tolerance design. *Parameter design* is that phase of the design process in which the "best" nominal values for the important design parameters are selected. Taguchi defines the best values for the nominals as those values that minimize the transmitted variability resulting from the noise factors. This is why we sometimes refer to parameter design as robust design. *Tolerance design* is the final phase of Taguchi's design process model, in which the tolerances on the nominal values of critical design factors determined during parameter design are analyzed. In tolerance design the evaluation of tolerances is considered to be an economic issue and the loss function model is used as a basis to strike the economic trade-off between increased manufacturing cost and associated reduction in quality loss due to functional variation.

Recognition of the Relative Roles of Factors in Influencing the Performance of the Product/Process

Taguchi suggests a sound engineering interpretation of the varying roles that the important system factors/variables play in influencing the product/process performance. Taguchi emphasizes the importance of evaluating quality performance under field conditions, as part of the design process, and the fact that functional variation in performance is influenced by noise factors that vary in the environment in which the product/process is functioning. For example, suppose that one decided to run a simple experiment to study the effects that four factors—carburetor design, engine displacement, tire pressure, and driving speed—may have, either singly or in concert, on the performance of an automobile as measured by fuel economy in miles per gallon. A simple two-level factorial design could be employed with the variable levels as defined below.

Carburetor design	Type A	Type B
Engine displacement (liters)	2.8	3.4
Tire pressure (psi)	22	28
Driving speed (mph)	45	55

A total of $2^4 = 16$ tests would be required and the purpose of the experiment might be to determine the "best" levels for each of the four factors (i.e., the levels that lead to the best fuel economy).

A more thoughtful consideration of the experiment above might lead one to believe that the notion of "best" levels does not have the same meaning for all four of the variables in question. In terms of the engineering design process, the selection of a "best" level for tire pressure and driving speed has very little meaning. In terms of product performance in the field, these two variables cannot be controlled to fixed levels. As a matter of fact, these two factors are likely to vary in a somewhat random fashion over time as the product performs in the field and the field environment (temperature, road terrain, etc.) varies. As tire pressure and driving speed change over time, their variation will be transmitted through the design and be felt in the output response of the product. As such, these two factors fit the definition of outer noise factors quite well.

The fact that some factors or variables influencing product performance in a substantial way cannot easily be controlled does not necessarily mean that we would not include them in an experiment aimed at understanding the nature of their influence and the influence of other, more readily controlled, factors. It does mean, however, that the structure of the experimental design might need to be adjusted from that defined by the more classical approach to the design of experiments.

Transfer Function Model: Classification of Variables

Figure 15.12 is a block diagram representation of a transfer function model for system performance. The transfer function describes the relationship between the inputs, defined by control factors and the outputs of the product/process, in the presence of noise factors varying in a generally uncontrollable fashion in the environment in which the product/process is functioning. When the state of knowledge permits, the transfer function might be derived from first prin-

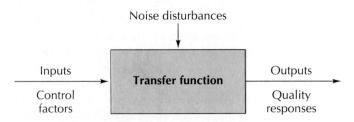

Figure 15.12 Transfer Function Model for Product/Process Performance

ciples and may be in the form of an explicit mathematical relationship. Often, however, such advanced knowledge is not available and hence we must rely on techniques such as design of experiments and empirical model building to develop the transfer function.

Taguchi's contribution to the foregoing transfer function model is a more rigorous and unique classification of the nature of the factors that can influence product/process performance. In formulating the engineering design problem, Taguchi has found it appropriate to classify the factors that can influence product/process performance into four categories.

1. **Signal Factors.**   These are the factors that may be adjusted by the user/operator to attain the target performance. The speed adjustment on a blender is a signal factor.
2. **Control Factors.**   The control factors are the product/process design parameters whose values are to be determined during the design process. The purpose of the design activity is to select the best levels of the control factors according to an appropriate design criterion.
3. **Noise Factors.**   The noise factors, as defined previously, are the uncontrollable factors and the factors that although controllable in principle, are controlled only at considerable effort/expense and are therefore considered as noise factors for the purposes of design.
4. **Scaling/Leveling Factors.**   These factors are special cases of control factors. They are factors that may easily be adjusted to achieve a desired functional relationship between a signal factor and the output response. For example, the gearing ratio in a steering mechanism can easily be adjusted to achieve a desired relationship between the turning radius and the steering angle.

The block diagram shown in Fig. 15.13 was adapted from Phadke[3] and provides a representation of a product or process in terms of the factors that govern its performance. The expression $f(x, m, n, r)$ is used to represent the relationship of the signal (m), control (x), noise (n), and scaling/leveling (r) factors to the response (y) of the product/process. Signal factors are those variables adjusted to attain the target/nominal performance. The control factors are those variables under the control of the designer. The selection of the nominal values for the control factors is the primary role of parameter design. Noise factors describe variables that are difficult or impossible to control but whose variation is transmitted through the design—as described by the transfer function—to the output.

The Concept of "Degree of Control"

Often, the matter of control of a given factor is an economic one. Factors that appear to be capable of being treated as control factors might be treated as noise

---

[3] M. S. Phadke, *Quality Engineering Using Robust Design*, Prentice Hall, Englewood Cliffs, N.J., 1989.

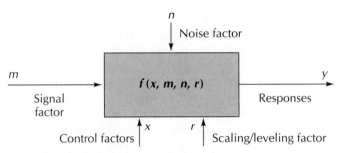

Figure 15.13 Block Diagram of a Product or Process

factors for the purpose of gaining economic advantage. This notion is particularly appealing in a simultaneous/concurrent engineering framework when product design and processing considerations are being dealt with jointly. The design for manufacturability concept can be served in a quantitative fashion by seeking levels for the control factors that minimize the transmitted variability to product performance of conditions present during processing—the manufacturing process.

Sometimes the identification of a factor as a noise factor is clear from the outset of the study. Only as product and process designs mature may it be possible to understand the economic ramifications of the control of certain factors. At other times, this is revealed only as the study progresses and data from experiments come to light. The opportunity to consider a factor as not needing to be actively controlled, or controlled at a lower level, may come to light only through the identification of nonlinear relationships that may exist between or among two or more variables.

The classification of factors in accordance with the roles they play in influencing the response/performance of the system is not meant to imply that the interpretation as to whether a certain factor is a control factor or a noise factor may not vary from one situation to another. The issue of the control of a certain factor is generally one of economics and therefore any factor might be thought of in terms of its *degree of control* rather than as controllable or not controllable. Consider the following example, which illustrates this idea.

Degree of Control: An Example.   Suppose that we are studying an injection-molding process because of a part shrinkage problem. A brainstorming session might lead to the following list of factors as having a potential influence on part shrinkage:

Cycle time   Mold temperature
Percent regrind versus virgin raw material   Cavity surface finish
Gate configuration   Raw material moisture content
Holding pressure   Raw material melt flow index
Screw speed   Mold cooling water temperature.

Although any or all of the factors above can influence part shrinkage, they are not all controlled or manipulated to the same degree or with the same ease economically. Some of these factors may be controlled at the engineering design process (e.g., the gate configuration or the cavity surface finish of the mold). Others are controllable at the molding machine (e.g., screw speed or cycle time). Still others are properties of the incoming raw material and are therefore under the control of the supplier (e.g., melt flow index). Finally, certain factors may best be categorized as defining part of the environment in which the process must function (e.g., mold cooling water temperature, percent regrind vs. virgin raw material).

It becomes readily apparent that it depends on who you are and what your point of view or objective is when it comes to determining whether a certain factor should be considered as a control or a noise factor. In dealing with an incoming raw material property, for example, one could either go back to the supplier and require a tightening of the specification to reduce the transmitted variability to the process output (part shrinkage), or one could treat that incoming material property as a noise factor and attempt to mitigate its influence on the output through adroit manipulation of other factors which are more easily controlled. In the illustration above, mold cooling water temperature and the virgin/regrind material mix are certainly factors that can be controlled, but if a part/die/process combination can be developed that is robust to variation in these factors, that would be economically attractive.

In a later chapter we examine in detail a case study dealing with rattles and closing effort for glove box assemblies in an instrument panel. In this case study several factors were examined as to their influence on closing effort, including latch striker angle, panel foam thickness, and striker positioning during assembly. The first of these is controlled through the design of the die that stamps out the striker. The second variable is controlled through the foam molding process. The third factor is controlled by the worker at the assembly process. Clearly, if there were a problem with glove box closing effort, which might manifest itself through warranty complaints, one might consider treating the striker position variable as a noise factor, since it is more difficult/expensive to control, and attempt to overcome the problems this factor is creating by the informed manipulation of other factors, such as striker angle, which are considerably easier to change/control.

Signal-to-Noise Ratio as a Measure of Product/Process Performance

Taguchi methods emphasize the importance of variation reduction and the relationship of variation reduction to the meaning of quality. Taguchi provides a definition of quality that can successfully be made operational upstream at the point of engineering design as a design criterion. The concept of signal-to-noise ratio as a means to evaluate system performance is an important contribution to a better understanding of the importance of variation reduction.

Furthermore, the loss function concept introduced in Chapter 2 and discussed in more detail in Chapter 9 is used to assess the specific economic consequences of variation.

In Chapter 1 we treated the concept of signal-to-noise ratio in a general way to examine quality design and improvement strategy in a global fashion. We will look at how Taguchi defines this measure as a more specific design of experiments response.

In assessing the results of experiments, Taguchi suggests the use of statistics known as *signal-to-noise ratios* (S/N). Instead of calculating an average response value and a variation in response value, Taguchi suggests that the signal-to-noise ratio be calculated at each design point. The combinations of the design variables that maximize the signal-to-noise ratio are then generally selected for further consideration as product or process parameter settings.

Maximizing the S/N ratios is presumed to be equivalent to minimizing the average/expected loss associated with striving to be on target with smallest variation. Three cases are of practical significance.

Case 1: The S/N ratio that Taguchi defines for "smaller is better" is

$$S/N = -10 \log_{10}(\text{mean-squared response})$$

$$= -10 \log_{10} \frac{\Sigma\, y_i^2}{n}.$$

It has been determined that for this case, minimizing the average loss is equivalent to minimizing $E(Y^2)$. Minimizing $\Sigma\, y_i^2/n$ is equivalent to maximizing $-10 \log_{10}(\Sigma\, y_i^2/n)$. Therefore, the S/N ratio is consistent with the loss function in this case.

Case 2: The S/N ratio that Taguchi defines for "larger is better" is

$$S/N = -10 \log_{10}(\text{mean square of the reciprocal response})$$

$$= -10 \log_{10} \left[\Sigma\, \frac{1/y_i^2}{n} \right].$$

For this case, minimizing the average loss is equivalent to minimizing $E(1/Y^2)$. Minimizing $\Sigma(1/y_i^2)/n$ is equivalent to maximizing $-10 \log_{10}[\Sigma(1/y_i^2)/n]$. Therefore, the S/N ratio is consistent with the loss function in this case also.

Case 3: One S/N ratio that Taguchi defines for "nominal/target (y_0) is best" is

$$S/N = 10 \log_{10} \frac{\text{mean-squared response}}{\text{variance}}$$

$$= 10 \log_{10} \frac{\bar{y}^2}{s^2},$$

where

$$\bar{y} = \frac{\Sigma\, y_i}{n} \quad \text{and} \quad s^2 = \frac{\Sigma(y_i - \bar{y}\,)^2}{n - 1}.$$

For this case, minimizing the average loss is equivalent to minimizing $E[(Y - y_0)^2]$. However, minimizing $\Sigma(y_i - y_0)^2/n$ is not necessarily equivalent to maximizing $10 \log_{10}[(\bar{y}^2/s^2)]$. Therefore, for this case the S/N ratio suggested by Taguchi is not necessarily consistent with the loss function.[4]

Parameter Design: The Robust Design Concept

Taguchi refers to parameter design as that phase of the design process in which the best nominal values for the important design parameters are selected. Taguchi defines the "best" values as those values that minimize the transmitted variability resulting from the noise factors. This is why we sometimes refer to parameter design as robust design.

Simple Example of Robust Design

Let us consider the following hypothetical example. Suppose that the setup of a certain injection molding machine is determined by the settings of five machine parameters and that these factors are likely to have some effect on part shrinkage. The five factors are

x_1: nozzle temperature.

x_2: holding pressure.

x_3: holding time.

x_4: mold temperature.

x_5: screw speed.

In addition to the five control factors above, three other factors are known to have some effect on part shrinkage:

n_1: mold cooling water temperature.

n_2: percent regrind versus virgin raw material.

n_3: ambient temperature at the machine.

The three factors above are somewhat difficult to control and therefore we would like to view them as noise variables. In the interest of running the process "on target with smallest variation," interest will naturally focus on the selection

---

[4] For more on this issue the reader is directed to R. G. Hunter, J. W. Sutherland, and R. E. DeVor, "A Methodology for Robust Design Using Models for the Mean, Variance, and Loss," *ASME Symposium on Quality and Performance: Design, Evaluation, and Improvements, ASME Winter Annual Meeting,* December 1989, PED-Vol. 42, pp. 25–42.

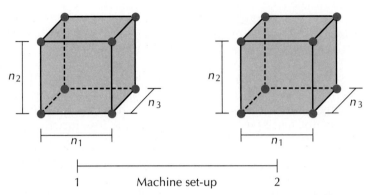

Figure 15.14 Robust Design Experiment for the Injection-Molding Example

of the five control variables that minimize part shrinkage and provide for the most consistently performing process in light of the presence of the three noise factors.

For the sake of simplicity in demonstrating the robust design concept, we will consider that the five control variables are collapsed into just one variable, the machine setup. Further, we will compare process performance across only two levels of the machine setup variable, machine setup 1 versus machine setup 2. It is proposed to run an experiment in which the conditions of the three noise variables are purposely varied for both of the machine setups. A two-level factorial design is proposed as a suitable matrix to define the variations of the noise variables. A graphical representation of the design of the experiment is shown in Fig. 15.14. From the figure it is seen that a total of 16 trials are required for the experiment proposed. The complete design matrix for this experiment is given in Table 15.3. In the table the values of shrinkage are in units of 0.001 inch.

To analyze the data from a robust design point of view, we need to determine measures of both the average performance and the consistency of perfor-

TABLE 15.3				Robust Design Experiment for the Molding Example				
	Machine Setup 1				Machine Setup 2			
Test	n_1	n_2	n_3	Shrinkage (y)	n_1	n_2	n_3	Shrinkage (y)
1	−	−	−	10	−	−	−	7
2	+	−	−	10	+	−	−	10
3	−	+	−	12	−	+	−	15
4	+	+	−	11	+	+	−	11
5	−	−	+	10	−	−	+	11
6	+	−	+	12	+	−	+	8
7	−	+	+	11	−	+	+	12
8	+	+	+	12	+	+	+	14

mance of the process at each of the two machine setup conditions. Calculating the average response at each of the two machine setup conditions, we find an identical value of $\bar{y} = 11.0$ (10^{-3} inch) for each machine setup. This means that as far as average performance is concerned, the two machine setups cannot be distinguished from each other. However, when we calculate the standard deviation in performance we find that machine setup 1 provides for much more consistent performance than does machine setup 2 over the variation in the outer noise variables. For machine setup 1 the standard deviation in performance is equal to $s_y = 0.93 \times 10^{-3}$ inch, while that for machine setup 2 is $s_y = 2.73 \times 10^{-3}$ inch. We would therefore favor machine setup condition 1 because it is more robust, that is, less sensitive to the noise variation in the process.

Inner and Outer Array Experimental Design Structures

The experimental design structure of Fig. 15.14 is sometimes referred to as an inner array/outer array structure. The inner array in that particular example considers only one factor at two levels, the control factor machine setup, and therefore contains only two test conditions. In Fig. 15.14 the outer array is a two-level factorial in the three noise factors and therefore is comprised of a total of eight test conditions. Taguchi's use of this design structure suggests that for each trial in the inner array we conduct the complete outer array of experiments. In this way we can see the variation in the response at each point in the inner array which results from the variation in the noise factors purposely created by the varying conditions of the outer array.

Taguchi has proposed the inner array/outer array structure as an experimental design structure to respond to his classification of variables. A simple example of the inner array/outer array experimental design structure is considered in Fig. 15.15. In this situation we have two control factors, each at two levels, and two noise factors, each also at two levels. The inner array would be defined by a 2^2 factorial and the outer array would also be a 2^2 factorial.

To examine the robustness of the product/process over the range of the two control factors as the two noise factors vary, we would conduct the outer array experiment for each of the four test conditions of the inner array. In this way we can see the variation in the response at each point in the inner array that results from the variation in the noise factors purposely created by the varying conditions of the outer array. Therefore, a total of 16 tests would need to be conducted.

Taguchi generally analyzes the results of such experiments by calculating the signal-to-noise ratio at each of the points in the inner array. Previously, the specific calculation of signal-to-noise ratio has been given for each of the three possible cases: smaller is better, larger is better, and target is best.

When the number of control and/or design factors becomes large and it is still desirable to examine each factor at only two levels, Taguchi suggests the use of *orthogonal arrays* for either an inner array or an outer array, or both. In the context of the work herein, these would be derived from the useful class of two-level fractional factorial designs, to be discussed later.

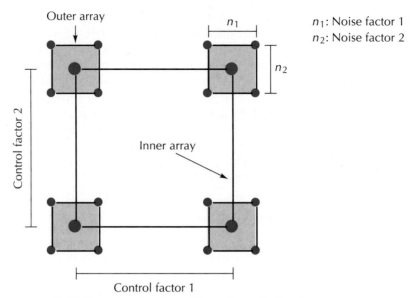

Figure 15.15 Example of an Inner/Outer Array Design Structure

   The inner/outer array design structure seems to have a great deal of appeal when one breaks factors down as being either strictly control factors or strictly noise factors. There are at least three disadvantages, however, of using only the inner/outer array structure and analysis advocated by Taguchi.

1. The use of this structure requires that the distinction between control and noise factors be made prior to conduct of the experiment. This is somewhat contrary to the notion of degree of control and constrains the problem unnecessarily.
2. The use of this structure generally leads to the need to run a rather large number of experiments since two separate designs are developed and then combined. As a result, there is a tendency to use very low resolution fractional factorials (to be discussed in Chapter 19).
3. The use of this structure makes it impossible to determine explicitly, and hence interpret, the interactions (nonlinearities) that may arise between control factors and noise factors. Yet it is these interdependencies that may be exploited to gain economic advantage.

   The comments above suggest that it may not only be possible but also desirable to combine control and noise factors in a single factorial design scheme. Figure 15.16 provides an example of a situation in which a control factor that is a design feature related to an automobile engine block cylinder bore has a strong interdependency/interaction effect with a factor that arises through processing of the cylinder, namely, the bore offset. Although bore offset (the relationship between the bore axis and the boring process axis) may be controlled through the casting process and fixturing of the block at the

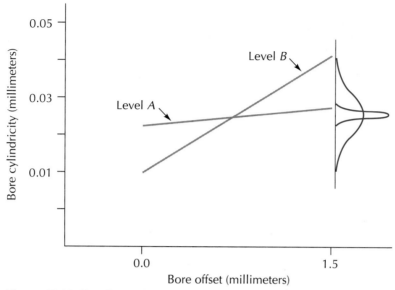

Figure 15.16 Two-Factor Interaction Between a Product Design/Control Factor and a Processing Noise Factor

boring station, the degree to which it needs to be controlled is the key issue. To control it to a point where it would not be a potential factor influencing final bore cylindricity would generally be considered cost prohibitive, particularly if an alternative design action is possible.

Figure 15.16 shows that there is a strong two-factor interaction effect between bore offset and a certain design feature of the bore. When the design feature takes on level B, the changes in the amount of bore offset seem to have a strong bearing on the bore quality, in a variability sense. On the other hand, when the design feature takes on level A, the influence of bore offset on bore quality variation is greatly reduced. While level A and level B provide about the same bore cylindricity on average, the variation in cylindricity is greatly affected by this design feature. Hence, thinking of bore offset as a noise variable, it would appear that by selecting the design feature nominal to be at level A, a more robust, less variable product/process design is attained.

15.6

Contrasting Philosophies for Dealing with Nuisance or Noise Factors

Dealing with noise factors in the conduct of experimental designs has been the subject of considerable study for some time. From the standpoint of enhancing comparative experiments two basic strategies emerged from the early work of Fisher: the techniques of blocking and randomization. For those nuisance/noise factors that are difficult to fix completely across the entire experiment, experimental subunits can be defined within which the nuisance factor can be controlled to a fixed level, and treatments are compared within the subunit, thereby blocking the impact of the noise factor in the response. The objective is to create a more sensitive comparison within the "block." At the same time, the experiment should give rise to variation from block to block that is representative of that encountered in the real world (i.e., outside the experiment). With the use of sound design of experiments practice, it is not only possible but generally useful to estimate directly the effects of the blocked factors. Such knowledge can ultimately lead to actions directed toward improved product design and/or process operation.

For those nuisance/noise variables that cannot be either fixed or blocked within the experiment (they may not even be identifiable), the technique of randomization ensures that their influence on the response is randomly distributed across the treatments, although the variation due to these factors does appear in the response. This measure guarantees the approximate validity of the statistical tests used to draw inferences on the data. Furthermore, the use of replication as a design strategy is also aimed at improving the sensitivity of the comparison of treatment means.

It is important to note that the experimental strategies above, as effective as they may be, are just that: design of experiments strategies, aimed at mitigating the forces of noise that could otherwise bias and/or desensitize the statistically based comparisons made through the experiment. The emphasis is clearly placed on enhancing the power of the statistical methods used to make inferences about the results of an experiment.

The strategic approach that Taguchi has taken in dealing with the forces of noise factors, as they may influence the performance response of interest, contains certain elements of the preceding discussion but at the same time assumes a fundamentally different posture. First, Taguchi is concerned about the influence of the nuisance/noise factors as they operate in the environment in which the product/process is functioning. In particular, his concern is focused on the extent to which this noise variation is transmitted through the product/process and hence influences the variation in the response from a field performance perspective. Hence the primary emphasis is on the functional environment in contrast to the experimental environment. The issue is whether the noise factor(s) are viewed as a nuisance to the experiment or as a nuisance to performance in the field. In fact, one might consider the question: Under what circumstances would a factor be a nuisance to the experiment but not a nuisance to field performance?

Second, Taguchi's use of designed experiments in evaluation of the impact of noise factors on field performance means that regardless of whether the noise factors are explicitly recognized (and therefore part of the design structure) or not, their influence on the response is realized in the experiment, and hence the concept of robust/parameter design is preserved for both cases. The ability to do this in an unbiased manner, of course, depends on the use of good experimental design practice, randomization, for example. In a sense, the inner/

outer array experimental design structure suggested by Taguchi may be thought of as a type of controlled replication where the influence of nuisance/noise factors, when varied over prescribed levels, can be observed and measured in the output performance.

But the purpose of this controlled replication is much different than that mentioned earlier in the context of the more traditional approach to experimental design. Taguchi's framework may be used to reveal interactions between the design/control factors and the noise factors that signal the presence of nonlinearities to be exploited to obtain a more consistently functioning (robust) product/process. In the design strategy that is proposed herein, judgment is reserved on what is a control variable and what is a noise variable (the degree-of-control notion) until the effects are estimated and their economic consequences can be considered more fully. Such a strategy adds flexibility to the entire process and enables the interactions between the control factors and the noise factors to be explicitly determined.

At the same time it is clear that Taguchi's use of inner/outer array experimental design structures does contain the element of blocking since the structure, although hierarchical in nature, is orthogonal and hence the variation induced by purposely varying the noise factors equally influences the treatment responses for the control variables in an average sense. It is rather the variation transmitted to the performance response as the noise factors vary that may be different for different levels of the control factors. All of this, of course, needs to be viewed more carefully in light of the confounding of variable effects evident in most highly fractionated designs.

Exercises

15.1. Suppose that you are asked to develop an experimental plan to evaluate possible differences in performance between two comparably priced subcompact cars. Propose an experiment(s) that may be used to determine if there is a detectable difference in the performance of the two cars. You are to develop a proposal for the experiments you will run. Discuss the important things you must take into consideration when you design comparative experiments for such situations.

15.2. Two high-performance automobiles (type A and type B) are being studied. Seven cars of type A and five cars of type B are randomly selected and tested. The driver(s) were asked to drive the automobiles for 5 minutes around a 2-mile oval test track at 60 miles per hour. The amount of fuel consumed (liters) for each trial is given in the table.

Car	1	2	3	4	5	6	7
Type A	1.65	1.52	1.62	1.73	1.78	1.68	1.76
Type B	1.60	1.42	1.70	1.38	1.53		

a. What kind of an analysis might you propose for the data?

b. What would you like to know about the process by which the data were collected?

15.3. In an effort to improve inventory control procedures, it has been suggested that marking (applying an inventory code number) items with a new "gun" may be faster than the presently used "stamp" method. To examine the time reduction claimed by the gun method, five different items of varying sizes, shapes, and inventory number placement locations were chosen, and the number of items that could be marked in 5 minutes were recorded using both methods. The data are shown in the table.

Item	Gun Method	Stamp Method
1	106	99
2	49	50
3	147	132
4	88	77
5	46	44

a. Discuss what important consideration was obviously taken into account as an experimental design strategy during the design of this experiment.

b. State the null and alternative hypotheses to be tested in this situation.

15.4. Explain why a process that is not in statistical control might not be a good candidate for experimentation. Suggest an alternative course of action that would be more effective in studying the process and explain your reasoning.

15.5. Explain the fundamental difference between experimentation and what is commonly referred to in the engineering world as testing. Why do you think experimentation is more useful than testing to the product design process?

15.6. Returning to Exercise 15.2, suppose it is learned that the tests were run under the following conditions: (1) All the type A cars were tested on Tuesday, and (2) all the type B cars were tested on Wednesday. Additionally, only two drivers were used for the tests, with one driver being used on Tuesday and the other on Wednesday. Express your concerns about the experiment.

15.7. A set of experiments has been planned to test the color intensity of batches of latex paint made on a certain machine. It has been decided to run five tests, one per day for an entire week, each time increasing the amount of dye added by 1%. Explain why this is a poor testing scheme and propose an alternative plan that will yield more useful data.

15.8. Randomization and blocking are two experimental design strategies used to deal with sources of variation in the experimental environment. Explain the difference between these two strategies.

15.9. An experiment is being planned to test an injection-molding process that makes snap-together toys that are put in boxes of children's cereal. Part quality, as measured by the frequency of various types of surface defects, is to be observed. Several variables that seem to be important are (1) machine, (2) raw material batch, (3) temperature of molten plastic, (4) injection pressure, (5) injection rate, (6) cooling time, and (7) size of the part being made. In designing a traditional experiment, which of these variables might be considered as variables to be blocked in the experiment? How would this be done, in principle? What would be the motivation for blocking these variables? If a Taguchi-like experiment were being planned, how would these variables be handled? Discuss the differences in these two approaches to handling the influence of these factors on the results of the experiments.

15.10. Assume that each of the following equations represents a certain relationship between the response of a process and the process variables. Sketch each function and suggest an appropriate experimental design (minimum number of levels of each factor) that would be required to verify this relationship.
   a. $\eta = 2x_1 + 5x_2 + 14$.
   b. $\eta = x_1^2 - 5$.
   c. $\eta = x_1 + 3x_2^2$.
   d. $\eta = 3x_1 + 10$.

15.11. Explain what is meant by the statement that one of the strengths of Taguchi's approach to quality is that "its center of gravity lies in the engineering design process."

15.12. For the process of cooking a pot of chili, define each of the following as a signal, a control, or noise factor:
   a. The list of ingredients.
   b. The amount of chili powder added "according to taste."
   c. The length of time the chili simmers.
   d. The consistency of the stove burner's temperature.
   e. The ability of the pot to distribute heat evenly.
   f. The brand of kidney beans used.
   g. The type of meat used (i.e., ground round, ground sirloin, etc.).

15.13. From your personal experience/knowledge, give two examples each to illustrate the three cases that Taguchi considers when maximizing the S/N ratio (i.e., "smaller is better," "larger is better," and "target/nominal is best").

15.14. Provide the geometrical representation of an inner/outer array where, the inner array is a 2^2 full factorial experiment and the outer array is a 2^3 full factorial experiment. How many tests are required for this experimental design?

15.15. In an experiment designed to test the performance of the punching process in a job shop that makes a variety of products out of sheet metal, consider the following variables:

a. Operator.
b. Sheet metal thickness.
c. Number of parts made per length of run.
d. Shift (first or second).
e. Ambient temperature.
f. Size of the part.
g. Die/punch tool material.
h. Press used (within range of acceptable tonnage).

Performance is measured as the fraction defective produced over the length of the run. For each of these variables, indicate whether or not you would choose to invoke randomization and/or blocking to deal with any or all of these variables as you design your experiment. Explain your decision in each case.

15.16. In considering the roles of various factors in influencing the response of a given process, discuss what is meant by "degree of control"; specifically, how is this concept invoked in the design and/or interpretation of an experiment?

15.17. In using the inner array/outer array structure for a designed experiment, how are the data generally analyzed according to the methods of Taguchi?

15.18. Suppose that an experiment is to be designed to study the coating used on the bottom surface of skis used for common recreational purposes by a broad class of users. Two different coatings are to be tested. It is desired to design an experiment and perform an associated statistical test to determine if there is a difference between the mean wear observed after a substantial period of use for the two coatings. The experiment designed should be as sensitive as possible. Describe the experiment to be performed.

15.19. Under Taguchi's concept of robust design, how might the experiment described in Exercise 15.18 be viewed? Explain and provide an alternative design of experiment.

15.20. Discuss the importance of statistical process control (SPC) in terms of its relationship to the design of experiments (DOE). How can SPC favorably affect DOE? How can DOE be used in aiding the work of SPC?

15.21. Explain why planned factorial experiments are preferred over one-variable-at-a-time approaches to experimentation.

15.22. Describe the purposes of each step in Taguchi's three-stage model of the design process.

15.23. Explain why experiments should be conducted in an iterative and sequential manner instead of conducting one large comprehensive experiment.

15.24. A chemical engineer is studying a new, fairly complicated process. The goal is to optimize the chemical yield for the process. Explain why the first experimental design that the engineer should conduct will consider the processing time and temperature at only two levels each.

15.25. A bank is studying the factors that influence the amount of time a customer must spend in the queue before being waited on by a teller. Some of the factors identified include (1) number of tellers on duty, (2) teller skill level, (3) time of day, (4) number of transactions per customer, and (5) transaction complexity. Design an experiment to study the system described. Indicate how you plan to deal with each variable.

16

Design and Interpretation of 2^k Factorial Experiments

16.1

Introduction

In Chapter 15 we discussed a number of considerations that collectively constitute good practice in the design of experiments. These include:

1. The ability to deal simultaneously with both qualitative (categorical) and quantitative factors/variables.
2. The ability of an experimental design to observe the effects of changing factors individually (all other factors held constant) but at the same time being able to do this at several different combinations of the other factors so that factor interdependencies can be properly revealed.
3. The ability of the experimental design to deal with the forces of nuisance variation, explicitly identifiable or not, through the techniques of blocking and randomization.
4. The ability of the experimental design to observe the forces of experimental error/system variability directly so that comparisons of performance can be made in light of the magnitude of system noise. Replication of tests provides us with the ability to do this.
5. Explicit recognition of the relationship between the proposed mathematical model and the experimental design that responds to that model.

6. Awareness of the notion of model building and the sequential and iterative nature of experimentation. In this regard, the ability to employ experiments that serve as building blocks, one upon the other, is important.

7. Recognition of the concepts of Taguchi, including parameter/robust design in the use of designed experiments for product/process design and improvement.

In addition, it would be useful if the experimental design were

- Easy to set up and carry out.
- Simple to analyze and interpret.
- Simple to communicate or explain to others.

We also suggested in Chapter 15 that a potentially useful class of experimental designs that exhibits many of these desirable properties is the class of two-level factorial designs. Over the next five chapters we study these designs and their sister designs, two-level fractional factorials, in considerable detail. We introduce the two-level factorial design through the use of a case study.

16.2

Process Study Using a Two-Level Factorial Design: Chemical Calibration for the Foam Molding Process[1]

In the manufacture of a new design of an automobile instrument panel that would be "soft to the touch," it was necessary to make some changes in certain raw materials. One such change was to reformulate the substrate foam padding material. This reformulation caused some processing difficulties mainly due to excessive foam voids. Foam voids are air pockets under the vinyl skin that are not completely filled with foam. With the help of c charts monitoring the frequency and location of foam voids, the study team made a number of improvements that reduced the average defect rate from 8.6 voids per sample of five parts to 0.46 void per sample of five parts over a period of several months. A summary of this study is shown in Fig. 16.1.

As can be seen from the foam molding study history of Fig. 16.1, the major improvement actions involved the maintenance of material-flow systems. These were primarily local faults that were detected and eliminated through application of the principles and methods of SPC. Although a significant reduction in the average rate of foam voids was achieved, the team of engineers wanted to learn more about the fundamental relationship between some of the variables

---

[1] R. J. Deacon and P. M. Belaire, "Quality Improvement: Design of Experiments Methodology," *IMPRO 88 Conference Proceedings*, Juran Institute, Chicago, 1988.

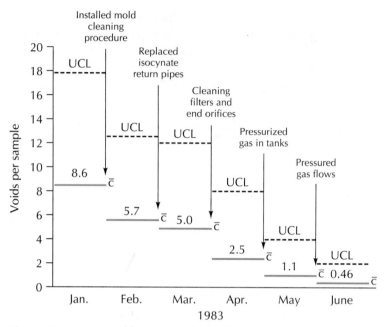

Figure 16.1 Foam Molding Improvement Summary

of the foam molding process and the occurrence of foam voids. This called for experimental investigation of the entire processing system. Since the system was now running "in control" with relatively low common-cause variability, it was considered an appropriate time to begin an experimental program.

In developing a strategy to experiment with the molding process, the team of engineers quickly realized that the experiments must be carried out in a series of at least three stages. They found out that two of the potential control variables of the foam molding process were themselves output responses from the foam calibration stand. Furthermore, four input variables of the foam calibration stand were responses associated with two chemicals at the chemical calibration stand upstream. A three-stage experimentation program was finally developed to study the entire foam molding system with each stage building information for the next stage. In the following sections we present an experimental study of the calibration of isocynate, in particular, the impingement pressure of the isocynate, one of the inputs to the foam calibration process. Figure 16.2 shows the three-stage process under investigation.

The experimental design formulated by the team for this first stage was a two-level, three-variable factorial design, simply designated as a 2^3 factorial design. The objective of this experiment was to determine the possible effects of three control variables on the isocynate weight and impingement pressure. The three variables were orifice size (diameter) of the mix head, pump setting for the isocynate tanks, and isocynate temperature. Two levels were chosen for

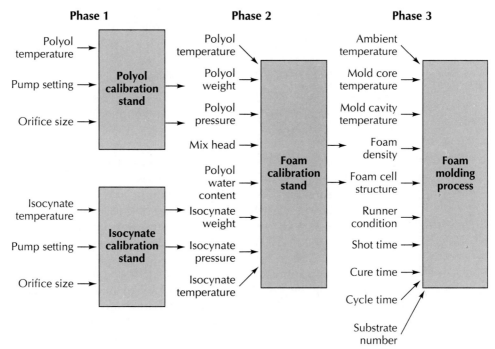

Figure 16.2 Foam Molding Process Experimental Design Flow Diagram

each variable. The high and low levels of the three variables are listed in Table 16.1. These particular levels were centered around the current operating conditions and represented a range that consumed what was believed to be about 30% of the feasible operating range.

To adopt a notation that will be the same for all two-level factorial designs, the variables are coded such that the high level will be denoted by $+1$, and the low level by -1. By so doing, regardless of the physical conditions represented by the two levels, the basic design of any two-level factorial becomes a simple tabulation of a systematic arrangement of plus and minus "ones."

TABLE 16.1 Variable Levels for the Isocynate Calibration Experiment			
Variable	Unit	Low Level	High Level
Orifice size, O	mm	1.30	1.50
Pump setting, P		4.00	4.50
Isocynate temperature, T	°C	22	30

16.3

Construction of the 2^3 Factorial Design

Standard Order for the Test Conditions

A complete 2^3 factorial design contains $2^3 = 8$ unique test conditions. One systematic way of writing down the eight test conditions in their coded forms is to proceed as follows:

1. For variable 1 (x_1), write down a column of the values $-1, +1, -1, +1, -1, +1, -1, +1$. The signs alternate each time.
2. For variable 2 (x_2), write down a column of the values $-1, -1, +1, +1, -1, -1, +1, +1$. The signs alternate in pairs.
3. For variable 3 (x_3), write down a column of the values $-1, -1, -1, -1, +1, +1, +1, +1$. The signs alternate in groups of four.

By writing down these three columns next to one another as shown in Table 16.2, we obtain the desired 2^3 factorial design matrix, which consists of the eight distinct sets of test conditions, in both coded and uncoded forms. The eight sets of test conditions are given by the eight rows, corresponding to test numbers 1 to 8. These constitute the recipes for running the tests. For instance, the actual calibration of the isocynate for test number 1 (denoted by $-1, -1, -1$) was to use an orifice of 1.30 millimeters in diameter with a pump set at 4.0 and temperature at 22°C. Table 16.2 is often referred to as the *design matrix*; it defines the tests associated with the experimental design.

In general, there are 2^k unique sets of test conditions for a 2^k factorial design (i.e., all possible combinations of k factors at two levels each). We can obtain the 2^k sets of coded test conditions by writing down the columns as follows:

1. For x_1, write down a column of $-1, +1, -1, +1, -1, +1, -1, +1, \ldots$. The signs alternate each time (i.e., $2^0 = 1$, one alternation every time). The column length is 2^k.
2. For x_2, write down a column of $-1, -1, +1, +1, -1, -1, +1, +1, \ldots$. The signs alternate in pairs (i.e., $2^1 = 2$, alternate in pairs).
3. For x_3, write down a column of $-1, -1, -1, -1, +1, +1, +1, +1, \ldots$. The signs alternate in groups of four (i.e., $2^2 = 4$, alternate in groups of four).
4. For x_4, the signs alternate in groups of eight ($2^3 = 8$).
5. Proceed in a similar way for $x_5, x_6, \ldots, x_k$. For x_k, write down a 2^{k-1} number of -1's, followed by a 2^{k-1} number of $+1$'s.

This method will yield all of the 2^k distinct sets of coded test conditions. We refer to this particular ordering of the tests as *standard order*. Such an arrangement of test conditions guarantees that all columns in the matrix are orthogonal to each other so that independent estimates of the factor effects may be obtained.

TABLE 16.2	Coded and Uncoded Test Conditions in Standard Order					
	Coded Test Conditions			Actual Test Conditions		
				O	P	T
Test	x_1	x_2	x_3	(mm)		(°C)
1	−1	−1	−1	1.30	4.0	22
2	+1	−1	−1	1.50	4.0	22
3	−1	+1	−1	1.30	4.5	22
4	+1	+1	−1	1.50	4.5	22
5	−1	−1	+1	1.30	4.0	30
6	+1	−1	+1	1.50	4.0	30
7	−1	+1	+1	1.30	4.5	30
8	+1	+1	+1	1.50	4.5	30

Geometric Representation of the 2^3 Factorial Design

If we consider the three variables as three mutually perpendicular coordinate axes x_1, x_2, and x_3, the 2^3 factorial design can be represented geometrically as a cube as shown in Fig. 16.3. The numbers encircled at the eight corner points of the cube represent the corresponding test numbers in standard order. The

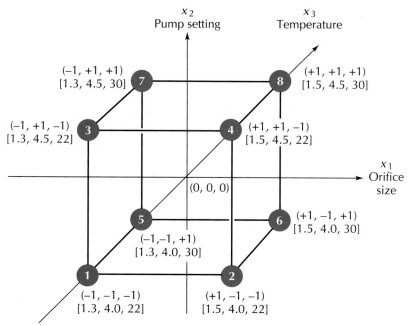

Figure 16.3 Geometric Representation of the 2^3 Factorial Design

TABLE 16.3	Correlation of Standard Test Ordering with Ambient Conditions			

Test	x_1	x_2	x_3	Ambient Condition
1	-1	-1	-1	Morning
2	$+1$	-1	-1	Morning
3	-1	$+1$	-1	Afternoon
4	$+1$	$+1$	-1	Afternoon
5	-1	-1	$+1$	Morning
6	$+1$	-1	$+1$	Morning
7	-1	$+1$	$+1$	Afternoon
8	$+1$	$+1$	$+1$	Afternoon

eight actual test conditions are given in brackets. This geometric configuration will be useful later when we calculate the main and interaction effects of the variables, and when we interpret the meaning of main effects and two-factor and three-factor interaction effects. The geometrical representation is also useful from the standpoint of communicating the design intent to others who are less informed.

Test Conduct: Random Order

In performing experiments, randomization of the test order should be exercised whenever possible. This is important since the standard order discussed previously is a very systematic ordering of the tests in terms of the pattern of the levels of the variables from test to test. Imagine that the 2^3 experiment was to be performed in 2 days and it was determined that two trials could be run in the morning and two in the afternoon. Suppose that due to changing ambient conditions from the morning to the afternoon, afternoon tests tend to produce uniformly higher isocynate impingement pressures. Since variable 2 (in standard order) has all (-1) level tests in the morning and all $(+1)$ level tests in the afternoon, as shown in Table 16.3, the effects of ambient condition and variable 2 would be hopelessly confused or confounded. To avoid or minimize such confounding problems, the test order should be randomized.

16.4

Calculation and Interpretation of Main Effects

After the tests in the design matrix were performed in a random order, the eight responses of impingement pressure in pounds per square inch (psi) were obtained. These responses will be used to calculate the main and interaction

TABLE 16.4 Test Results for Isocynate Calibration Experiment

Test	x_1	x_2	x_3	Test Order	Response, y (psi)
1	−1	−1	−1	6	1550
2	+1	−1	−1	8	1925
3	−1	+1	−1	1	2150
4	+1	+1	−1	2	2350
5	−1	−1	+1	5	1525
6	+1	−1	+1	3	1800
7	−1	+1	+1	4	2175
8	+1	+1	+1	7	2200

effects on impingement pressure for the three variables under study. These responses are shown in the right-hand column of Table 16.4. Note that the column test order in the table shows the order in which the tests were actually run.

Given the results in Table 16.4, how could we evaluate the individual and joint influences of orifice size, pump setting, and isocynate temperature on impingement pressure of the isocynate? To help us see what these variable effects are—what they mean and how they may be calculated—we employ the geometrical representation of the experimental design. The cube in Fig. 16.4

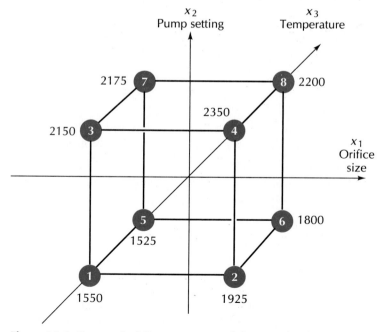

Figure 16.4 Geometrical Representation of the Test Results

depicts the 2^3 design geometrically. The response (impingement pressure in psi) for each test is given at the appropriate corner of the cube.

Calculation of the Main Effect of Orifice Size

We might first ask the question: How is the isocynate impingement pressure influenced by a change in any one of the three variables while the other two are being held constant? Suppose that we consider this question for orifice size. A glance at the cube in Fig. 16.4 indicates that there are four comparisons or contrasts of test results that indicate how impingement pressure changes when orifice size is changed from low (small) to high (large), with pump setting and isocynate temperature each being held at a constant level.

The following pairs of tests may be compared to see the effect of orifice size:

- y_2 and y_1
- y_4 and y_3
- y_6 and y_5
- y_8 and y_7.

The differences in the results within each of the four pairs of tests just defined reflect the effect of orifice size alone on impingement pressure. The differences in units of psi are

Individual Contrasts

Tests 1 and 2: $y_2 - y_1 = 1925 - 1550 = 375$ psi

Tests 3 and 4: $y_4 - y_3 = 2350 - 2150 = 200$ psi

Tests 5 and 6: $y_6 - y_5 = 1800 - 1525 = 275$ psi

Tests 7 and 8: $y_8 - y_7 = 2200 - 2175 = 25$ psi.

The average effect of orifice size, designated by E_1, is defined to be the average of the four differences or contrasts above. Note that this average effect is also commonly referred to as a *main effect* or *location effect*. That is,

$$E_1 = (\tfrac{1}{4})[(y_2 - y_1) + (y_4 - y_3) + (y_6 - y_5) + (y_8 - y_7)]$$
$$= (\tfrac{1}{4})[(375) + (200) + (275) + (25)]$$
$$= 218.75 \text{ psi.}$$

Geometrical Representation of the Main Effect of Orifice Size

Geometrically, the average (main) effect of orifice size, E_1, is the difference between the average test result on plane II (high level of orifice size) and the average test result on plane I (low level of orifice size), as shown in Fig. 16.5.

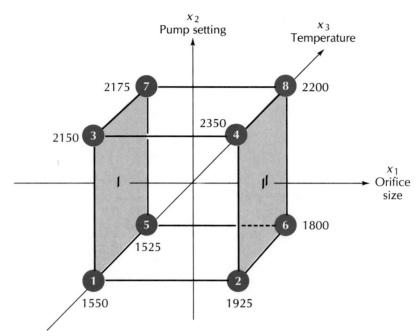

Figure 16.5 Geometric Representation of the Main Effect of Orifice Size

This can be seen by rearranging the previously defined average effect equation for E_1 as

$$E_1 = \frac{y_2 + y_4 + y_6 + y_8}{4} - \frac{y_1 + y_3 + y_5 + y_7}{4}.$$

It is noted that:

- All tests on plane II have one thing in common—a high level of orifice size
- All tests on plane I have one thing in common—a low level of orifice size.

The main effect of orifice size tells us that on the average, over the ranges of the variables studied in this investigation, the effect of changing the orifice size from its low level to its high level is to increase the impingement pressure by 218.75 psi. Notice, however, that the individual differences (375, 200, 275, and 25 psi) are actually quite different. The average effect, therefore, must be interpreted with considerable caution since this effect is not particularly consistent over the four unique combinations of pump setting and isocynate temperature (x_2 and x_3), as shown in Fig. 16.6.

Main Effect of Pump Setting, E_2

Returning to Table 16.4, consider tests 1 and 3. For these two tests, the orifice size (x_1) and isocynate pressure (x_3) are the same, but for test 1 the pump setting

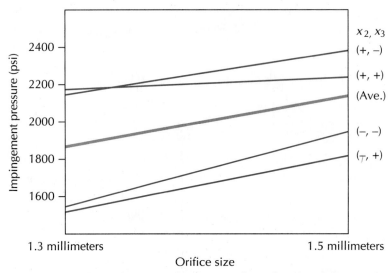

Figure 16.6 Individual Contrasts and Main Effect of Orifice Size (Heavy Line)

(x_2) is at the low level, and for test 3, the pump setting is at the high level. The difference in the two test results can be attributed to the effect of pump setting alone. Similarly, for tests 2 and 4, 5 and 7, and 6 and 8, the orifice size and isocynate temperature are the same within each pair, but the pump setting is at different levels.

Again, referring to Fig. 16.4, compare or contrast the four pairs of tests to see the effect of pump setting. The average effect of pump setting, E_2, can be obtained by taking the average of the four individual contrasts:

Individual Contrasts

Tests 1 and 3: $y_3 - y_1 = 2150 - 1550 = 600$ psi
Tests 2 and 4: $y_4 - y_2 = 2350 - 1925 = 425$ psi
Tests 5 and 7: $y_7 - y_5 = 2175 - 1525 = 650$ psi
Tests 6 and 8: $y_8 - y_6 = 2200 - 1800 = 400$ psi.

The average (main) effect of x_2 is, therefore,

$$E_2 = (\tfrac{1}{4})(600 + 425 + 650 + 400)$$
$$= 518.75 \text{ psi.}$$

Geometrically, the average effect of a pump setting is the difference between the average result on plane IV (high level of pump setting) and the average result on plane III (low level of pump setting) as shown in Fig. 16.7. The average effect of pump setting is 518.75 psi, which tells us that, on the

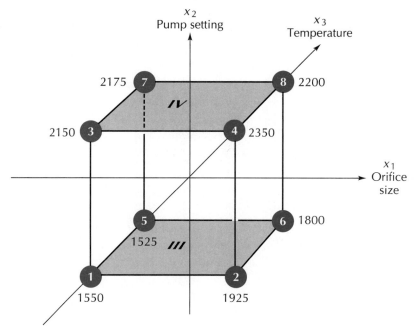

Figure 16.7 Main Effect of Pump Setting

average, an increase in pump setting from its low level to high level gives rise to an increase in impingement pressure of 518.75 psi.

Main Effect of Isocynate Temperature, E_3

Similarly, the average effect of isocynate temperature, E_3, is the corresponding comparison between plane VI (high level of isocynate temperature) and plane V (low level of isocynate temperature), as indicated in Fig. 16.8. This average or main effect is calculated as follows:

$$E_3 = (\tfrac{1}{4})[(y_5 - y_1) + (y_6 - y_2) + (y_7 - y_3) + (y_8 - y_4)]$$
$$= (\tfrac{1}{4})[(-25) + (-125) + (25) + (-150)]$$
$$= -68.75 \text{ psi.}$$

Interpretation of the Main Effects

We have defined the average or main effect of a variable as the amount of change we witness in the response, on the average, when only that variable changes from its low to high level. It is an average effect because within a 2^k experimental design there are 2^{k-1} individual comparisons or contrasts that measure how much the response changes when any one variable changes. The average of all such contrasts across the entire experimental design is called the

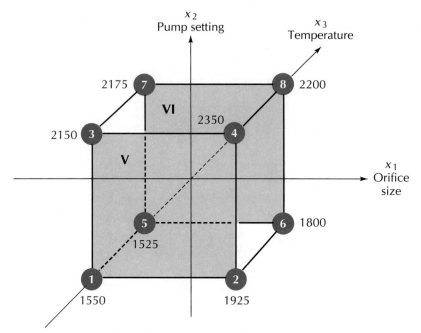

Figure 16.8 Main Effect of Isocynate Temperature

average effect, main effect, or *location effect,* of that factor. Both the sign and magnitude of an average (main) effect mean something:

- The sign tells us of the direction of the effect, that is, if the response increases or decreases.
- The magnitude tells us of the strength of the effect.

Summary: Main Effects of the Isocynate Calibration Experiment

The relative importance of the three main effects of the control variables on impingement pressure are shown graphically in Fig. 16.9. The figure seems to show that pump setting has the largest average effect among the three factors. But an average effect is just that—an average of individual contrasts taken over all possible combinations of levels of the other factors. If the individual contrasts comprising this average are quite similar in magnitude and in sign, the effect of that particular variable tends to be independent of the levels of other variables. On the other hand, if the individual contrasts comprising this average are quite different in magnitude and/or sign, the effect of that particular variable tends to be dependent on the levels of one or more of the other variables. In this case, the average (main) effects may not mean much, and we must look at what happens to the response of interest when two or more variables change simultaneously.

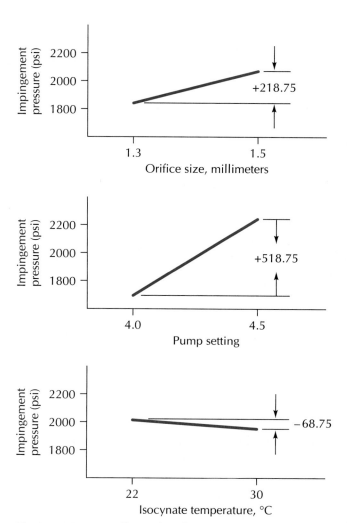

Figure 16.9 Main Effects of Orifice Size, Pump Setting, and Isocynate Temperature

---

16.5

Calculation and Interpretation of Interaction Effects

In calculating the main effect of pump setting, we saw that the amount of change in impingement pressure with change in pump setting seemed to depend on the particular levels of orifice size and isocynate temperature. Table 16.5 shows this. What is the particular nature of the dependency of the effect of pump setting on orifice size and isocynate temperature? Looking more closely at the table, we see that:

1. The average effect of pump setting is very different, depending on the level of orifice size:

    a. *Orifice size = 1.30 millimeters* (averaged over both levels of isocynate temperature).

$$E_{\text{pump}} = \frac{600 + 650}{2} = 625 \text{ psi}$$

    b. *Orifice size = 1.50 millimeters* (averaged over both levels of isocynate temperature).

$$E_{\text{pump}} = \frac{425 + 400}{2} = 412.50 \text{ psi}$$

2. The average effect of pump setting is about the same, regardless of the level of isocynate temperature:

    a. *Isocynate temperature = 22 degrees Celsius* (averaged over both levels of orifice size).

$$E_{\text{pump}} = \frac{600 + 425}{2} = 512.5 \text{ psi}$$

    b. *Isocynate temperature = 30 degrees Celsius* (averaged over both levels of orifice size).

$$E_{\text{pump}} = \frac{650 + 400}{2} = 525 \text{ psi}$$

In short, the effect of pump setting on impingement pressure seems to be

- Indifferent to the magnitude of isocynate temperature.
- Quite dependent on the magnitude of orifice size.

The degree of dependency of the pump setting effect on the particular levels of orifice size and isocynate temperature may be appreciated via the *two-way diagrams* in Fig. 16.10. Figure 16.10(a) indicates that the change in impingement pressure with a change in pump setting is the same, regardless of the level of isocynate temperature. There appears, therefore, to be no dependency

TABLE 16.5   Interdependency Among Variable Effects

Orifice Size (mm)	Isocynate Temperature (°C)	Effect of Pump Setting (psi)
1.30	22	600
1.50	22	425
1.30	30	650
1.50	30	400

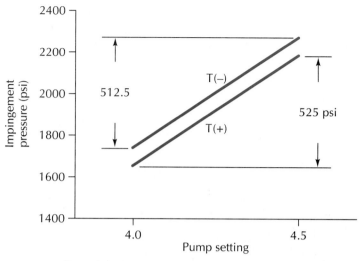

(a) Effect of pump setting for varying isocynate temperatures.

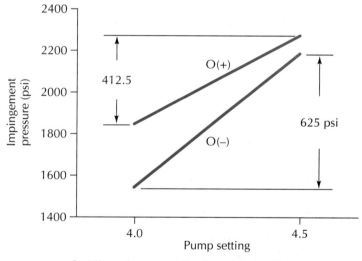

(b) Effect of pump setting for varying orifice sizes.
Figure 16.10 Two-Way Diagrams of Interaction Effects

or interaction between isocynate temperature and pump setting when averaged over orifice size. Again, however, we must interpret this interaction effect with caution since it too is an average effect, averaged over the high and low levels of orifice size. Figure 16.10(b) suggests that the change in impingement pressure with a change in pump setting will depend on the level of orifice size. The average impingement pressure, as pump setting is raised from low to high, increases by 412.5 psi with a larger orifice size, and 625 psi with a smaller orifice size. Since the pump effect depends on the size of the orifice, there is a certain interaction between orifice size and pump setting.

Still another way to see or appreciate the nature of the interaction that may exist between two factors is to construct another type of *two-way diagram*. Referring to the geometrical representation of the 2^3 factorial in Fig. 16.11(a), we can examine the interaction between orifice size and pump setting by

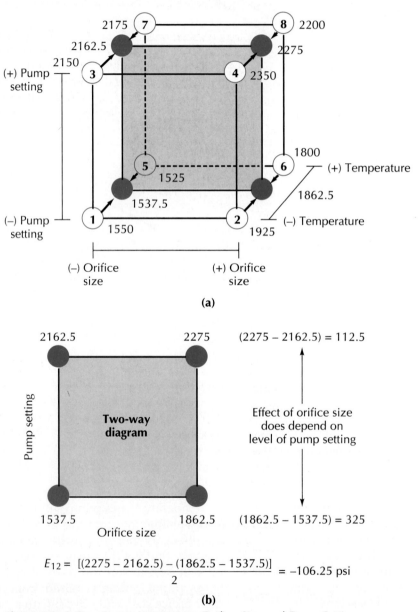

(a)

$$E_{12} = \frac{[(2275 - 2162.5) - (1862.5 - 1537.5)]}{2} = -106.25 \text{ psi}$$

(b)

Figure 16.11 Two-Factor Interaction: Orifice Size and Pump Setting

compressing the cube along the isocynate temperature axis. By compressing the cube we mean that we average the response values for a given orifice size and pump setting combination [say $(-)$, $(-)$] across the high and low levels of isocynate temperature. The result is that the cube becomes a square, as shown in Fig. 16.11(b). Similarly, the orifice–temperature and pump–temperature interactions can be represented by their respective squares, called two-way diagrams. Figures 16.12 and 16.13 illustrate these two interactions.

To characterize the orifice size–pump setting interaction numerically, let us examine Fig. 16.11(b). As is evident, the effect of orifice size is 112.5 psi when the pump setting is at its high level, while the effect of the orifice size is 325 psi when the pump setting is at its low level. Earlier, we said that the average of these two effects was the average effect of the orifice size:

$$E_1 = \frac{112.5 + 325.0}{2}$$

$$= 218.75 \text{ psi.}$$

The difference between these two effects (112.5 and 325.0 psi) divided by 2 measures the dependency of the effect of x_1 on the level of x_2, that is, the interaction effect between the orifice size and the pump setting. Thus the orifice size–pump setting interaction (the "12" interaction) is

$$E_{12} = \frac{(2275 - 2162.5) - (1862.5 - 1537.5)}{2}$$

$$= \frac{112.5 - 325}{2}$$

$$= -106.25.$$

In a like manner, the two-factor interactions orifice size–isocynate temperature and pump setting–isocynate temperature may also be calculated (refer to Figs. 16.12 and 16.13) as

$$E_{13} = [(\text{effect of orifice size at high level for temperature}) -$$

$$(\text{effect of orifice size at low level for temperature})]/2$$

$$= \frac{(2000 - 1850) - (2137.5 - 1850)}{2}$$

$$= \frac{150 - 287.5}{2}$$

$$= -68.75 \text{ psi}$$

$$E_{23} = [(\text{effect of temperature at high level for pump setting}) -$$

$$(\text{effect of temperature at low level for pump setting})]/2$$

$$= \frac{(2187.5 - 2250) - (1662.5 - 1737.5)}{2}$$

$$= \frac{-62.5 - (-75.0)}{2}$$

$$= 6.25 \text{ psi.}$$

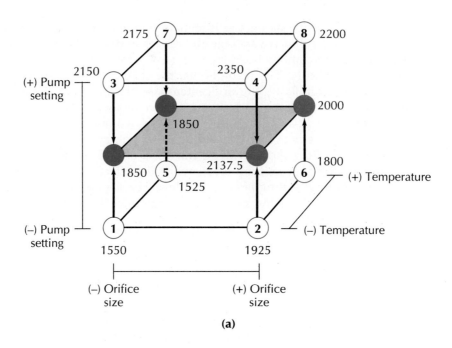

(a)

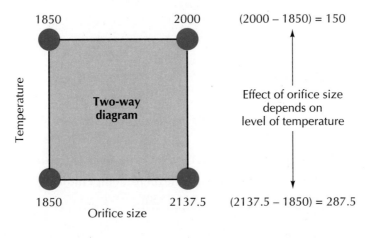

$$E_{13} = \frac{[(2000 - 1850) - (2137.5 - 1850)]}{2} = -68.75 \text{ psi}$$

(b)

Figure 16.12 Two-Factor Interaction: Temperature and Orifice Size

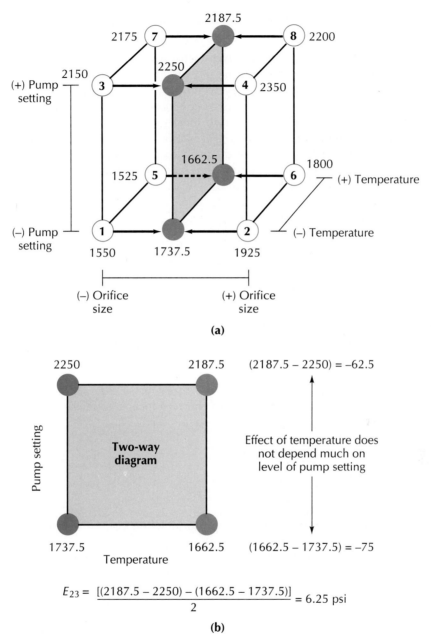

Figure 16.13 Two-Factor Interaction: Pump Setting and Temperature

Three-Factor Interaction: Orifice Size–Pump Setting–Isocynate Temperature

In our previous discussion of two-factor interactions we "compressed the cube" (i.e., averaged over the variable not under examination). Suppose that we look at the two-factor interaction between orifice size and pump setting locally; that is, we do not compress the cube along the isocynate temperature axis. Figure 16.14(b) and (c) shows these two local two-factor interactions, one at the low level of temperature and the other at the high level of temperature. These two squares are each a two-way diagram that can be used to calculate two-factor interactions of orifice size and pump setting E'_{12} and E''_{12}, as shown in Fig. 16.14.

Again, taking the average difference of the two contrasts for orifice size, at high and low pump settings, we have

$$E'_{12} = \tfrac{1}{2}[(2350 - 2150) - (1925 - 1515)]$$
$$= -87.5 \, \text{psi}$$
$$E''_{12} = \tfrac{1}{2}[(2200 - 2175) - (1800 - 1525)]$$
$$= -125.0 \, \text{psi}.$$

Note that the nature of the two-factor interaction between orifice size and pump setting is not too different across the two levels of isocynate temperature. This means that the two-factor interaction between orifice size and pump setting does not depend much on isocynate temperature, or the three-factor interaction of orifice size, pump setting, and isocynate temperature is not likely to be important. An estimate of the three-factor interaction can be obtained by finding the average difference in the two local two-factor interactions, E'_{12} and E''_{12}:

$$E_{123} = \frac{E''_{12} - E'_{12}}{2}$$
$$= \frac{(-125) - (-87.5)}{2} = -18.75 \, \text{psi},$$

which is relatively small in magnitude when compared to the other variable effects.

Previously, we found that $E_{12} = -106.25$ psi. But again this is an average. It is the average of the 12 interaction when 3 is high $(+)$ and the 12 interaction when 3 is low $(-)$. If these two local two-factor interactions are of the same sign and comparable in magnitude, regardless of the levels of 3, then E_{123} may not be important. If, however, the two local two-factor interaction effects were quite different in magnitude and/or different in sign, a three-factor (and higher order, in general) interaction might be at work.

Again, the three-factor interaction is an average difference in two local two-factor interactions. Although this procedure can be extended for the calculation of interaction effects of four factors, five factors, and so on, clearly a simpler method is needed. Although the geometrical representation and graphical analysis of the experimental design has been quite useful in our understanding of

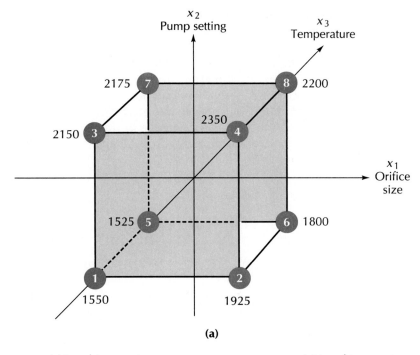

(a)

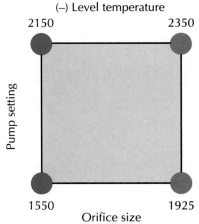

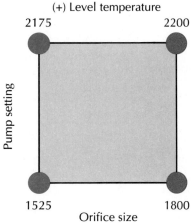

(−) Level temperature

2150   2350

Pump setting

1550   1925

Orifice size

$$E'_{12} = \frac{[(2350 - 2150) - (1925 - 1550)]}{2}$$

$$= -87.5 \text{ psi}$$

(b)

(+) Level temperature

2175   2200

Pump setting

1525   1800

Orifice size

$$E''_{12} = \frac{[(2200 - 2175) - (1800 - 1525)]}{2}$$

$$= -125 \text{ psi}$$

(c)

Figure 16.14 Three-Factor Interaction

the design and the calculation of effects, we will need a simpler and more general method for effect calculations. One such general method is presented in Section 16.6.

Through the use of a geometrical representation of this 2^3 factorial experiment for the isocynate calibration study, we have been able to see how we can calculate variable effects (both main and interaction) and, hopefully, how we might interpret them. But proper interpretation of the existence of real effects cannot be made without any knowledge of the experimental error. Some methods for the analysis of experimental results and statistical interpretation of the effects are discussed in Chapter 17.

16.6
Generalized Method for the Calculation of Effects

The purpose of going through the average effects and interactions in a rather detailed and descriptive manner, using the geometrical representation of the design, is to provide a basic understanding and better appreciation of the meaning of these effects. However, the usefulness of this approach for calculation purposes is somewhat limited since extension to more than three variables is cumbersome. A simplified calculation procedure, which is easily extended for analyzing two-level factorial designs in any number of variables, is therefore needed and is now described.

Let us refer again to the design matrix for the chemical calibration example (Table 16.4).

Test	x_1	x_2	x_3	y (psi)
1	-1	-1	-1	1550
2	$+1$	-1	-1	1925
3	-1	$+1$	-1	2150
4	$+1$	$+1$	-1	2350
5	-1	-1	$+1$	1525
6	$+1$	-1	$+1$	1800
7	-1	$+1$	$+1$	2175
8	$+1$	$+1$	$+1$	2200

We now notice that if we are to multiply the column of ± 1 values associated with x_1 by the column of data y and then divide the sum by $N/2 = 8/2 = 4$, we would have the main effect of orifice size (N = total number of tests, $2^3 = 8$):

x_1		y
-1	$\times$	1550
$+1$	$\times$	1925
-1	$\times$	2150
$+1$	$\times$	2350
-1	$\times$	1525
$+1$	$\times$	1800
-1	$\times$	2175
$+1$	$\times$	2200
Sum $=$		875
Sum/4 $=$		875/4
E_1 $=$		218.75.

This result should not be too surprising; it is simply a rearrangement of the individual results that form the four contrasts that were previously averaged.

A similar application of the signs of the x_2 and x_3 columns to the data followed by a summing operation and, finally, dividing by 4, produces the main effects of pump setting and isocynate temperature, respectively.

x_2		y		x_3		y
-1	$\times$	1550		-1	$\times$	1550
-1	$\times$	1925		-1	$\times$	1925
$+1$	$\times$	2150		-1	$\times$	2150
$+1$	$\times$	2350		-1	$\times$	2350
-1	$\times$	1525		$+1$	$\times$	1525
-1	$\times$	1800		$+1$	$\times$	1800
$+1$	$\times$	2175		$+1$	$\times$	2175
$+1$	$\times$	2200		$+1$	$\times$	2200
Sum $=$		2075		Sum $=$		-275

$$E_2 = \frac{2075}{4} = 518.75 \text{ psi} \qquad E_3 = \frac{-275}{4} = -68.75 \text{ psi.}$$

Again, this result is not a surprise to us, since in the geometrical representation, the average effect was simply an average of four contrasts, each contrast being the difference between a test result at a high level $(+1)$ for that factor and one at a low level (-1) for that factor.

Although the simplified and generalized method for calculating the average (main) effects via the design matrix is somewhat obvious, it may not be immediately obvious how to extend this to the calculation of interaction effects. To see how we might handle the calculation of an interaction effect, we can examine the mathematical model associated with a 2^2 factorial:

$$y = b_0 + b_1 x_1 + b_2 x_2 + b_{12} x_1 x_2 + \epsilon.$$

The coefficients b_1 and b_2 correspond to the average or main effects of variables 1 and 2, respectively, while the coefficient b_{12} corresponds to the interaction effect between variables 1 and 2.

Given y, b_0, b_1, b_2, and particular values for x_1 and x_2, the key to obtaining a solution for b_{12} is to form the cross product x_1x_2. Similarly, the associated mathematical model for a 2^3 factorial design is

$$y = b_0 + b_1x_1 + b_2x_2 + b_3x_3 + b_{12}x_1x_2 + b_{13}x_1x_3$$
$$+ b_{23}x_2x_3 + b_{123}x_1x_2x_3 + \epsilon.$$

The estimates of the b's are actually exactly one-half the value of the E's because the coefficient b measures the incremental change in y per *one* unit of change in the corresponding x (e.g., from 0 to 1), while the effect E measures the incremental change in y for a *two*-unit change in x (from -1 to $+1$).

To obtain the interaction effects (E_{12}, E_{13}, E_{23}, and E_{123}) we may form the cross-product columns x_1x_2, x_1x_3, x_2x_3, and $x_1x_2x_3$, respectively, as shown in Table 16.6. This matrix will be referred to as the *calculation matrix*. The cross-product columns of $\pm$ signs are simply the inner products of the individual columns [e.g., $x_1x_2 = (x_1)(x_2)$].

Note that Table 16.6 also contains a column of all $+$ signs, denoted I. This column of $+$ signs, when multiplied by the y column, summed, and divided by 8 provides an estimate of b_0, the mean response.

The two- and three-variable interactions can now easily be calculated. For instance, to calculate the interaction between orifice size and pump setting, E_{12}, we multiply the column x_1x_2 by the column y, sum algebraically, and then divide the sum by $2^3/2 = 4$.

x_1x_2	y
$(+1) \times (1550)$	$+1550$
$(-1) \times (1925)$	-1925
$(-1) \times (2150)$	-2150
$(+1) \times (2350) \quad =$	$+2350$
$(+1) \times (1525)$	$+1525$
$(-1) \times (1800)$	-1800
$(-1) \times (2175)$	-2175
$(+1) \times (2200)$	$+2200$
Sum $\quad =$	-425

Dividing the sum by 4, we obtain the answer that was determined previously, namely $E_{12} = -106.25$ psi, for this two-variable interaction. Similarly, all other interaction effects can be calculated the same way using appropriate interaction columns.

The foregoing algebraic method using a calculation matrix is a general method because it can easily be extended for the calculation of all the effects for two-level factorial designs of any number of variables, k. Since the method is systematic, it lends itself readily to the computer.

		Main Effects			Interactions				
Test	I	x_1	x_2	x_3	x_1x_2	x_1x_3	x_2x_3	$x_1x_2x_3$	y (psi)
1	+	−1	−1	−1	+1	+1	+1	−1	1550
2	+	+1	−1	−1	−1	−1	+1	+1	1925
3	+	−1	+1	−1	−1	+1	−1	+1	2150
4	+	+1	+1	−1	+1	−1	−1	−1	2350
5	+	−1	−1	+1	+1	−1	−1	+1	1525
6	+	+1	−1	+1	−1	+1	−1	−1	1800
7	+	−1	+1	+1	−1	−1	+1	−1	2175
8	+	+1	+1	+1	+1	+1	+1	+1	2200

TABLE 16.6   Calculation Matrix for 2^3 Design

16.7

Mathematical Model of the Response

Recall that the objective of the isocynate calibration experiment was to learn about how isocynate impingement pressure can be controlled to prepare for the experiments to study the foam calibration stand, and ultimately, the foam molding machine to discover the root causes of foam voids. Since the impingement pressure is one of the control variables for foam calibration, the team of engineers wanted to be able to calibrate with precision the isocynate to produce the impingement pressure at some appropriate levels for foam experiments. In other words, they hoped to know, from the isocynate experiment, how to set some or all of the three factors—orifice, pump, and temperature—to produce impingement pressure at some target value. We will now evaluate the results of the isocynate experiment to see if a solution could be developed to meet the intended objective, by way of a fitted mathematical model.

First, this 2^3 factorial design was used with the following mathematical model in mind:

$$y = b_0 + b_1x_1 + b_2x_2 + b_3x_3 + b_{12}x_1x_2 + b_{13}x_1x_3$$
$$+ b_{23}x_2x_3 + b_{123}x_1x_2x_3 + \epsilon,$$

where all the b's, except b_0, can be substituted with their corresponding $\frac{1}{2}E$'s, and $\hat{b}_0$ is an average of all eight observed responses:

$$\hat{b}_0 = (\tfrac{1}{8})(y_1 + y_2 + y_3 + y_4 + y_5 + y_6 + y_7 + y_8),$$

which could also be thought of as the response value at the center of the cube. Finally, the ϵ represents the random noise or experimental error associated with the response y, and it has an expected value of zero. Additionally, let us assume that the model adequately represents the functional relationship between y and the x's.

From the results of the isocynate experiment, our estimate of the mean impingement pressure is given by

$$\hat{b}_0 = (\tfrac{1}{8})(1550 + 1925 + 2150 + 2350 + 1525 + 1800 + 2175 + 2200)$$
$$= 1959.375 \text{ psi}$$

and the other coefficients are estimated by

$$\hat{b}_1 = \frac{E_1}{2} = \frac{218.75}{2} = 109.375$$

$$\hat{b}_2 = \frac{E_2}{2} = \frac{518.75}{2} = 259.375$$

$$\hat{b}_3 = \frac{E_3}{2} = \frac{-68.75}{2} = -34.375$$

$$\hat{b}_{12} = \frac{E_{12}}{2} = \frac{-106.25}{2} = -53.125$$

$$\hat{b}_{13} = \frac{E_{13}}{2} = \frac{-68.75}{2} = -34.375$$

$$\hat{b}_{23} = \frac{E_{23}}{2} = \frac{6.25}{2} = 3.125$$

$$\hat{b}_{123} = \frac{E_{123}}{2} = \frac{-18.75}{2} = -9.375.$$

It is noted that the seven $\hat{b}$'s, or equivalently the E's, are quite different from one another in both magnitude and sign. This shows at least that not all effects are equally important. In fact, some of the smaller effects, positive or negative, could be just sample reflections of the experimental error, while their true effects are zero. The importance of an estimated effect should be judged by comparing it with the experimental error if it is known or can be estimated. We will address this issue in detail in Chapter 17.

Building the Math Model: Significance of Variable Effects

A closer examination of the calculated effects seems to show that some of them are quite large (259.375, 109.375, −53.125) while others are fairly small (−9.375, 3.125). The large sample effects may be important in affecting impingement pressure, while the smaller effects may result only from the forces of error/ random variation and therefore could be ignored. Rigorous procedures for assessing the relative importance of the variable effects in light of random variation are presented in Chapter 17. To show how the estimated effects may be used through the mathematical model, we will tentatively assume that E_3, E_{13}, E_{23}, and E_{123} are small enough to be considered as completely explainable by experimental error. Equivalently, we assume that b_3, b_{13}, b_{23}, and b_{123} are zero and their associated terms can be dropped from the postulated model.

Tentatively, our fitted model could be expressed with some rounding off as

$$\hat{y} = 1959 + 109x_1 + 259x_2 - 53x_1x_2, \tag{16.1}$$

where $\hat{y}$ is the predicted impingement pressure. This fitted model should be subjected to further analysis and confirmatory tests to check its adequacy as a prediction equation of the impingement pressure for any values of x_1 and x_2 within their high and low levels. Complete procedures for model building are discussed in Chapter 18.

Use of the Fitted Model

Assuming that the fitted model is judged to be adequate, we can use it to represent or approximate the true response surface of the impingement pressure in the space of orifice size and pump setting as shown in Fig. 16.15. Note that the surface shown with heavy lines is not a flat plane because of the interaction term, E_{12}, in the fitted model.

Once this estimated response surface model has been properly validated, it can be used for the selection of the levels or settings of x_1 and x_2 for any given response target (y_0). Conversely, given any combination of x_1 and x_2, the surface can be used to predict the impingement pressure, $\hat{y}$. In either case, some confirmatory tests should be run to check the agreement between the test results and the predicted values from the response surface. In the following discussions, we assume that the model is valid within the domain of the experiment.

Since x_3, the isocynate temperature, was found to have no significant effect on impingement pressure, we can take x_3 out of our design matrix and rewrite a design matrix of x_1 and x_2 alone, with each unique test condition replicated as shown in Table 16.7. There are at least three useful observations that can be made from the resulting math model and its geometrical interpretation in Fig. 16.15.

TABLE 16.7   Design Matrix for x_1 and x_2

Test	x_1	x_2	x_1x_2	Replicated Responses		Average Response	Replicate Difference
1	−1	−1	+1	1550	1525	1537.5	−25
2	+1	−1	−1	1925	1800	1862.5	+125
3	−1	+1	−1	2150	2175	2162.5	−25
4	+1	+1	+1	2350	2200	2275.0	+150

Observation 1. An increase in either the orifice size or the pump setting, or both, increases the average impingement pressure. It is, however, the pump setting that has the most influence on the impingement pressure (two and a half times greater than the effect of orifice size). This could also be seen from the relative values of $\hat{b}_1$ and $\hat{b}_2$ in the fitted model (109 versus 259). These average effects of x_1 and x_2 are also called location effects because they change the level or location of the response.

Observation 2. Because of the presence of the interaction effect, $\hat{b}_{12}$, the slope of the response surface in Fig. 16.15 along either the x_1 or x_2 axes is changing as a function of either x_2 or x_1, respectively. For example, for $x_1 = -1$ (1.3 mm orifice size), the pump setting effect on pressure is given by

$$\hat{y} = 1959 + 109(-1) + 259x_2 - 53(-1)x_2$$
$$\hat{y} = 1850 + 312x_2.$$

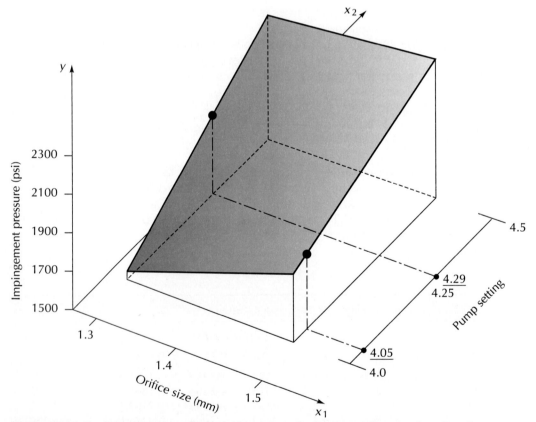

Figure 16.15 Predicted Response Surface of Isocynate Impingement Pressure As a Function of Orifice Diameter and Pump Setting

For $x_1 = +1$ (1.5-mm orifice size), however, the pump setting effect on pressure is given by

$$\hat{y} = 1959 + 109(+1) + 259 - 53(+1)x_2$$
$$\hat{y} = 2068 + 206x_2.$$

Note that the increase in pressure with increase in pump setting is 50% greater for the small orifice size than it is for the large orifice size.

Observation 3. Suppose that it was desired for further experimentation on the foam molding process to set the impingement pressure at 1900 psi. What values of orifice size and pump setting could/should be used? The model above offers an answer to this question. If the small (-1 level) orifice size is used, the pump setting should be

$$1900 = 1850 + 312x_2$$

$$x_2 = 0.16 \quad \text{(a pump setting of 4.29 in uncoded units).}$$

On the other hand, if the large orifice size ($+1$ level) is used, the pump setting should be

$$1900 = 2068 + 206x_2$$

$$x_2 = -0.81 \quad \text{(a pump setting of 4.05 in uncoded units).}$$

The two points are shown on the response surface in Fig. 16.15.

Finding Pump Setting Levels for the Next Experiment. For the next experiment on the foam molding process, suppose that it is desired to set the high and low levels of the impingement pressure at 2050 and 1600 psi, respectively. Further suppose that an orifice setting of 1.3 millimeter (-1) is used. The pump settings can be determined using the response model. For the impingement pressure $\hat{y}$ to be equal to an average of 2050 psi, the fitted model (Eq. 16.1) may be written as

$$2050 = 1959 + (109)(-1) + (259)(x_2) - (53)(-1)(x_2)$$

or

$$x_2 = \frac{200}{306} = 0.654,$$

which should be decoded to the actual units:

Pump setting $= (4.25) + (0.654)(4.50 - 4.25) = 4.4135.$

Similarly for $\hat{y} = 1600$, using Eq. 16.1 $x_2 = -250/306 = -0.817$, which is converted to the actual units of pump setting $= 4.04575$.

The foregoing solution for the pump settings may be determined graphically using the predicted response surface as illustrated in Fig. 16.16. The two levels of the selected impingement pressure are indicated by the [+] and [−] on the vertical axis in Fig. 16.16. By projecting these two points first to the

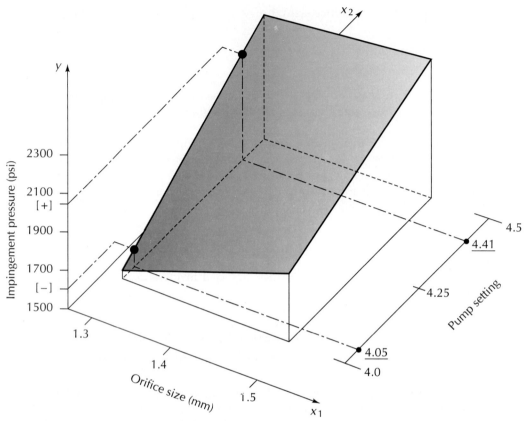

Figure 16.16 Use of Response Surface to Select Levels of Orifice Size and Pump Setting

surface and then down to the orifice size–pump setting plane, the desired levels of the orifice size and the pump setting can be read from the x_1 and x_2 axes. If somewhat larger orifice sizes are preferred for other considerations, the response surface of Fig. 16.16 may be used as an aid in choosing a particular combination of orifice size and pump setting for each impingement pressure selected.

Exercises

16.1. A 2^3 factorial design was conducted to study the influence of temperature (x_1), pressure (x_2), and cycle time (x_3) on the occurrence of splay in an injection-molding process. For each of the eight unique trials, 50 parts were made and the response observed was the number of incidences of the occurrence of splay on the part surface across all 50 parts. The design matrix and test responses are given in the table.

Test	1	2	3	y
1	−	−	−	12
2	+	−	−	15
3	−	+	−	24
4	+	+	−	17
5	−	−	+	24
6	+	−	+	16
7	−	+	+	24
8	+	+	+	28

    a. Draw the geometrical representation of this experiment and label each corner of the cube with the corresponding test conditions (e.g., +, −, +) as well as the test response.

    b. Calculate the main effects of all three variables.

    c. Calculate all the two- and three-factor interaction effects.

    d. Write down the appropriate mathematical model, assuming that all estimated effects are to be included in the model.

16.2. Suppose that it is proposed to run the tests outlined in Exercise 16.1 using the test order specified. Additionally, it is planned to run two tests per shift, and the operation is to be run for two shifts per day. If the eight tests are performed in this manner, what possible problem(s) could arise in the ultimate interpretation of the results? How can this situation be rectified?

16.3. A 2^2 factorial experiment was conducted to determine the effects that tool surface coating (x_1) and the application of cutting fluid (x_2) might have on the surface finish of a turned part. The low level for each of the two variables represented the absence of the technology in question (i.e., no surface coating, no cutting fluid). The high level represented the technology at some commercially recommended level. The response is the part surface finish in microinches (R_a value).

x_1	x_2	R_a
−	−	44
+	−	32
−	+	55
+	+	20

    a. Draw the geometrical representation of this experiment, labeling the corners with the corresponding test conditions (e.g., +, −).

    b. Calculate all the variable effects.

    c. Sketch a graphical representation of the two-factor interaction effect using a two-way diagram and explain the nature of the interaction effect in words.

16.4. A 2^3 factorial design was constructed to test the effect of three different variables on the alertness of students in an early morning class. The variables tested, the design matrix, and the responses are shown in the tables.

Variable	Low Level	High Level
1. Hours of sleep	4	8
2. Ounces of coffee	6	12
3. Number of donuts	1	2

Test	1	2	3	Alertness (%)
1	−	−	−	56
2	+	−	−	72
3	−	+	−	68
4	+	+	−	89
5	−	−	+	43
6	+	−	+	62
7	−	+	+	59
8	+	+	+	75

a. Calculate estimates of all the main and interaction effects, using the algebraic method based on the calculation matrix for this design.

b. Interpret the two-factor interaction effects geometrically by making two-way diagrams for all interaction effects. Briefly summarize in words the information that each graph yields.

c. Explain the physical meaning of a three-factor interaction effect. Use any graphical summaries that you think are useful to your explanation.

16.5. Draw the geometrical representation of a 2^4 factorial experiment and label the diagram in appropriate places with the corresponding test conditions.

16.6. A 2^3 factorial design was used to study the possible effects of temperature, time, and pressure on the output of a chemical process. The data, including experimental results, are shown in the tables.

Variable	Low Level	High Level
1. Temperature (°F)	80	100
2. Time (min)	5	7
3. Pressure (psi)	120	140

Test	1	2	3	Response
1	−	−	−	10
2	+	−	−	20
3	−	+	−	4
4	+	+	−	10
5	−	−	+	8
6	+	−	+	18
7	−	+	+	6
8	+	+	+	12

a. Draw the geometrical representation of this experiment and label each corner of the cube with the corresponding test conditions (e.g., +, −, +) as well as the test responses.

b. Use the geometrical representation you created in part (a) to determine the value of the main effects and the two-factor interaction effects.

c. Write down the calculation matrix and use it to obtain all main and interaction effect estimates. [Compare with the effect estimates obtained in part (b).]

16.7. A 2^3 factorial experiment was designed and conducted to determine the effects of packing material, distance shipped, and home shipments per load used on the average amount of damage sustained per load by a moving company. The variable levels used and the results of the experiment are shown in the first table.

Variable	Low Level	High Level
1. Packing material	Shredded paper	Foam "peanuts"
2. Distance traveled	Less than 500 miles	More than 1000 miles
3. Home shipments per load	One	Two or more

The response observed was the number of damaged items per 100 packed boxes, shown in the second table.

Test	1	2	3	Response
1	−	−	−	2.55
2	+	−	−	3.95
3	−	+	−	1.55
4	+	+	−	2.50
5	−	−	+	1.15
6	+	−	+	2.15
7	−	+	+	2.55
8	+	+	+	3.75

   a. Write out the calculation matrix and use it to calculate all the main and interaction effects.
   b. Interpret the interaction effects geometrically by making two-way diagrams for all interaction effects. Briefly summarize the information each graph yields.
   c. Write down the appropriate mathematical model in terms of the coded variables, assuming that all effects may be important.
   d. Determine the predicted value, $\hat{y}$, for each test condition.

16.8. A 2^2 factorial design experiment was conducted, and the results are shown in the table.

Test	1	2	Response
1	−	−	45.0
2	+	−	30.5
3	−	+	39.2
4	+	+	28.9

   a. Draw the geometrical representation of this experiment and label your diagram with the corresponding test conditions (e.g., +, −).
   b. Calculate all the main and interaction effects.
   c. Draw the two-way diagram for the interaction effect, and interpret this effect.

16.9. A chemical engineer performed the experiments shown in the first table, randomizing the order of the tests within each week. The response of interest is the chemical yield, and the run order is given in parentheses.

Test	Temp.	Catalyst	pH	Week 1	Week 2
1	−	−	−	64.0 (5)	61.2 (10)
2	+	−	−	72.3 (8)	75.5 (11)
3	−	+	−	60.2 (1)	61.4 (13)
4	+	+	−	65.3 (3)	68.7 (16)
5	−	−	+	62.8 (7)	66.2 (9)
6	+	−	+	83.6 (4)	84.7 (14)
7	−	+	+	68.9 (6)	71.7 (12)
8	+	+	+	76.5 (2)	78.1 (15)

The variable levels are given in the second table.

Level	Temperature (°F)	Catalyst	pH
−	160	A	6
+	180	B	7

   a. Sketch the geometrical representation of the 2^3 design, labeling each corner with the corresponding test conditions.

b. Calculate the average response for each test condition, and add these values to the drawing in part (a).

c. Write out the calculation matrix and calculate all the effects.

d. Interpret the three-factor interaction using a pair of two-way diagrams associated with any one of the two-factor interactions.

16.10. A reexamination of the experiment performed in Exercise 16.9 revealed that all the tests performed in week 1 were performed by one operator, and the tests performed in week 2 were performed by another operator. Another interpretation of the experiment may therefore be that it is a 2^4 factorial, varying temperature, catalyst, pH, and operator.

a. Write out the design matrix in standard order for the 2^4 design.

b. Write out the calculation matrix for the 2^4 design.

c. Calculate all the effects.

d. Comment on the advisability of the manner in which the 2^4 factorial experiment was performed.

16.11. The results of a 2^3 factorial design are given in standard order in the table.

Test	Response, y
1	48.8
2	91.6
3	40.8
4	60.4
5	48.8
6	89.2
7	40.4
8	60.0

Let x_1, x_2, and x_3 correspond to the coded variables.

a. Write out the calculation matrix and use it to calculate estimates of all the main and interaction effects.

b. Interpret the interaction effects geometrically by making two-way diagrams for all interaction effects. Briefly summarize the information each graph yields.

16.12. A 2^2 factorial experiment was conducted to test the effect of lighting and watchman on the average number of car burglaries per month from four different parking garages, with data as shown in the tables.

Variable	Low Level	High Level
1. Lighting	Poor	Good
2. Watchman	No	Yes

Test	1	2	y
1	−	−	2.80
2	+	−	1.00
3	−	+	2.40
4	+	+	0.75

    a. Draw the geometrical representation of this experiment and label your diagram with the corresponding test conditions (e.g., +, −).

    b. Calculate all the main and interaction effects.

    c. Interpret the results.

16.13. A 2^4 factorial experiment was conducted and the results are shown in the table.

Test	1	2	3	4	y
1	−	−	−	−	34.62
2	+	−	−	−	41.35
3	−	+	−	−	23.08
4	+	+	−	−	18.27
5	−	−	+	−	26.92
6	+	−	+	−	33.65
7	−	+	+	−	31.73
8	+	+	+	−	21.15
9	−	−	−	+	30.77
10	+	−	−	+	37.50
11	−	+	−	+	24.04
12	+	+	−	+	23.08
13	−	−	+	+	16.35
14	+	−	+	+	13.46
15	−	+	+	+	25.00
16	+	+	+	+	19.23

    a. Sketch the geometrical representation of this design, and and place the test results on this representation.

    b. Write out the calculation matrix and calculate all the main effects and interaction effects.

16.14. Suppose that the investigator decided to ignore the presence of variable 4 in Exercise 16.13 and treat the design as a 2^3 factorial with replication.

    a. How does this change the estimates of the effects involving variables 1, 2, and 3?

    b. How does this change the actual interpretation of the results (i.e., how does the interpretation change between the 2^4 and 2^3 designs)?

16.15. A 2^3 factorial design was constructed to study the effects of lighting, ambient temperature, and years of service on the average number of injuries per month in three different factories. The results are shown in the tables.

Variable	Low Level	High Level
1. Lighting	Poor	Well-lit
2. Average ambient temperature (°F)	68	75
3. Years of service	Below 5	Above 5

Test	1	2	3	y
1	−	−	−	13.7
2	+	−	−	11.0
3	−	+	−	10.2
4	+	+	−	6.4
5	−	−	+	17.3
6	+	−	+	12.6
7	−	+	+	13.5
8	+	+	+	11.1

a. Draw the geometrical representation of this experiment and label each corner of the cube with the corresponding test conditions (e.g., +, −, +) as well as the test response.

b. Use the geometrical representation you created in part (a) to determine the value of the main effects and the two-factor interaction location effects.

c. Write down the calculation matrix and use it to obtain all main and interaction effects. [Compare with the effects obtained in part (b).]

16.16. A 2^3 factorial experiment was designed to test the effect of three different variables on the ultimate tensile strength of a weld.[2] The three variables and the levels used are shown in the tables together with the test results. Note that this experiment was replicated.

Variable	Low Level	High Level
1. Ambient temperature (°F)	0	70
2. Wind velocity (mph)	0	20
3. Bar size (in.)	4/8	11/8

Test	1	2	3	y_1	y_2
1	−	−	−	84.0	91.0
2	+	−	−	90.6	84.0
3	−	+	−	69.6	86.0
4	+	+	−	76.0	98.0
5	−	−	+	77.7	80.5
6	+	−	+	99.7	95.5
7	−	+	+	82.7	74.5
8	+	+	+	93.7	81.7

[2] S. M. Wu, "Analysis of Rail Steel Bar Welds by Two-Level Factorial Design," *Welding Journal Research Supplement*, 1964, pp. 1795–1835.

    a. Sketch the geometrical representation of the 2^3 design, labeling each corner with the corresponding test conditions.

    b. Calculate the average response for each test condition, and add these values to the drawing in part (a).

    c. Write out the calculation matrix, and calculate all the effects.

16.17. An investigator performed a 2^3 factorial experiment, and calculated the average of all eight tests and the main effect of variable 1 before leaving the office one evening. Upon returning to work the next day, the investigator was only able to find the six test results shown in the table, and the numerical values for the average (25.00375) and E_1 (10.2675).

1	2	3	y
−	+	−	13.76
−	−	−	30.72
+	−	−	37.81
−	−	+	12.40
+	+	−	25.69
+	−	+	21.79

    a. Write down the test recipes for the two missing tests.

    b. Calculate the remaining main and interaction effects.

16.18. A 2^2 factorial design was run to study the effects of product model (x_1) and fixture type (x_2) on the productivity of an assembly process. The responses, parts assembled per hour, are shown below. The next day, the same experiment was repeated with a completely different set of workers doing the assembly.

Day	Test	1	2	y
1	1	−	−	59.0
	2	+	−	76.3
	3	−	+	28.9
	4	+	+	51.4
2	5	−	−	20.3
	6	+	−	41.6
	7	−	+	48.3
	8	+	+	73.1

    a. Calculate E_1, E_2, and E_{12} for each day's experiment.

    b. Compare the results of the two experiments, and offer an explanation for possible differences observed.

    c. Calculate the main effect of worker set on assembly time.

    d. Calculate the two-factor interaction of fixture type by worker set.

    e. Comment on the way in which this experiment was performed. Does the manner in which the experiment was performed affect the interpretation of the results?

16.19. A 2^3 factorial design is constructed in an attempt to determine the effects of the amount of rest, the amount of preparation, and the exam room's temperature on the ability of a student to score well on an exam. The variable levels, design matrix, and responses are shown in the tables.

Variable	Low Level	High Level
1. Sleep (hr)	4	8
2. Preparation (hr)	2	6
3. Temperature (°F)	65	75

Test	1	2	3	Score
1	−	−	−	60
2	+	−	−	70
3	−	+	−	85
4	+	+	−	95
5	−	−	+	55
6	+	−	+	65
7	−	+	+	80
8	+	+	+	95

   a. Which factor has the largest average effect on exam score?
   b. Calculate and interpret the three two-factor interaction effects.
   c. For a person who has prepared very little for the exam, what action should be taken if they wish to score well? Explain.

16.20. While in the midst of stealing the results of a 2^3 factorial experiment, a disgruntled employee is interrupted and obtains only the seven test results shown in the table.

1	2	3	y
+	−	−	157.0
−	−	−	161.3
+	+	+	141.1
−	+	−	152.1
+	+	−	160.9
−	−	+	131.6
−	+	+	139.2

   a. What is the recipe for the missing test?
   b. What assumption could reasonably be made to permit calculation of all main and interaction effects?
   c. Using the assumption of part (b), calculate all the main and interaction effects.

16.21. For each diagram shown, determine whether a two-factor interaction exists. Calculate the interaction effects in each case.

a.

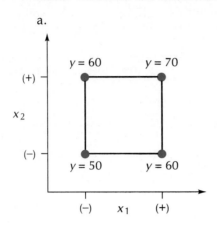

b.

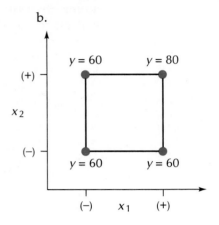

c.

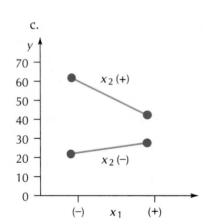

d.

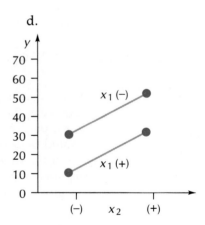

e.

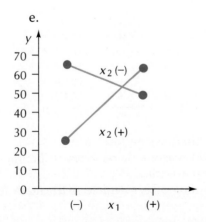

16.22. A 2^2 factorial design has produced the following model relating machined part surface finish to feed rate (x_1) and nose radius (x_2) of the tool (x_1, x_2 are in coded units of $\pm$ 1):

$$\hat{y} = 90 + 10x_1 - 15x_2 + 5x_1x_2.$$

a. What is the average surface finish across the experiment?
b. What are the values for E_1, E_2, and E_{12} as obtained from the experiment?
c. Sketch the response surface in the x_1, x_2, y coordinate space.
d. What value for the surface finish does the model predict when the feed rate and nose radius are at their high levels for the experiment?
e. If a surface finish of 85 is desired, what value of feed rate (coded units) should be used if the nose radius takes on the low level used in the experiment?

16.23. Apply the responses $y = 5$, $y = 10$, and $y = 15$ to the graphical representation shown (a response may be repeated or not used at all) to satisfy the following four different situations:

a. Positive main effects of x_1 and x_2 and interaction effect = 0.
b. Positive main effect of x_1, main effect of $x_2 = 0$, and interaction effect = 0.
c. Main effects of x_1 and $x_2 = 0$, interaction effect $\neq$ 0.
d. Positive main effect of x_1, negative main effect of x_2, and interaction effect $\neq$ 0.

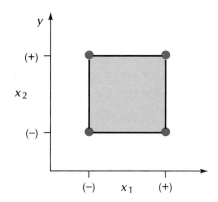

16.24. A 2^4 factorial experiment was conducted producing the following test results, in standard order: 10, 12, 12, 14, 8, 7, 9, 16, 8, 9, 17, 12, 10, 11, 6, 16.

a. Calculate the value of the two-factor interaction E_{34}.
b. Calculate the value of the three-factor interaction E_{124}.
c. Show the two-factor interaction, E_{23}, in graphical form (i.e., construct the two-way diagram to interpret this interaction).

16.25. The incidence of splay defects (measured in percent of part coverage) is being studied as a function of temperature and holding pressure for an injection molding process. The results of a 2^2 factorial experiment are depicted graphically as shown.

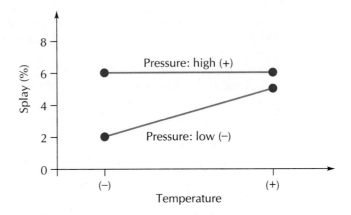

a. Write down the design matrix, and indicate the numerical value of the four test results.
b. Calculate E_1, E_2, and E_{12}.
c. Interpret, in your own words, the two-factor interaction (i.e., what is the specific nature of the pressure–temperature dependency?).

Analysis of Two-Level Factorial Designs

17.1
Introduction

This chapter takes an in-depth look at the analysis of the results of two-level factorial designs. The emphasis is placed on ways to evaluate the level of noise (experimental error) inherent in the environment in which the tests were conducted, and how this noise influences the *effect estimates*. We would like to distinguish those effects that are truly important from those that are not, in light of the experimental error in the testing environment. It will be seen that experimental replication provides one method by which the important variable effects may be identified. For unreplicated experiments, two techniques are presented that may also be used to distinguish significant effects from those that have arisen solely due to the forces of error.

Chapter 16 presented a detailed account of the calculation and interpretation of variable effects for a chemical calibration problem. At one point we simply assumed certain main and interaction effects to be unimportant because their calculated magnitudes were relatively small. We know that if the entire experiment were repeated (another run for each of the eight unique test conditions), the results and hence the effect estimates would be somewhat different. The question is, how different? To answer this question in a somewhat general way, consider the following.

If we ran the experiment five times and for E_{pump} (the main effect of pump setting) got the results

533 psi,   496 psi,   524 psi,   509 psi,   477 psi,

our attitude would be to consider the effect of pump setting as important—real. It would be considered important because all five effect values are comparable and their average is quite different from zero relative to the variability among them. However, if we ran the experiment five times and for E_{pump} got the results

150 psi, -420 psi,   50 psi,   525 psi, -295 psi,

our attitude would be quite different concerning the importance of the pump effect because these five values are widely different and average out close to zero. Figure 17.1 is a graphical depiction of these two hypothetical situations.

In the second case [Fig. 17.1(b)], some individual effect estimates are quite different from zero, but as a group their average is nearly zero. In some cases increasing pump setting increases impingement pressure, while in other cases it decreases pressure. On the other hand, for the distribution of sample effects that covers the effect estimates in Fig. 17.1(a), zero looks to be a quite implausible value to arise as a sample effect. That is, the sample effects, as a group, appear to be distinctly different from zero.

It is essential to have some measure of the effect variability before any judgment about the importance of an effect can be made. Since the variations in effects are the result of experimental errors in the individual responses, it is best to have an estimate of the variance of the responses and then develop from it a variance estimate for the effects. This requires replicated runs of at least some of the test settings. Unfortunately, often we can only run the experiment once (without any replications), particularly when the design calls for a large number of tests. For these unreplicated designs, we will have to develop an estimate of the effect variance from the observed effects directly, usually with additional assumptions. In this chapter we discuss methods of judging the importance of effects for both replicated and unreplicated experiments.

17.2
Analysis of Replicated 2^k Experiments

To judge the relative importance of a variable effect based on our calculated main and interaction effects, we can make use of test replications to estimate the error we expect to see in an effect estimate. In particular, we will

1. Estimate the experimental error via the variance of individual observations.
2. Estimate the variance associated with the main or interaction effects, which are simply a linear combination (sum) of individual observations.
3. Construct the statistical distribution of effect estimates and examine the range of estimates that could reasonably arise, say, if the true effect were really zero. (Due to the central limit theorem, the distribution of effect estimates is generally assumed to be normal, as in the two illustrations of Fig. 17.1.)

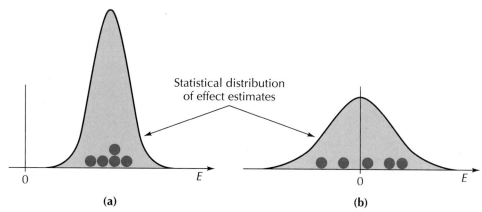

Figure 17.1 Graphical Depiction of the Distribution of Sample Effects

Glove Box Door Alignment Experiment

To present the method of analysis for replicated two-level factorial experiments, we will use an automobile glove box door alignment study. An automobile assembly plant using an automated fixture to center a glove box door to the instrument panel was experimenting with ways to improve the door parallelism during assembly. The actual assembly operation at the time was producing an average deviation of 2 millimeters off-parallel with some assemblies exceeding 4 millimeters. The purpose of the study was to find better ways to set up the fixtures to improve the door parallelism measure to within ± 1.5 millimeters. At one phase of this study, the following four variables were examined over two levels each to investigate their possible effects on parallelism in millimeters.

Variable	Low ($-$)	High ($+$)
x_1: RH cowl fore/aft movement	Nominal	-5 mm
x_2: Center brace attachment sequence	Before	After
x_3: Plenum gasket	No	Yes
x_4: Evaporator case setup, fore/aft	Nominal	-5 mm

To examine the possible influence of these factors on parallelism, the engineers decided to use a 2^4 design with each test replicated once. The total of 32 tests were conducted in a random order. The results are given in Table 17.1.

To obtain the effect estimates, we first augment the design matrix of Table 17.1, with columns for the two-, three-, and four-factor interactions. These new columns are formed from the design matrix by taking all possible products of the main variable columns two at a time, three at a time, and four at a time. The resulting matrix is referred to as the *calculation matrix* and also contains a column of all "$+1$" values for the average response. This additional column is generally denoted as the I or identity column. The calculation matrix should be

TABLE 17.1 2^4 Design and Results of the Glove Box Door Experiment (Test Order in Parentheses)

					Parallelism (mm)	
					Run 1	Run 2
Test	x_1	x_2	x_3	x_4	y_{i1}	y_{i2}
1	−	−	−	−	−1.44 (7)	−0.08 (28)
2	+	−	−	−	−1.79 (10)	−1.01 (24)
3	−	+	−	−	0.39 (14)	0.17 (32)
4	+	+	−	−	−0.50 (2)	−0.24 (21)
5	−	−	+	−	−0.20 (9)	0.17 (27)
6	+	−	+	−	−0.79 (6)	−0.64 (30)
7	−	+	+	−	1.22 (13)	0.28 (20)
8	+	+	+	−	0.21 (8)	0.28 (18)
9	−	−	−	+	−0.40 (1)	−0.65 (31)
10	+	−	−	+	−0.63 (15)	−1.19 (25)
11	−	+	−	+	0.47 (3)	0.44 (17)
12	+	+	−	+	−0.01 (5)	−0.03 (23)
13	−	−	+	+	1.29 (12)	0.64 (29)
14	+	−	+	+	−1.17 (4)	0.14 (19)
15	−	+	+	+	0.48 (16)	1.06 (22)
16	+	+	+	+	0.40 (11)	0.34 (26)

TABLE 17.2 Calculation Matrix of the Glove Box Door Experiment

Test	I	1	2	3	4	12	13	14	23	24	34	123	124	134	234	1234	$\bar{y}_i$	d_i
1	+	−	−	−	−	+	+	+	+	+	+	−	−	−	−	+	−0.76	−1.36
2	+	+	−	−	−	−	−	−	+	+	+	+	+	+	−	−	−1.40	−0.78
3	+	−	+	−	−	−	+	+	−	−	+	+	+	−	+	−	0.28	0.22
4	+	+	+	−	−	+	−	−	−	−	+	−	−	+	+	+	−0.37	−0.26
5	+	−	−	+	−	+	−	+	−	+	−	+	−	+	+	−	−0.02	−0.37
6	+	+	−	+	−	−	+	−	−	+	−	−	+	−	+	+	−0.72	−0.15
7	+	−	+	+	−	−	−	+	+	−	−	−	+	+	−	+	0.75	0.94
8	+	+	+	+	−	+	+	−	+	−	−	+	−	−	−	−	0.25	−0.07
9	+	−	−	−	+	+	+	−	+	−	−	−	+	+	+	−	−0.53	0.25
10	+	+	−	−	+	−	−	+	+	−	−	+	−	−	+	+	−0.91	0.56
11	+	−	+	−	+	−	+	−	−	+	−	+	−	+	−	+	0.46	0.03
12	+	+	+	−	+	+	−	+	−	+	−	−	+	−	−	−	−0.02	0.02
13	+	−	−	+	+	+	−	−	−	−	+	+	+	−	−	+	0.97	0.65
14	+	+	−	+	+	−	+	+	−	−	+	−	−	+	−	−	−0.52	−1.31
15	+	−	+	+	+	−	−	−	+	+	+	−	−	−	+	−	0.77	−0.58
16	+	+	+	+	+	+	+	+	+	+	+	+	+	+	+	+	0.37	0.06

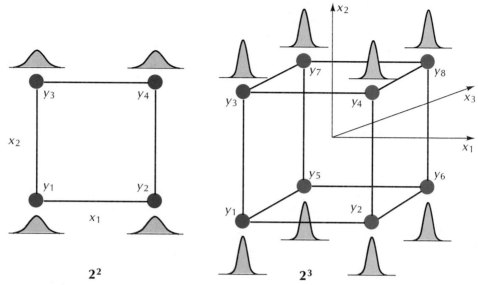

Figure 17.2 Distribution of Responses on Factorial Designs

a square matrix; that is, the number of columns/effects is equal to the number of rows/test conditions. The calculation matrix with experimental results of the 2^4 design used for the glove box door study is shown in Table 17.2. The last two columns in Table 17.2 are the average response, $\bar{y}_i = (y_{i1} + y_{i2})/2$, and difference $d_i = (y_{i1} - y_{i2})$ for each of the $i = 1, 2, 3, \ldots, 16$ unique test conditions. The differences in the last column of Table 17.2 will be used later to estimate the experimental error variance for an individual test observation.

Estimation of Variance of an Individual Observation

Let us assume that the amount of error in an observation is the same for all observations (i.e., that the true variance is the same for all observations) and that the observations are independent. In Chapter 18 we check this assumption of equal variance through a statistical test. Figure 17.2 depicts this situation for 2^2 and 2^3 designs. When we calculate location effects (main and interaction effects on the mean response) we are really determining the extent to which the mean level of the response distribution is changing from one level of a variable to another. However, if the variance of the response distribution is dependent on one of the variables, we say that there exists a significant dispersion effect on the response. Dispersion effects or effects on variation of the response have recently received some attention.[1]

---

[1] G. E. P. Box and R. D. Meyer, "An Analysis for Unreplicated Fractional Factorials," and "Dispersion Effects from Fractional Designs," *Technometrics*, Vol. 28, 1986, pp. 11–18 and 19–27.

We will now describe a procedure for the estimation of the variance of an individual observation under the assumption of a common variance for all of the responses. For test condition 1 of the glove box door parallelism experiment, the two replicated observations are -1.44 and -0.08 millimeters. The sample variance for this test condition, designated as s_1^2, can be calculated as

$$s_1^2 = \frac{(y_{11} - \bar{y}_1)^2 + (y_{12} - \bar{y}_1)^2}{2 - 1}$$
$$= [-1.44 - (-0.76)]^2 + [-0.08 - (-0.76)]^2$$
$$= 0.9248.$$

In this example, we can calculate 16 independent sample variances $s_1^2, s_2^2, \ldots, s_{16}^2$, one for each unique test condition of the design. The 16 sample variances have been calculated to be

$$
\begin{aligned}
s_1^2 &= 0.92480 & s_9^2 &= 0.03125 \\
s_2^2 &= 0.30420 & s_{10}^2 &= 0.15680 \\
s_3^2 &= 0.02420 & s_{11}^2 &= 0.00045 \\
s_4^2 &= 0.03380 & s_{12}^2 &= 0.00020 \\
s_5^2 &= 0.06845 & s_{13}^2 &= 0.21125 \\
s_6^2 &= 0.01125 & s_{14}^2 &= 0.85805 \\
s_7^2 &= 0.44180 & s_{15}^2 &= 0.16820 \\
s_8^2 &= 0.00245 & s_{16}^2 &= 0.00180.
\end{aligned}
$$

Since we are assuming that there is a common variance for all 32 observations, each of the sample variances calculated is an independent estimate of that common variance, σ_y^2. A more precise, single estimate of this common variance may be obtained by combining, or pooling, the 16 independent estimates of the variance. The pooled sample variance, s_p^2 is, in general, given by

$$s_p^2 = \frac{\nu_1 s_1^2 + \nu_2 s_2^2 + \cdots + \nu_m s_m^2}{\nu_1 + \nu_2 + \cdots + \nu_m} = \frac{\sum\limits_{i=1}^{m} \nu_i s_i^2}{\sum\limits_{i=1}^{m} \nu_i}, \tag{17.1}$$

where ν_1, ν_2, and so on, are the degrees of freedom associated with the variance estimates s_1^2, s_2^2, and so on, and m is the number of independent estimates of the common variance σ_y^2 that are available. Determining the pooled variance amounts simply to calculating s_p^2 as a weighted average of the individual s_i^2, where the weights are the degrees of freedom associated with each estimate, s_i^2. When all tests are replicated n times each (i.e., $n_1 = n_2 = \cdots = n_m = n$), the formula for s_p^2 reduces to

$$s_p^2 = \frac{s_1^2 + s_2^2 + \cdots + s_m^2}{m}. \tag{17.2}$$

If the number of replicates is equal to two for all tests (i.e., $n = 2$), a convenient form for s_i^2 may be developed in terms of the difference $d_i = (y_{i1} - y_{i2})$:

$$s_i^2 = \frac{\left(y_{i1} - \dfrac{y_{i1} + y_{i2}}{2}\right)^2 + \left(y_{i2} - \dfrac{y_{i1} + y_{i2}}{2}\right)^2}{2 - 1} = \left(\frac{d_i}{2}\right)^2 + \left(\frac{d_i}{2}\right)^2 = \frac{d_i^2}{2}. \tag{17.3}$$

Using this relationship, we can express the pooled variance for this particular case as

$$s_p^2 = \sum_{i=1}^{m} \frac{d_i^2/2}{m}. \tag{17.4}$$

For the case at hand, therefore, using Eq. (17.2) the pooled variance is

$$s_p^2 = \frac{s_1^2 + s_2^2 + \cdots + s_{16}^2}{16} = \frac{0.9248 + \cdots + 0.0018}{16}$$

$$= 0.20242,$$

or alternatively, using Eq. (17.4) and the d_i values in Table 17.2, is

$$s_p^2 = \frac{d_1^2/2 + \cdots + d_{16}^2/2}{16} = \frac{(-1.36)^2 + \cdots + (0.06)^2}{32}$$

$$= 0.20242.$$

Estimation of the Variances Associated with the Main Effects and Interaction Effects

Given the data of Table 17.2, we find that the main effect of variable 1, E_1, is

$$E_1 = (\tfrac{1}{8})\, [(\bar{y}_2 - \bar{y}_1) + (\bar{y}_4 - \bar{y}_3) + \cdots + (\bar{y}_{14} - \bar{y}_{13}) + (\bar{y}_{16} - \bar{y}_{15})].$$

Since each $\bar{y}_i$ is an average of two observations y_{i1} and y_{i2}, E_1 can be written as

$$E_1 = \frac{\dfrac{y_{2,1} + y_{2,2}}{2} - \dfrac{y_{1,1} + y_{1,2}}{2} + \cdots + \dfrac{y_{16,1} + y_{16,2}}{2} - \dfrac{y_{15,1} + y_{15,2}}{2}}{8}$$

or

$$E_1 = \frac{y_{2,1} + y_{2,2} - y_{1,1} - y_{1,2} + \cdots + y_{16,1} + y_{16,2} - y_{15,1} - y_{15,2}}{16}.$$

Thus we see that an effect estimate is a linear combination of $(n \times m = 32)\ y_{ij}$.

By using the calculation matrix and the algebraic method of effect calculation, all the estimated effects for the glove box parallelism experiment were calculated, and are (in millimeters)

$$\text{average} = -0.087$$

E_1	$= -0.654$	$E_{13} = -0.117$	E_{123}	$=$	0.172
E_2	$= 0.794$	$E_{14} = -0.031$	E_{124}	$=$	0.101
E_3	$= 0.638$	$E_{23} = -0.191$	E_{134}	$=$	-0.138
E_4	$= 0.322$	$E_{24} = -0.154$	E_{234}	$=$	-0.104
E_{12}	$= 0.147$	$E_{34} = 0.009$	E_{1234}	$=$	$0.121.$

Under the assumption that the observations y_{ij} are independent, each with an unknown but common variance σ_y^2, the variance of E_1, $\text{Var}(E_1)$, is

$$\text{Var}(E_1) = \text{Var}\left(\frac{y_{2,1} + y_{2,2} - y_{1,1} - y_{1,2} + \cdots + y_{16,1} + y_{16,2} - y_{15,1} - y_{15,2}}{16}\right)$$

$$= \left(\frac{1}{256}\right)[\text{Var}(y_{2,1}) + \text{Var}(y_{2,2}) + \cdots + \text{Var}(y_{15,1}) + \text{Var}(y_{15,2})]$$

$$= \left(\frac{1}{256}\right)(\sigma_y^2 + \sigma_y^2 + \cdots + \sigma_y^2 + \sigma_y^2)$$

$$= \frac{32}{256}\sigma_y^2 = \frac{4\sigma_y^2}{32} = \frac{\sigma_y^2}{8},$$

where the variance of an observation σ_y^2 is to be estimated by the sample (pooled) variance, s_p^2.

Since each effect is a linear combination of the same set of 32 responses, the variances of all other main effects and the interaction effects are the same as the variance of E_1:

$$\text{Var}(E_1) = \text{Var}(E_2) = \text{Var}(E_3) = \text{Var}(E_4) = \text{Var}(E_{12}) = \text{Var}(E_{13})$$
$$= \text{Var}(E_{14}) = \text{Var}(E_{23}) = \text{Var}(E_{24}) = \text{Var}(E_{34}) = \text{Var}(E_{123})$$
$$= \text{Var}(E_{124}) = \text{Var}(E_{134}) = \text{Var}(E_{234}) = \text{Var}(E_{1234})$$
$$= \frac{\sigma_y^2}{8}.$$

Substituting for σ_y^2, the pooled sample variance, s_p^2, we obtain the sample variance of the effects for this case as

$$s_{\text{effect}}^2 = \frac{s_p^2}{8}.$$

In general, an effect estimate is determined from the equation

$$E = \frac{2}{N}(\pm y_1 \pm y_2 \pm \cdots \pm y_N), \tag{17.5}$$

where N is the total number of tests conducted, including all replicates. Therefore, the variance of an effect estimate is given as

$$\text{Var}(E) = \text{Var}\left[\frac{2}{N}(\pm y_1 \pm y_2 \pm \cdots \pm y_N)\right] = \frac{4}{N^2}(N\sigma_y^2) = \frac{4\sigma_y^2}{N}. \tag{17.6}$$

Thus, in general, the sample variance of an effect is calculated by

$$s_{\text{effect}}^2 = \frac{4s_p^2}{N}.$$

For the glove box door experiment, $N = 32$ (16 unique test conditions replicated once) and $s_p^2 = 0.20242$. The estimated variance of an effect is therefore

$$s_{\text{effect}}^2 = \frac{4s_p^2}{32}$$

$$= \frac{s_p^2}{8}$$

$$= \frac{0.20242}{8}$$

$$= 0.0253.$$

The square root of s_{effect}^2 is sometimes called the standard error (s.e.) of an effect, which will be used for significance tests applied on the effects. For our example,

$$\text{s.e.} = s_{\text{effect}} = 0.159 \text{ mm}.$$

The variance of the average response is the familiar form:

$$\text{Var(average)} = \text{Var}\left(\frac{y_1 + y_2 + \cdots + y_N}{N}\right) = \frac{1}{N^2}(N\sigma_y^2) = \frac{\sigma_y^2}{N}. \tag{17.7}$$

Thus the sample variance of the average, s_{ave}^2, is s_p^2/N. For our example, the numerical value for s_{ave} is 0.0795, which is observed to be one-half of the standard error for an effect.

17.3

Determination of Statistically Significant Effects

Hypothesis Testing Formulation

Recall that based on the test results from the 2^4 glove box door alignment experiment, the following effect estimates were calculated:

E_1	$= -0.654$	$E_{13} = -0.117$	E_{123}	$=$	0.172
E_2	$= 0.794$	$E_{14} = -0.031$	E_{124}	$=$	0.101
E_3	$= 0.638$	$E_{23} = -0.191$	E_{134}	$=$	-0.138
E_4	$= 0.322$	$E_{24} = -0.154$	E_{234}	$=$	-0.104
E_{12}	$= 0.147$	$E_{34} = 0.009$	E_{1234}	$=$	$0.121.$

It is desired at this point to assess which of these estimates, if any, are important, that is, statistically significant.

In Chapter 5 the concept of hypothesis testing was developed in a general context and was also used to formulate the Shewhart control chart model. The hypothesis-testing idea may also be applied in the situation at hand to judge whether any of the calculated effect estimates are important.

In conducting a hypothesis test on the effect estimates above, we will assume that the true effect values are zero, that is, $\mu_{effect} = 0.0$. This assumption serves as our null hypothesis (H_0) and is depicted in Fig. 17.3. If a calculated effect estimate is not consistent with H_0, we will reject H_0 and accept the alternative hypothesis (H_1) that the true value for that effect is nonzero. Let us also assume that the effect estimates are normally distributed and that their variability is characterized by σ^2_{effect}. The assumption of normality is not unreasonable since the estimates are a linear combination of N responses [refer to Eq. (17.5)], and such linear combinations are approximately normally distributed due to the Central Limit Theorem.

As we saw in Chapter 5, to calculate the probability of obtaining a given value for E_i or one larger (or smaller), requires that we perform a transformation to the standard normal distribution,

$$Z = \frac{E_i - \mu_{effect}}{\sigma_{effect}}.$$

Generally, we do not know the true variance of an effect, σ^2_{effect}, but we may estimate it with s^2_{effect}. We might, therefore, consider performing the following transformation:

$$\frac{E_i - \mu_{effect}}{s_{effect}}.$$

The quantity given above no longer follows the standard normal distribution. The question is: What distribution does it follow?

Student's t Distribution

When samples are drawn from a normal distribution, the sample means also follow the normal distribution. If the population variance, σ^2_y, is unknown and is estimated by the sample variance, s^2_y, the quantity

$$\frac{\bar{y} - \mu_{\bar{y}}}{s_{\bar{y}}} = \frac{\bar{y} - \mu_{\bar{y}}}{s_y / \sqrt{n}}$$

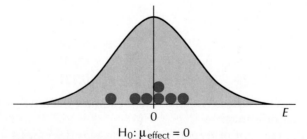

$$H_0: \mu_{effect} = 0$$

Figure 17.3 Distribution of the Effect Estimates Under the Null Hypothesis

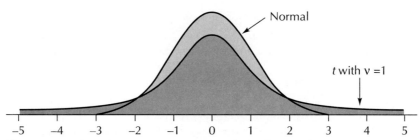

Figure 17.4 t Distribution ($v = 1$ Degree of Freedom) and Standard Normal Distribution

is said to be distributed according to Student's t distribution with $v = n - 1$ degrees of freedom. The t distribution is symmetric about its mean of zero and has varying spread, depending on the amount of information (degrees of freedom) that forms the estimate of the variance $\sigma_{\bar{y}}^2$, in our case s_p^2. The t distribution has a wider spread than the standard normal distribution due to the uncertainty introduced by replacing the population variance with the sample variance. The spread in the t distribution decreases as the sample size increases (i.e., the degrees of freedom increases), and in fact it approaches the standard normal distribution as the sample size goes to infinity. Figure 17.4 shows a t distribution with $v = 1$ degree of freedom along with a standard normal distribution. Table A.3 gives the t values associated with various α-risk levels for several different degrees of freedom, v.

To make certain there is no confusion here, the t distribution arises only in cases where the population is normal and the variance is unknown, regardless of the size of the sample. Remember, it is the quantity $(\bar{y} - \mu_{\bar{y}})/s_{\bar{y}}$ that follows the t distribution. The sole purpose of the t distribution is to account for the uncertainty introduced when estimating the population variance with the sample variance.

Return to Glove Box Alignment Case Study

Returning to our hypothesis test on the estimated effects, we may calculate the t statistic for each estimate, under $H_0: \mu_{\text{effect}} = 0$, using the transformation

$$t = \frac{E_i - \mu_{\text{effect}}}{s_{\text{effect}}}$$
$$= \frac{E_i - 0.0}{0.159}.$$

Note that the estimate of the true standard deviation of the effects used in this transformation is the standard error of the effects calculated previously, $s_{\text{effect}} = 0.159$ mm. It should be noted further that this estimate of the standard deviation has $v = 16$ degrees of freedom (each unique test condition supplies an estimate of variance with 1 degree of freedom, and the pooled variance based on these 16 estimates then has 16 degrees of freedom). The degrees of

Effect	Effect Estimate	Associated t Value	Effect	Effect Estimate	Associated t Value
		TABLE 17.3			

TABLE 17.3   Effect Estimates and Associated t Values for the Glove Box Door Alignment Study

Effect	Effect Estimate	Associated t Value	Effect	Effect Estimate	Associated t Value
E_1	-0.654	-4.1132	E_{24}	-0.154	-0.9686
E_2	0.794	4.9937	E_{34}	0.009	0.0566
E_3	0.638	4.0126	E_{123}	0.172	1.0818
E_4	0.322	2.0252	E_{124}	0.101	0.6352
E_{12}	0.147	0.9245	E_{134}	-0.138	-0.8679
E_{13}	-0.117	-0.7358	E_{234}	-0.104	-0.6541
E_{14}	-0.031	-0.1950	E_{1234}	0.121	0.7610
E_{23}	-0.191	-1.2013			

freedom associated with a particular t statistic are always defined by the degrees of freedom associated with the variance estimate. The t values associated with the effect estimates are given in Table 17.3. For $E_1 = -0.654$,

$$t = \frac{-0.654 - 0}{0.159}$$

$$= -4.1132.$$

From Table A.3 it is seen that for a t distribution with $\nu = 16$ degrees of freedom, values of t such that the probability of occurrence of that value or one more extreme (larger or smaller) is 0.025 are

$$t_{16,0.025} = -2.120$$

$$t_{16,0.975} = 2.120,$$

where the subscripts on t represent the degrees of freedom and the area under the t distribution to the left of that value, respectively. Thus any calculated t value with an absolute value greater than 2.120 is judged to be too large to have arisen under H_0. The risk that we will reject H_0 here, when it is true, is 0.05.

Upon examining Table 17.3, it is seen that the t values associated with E_1, E_2, and E_3 are greater in absolute value than $t_{16,0.975} = 2.120$; thus we must reject the null hypothesis for those effects. We, therefore, conclude that the true means of the effect values for E_1, E_2, and E_3; namely μ_{E_1}, μ_{E_2}, and μ_{E_3}, are nonzero. Figure 17.5 depicts the situation at hand. It should be noted that we cannot infer from the results that the other effects are unimportant, only that we have no reason to doubt the null hypothesis. Their true effects may or may not be equal to zero.

Calculation of Confidence Intervals for the True Mean Effects

Through the hypothesis-testing procedure described previously, the main effects of variables 1, 2, and 3 have been identified as important. Alternatively,

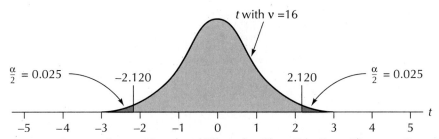

Figure 17.5 t Test for Significance of Effects for Glove Box Door Alignment Experiment

we might say that the true mean effect values μ_{E1}, μ_{E2}, and μ_{E3} are nonzero. The effect estimates obtained from the experiment, $E_1 = -0.654$, $E_2 = 0.794$, and $E_3 = 0.638$, are simply values that have been drawn from distributions centered at μ_{E1}, μ_{E2}, and μ_{E3}, respectively. Although the hypothesis test above has led us to believe that these three μ_{Ei} are nonzero, one might ask more specifically: What value or values for μ_{Ei} are likely given the data?

Unfortunately, we cannot determine exact values for the true mean effect values. We can, however, estimate the values for them using the effect estimates. Specifically, $E_1 = -0.654$ is an estimate of μ_{E1}, $E_2 = 0.794$ is an estimate of μ_{E2}, and $E_3 = 0.638$ is an estimate of μ_{E3}. We say that such effect estimates are point estimates of the true mean effect values. In short, the effect estimates are the best single values that we have for the true mean effects.

Although we recognize that the calculated effect value is the best estimate that we have for a true mean of an effect, we also recognize that the true mean of an effect may be larger or smaller than the estimate. We might therefore ask the question: How small (or large) could the true mean of an effect be? To answer this question, let us consider posing a succession of null hypotheses where μ_{effect} takes on values in the vicinity of E_i. The smallest and largest values for μ_{effect} for which the null hypothesis cannot be rejected (i.e., for which the observed effect estimate is a plausible realization) are given by

$$Z_{1-\alpha/2} = \frac{E_i - \mu_{\text{low}}}{\sigma_{\text{effect}}}; \qquad Z_{\alpha/2} = \frac{E_i - \mu_{\text{high}}}{\sigma_{\text{effect}}}.$$

With a bit of algebra, we can rewrite the above as

$$\mu_{\text{low}} = E_i - Z_{1-\alpha/2}\,(\sigma_{\text{effect}}) \qquad \mu_{\text{high}} = E_i - Z_{\alpha/2}\,(\sigma_{\text{effect}})$$

or, simplifying,

$$\mu_{\text{low}}, \mu_{\text{high}} = E_i \pm Z_{1-\alpha/2}\,(\sigma_{\text{effect}}). \qquad (17.8)$$

The situation depicted mathematically by Eq. (17.8) is illustrated graphically in Fig. 17.6. Equation (17.8) is termed an interval estimate for the true mean of an effect μ_{effect}, and is centered at the point estimate, E_i. As we saw in our development of the hypothesis-testing idea, we must be willing to accept some risk in the expression of our interval estimate. This risk arises from the fact that the calculated effect estimate, E_i, may not be a typical realization from the

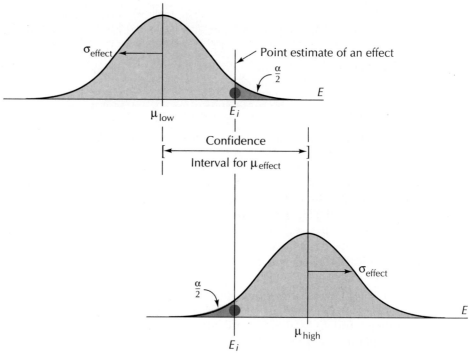

Figure 17.6 Confidence Interval Idea

distribution of effect estimates centered at μ_{effect}. This risk is expressed in the present context by referring to the interval estimate as being a *confidence interval*. Thus Eq. (17.8) should be properly referred to as a $100(1 - \alpha)\%$ confidence interval for the true mean of an effect.

In developing the confidence interval of Eq. (17.8) it may be observed that σ_{effect} was assumed to be known. As we saw previously, we may replace σ_{effect} with its estimate, s_{effect}, to produce the following form of the confidence interval:

$$\mu_{\text{low}}, \mu_{\text{high}} = E_i \pm t_{v,1-\alpha/2}\, s_{\text{effect}}. \tag{17.9}$$

Note that the replacement of σ_{effect} with s_{effect} forces us to replace $Z_{1-\alpha/2}$ with $t_{v,1-\alpha/2}$.

Confidence intervals may now be developed for all the true mean effects based on the calculated effect estimates ($E_1, E_2, \ldots, E_{1234}$) and the standard error of the effects ($s_{\text{effect}} = 0.159$, $v = 16$). For a $100(1 - \alpha)\%$ confidence interval with $\alpha = 0.05$ (95% confidence interval), $t_{16,0.975}$ is 2.120. The 95% confidence interval for μ_{effect} is, therefore,

$$E_i \pm t_{16,0.975} s_{\text{effect}}$$
$$E_i \pm (2.120)(0.159)$$
$$E_i \pm 0.337.$$

TABLE 17.4 95% Confidence Intervals for True Mean Effects of the Glove Box Door Alignment Study Based on Replicated Experiment	
Main Effects	*95% Confidence Interval*
RH cowl fore/aft (E_1)	-0.654 ± 0.337 mm*
Center brace (E_2)	0.795 ± 0.337 mm*
Plenum gasket (E_3)	0.638 ± 0.337 mm*
Evaporator case (E_4)	0.322 ± 0.337 mm
Two-Variable Interactions	*95% Confidence Interval*
RH cowl $\times$ center brace (E_{12})	0.147 ± 0.337 mm
RH cowl $\times$ plenum gasket (E_{13})	-0.117 ± 0.337 mm
RH cowl $\times$ evaporator case (E_{14})	-0.031 ± 0.337 mm
Center brace $\times$ plenum gasket (E_{23})	-0.191 ± 0.337 mm
Center brace $\times$ evaporator case (E_{24})	-0.154 ± 0.337 mm
Plenum gasket $\times$ evaporator case (E_{34})	0.009 ± 0.337 mm
Three-Variable Interaction	*95% Confidence Interval*
Cowl $\times$ brace $\times$ plenum (E_{123})	0.172 ± 0.337 mm
Cowl $\times$ brace $\times$ evaporator (E_{124})	0.101 ± 0.337 mm
Cowl $\times$ plenum $\times$ evaporator (E_{134})	-0.138 ± 0.337 mm
Brace $\times$ plenum $\times$ evaporator (E_{234})	-0.104 ± 0.337 mm
Four-Variable Interaction	*95% Confidence Interval*
Cowl $\times$ brace $\times$ plenum $\times$ evaporator (E_{1234})	0.121 ± 0.337 mm

* Confidence interval shows significant effect.

As an example, the 95% confidence interval for the true mean effect of variable 1, the fore/aft movement of the right-hand cowl, on parallelism (in millimeters), is

$$E_1 \pm 0.337$$
$$-0.654 \pm 0.337$$
$$-0.991, -0.317.$$

Thus we are 95% confident that the true mean effect of variable 1 lies on the interval from -0.991 to -0.317. Table 17.4 gives the 15 effect estimates together with the corresponding 95% confidence intervals for their true effect values.

A confidence interval may also be used to judge the statistical significance of variable effects. We will reject the null hypothesis, H_0: true mean effect = 0, if the confidence interval does not include "zero." In other words, we will conclude that an effect is statistically nonzero if the confidence interval values are strictly either positive or negative. From Table 17.4 it would appear that only three of the estimated effects are important based on a 95% confidence level for their true mean effect values. The confidence intervals associated with significant variable effects, E_1, E_2, and E_3, are identified by an asterisk in Table

17.4. It should come as no surprise that the conclusions made as a result of hypothesis testing are exactly the same as those made based on this confidence interval approach. The two approaches essentially supply identical information concerning the *significance* of variable effects.

Graphically, we can present these confidence intervals as shown in Fig. 17.7. From the figure it is seen that only the first three confidence intervals do not include zero, which was the hypothesized true mean for all the main and interaction effects under H_0.

In addition to the confidence intervals developed for each of the true mean effects above, one may also determine such an interval for the mean response. Such an interval may be of special concern if it is desired to ascertain if the average response ($\hat{b}_0$) is to be included in a math model such as that developed in Chapter 16. Therefore, we may wish to check to see if the calculated estimate for the average or mean response could have been a sample observation drawn from a distribution with a true mean equal to zero. Our calculated average parallelism for the 32 responses is -0.087 millimeter. The corresponding 95% confidence interval for the mean is then

$$\text{average} \pm t_{16,0.975} \frac{s_p}{\sqrt{32}}$$

$$-0.087 \pm 0.1685,$$

which shows that the true mean parallelism is not significantly different from zero. Hence if it were desired to write down a math model for parallelism, based on the results of the experiment, the $\hat{b}_0$ term would be left out.

17.4

Analysis of the Results of Unreplicated Two-Level Factorials

There are times when it is neither feasible nor desirable (due to perhaps very limited resources or due to a large number of variables) to include replication in a two-level factorial design. In such cases it is not possible to obtain an independent estimate of the experimental error for judging the relative importance of the main and interaction effects. There are, however, at least two methods that can be employed to aid in making such decisions. These two methods are

1. The use of normal probability plots of the effect estimates to isolate the important effects from those that appear to be driven by the forces of experimental error alone.
2. The use of higher-order interaction effect estimates to estimate directly the variance of the effects for significance testing.

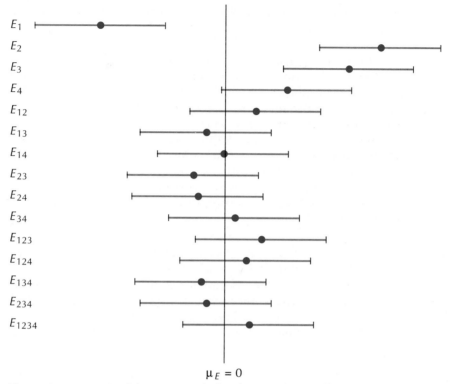

Figure 17.7 95% Confidence Intervals for the True Mean Effects (Only the Intervals Based on E_1, E_2, and E_3 Do Not Include Zero)

Normal Probability Plot to Analyze Unreplicated Designs

We saw in Chapter 3 that the normal probability plot is a graphical method to check whether or not a set of data could be considered as arising from a normal distribution. If a normal probability plot is constructed for data drawn from a single normal distribution, the data will fall along/about a straight line. In addition, if all the data in a sample do come from the same normal distribution with the exception of a few data points, the normal probability plot will show these exceptional points to distribute themselves off a line that passes through the other points.

To use a normal probability plot to judge the importance of effects estimated from two-level factorial experiments, we make use of two important facts surrounding the situation at hand:

1. Since effect estimates are linear combinations of the individual response values, they tend to be normally distributed due to the Central Limit Theorem.
2. Our purpose in making a test of significance for the effect estimates is to sort them into two piles: those that are insignificant (their true mean values

are zero) and those that are significant (their true mean values are non-zero).

In addition, the variance of the effect values should be equal, an assumption we have made earlier. The first fact allows us to use normal probability plots as a means to evaluate the effect estimates, while the second, when taken with the first, means that all the insignificant effect estimates should fall on a straight line on the plot centered near zero.

We will illustrate this method using the data from the glove box door parallelism study. We will assume now that we have no way of estimating the variance of an individual response value.

Following the procedure outlined in Chapter 3, we need to first arrange the effect estimates in ascending order and estimate their corresponding cumulative probabilities using

$$p_i = \frac{100(i - 1/2)}{2^4 - 1}, \qquad i = 1, 2, \ldots, (2^4 - 1).$$

Table 17.5 contains all the information required to make the normal probability plot, and Fig. 17.8 gives the resulting normal plot. In the figure, a straight line has been drawn through a large number of the effect estimates, passing near the (0, 50%) coordinate such that their scatter about this line is fairly random.

TABLE 17.5   Data Points for the Normal Probability Plot (Glove Box Door Parallelism Study, Unreplicated Average Responses)

Order Number, i	Effect Estimates, E_i	Identity of Effects	Cumulative Probability, p_i (%)
1	−0.654	1	3.3
2	−0.191	23	10.0
3	−0.154	24	16.7
4	−0.138	134	23.3
5	−0.117	13	30.0
6	−0.104	234	36.7
7	−0.031	14	43.3
8	0.009	34	50.0
9	0.101	124	56.7
10	0.121	1234	63.3
11	0.147	12	70.0
12	0.172	123	76.7
13	0.322	4	83.3
14	0.638	3	90.0
15	0.794	2	96.7

It is seen that there are three effect estimates that fall well off this line. We conclude, therefore, that effects 1, 2, and 3 (circled and marked in Fig. 17.8) are important; that is, they are not likely to be arising from a normal distribution with a mean of zero. This conclusion agrees with the one we reached by the method for replicated experiments in the preceding section. This example demonstrates the effectiveness of the use of normal probability plots for analyzing the results of unreplicated experiments.

To ensure an effective application of the normal probability plot for the identification of significant effects, it is recommended that it be used for two-

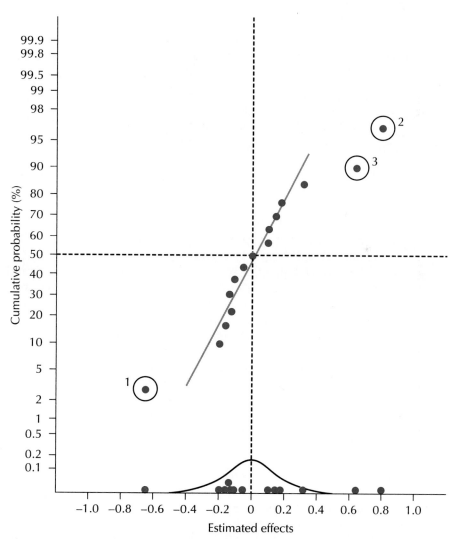

Figure 17.8 Normal Probability Plot of the Sample Effects: Glove Box Door Parallelism Experiment

level designs with at least four variables. With less than four variables, (e.g., a 2^3 factorial design), if several of the true mean effects are nonzero, it becomes very difficult to decide which set of estimates have arisen from a normal distribution with mean zero.

Being a graphical method, the use of normal probability plots always involves a certain degree of subjective judgment. Sometimes, there may be a few points that are not easy to be ruled as off-line or on-line. If we follow the principle of parsimony in model building, we would first only select the points that are obviously well off the line and conduct further math model checks (to be discussed later). If these checks indicate some model inadequacies or a general inability to confirm model predictions through additional testing, other effect estimates close to the line might be included in the model.

Use of Higher-Order Interaction Effects to Estimate Error and Assess the Importance of Variable Effects

The interpretation of the normal plot of estimated effects in Fig. 17.8 suggests that we might be able to obtain an estimate of the standard error of estimated effects from the higher-order interaction effects. Why is this so? The answer comes as we reexamine the normal plot. We see two things:

1. The majority of the insignificant effects that comprise the straight line about zero are interaction effect estimates and in fact all the third- and higher-order interactions are randomly distributed about this line. This seems to suggest that at some point we might be able to assume that interactions of a certain order and above may be neglected—their true mean value assumed to be zero.
2. When we collapse the effect estimates that fall on the straight line about zero to a "pile" in a dot diagram form, we can see the amount of scatter (the magnitude of the standard error) we expect to be associated with effect estimates. We actually see a data-based representation of the reference distribution of effects, as shown at the bottom of Fig. 17.8.

Based on the two observations above, let us assume that all third- and higher-order interaction effects are negligible. This means that we are declaring that the true mean effects of interactions 123, 124, 134, 234, and 1234 are zero. With this assumption we can estimate the variance of an effect directly from the formula

$$s_{\text{effect}}^2 = \sum_{\substack{\text{higher-order} \\ \text{interactions}}} \frac{(E_i - \mu_{Ei})^2}{\text{no. of higher-order interactions}}. \tag{17.10}$$

This is a sample variance formula where we can substitute 0 for the true mean effect μ_{E_i} that is assumed to be known. The number of higher-order interactions available in this case is $n = 5$ and we divide by $n = 5$ instead of $n - 1 = 4$, since the mean μ_{E_i} is assumed to be known.

Recall that the effect estimates for the third- and higher-order interactions were

$$E_{123} = +0.172 \qquad E_{234} = -0.104 \qquad E_{1234} = +0.121.$$
$$E_{124} = +0.101 \qquad E_{134} = -0.138$$

From Eq. (17.10) with $\mu_{E_i} = 0$ we obtain

$$s_{\text{effect}}^2 = \frac{[(0.172 - 0)^2 + (0.101 - 0)^2 + (-0.104 - 0)^2 + (-0.138 - 0)^2 + (0.121 - 0)^2]}{5}$$

$$= 0.0168572.$$

Therefore, the estimated standard error of an effect is

$$s_{\text{effect}} = \text{s.e.} = 0.1298.$$

When we use this standard error to construct a 95% confidence interval with $t_{5,0.975} = 2.571$, we have

Effect estimate $\pm$ (2.571)(0.1298)

$$E_i \pm 0.334.$$

With this confidence interval, our conclusions about the importance of the effects are

- The main effects of variables 1, 2, and 3 are statistically significant.
- The remaining effects are not likely to be important.

Again, the same conclusion was reached previously by the other methods. It is interesting to note that the half width of the 95% confidence interval calculated from the replicated responses was 0.337, which is about the same as the one above.

Summary

In this chapter we have presented techniques for the assessment of important variable effects for both replicated and unreplicated two-level factorial designs. In identifying the important effects, our ultimate goal will be to develop a mathematical model for the process that can then be used for design and/or control purposes. In the next chapter we will discuss the building of such models.

The effects we have been studying up to now describe how variables and interactions between variables affect the centering or location of a distribution of responses. For this reason we term the types of effects described in this chapter as "location effects." In developing techniques to judge location effect importance, we have relied on the assumption that the variance of the observations is constant. Accordingly, we have assumed that none of the variables being studied in the experiment have any effect on the level of noise associated with the response. In the next chapter we check for the possible presence of "dispersion" or variability effects.

Exercises

17.1. The yield of a chemical process has been studied as a function of three factors, at the following experimental levels:

Variable	Low Level	High Level
1. Temperature (°C)	80	120
2. Pressure (psi)	50	70
3. Reaction time (min)	5	15

A 2^3 factorial design was performed. Each test, or unique combination of the three variables, was performed three times. The table provides the results of the tests.

				Trial			
Test	Temperature	Pressure	Time	1	2	3	Average
1	80	50	5	61.43	58.58	57.07	59.03
2	120	50	5	75.62	77.57	75.75	76.31
3	80	70	5	27.51	34.03	25.07	28.87
4	120	70	5	51.37	48.49	54.37	51.41
5	80	50	15	24.80	20.69	15.41	20.30
6	120	50	15	43.58	44.31	36.99	41.63
7	80	70	15	45.20	49.53	50.29	48.34
8	120	70	15	70.51	74.00	74.68	73.07

a. Calculate all the main and interaction effects.
b. Calculate the sample pooled variance, s_p^2, based on the replication within each test run.
c. Calculate the sample variance of an effect based on the sample pooled variance.
d. Using a 95% confidence interval, find the statistically significant effects.

17.2. A 2^2 factorial design was conducted to study the effect of welding temperature (degrees Fahrenheit) and flux quantity (inches per minute) on the resulting weld strength. The table describes the conditions that were examined in the experiment and the associated measured weld strengths (note that each unique combination of temperature and flux quantity was tested three times). The table also shows the average and sample variance for the weld strength calculated for each unique combination of the temperature (variable 1) and flux quantity (variable 2).

Test	Temperature	Flux Quantity	Trial 1	Trial 2	Trial 3	Average	Sample Variance
1	300	10	3088	2724	3586	3132.67	187,257.33
2	500	10	799	1202	1079	1026.67	42,656.33
3	300	20	813	1124	1101	1012.67	30,032.33
4	500	20	2806	2946	3336	3029.33	75,433.33

a. Calculate the effect estimates (i.e., the average, E_1, E_2, and E_{12}).
b. Based on the replicated trials for each unique test combination, develop a 95% confidence interval to determine which effect estimates are significant.
c. Prepare graphical aids that help to explain *all* significant effects (other than the average). Briefly explain/interpret the meaning of all significant effects.

17.3. An unreplicated 2^3 factorial design was conducted. Three variables (x_1, x_2, and x_3) were considered, with the following results:

$$\begin{aligned}
\text{average} &= 60.0 & E_{12} &= 3.0 \\
E_1 &= -12.0 & E_{13} &= 23.0 \\
E_2 &= -3.0 & E_{23} &= -2.0 \\
E_3 &= -18.0 & E_{123} &= 1.0.
\end{aligned}$$

a. Determine which effect estimates are significant by plotting them on normal probability paper. Show all your calculations.
b. Write down what appears to be the most appropriate math model for the response.

17.4. A 2^3 factorial design was conducted. The design matrix and data are shown in the table.

Test	1	2	3	y
1	−	−	−	12
2	+	−	−	15
3	−	+	−	24
4	+	+	−	17
5	−	−	+	24
6	+	−	+	16
7	−	+	+	24
8	+	+	+	28

a. Calculate all the effects.
b. Plot the effect estimates on normal probability paper. Based on the

normal probability plot, indicate which effects seem to be significant.

17.5. A replicated 2^2 factorial experiment was performed, the results of which are shown in the table.

x_1	x_2	y_{i1}	y_{i2}
−	−	0.50	0.46
+	−	0.42	0.37
−	+	0.33	0.30
+	+	0.11	0.16

a. Calculate the effect estimates (i.e., the average, E_1, E_2, and E_{12}).

b. Based on the replicated trials for each unique test combination, develop a 90% confidence interval to determine which effect estimates are significant.

17.6. A local nightclub owner has performed a two-level factorial experiment to investigate the effects of five factors on the business (sales) at his club during a single evening (measured in hundreds of dollars).

Variable	Low Level	High Level
1. Price of beverage	$0.50	$1.00
2. Jukebox	Off	On
3. Bartender	Sam	Woody
4. Waitress	Diane	Carla
5. Mirror-ball	Off	On

The following effects have been calculated based on the 32 combinations of the five variables:

average $=$	13.00	$E_{14} =$	-0.50	$E_{123} =$	0.24	$E_{245} = \quad 0.37$
$E_1 =$	-4.30	$E_{15} =$	-0.20	$E_{124} =$	0.57	$E_{345} = \quad 0.82$
$E_2 =$	0.03	$E_{23} =$	0.29	$E_{125} =$	0.38	$E_{1234} = -0.60$
$E_3 =$	-0.50	$E_{24} =$	0.26	$E_{134} =$	-1.00	$E_{1235} = \quad 0.13$
$E_4 =$	0.75	$E_{25} =$	-0.50	$E_{135} =$	-1.40	$E_{1245} = -0.40$
$E_5 =$	7.80	$E_{34} =$	-0.60	$E_{145} =$	0.35	$E_{1345} = \quad 0.22$
$E_{12} =$	-0.40	$E_{35} =$	0.01	$E_{234} =$	0.00	$E_{2345} = -0.50$
$E_{13} =$	-0.80	$E_{45} =$	6.81	$E_{235} =$	-0.20	$E_{12345} = \quad 0.48.$

a. Plot the effects on normal probability paper.

b. Based on the normal probability plot, indicate which effects appear to be significant.

17.7. Referring to Exercise 17.6:

• a. Use the third- and higher-order interactions to calculate the sample variance of an effect.

b. Develop 95% confidence intervals for each of the true mean effects. Which effects appear to be statistically significant?

c. Are the results here consistent with those of Exercise 17.6? Explain.

17.8. Based on the significant effects determined in Exercise 17.6 or 17.7, what strategy would you take as the owner to maximize your revenues? Be specific. Provide graphical interpretations for all important main and interaction effects.

17.9. After completing the analysis for the experiment of Exercise 17.6, it was found that the jukebox was on for all 32 tests. Reanalyze the data of this experiment, given this new information and the additional analysis capability it provides. Are the results here consistent with the results of Exercise 17.6? Comment.

17.10. A 2^3 factorial design was conducted to study the effect of three variables on a response of interest. The table describes the conditions that were examined in the experiment and the associated response (note that each unique combination of the variables was tested four times). Also evident from an examination of the table is that fact that the average and sample variance for the response have been calculated for each unique combination of the variables.

x_1	x_2	x_3	Trial 1	Trial 2	Trial 3	Trial 4	Average	Sample Variance
−	−	−	13	28	12	31	21	98
−	+	−	9	18	30	23	20	78
+	+	+	38	33	53	32	39	94
+	+	−	57	55	39	45	49	72
−	−	+	52	34	49	41	44	66
−	+	+	30	42	43	33	37	42
+	−	−	41	53	60	62	54	90
+	−	+	38	41	50	27	39	90

The following effect estimates have been calculated:

$$
\begin{aligned}
\text{average} &= 37.875 & E_{12} &= 0.75 \\
E_1 &= 14.75 & E_{13} &= -16.25 \\
E_2 &= -3.25 & E_{23} &= \ ? \\
E_3 &= \ ? & E_{123} &= 2.75.
\end{aligned}
$$

a. Calculate E_3 and E_{23}.

b. Based on the replicated trials for each unique combination, calculate an estimate of the standard error of an effect. Using a 95% confidence interval, which effect estimates are statistically significant?

c. Graphically explain the numerical value obtained for E_{13} by using a two-way diagram. Explain in your own words the meaning of the graphical display.

17.11. A fisherman has conducted a 2^5 factorial design to determine the effects of five factors on the number of fish he catches in a four-hour period.

The variables studied and their levels for experiment are shown in the table.

Variable	Low Level	High Level
1. Fishing location	Pier	Boat
2. Bait type	Worms	Night crawlers
3. Time of day	Day	Night
4. Weather	No rain	Rain
5. Hook size	Small	Large

The experimental trials were all performed at Jack's Fish Farm in northern Wisconsin, stocked with bluegills. The following table gives the results of the experiment.

Test	x_1	x_2	x_3	x_4	x_5	y
1	−	−	−	−	−	25
2	+	−	−	−	−	26
3	−	+	−	−	−	31
4	+	+	−	−	−	34
5	−	−	+	−	−	24
6	+	−	+	−	−	26
7	−	+	+	−	−	32
8	+	+	+	−	−	36
9	−	−	−	+	−	30
10	+	−	−	+	−	36
11	−	+	−	+	−	40
12	+	+	−	+	−	43
13	−	−	+	+	−	30
14	+	−	+	+	−	34
15	−	+	+	+	−	40
16	+	+	+	+	−	42
17	−	−	−	−	+	2
18	+	−	−	−	+	21
19	−	+	−	−	+	26
20	+	+	−	−	+	43
21	−	−	+	−	+	4
22	+	−	+	−	+	20
23	−	+	+	−	+	25
24	+	+	+	−	+	44
25	−	−	−	+	+	10
26	+	−	−	+	+	27
27	−	+	−	+	+	33
28	+	+	−	+	+	51
29	−	−	+	+	+	9
30	+	−	+	+	+	27
31	−	+	+	+	+	32
32	+	+	+	+	+	53

a. Calculate all effect estimates.
b. Assuming that the third- and higher-order interactions are negligible, determine an estimate for the variance of an effect and construct 95% confidence intervals based on the effect estimates.
c. Based on part (b), which effects appear to be important?
d. Construct two-way diagrams for all important two-factor interactions. Comment.

17.12. Using the data presented in Exercise 17.11:
a. Calculate all effect estimates.
b. Construct a normal probability plot of the effect estimates.
c. Interpret the normal probability plot. Which effects appear to be significant?
d. Construct two-way diagrams for all two-factor interactions. Comment.

17.13. A 2^3 factorial design was used to study the possible effects of temperature, time, and pressure on the yield of a chemical process. The data, including experimental results, are shown in the tables.

Variable	Low Level	High Level
1. Temperature (°F)	80	100
2. Time (min)	5	7
3. Pressure (psi)	120	140

				Response	
Test	1	2	3	Trial 1	Trial 2
1	−	−	−	9	11
2	+	−	−	18	22
3	−	+	−	4	4
4	+	+	−	9	11
5	−	−	+	7	9
6	+	−	+	17	19
7	−	+	+	5	7
8	+	+	+	10	14

a. Calculate all the effects.
b. Using the replicated results, determine an estimate for the variance of the effects.
c. Evaluate the statistical significance of each of the effects by calculating a t value for each effect estimate and comparing it to the critical t value at a 5% significance level.
d. Based on your interpretation of the results, what action(s) might be taken to increase the yield?

17.14. Using the data presented in Exercise 17.13:
a. Calculate all the effect estimates.

     b. Make a normal plot of the effect estimates, and interpret the plot.

     c. Based on your interpretation of the results, what action(s) might be taken to increase the yield?

17.15. A 2^2 factorial experiment is conducted with each unique test condition being performed three times (giving 12 tests total). $100(1 - \alpha)\%$ confidence intervals are to be constructed for the true mean effects based on the effect estimates. How many degrees of freedom are associated with the Student's t statistic value to be used in construction of the confidence intervals?

17.16. A 2^2 factorial design was conducted and each unique test condition was run four times. The sample variance of an effect estimate was calculated and found to be $s_{effect}^2 = 9$. The calculated effect estimates were: average $= 7.5$, $E_1 = -6.0$, $E_2 = 8.0$, and $E_{12} = -5.5$. Using 95% confidence intervals, determine which of the effect estimates appear to be significant.

17.17. A 2^5 factorial experiment is conducted. It is assumed that the third- and higher-order interactions can be neglected, and these interactions are used to estimate the variance of an effect. How many degrees of freedom are associated with the t value that will be used to test the significance of the variable effects?

17.18. A 2^3 factorial design is to be conducted. The variance of an individual observation is $\sigma_y^2 = 4$. Suppose that we want the length of a 95% confidence interval to be 1.8 or smaller. What is the minimum number of runs at each distinct test condition that will satisfy the foregoing requirement, assuming that each test will be run an equal number of times?

17.19. The results of a 2^3 factorial design are given in standard order in the table.

Test	Response, y
1	48.8
2	91.6
3	40.8
4	60.4
5	48,8
6	89.2
7	40.4
8	60.0

Let x_1, x_2, and x_3 correspond to the coded variables.

     a. Calculate the effects.

     b. Using a normal probability plot, which effects appear to be significant?

17.20. Referring to Table A.4 in the Appendix, sketch the following distributions using the same horizontal scale.

     a. A standard normal distribution.

     b. A t distribution with 1 degree of freedom.

     c. A t distribution with 8 degrees of freedom.

Model Building for Design and Improvement Using Two-Level Factorial Designs

18.1

Introduction

In Chapter 16 we examined how a performance measure or response could be effectively studied through the use of two-level factorial designs. It was seen that the dependence of an experimental response on one or more variables, both singly and in concert with each other, could be characterized through the calculation of effects. In Chapter 17 we examined several techniques for separating the effects that are truly important from those that are not, in light of the experimental error in the testing environment. In developing these techniques, some assumptions about the character of the noise were made that need to be examined further to judge their appropriateness.

In this chapter we also consider two-level factorial designs, focusing on the development of empirical models for design, prediction, and control purposes. The development of such models requires that we check the appropriateness and adequacy of the prediction model. It is, at least in considerable part, the building of such models for quality/productivity design and improvement that experiments are designed, conducted, and analyzed in the first place.

General Procedures for Model Building

It is useful to think of the model-building activity as an iterative procedure evolving in four distinct stages.

1. Postulation of a tentative general model form.
2. Design of an appropriate experiment.
3. Fitting the general model form to the data.
4. Diagnostic checking of the fitted model.

Postulation of a tentative model may be based on prior knowledge of the underlying behavior of the response over the ranges of the variables under study. When little is known in this regard and/or the variable level ranges are not large, low-order polynomial forms may be useful.

Given a tentative model form, an experiment must be designed that has the capacity to estimate the parameters/coefficients of the variable terms in the model. When low-order polynomial forms are proposed, two- and three-level factorials serve this purpose well.

Once the data are at hand, the model may be fit to the data. For two-level factorials and their associated polynomial forms, the effect estimates E_i are employed. These estimates can be shown to be the least squares estimates of the model coefficients, when suitably scaled, and as a result have several desirable properties.

The fact that the initially postulated model is fit to the data is no guarantee that the model will provide good predictions of the process behavior. If the model postulation is misguided, the predictions will be poor even if the experiment is well designed and carried out. Remember, the experimental design is in response to the model postulated, good or bad. In diagnostically checking the model, two basic questions are addressed:

1. Are all the terms initially included in the model necessary?
2. Are the terms initially included in the model sufficient?

The first question may be dealt with given the experimental results at hand. Dealing with the second may require additional experimentation.

18.2
Glove Box Door Alignment Study Revisited

Review of Previous Analysis

In Chapter 17 a study was presented that focused on improving the parallelism of a glove box door with respect to the instrument panel. Four variables were studied over two levels each to investigate their possible effects on parallelism. These variables and their low and high levels are given in the table.

Variable		Low (−)	High (+)
x_1:	RH cowl fore/aft movement	Nominal	−5 mm
x_2:	Center brace attachment sequence	Before	After
x_3:	Plenum gasket	No	Yes
x_4:	Evaporator case setup, fore/aft	Nominal	−5 mm

A replicated 2^4 design was performed, and the parallelism was recorded for each of the 32 tests. The results of the experiment are given in Table 18.1.

Selection of this experimental design means that the tacit assumption has already been made that a model of the form of Eq. (18.1) may provide for an adequate representation of the phenomenon under study:

$$
\begin{aligned}
y = {} & b_0 + b_1 x_1 + b_2 x_2 + b_3 x_3 + b_4 x_4 + b_{12} x_1 x_2 + b_{13} x_1 x_3 \\
& + b_{14} x_1 x_4 + b_{23} x_2 x_3 + b_{24} x_2 x_4 + b_{34} x_3 x_4 + b_{123} x_1 x_2 x_3 \\
& + b_{124} x_1 x_2 x_4 + b_{134} x_1 x_3 x_4 + b_{234} x_2 x_3 x_4 \\
& + b_{1234} x_1 x_2 x_3 x_4 + \epsilon,
\end{aligned} \tag{18.1}
$$

where the ϵ are the deviations between the data y and the mean response η, represented here by the polynomial form. It is typically assumed that ϵ should behave as a normally and independently distributed variable with zero mean

TABLE 18.1 2^4 Design and Results of the Glove Box Door Experiment

Test	x_1	x_2	x_3	x_4	Run 1 y_{i1}	Run 2 y_{i2}	$\bar{y}_i$	d_i	s_i^2
1	−	−	−	−	−1.44	−0.08	−0.760	−1.36	0.92480
2	+	−	−	−	−1.79	−1.01	−1.400	−0.78	0.30420
3	−	+	−	−	0.39	0.17	0.280	0.22	0.02420
4	+	+	−	−	−0.50	−0.24	−0.370	−0.26	0.03380
5	−	−	+	−	−0.20	0.17	−0.015	−0.37	0.06845
6	+	−	+	−	−0.79	−0.64	−0.715	−0.15	0.01125
7	−	+	+	−	1.22	0.28	0.750	0.94	0.44180
8	+	+	+	−	0.21	0.28	0.245	−0.07	0.00245
9	−	−	−	+	−0.40	−0.65	−0.525	0.25	0.03125
10	+	−	−	+	−0.63	−1.19	−0.910	−0.56	0.15680
11	−	+	−	+	0.47	0.44	0.455	0.03	0.00045
12	+	+	−	+	−0.01	−0.03	−0.020	0.02	0.00020
13	−	−	+	+	1.29	0.64	0.965	0.65	0.21125
14	+	−	+	+	−1.17	0.14	−0.515	−1.31	0.85805
15	−	+	+	+	0.48	1.06	0.770	−0.58	0.16820
16	+	+	+	+	0.40	0.34	0.370	0.06	0.00180

and fixed variance σ_y^2. Since the ϵ's may be thought of as an aggregation of many small random errors, it is not unreasonable to expect the above assumptions to be met if the model η is adequate.

Based on the results in Table 18.1, the following effect estimates were calculated:

$$
\begin{array}{llll}
\text{average} = -0.087 \\
E_1 = -0.654 & E_{13} = -0.117 & E_{123} = 0.172 \\
E_2 = 0.794 & E_{14} = -0.031 & E_{124} = 0.101 \\
E_3 = 0.638 & E_{23} = -0.191 & E_{134} = -0.138 \\
E_4 = 0.322 & E_{24} = -0.154 & E_{234} = -0.104 \\
E_{12} = 0.147 & E_{34} = 0.009 & E_{1234} = 0.121.
\end{array}
$$

To judge the relative importance of these effects, we may use the replicated responses from the experiment. As noted in Chapter 17, if we assume that the responses are all drawn from distributions with the same variance, the sample variances determined for each pair of replicated tests may be pooled to provide for a more precise estimate of this common variance,

$$
s_p^2 = \frac{s_1^2 + s_2^2 + \cdots + s_{16}^2}{16} = \frac{0.9248 + \cdots + 0.0018}{16}
$$

$$
= 0.20243.
$$

Recall that the variance of an effect for a two-level factorial design is given by

$$
\sigma_{\text{effect}}^2 = \frac{4\sigma_y^2}{N}.
$$

For this experiment the sample variance of an effect is

$$
s_{\text{effect}}^2 = \frac{4s_p^2}{32}
$$

$$
= \frac{s_p^2}{8}
$$

$$
= \frac{0.20243}{8}
$$

$$
= 0.0253.
$$

A 95% confidence interval may be developed for the true mean value of each effect estimate,

$$
E_i \pm t_{16,0.975}\, s_{\text{effect}} = E_i \pm (2.120)(0.159) = E_i \pm 0.337.
$$

Table 18.2 summarizes the 95% confidence intervals based on each of the effect estimates, and also includes a confidence interval for the mean response. It may be seen in Table 18.2 that the confidence intervals for μ_{E_1}, μ_{E_2}, and μ_{E_3} do not include zero (bold type in the table). Therefore, it may be concluded that the true mean effects of variables 1, 2, and 3 differ significantly from zero.

TABLE 18.2	Confidence Intervals for Variable Effects of the Glove Box Door Alignment Study		
Effect	95% Confidence Interval	Effect	95% Confidence Interval
Mean	-0.087 ± 0.169	E_{23}	-0.191 ± 0.337
$\boldsymbol{E_1}$	$\boldsymbol{-0.654 \pm 0.337}$	E_{24}	-0.154 ± 0.337
$\boldsymbol{E_2}$	$\boldsymbol{0.794 \pm 0.337}$	E_{34}	0.009 ± 0.337
$\boldsymbol{E_3}$	$\boldsymbol{0.638 \pm 0.337}$	E_{123}	0.172 ± 0.337
E_4	0.322 ± 0.337	E_{124}	0.101 ± 0.337
E_{12}	0.147 ± 0.337	E_{134}	-0.138 ± 0.337
E_{13}	-0.117 ± 0.337	E_{234}	-0.104 ± 0.337
E_{14}	-0.031 ± 0.337	E_{1234}	0.121 ± 0.337

Development of a Prediction Model Based on the Experiment Results

In our examination of two-level factorial designs we have seen that important variables and interactions can be distinguished from those that are not important using several techniques. One of the primary motivations in our study of two-level factorial designs was to mathematically describe the relationships between the experimental response and the variables manipulated during the experiment. Let us consider, therefore, how such a relationship or model can be developed.

As we have said earlier, in analyzing a 2^k factorial design we may express the experimental response using a model of the form of Eq. (18.1). It should be noted that for a 2^k factorial design there is a model coefficient associated with each of the $2^k - 1$ main and interaction effects, plus the mean response b_0. For a 2^k factorial design, estimates of the model coefficients may be obtained directly from the effect estimates, which are, in fact, the least squares estimates. The coefficient estimates are simply the effect estimates divided by 2:

$$\hat{b}_i = \frac{E_i}{2}.$$

Thus a calculated effect, E_i, is an estimate of the true mean effect, and $\hat{b}_i$ is an estimate of the model coefficient, b_i.

A common question that is asked with respect to obtaining estimates of the model coefficients from the effect estimates is: "Why is the effect estimate divided by 2?" To understand the reason behind this, consider the graphical representation of the meaning of the main effect of variable 1, which is illustrated in Fig. 18.1. As can be seen from the figure, changing the level of the RH cowl movement from the nominal value (-1 level) to -5 millimeters ($+1$ level) produces a change in the parallelism of -0.6544 millimeters. In other words, a two-unit change in variable x_1 (from -1 to $+1$) produces a -0.6544 millimeter change in the response. The dependency of the parallelism on x_1

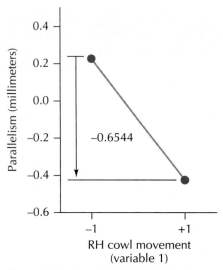

Figure 18.1 Graphical Aid for the Interpretation of the Main Effect of Variable 1

may also be expressed via the slope of the line in Fig. 18.1. This slope, $\hat{b}_1$, is given by

$$\hat{b}_1 = \frac{-0.6544}{1 - (-1)} = -0.327$$

$$= \frac{E_1}{2}.$$

If estimates for all the model coefficients are inserted into Eq. (18.1), an equation is produced that may be used to predict the response as a function of the independent variables.

$$\begin{aligned}
\hat{y} = \hat{b}_0 &+ \hat{b}_1 x_1 + \hat{b}_2 x_2 + \hat{b}_3 x_3 + \hat{b}_4 x_4 + \hat{b}_{12} x_1 x_2 + \hat{b}_{13} x_1 x_3 \\
&+ \hat{b}_{14} x_1 x_4 + \hat{b}_{23} x_2 x_3 + \hat{b}_{24} x_2 x_4 + \hat{b}_{34} x_3 x_4 + \hat{b}_{123} x_1 x_2 x_3 \\
&+ \hat{b}_{124} x_1 x_2 x_4 + \hat{b}_{134} x_1 x_3 x_4 + \hat{b}_{234} x_2 x_3 x_4 \\
&+ \hat{b}_{1234} x_1 x_2 x_3 x_4,
\end{aligned} \tag{18.2}$$

where $\hat{y}$ is the predicted response. Examining Eq. (18.2), we see that there are terms associated with all the variables and their interactions. We know from Table 18.2, however, that only a few of the coefficients associated with these terms are important. If we drop the insignificant terms from the model, we obtain

$$\hat{y} = \hat{b}_1 x_1 + \hat{b}_2 x_2 + \hat{b}_3 x_3. \tag{18.3}$$

Recall that the mean response $\hat{b}_0$ could not be distinguished from zero and therefore does not appear in the model. If the calculated values for the $\hat{b}$'s are inserted into Eq. (18.3), the following equation is produced:

$$\hat{y} = -0.327 x_1 + 0.397 x_2 + 0.319 x_3. \tag{18.4}$$

This equation provides us with a tool that can be used to study the response as a function of variables 1, 2, and 3 (RH cowl movement, attachment sequence, and plenum gasket). However, before we exercise this model for quality design and improvement purposes, we should examine the model more closely to determine if it is an adequate representation of the phenomenon under study. If it is not, some improvement may be possible given the existing data. However, it may be necessary to conduct more experiments to develop an adequate model.

18.3

Diagnostic Checking of the Fitted Model

Calculation of Predicted Responses and Model Residuals

In the case of our glove box door parallelism study, we identified the main effects of variables 1, 2, and 3 as the only important effects and therefore we obtained the following model for the prediction of the door parallelism [Eq. (18.4)]:

$$\hat{y} = -0.327x_1 + 0.397x_2 + 0.319x_3.$$

This equation may be used to calculate a predicted response, $\hat{y}_i$, for each of the test conditions. If this model contains all the necessary terms to adequately predict the response, y, the model residual errors (the differences between the data and the model predictions), $e_{ij} = (y_{ij} - \hat{y}_i)$, should contain no structured variation at all; that is, they should behave in the following way:

1. They should be centered about a mean of zero.
2. They should tend to be normally distributed.
3. They should have a scatter that does not vary as a function of the predicted response ($\hat{y}$'s).
4. They should not be correlated with the independent variables x, time order of the testing, or any other variables related to the experiment. They should be truly random errors.

In short, the model residuals (e's) should have the same properties as we assumed for the errors (ϵ's). If the model residuals do not demonstrate these properties, this may be an indication that the assumptions made with regard to the errors (ϵ's) do not hold—perhaps because the model η is inadequate.

To check to see if the residuals behave as above, the procedure outlined below should be followed:

- Use the model to predict the parallelism for each of the tests performed.
- Compare the model prediction with the observed data and calculate the residual errors, e_{ij}, for all i and j.

- Perform a series of tests and checks on the residuals to see if they do behave as they should if the predicting model is adequate.

At each of the 16 unique combinations of x_1, x_2, x_3, and x_4 in the experimental design, we can predict the response, $\hat{y}$, using Eq. (18.4). Since only x_1, x_2, and x_3 are important, the 16 test conditions will produce only eight unique predicted responses. For the first test (in standard order), the levels for x_1, x_2, and x_3 are all -1. Therefore, the predicted response is

$$\hat{y}_1 = -0.327(-1) + 0.397(-1) + 0.319(-1) = -0.389.$$

For test 1, there are two observed responses, $y_{11} = -1.44$ and $y_{12} = -0.08$. The model residual errors are then $e_{1j} = (y_{1j} - \hat{y}_1)$, $j = 1, 2$; $e_{11} = -1.051$ and $e_{12} = 0.309$. Table 18.3 gives the predicted responses and residual errors associated with the prediction model over the entire experiment.

Residual Analysis

We may now perform a series of checks on the residuals, e_{ij}. These should include:

- A normal probability plot of the residuals.
- A plot of the residuals versus the time/run order of the tests.
- A plot of the residuals versus the predicted response.
- Plots of the residuals versus the independent variables.

TABLE 18.3   Model Predictions and Residuals of the Parallelism Prediction Model (Glove Box Door Study)

Test	x_1	x_2	x_3	x_4	Predicted Response, $\hat{y}_i$	Observed Response, y_{i1}	Run Order	Model Residual, e_{i1}	Observed Response, y_{i2}	Run Order	Model Residual, e_{i2}
1	−	−	−	−	−0.389	−1.440	(7)	−1.051	−0.080	(28)	0.309
2	+	−	−	−	−1.043	−1.790	(10)	−0.747	−1.010	(24)	0.033
3	−	+	−	−	0.405	0.390	(14)	−0.015	0.170	(32)	−0.235
4	+	+	−	−	−0.249	−0.500	(2)	−0.251	−0.240	(21)	0.009
5	−	−	+	−	0.249	−0.200	(9)	−0.449	0.170	(27)	−0.079
6	+	−	+	−	−0.405	−0.790	(6)	−0.385	−0.640	(30)	−0.235
7	−	+	+	−	1.043	1.220	(13)	0.177	0.280	(20)	−0.763
8	+	+	+	−	0.389	0.210	(8)	−0.179	0.280	(18)	−0.109
9	−	−	−	+	−0.389	−0.400	(1)	−0.011	−0.650	(31)	−0.261
10	+	−	−	+	−1.043	−0.630	(15)	0.413	−1.190	(25)	−0.147
11	−	+	−	+	0.405	0.470	(3)	0.065	0.440	(17)	0.035
12	+	+	−	+	−0.249	−0.010	(5)	0.239	−0.030	(23)	0.219
13	−	−	+	+	0.249	1.290	(12)	1.041	0.640	(29)	0.391
14	+	−	+	+	−0.405	−1.170	(4)	−0.765	0.140	(19)	0.545
15	−	+	+	+	1.043	0.480	(16)	−0.563	1.060	(22)	0.017
16	+	+	+	+	0.389	0.400	(11)	0.011	0.340	(26)	−0.049

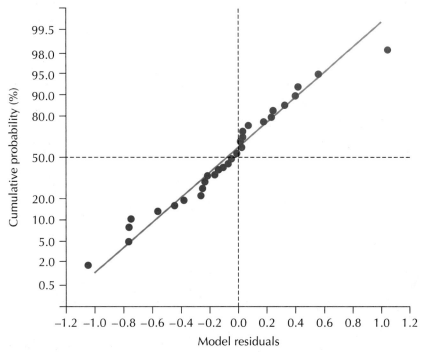

Figure 18.2 Normal Probability Plot of the Residuals

Figure 18.2 shows a normal probability plot of the residuals. Most of the residuals seem to fall on a straight line on the normal probability paper, with the exception of two or three points, and are centered approximately at zero. These off-line points should be checked for possible causes, such as unusual events surrounding the environment of their respective tests, possible deviations from designed test conditions, and possible errors in the measurements and recording of these tests. On the whole, the plot looks as if the residual errors are quite normal.

Figures 18.3 through 18.9 give some additional plots of the residuals to further reveal possible inadequacies of the prediction model for the glove box door study. Even though $\hat{b}_4$ was not included in the model, a residual plot versus x_4 should be made to reveal the possible presence of structure in the data with respect to x_4. The residuals seem to be random and appear to have a consistent amount of scatter, as none of the residual plots of Figs. 18.3 to 18.9 show any obvious relationship with the time order of running the tests or the levels of the independent variables. The residuals are centered about zero and there does not appear to be any structure within the residuals. Figure 18.5 examines the extent to which the magnitude of replication error may be related to the magnitude of the response. This check addresses the issue of constant variance, which will be examined in more detail later in this chapter.

(*Continued on page 625.*)

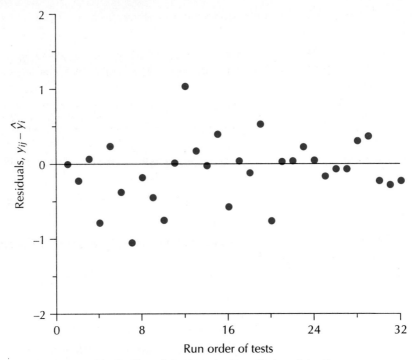

Figure 18.3 Residuals Plotted Against the Run Order of the Tests

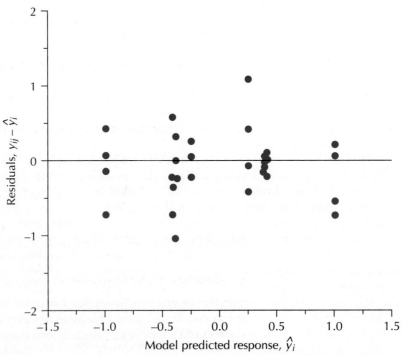

Figure 18.4 Residuals Plotted Against the Predicted Responses

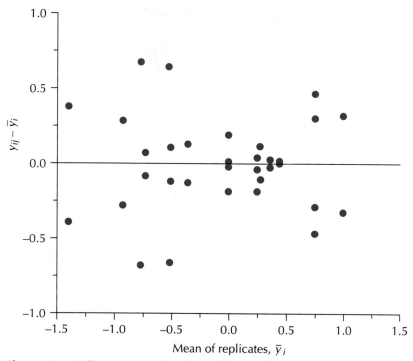

Figure 18.5 Difference Between the Replicated Response and the Mean of the Replicates Plotted Against the Mean of the Replicates

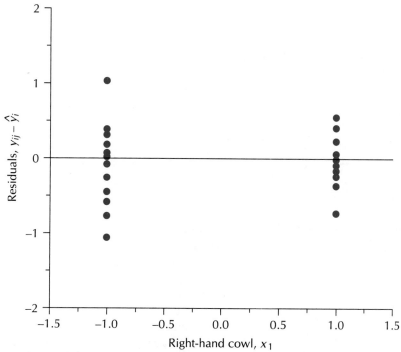

Figure 18.6 Residuals Versus Right-Hand Cowl Movement Level: Nominal (-1), -5 Millimeters $(+1)$

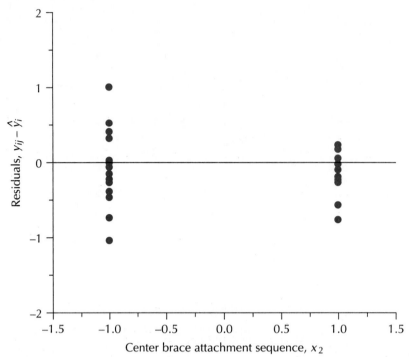

Figure 18.7 Residuals Versus Center Brace Attachment Level: Before (-1), After ($+1$)

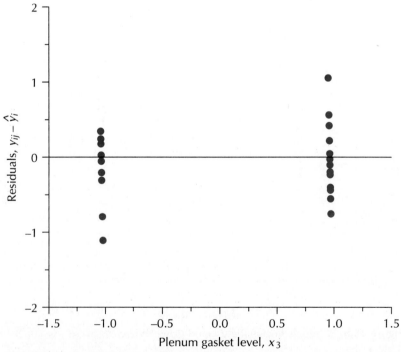

Figure 18.8 Residuals Versus Plenum Gasket Level: No (-1), Yes ($+1$)

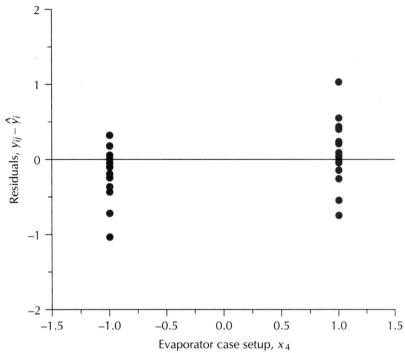

Figure 18.9 Residuals Versus Evaporator Case Setup Level: Nominal (-1), -5 Millimeters $(+1)$

Some Nonrandom Patterns of Residual Plots

Residual plots are used primarily to check for model inadequacies. Sometimes they may also reveal other concerns, such as the assumption of equal variance of the responses or problems with the actual conduct or experimental environment of the tests. To assist in the identification of residual plots with nonrandom patterns, several residual plots with common nonrandom patterns are shown in Fig. 18.10. It is noted that all 10 nonrandom patterns of Fig. 18.10 have a zero mean, except pattern (b). A brief explanation of some typical causes of each nonrandom pattern follows.

    a. The residuals form a linear trend (upward or downward). If it is a trend over time, this might be due to some consistent and gradual buildup effect of certain external factors (which are not being studied), such as the moisture content of the raw material that was continuously decreasing. If the trend is a function of the predicted response $\hat{y}$, it would indicate that the model is inadequate due to perhaps a bias in the constant term in the model.

    b. The model has a constant bias (positive or negative). Possible causes include error in the estimation of the mean and missing effects in the model. Also check the response data, particularly for one or two possible "outliers."

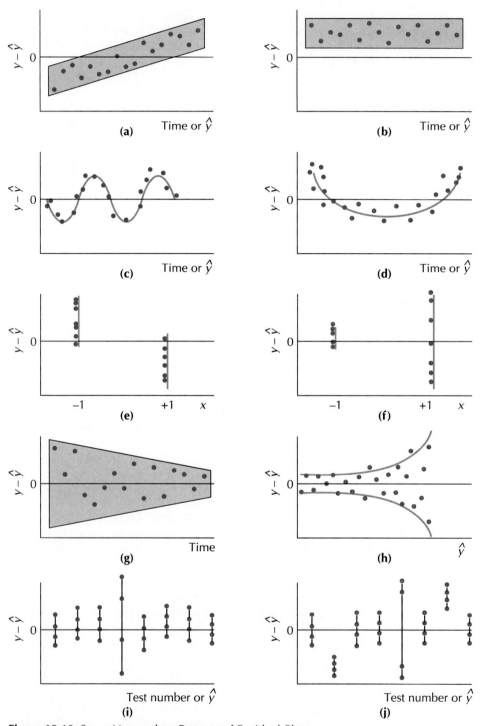

Figure 18.10 Some Nonrandom Patterns of Residual Plots

   c. The residuals have a periodic trend. If it is seen from the time plot, it is probably caused by an unknown cyclic factor such as day–night–day. If it shows up on the $\hat{y}$ plot, it means that a higher-order model would be needed.

   d. A quadratic trend (concave or convex). This may be revealing the presence of higher-order effects.

   e. The model is missing a main or interaction effect associated with an x_i. If the control variable x_i is not in the model, consider possible inclusion of it. If x_i is one of the variables already included in the model, check for inclusion of some interaction effects involving x_i.

   f. The model is not uniform in variability. This is a strong indication of a dispersion effect of x_i; that is, the variation level varies as a function of x_i.

   g. Instability in the experiment process. This would indicate certain instability of the experimental environment for the tests that run earlier in the series. A learning effect of the experimenter or the transient state of certain new experimental equipment are examples of causes for this pattern. The use of data transformations such as a log transformation or adding a time-dependent term in the model could save the experiment in this case if the instability was well behaved.

   h. There is an increasing response variability. The variability of the response is proportional to the magnitude of the response. The response data may have to be transformed to obtain a constant variability in the transformed responses. The use of log transformation often works well.

   i. Outlier responses among the replicates. If the replicated responses are fairly uniform among all test conditions with the exception of one or two whose responses vary widely, check to verify the proper settings of these test conditions. Perhaps these test conditions are much more difficult to set up accurately each time.

   j. Uneven residual variability. When several test conditions have widely different distributions for their respective residuals, a number of things might be at work. Possible causes include inaccurate measurement method, missing interaction effects in the model, and/or the presence of dispersion effects.

18.4

Checking the Assumption of Common Variance of the Responses

Several of the residual plots described previously either directly or indirectly focus on checking to see if the variability of the responses is constant. We have repeatedly relied on the assumption of constant variance throughout our analysis of two-level factorials. When an experiment is replicated as in the case study above, we are afforded the opportunity to check the assumption of constant variance through a procedure that is a bit more rigorous than simply checking residual plots.

Recall from Table 18.1 that variance estimates for the response, door parallelism, were obtained for all 16 of the unique test conditions. These variance estimates are repeated in Table 18.4.

At this point, we might ask the question: Is it reasonable to expect that all these sample variances are estimates of the same common variance? In other words, "Is the variation among these sample variances greater than would be expected if only random error variation was at work?" To answer this question, we must examine how sample variances are distributed (i.e., arise in a frequency sense due to chance variation forces alone).

Sampling Distribution of the Sample Variance, s^2

Up to this point, we have concerned ourselves with evaluating hypotheses concerning the true mean of a distribution. For example, the true mean foundry workers' wage rate is \$16, the true effect value for a factorial experiment is 0, and so on. Hypotheses may also be made concerning the true variance of a distribution. The procedure to test a hypothesis concerning the variance is exactly the same as that described previously for the mean.

Let us assume that a set of individual measurements is normally distributed, and that the variance of these individuals, σ_y^2, is known or hypothesized to be a specific value. If a sample of size n is drawn from this distribution and is used to calculate the sample variance, s_y^2, the quantity

$$\frac{(n-1)s_y^2}{\sigma_y^2} \tag{18.5}$$

is distributed according to the *chi-square* (χ^2) *distribution*. The chi-square distribution defines the distribution of sample variances and is dependent on the degrees of freedom, $v = n - 1$, used to calculate the sample variance. The probability density functions for several chi-square distributions (different val-

TABLE 18.4   Variance Estimates from Glove Box Door Alignment Study

Test	s_i^2	Test	s_i^2
1	0.92480	9	0.03125
2	0.30420	10	0.15680
3	0.02420	11	0.00045
4	0.03380	12	0.00020
5	0.06845	13	0.21125
6	0.01125	14	0.85805
7	0.44180	15	0.16820
8	0.00245	16	0.00180

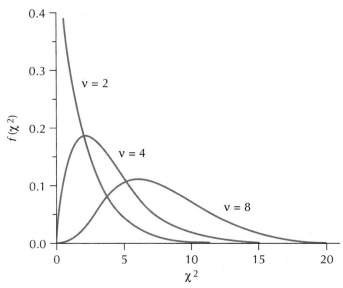

Figure 18.11 Chi-Square Distributions for $v = 2$, 4, and 8

ues for v) are illustrated in Fig. 18.11. Table A.5 provides a table of chi-square values beyond which the probability of the random variable χ^2 is α for various degrees of freedom.

Returning to the problem at hand, suppose that based on prior knowledge we believe that the parallelism response has a variance, σ_y^2, of 0.2. We may use the χ^2 distribution to judge whether any of the sample variances in Table 18.5 provides evidence that the true variance is not 0.2 for a given level of risk. Using the fact that the quantity $(n - 1)s_y^2/\sigma_y^2$ is χ^2 distributed, we may calculate χ^2 values for each of the 16 sample variances, as shown in Table 18.5. (Note that since two responses were used to obtain each of the sample variances, the value for n is 2.)

TABLE 18.5   Sample Variances and Calculated χ^2 Values

Test	s_i^2	χ^2	Test	s_i^2	χ^2
1	0.92480	4.62400	9	0.03125	0.15625
2	0.30420	1.52100	10	0.15680	0.78400
3	0.02420	0.12100	11	0.00045	0.00225
4	0.03380	0.16900	12	0.00020	0.00100
5	0.06845	0.34225	13	0.21125	1.05625
6	0.01125	0.05625	14	0.85805	4.29025
7	0.44180	2.20900	15	0.16820	0.84100
8	0.00245	0.01225	16	0.00180	0.00900

These calculated χ^2 values may be referred to the χ^2 distribution with $\nu = 1$ degree of freedom. Referring to Table A.5, it is seen that for 1 degree of freedom, χ^2 values of approximately 0.001 and 5.02 bracket 95% of all sampled χ^2 values. Since none of the calculated χ^2 values fall outside this range, we will not reject the hypothesis that the true common variance is 0.20 at a significance level of 0.05 (5%). It should be noted that this significance level places 0.025 (2.5%) in each tail of the chi-square distribution.

Test for the Homogeneity of the Variance

The following widely used procedure, known as *Bartlett's test*,[1] may be used to test for equality of the variance (i.e., the null hypothesis below):

$$H_0: \quad \sigma_1^2 = \sigma_2^2 = \cdots = \sigma_m^2 = \sigma_y^2$$
$$H_1: \quad \text{at least one } \sigma_i^2 \neq \sigma_j^2, \quad i \neq j.$$

The procedure computes a statistic whose sampling distribution is closely approximated by a chi-square distribution with $(m - 1)$ degrees of freedom. In this case $m = 2^k$ is the number of random samples of size n_i drawn from independent normal populations. For m sample variances calculated based on a total of N responses ($N = n_1 + n_2 + \cdots + n_m$), the test statistic is given by

$$\chi_{\text{calc}}^2 = \frac{M}{c}, \tag{18.6}$$

where

$$M = (N - m) \ln s_p^2 - \sum_{i=1}^{m} (n_i - 1) \ln s_i^2$$

$$c = 1 + \frac{1}{3(m - 1)} \left[\left(\sum_{i=1}^{m} \frac{1}{n_i - 1} \right) - \frac{1}{N - m} \right]$$

$$s_p^2 = \frac{\sum_{i=1}^{m} (n_i - 1)s_i^2}{N - m}.$$

The value for M will be large if the sample variances, s_i^2, differ greatly in magnitude, and will be zero if all the sample variances are exactly equal. We will reject H_0 if χ_{calc}^2 is too large; that is, we reject H_0 if

$$\chi_{\text{calc}}^2 > \chi_{m-1,\alpha}^2,$$

where $\chi_{m-1,\alpha}^2$ places α in the upper tail of the chi-square distribution with $m - 1$ degrees of freedom. Bartlett's test is sensitive to the normality assumption and should not be applied in cases where the distribution of the responses is not normal. Furthermore, it has been suggested that the test should be viewed with some caution if any of the n_i are less than 4.

---

[1] M. S. Bartlett, "The Use of Transformations," *Biometrics*, Vol. 3, 1947, pp. 39–52.

We now apply Bartlett's test to the sample variances obtained for the glove box door alignment study to test the common variance assumption. Using the sample variances of Table 18.5, the following may be calculated:

$$s_p^2 = \frac{(2-1)0.92480 + (2-1)0.30420 + \cdots + (2-1)0.00180}{32-16} = 0.20243$$

$$M = (32-16)\ln(0.20243)$$
$$- [(2-1)\ln 0.92480 + (2-1)\ln 0.30420 + \cdots + (2-1)\ln 0.00180]$$
$$= 28.1695$$

$$c = 1 + \frac{1}{3(16-1)}\left(\frac{1}{2-1} + \frac{1}{2-1} + \cdots + \frac{1}{2-1} - \frac{1}{32-16}\right)$$
$$= 1.3542$$

and the test statistic is

$$\chi_{\text{calc}}^2 = \frac{28.1695}{1.3542} = 20.8016.$$

Since $\chi_{15,0.05}^2 = 25.0$, we cannot reject the null hypothesis that the variances are equal. It may be noted that since $n = 2$ for this example, this conclusion must be viewed with caution; however, the conclusion is consistent with the interpretations we made based on our study of the residual plots.

18.5

Using the Fitted Model for Quality Improvement

Recall that the objective of the glove box door alignment study was to find a fixturing method for the assembly of the door to the instrument panel so that the amount of off-parallelism would be within ± 1.5 millimeters. Therefore, we would like to find a solution that produces a zero off-parallelism on the average. To find such a solution, we will use the model that was developed previously for the parallelism response.

The fitted model we developed for the parallelism response is

$$\hat{y} = -0.327x_1 + 0.397x_2 + 0.319x_3.$$

All residual checks and modeling assumption checks have shown this model to be an adequate representation of the phenomenon under study. Some confirmatory tests were conducted and showed no reason to doubt the legitimacy of the model.

From the fitted model it is clear that the parallelism is a simple additive function of the linear effects of RH cowl movement (variable 1), brace attachment sequence (variable 2), and plenum gasket (variable 3). The levels for these variables that were used during the conduct of the experiments are summarized as follows:

Variable	Low Level (-1)	High Level ($+1$)
x_1: RH cowl movement	0 mm	-5 mm
x_2: Brace attachment sequence	Before	After
x_3: Plenum gasket	No	Yes

An examination of this table shows that while variable 1 is a continuous variable that can take on any value between -1 (0 millimeters) and $+1$ (-5 millimeters), variables 2 and 3 are discrete variables that can only take on values of -1 or $+1$. The four combinations of the levels for variables 2 and 3 were substituted into Eq. (18.4) (repeated above) to produce the following simplified equations:

$$1.\ x_2 = -1 \quad x_3 = -1 \quad \hat{y}_1 = -0.716 - 0.327x_1 \tag{18.7}$$
$$2.\ x_2 = +1 \quad x_3 = -1 \quad \hat{y}_2 = 0.078 - 0.327x_1 \tag{18.8}$$
$$3.\ x_2 = -1 \quad x_3 = +1 \quad \hat{y}_3 = -0.078 - 0.327x_1 \tag{18.9}$$
$$4.\ x_2 = +1 \quad x_3 = +1 \quad \hat{y}_4 = 0.716 - 0.327x_1. \tag{18.10}$$

Recall that it is desired to select levels for the three important variables that produce a zero off-parallelism on the average, or equivalently, values for x_1, x_2, and x_3 that make $\hat{y} = 0$. It is clear that for both Eqs. (18.7) and (18.10), no value for x_1 on the interval from -1 to $+1$ is capable of satisfying this requirement. Potentially, values for x_1 outside this interval may give a zero off-parallelism on the average; however, it is not advisable to use the model to extrapolate outside the experimental range, at least without exhausting other alternatives first.

Equations (18.8) and (18.9) suggest that a value for variable 1 (between -1 and $+1$) does exist that produces zero off-parallelism on the average. The behavior of the predicted response as a function of x_1 is illustrated in Fig. 18.12 for these two equations. To find the value of x_1 that gives a predicted response of 0 for $x_2 = +1$ and $x_3 = -1$, Eq. (18.8) may be set equal to zero and x_1 solved for:

$$\hat{y} = 0.0 = 0.078 - 0.327x_1$$
$$x_1 = \frac{-0.078}{-0.327}$$
$$= 0.239.$$

To find the value of x_1 that gives a predicted response of 0 for $x_2 = -1$ and $x_3 = +1$, Eq. (18.9) may be set equal to zero and x_1 solved for:

$$\hat{y} = 0.0 = -0.078 - 0.327x_1$$
$$x_1 = \frac{0.078}{-0.327}$$
$$= -0.239.$$

It remains to transform these coded (-1, $+1$ in the experimental design) variable values back to the original variable units.

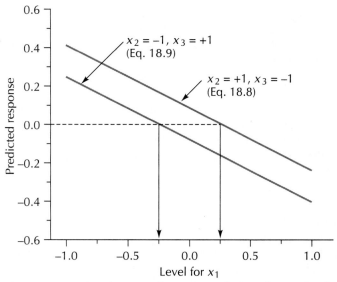

Figure 18.12 Identification of x_1 Values That Produce a Predicted Parallelism of 0.0 Millimeter

Transformations to and from the Coded Variable Space

To find the level of cowl movement corresponding to the values of x_1 calculated above, which are in coded form, recall that the low level of cowl movement was 0 millimeters ($x_1 = -1$) and the high level was -5 millimeters ($x_1 = +1$). To find the desired cowl movement values, we may use the following transformation:

$$x = \frac{r - (r_{-1} + r_{+1})/2}{(r_{+1} - r_{-1})/2} \tag{18.11}$$

or

$$r = \frac{r_{-1} + r_{+1}}{2} + x\left(\frac{r_{+1} - r_{-1}}{2}\right), \tag{18.12}$$

where x is the coded level associated with the actual level for a variable, r. It should be noted that r_{-1} is the low level for the variable (e.g., 0 millimeters for cowl movement) and r_{+1} is the high level for the variable (e.g., -5 millimeters for cowl movement). Using Eq. (18.12), we may obtain the desired values for cowl movement:

$$r = \frac{0 + (-5)}{2} + 0.239\left(\frac{-5 - 0}{2}\right) = -3.10 \text{ mm} \qquad \text{when } (x_2, x_3) = (+1, -1)$$

$$r = \frac{0 + (-5)}{2} + (-0.239)\left(\frac{-5 - 0}{2}\right) = -1.90 \text{ mm} \qquad \text{when } (x_2, x_3) = (-1, +1).$$

Thus if the center brace is attached before ($x_2 = -1$) and the gasket is present ($x_3 = +1$), a cowl movement position of -1.90 millimeters ($x_1 = -0.239$) gives a predicted off-parallelism value of 0. If the center brace is attached after ($x_2 = +1$) and the gasket is absent ($x_3 = -1$), a cowl movement position of -3.10 millimeters ($x_1 = 0.239$) also gives a predicted off-parallelism value of 0.

A final note of caution is in order at this point. It must be remembered that any solution derived from the fitted model is only a prediction. The final validation of such a prediction should be made through the results of confirmatory tests before action is taken (a system change is made) "on the floor."

Contour Plots

It may be useful to look at a more generalized model analysis. Often, two or more of the factors under study vary continuously and therefore a contour plot of the response surface is quite enlightening. Let us again consider the model developed previously for the parallelism response, but limit our discussion to the case where the center brace is attached after (i.e., $x_2 = +1$). For this situation, the predicted parallelism is given by the equation

$$\hat{y} = 0.397 - 0.327x_1 + 0.319x_3. \tag{18.13}$$

Previously, we considered a case where variable x_3 (plenum gasket presence) could only take on two levels: $x_3 = -1$ (no) and $x_3 = +1$ (yes). Let us now examine what happens if we redefine this variable as the plenum gasket thickness. The levels under which the experiment were performed now become 0 millimeters ($x_3 = -1$) and 2 millimeters ($x_3 = +1$), since the gasket used in the experiment had a thickness of 2 millimeters.

A convenient way to simultaneously examine the effect of variable 1 (cowl movement) and variable 3 (gasket thickness) on the predicted response is a contour plot. Figure 18.13 displays a contour plot for predicted parallelism based on Eq. (18.13). As can be seen from Fig. 18.13, there are contours associated with predicted responses of -0.2, 0.0, 0.2, 0.4, 0.6, 0.8, and 1.0 millimeters of parallelism. A given contour level essentially defines all the combinations of variables x_1 and x_3 that produce a predicted response equal to that response value.

Contours like those illustrated in Fig. 18.13 may be constructed by expressing Eq. (18.13) in terms of x_3:

$$x_3 = \frac{\hat{y} - 0.397 + 0.327x_1}{0.319}. \tag{18.14}$$

To determine combinations of x_1 and x_3 that produce a predicted response, $\hat{y}$, equal to 0.0 (zero off-parallel), we may calculate the values of x_3 corresponding to the following values of x_1 using Eq. (18.14) with $\hat{y} = 0$:

x_1	0.00	0.20	0.40	0.60	0.80	1.00
x_3	-1.24	-1.04	-0.83	-0.63	-0.42	-0.22

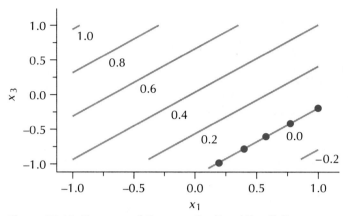

Figure 18.13 Contours of Constant Predicted Parallelism

These points were used to construct the $\hat{y} = 0$ contour shown in Fig. 18.13. The table indicates that values of $x_1 = 1.00$ and $x_3 = -0.22$ produce a predicted response of 0.0 for the case where $x_2 = +1$ (center brace is attached after). The actual variable levels corresponding to the coded levels for variables 1 and 3 may also be found using Eq. (18.12):

$$\text{cowl movement} = \frac{0 + (-5)}{2} + 1\left(\frac{-5-0}{2}\right) = -5.00 \text{ mm}$$

$$\text{gasket thickness} = \frac{0+2}{2} + (-0.22)\left(\frac{2-0}{2}\right) = 0.78 \text{ mm}.$$

Thus a cowl movement of -5.00 millimeters and a gasket thickness of 0.78 millimeter produce a predicted off-parallelism value of zero on the average. Many other solutions are possible. It should be noted again, however, that any solution derived from the fitted model is only a prediction, and final validation of any prediction should be made through the results of confirmatory tests.

18.6

Another Case Study: Surface Finish of a Machined Part

In this chapter we have introduced and exercised a four-stage model building procedure using two-level factorial designs. Once a model has been developed and checked for potential inadequacies using this procedure, it may then be used for the purposes of quality and productivity design and improvement, as was seen in the example above. In this section we examine another case study concerned with finding variable settings that produce a desired surface finish for a machined part.

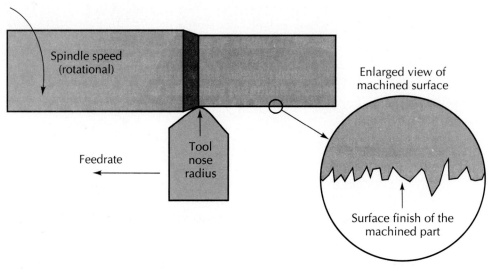

Figure 18.14 Variables Considered in the Surface Finish Case Study

Background of Surface Finish Case Study

A manufacturer was concerned with the surface finish of a cylindrical part produced through a single-point turning machining operation. The specifications for the part called for a nominal surface finish of 100 microinches, and although the process producing the parts had been stabilized through the use of $\overline{X}$ and R charts, the average surface finish of the parts produced by the process was approximately 120 microinches. To identify how the process could be centered at the target value of 100 microinches, a study team was formed to examine the problem.

The study team identified three variables of interest: spindle speed, feedrate, and tool nose radius. These variables are defined in Fig. 18.14. The study team decided to perform a 2^3 factorial design to study the effects of the three variables on, and ultimately to develop a model for, the machined surface finish. In selecting the levels for these three variables for the experiment, the study team selected levels spaced about the current operating conditions for the process (spindle speed = 2500 revolutions per minute, feedrate = 0.008 inches

TABLE 18.6 Variable Levels for the Surface Finish Experiment		
Variable	Low (−1) Level	High (+1) Level
x_1: Spindle speed (rev/min)	2400	2600
x_2: Feedrate (in./rev)	0.005	0.010
x_3: Nose radius (in.)	1/64	1/32

TABLE 18.7	2^3 Design and Results of the Surface Finish Experiment				
Test	x_1	x_2	x_3	Surface Finish (μin.)	Run Order
1	−	−	−	52	4
2	+	−	−	63	2
3	−	+	−	213	8
4	+	+	−	206	5
5	−	−	+	31	1
6	+	−	+	28	7
7	−	+	+	110	3
8	+	+	+	105	6

per revolution, and nose radius $= \frac{1}{64}$ inch). The high and low levels selected for the three variables for the experiment are summarized in Table 18.6. The results of the 2^3 experiment are given in Table 18.7.

Calculation of Effect Estimates and Determination of Significant Effects

Given the experimental results of Table 18.7, the effect estimates were computed using the calculation matrix and the algebraic method of effect calculation. The calculation matrix along with the response values is given in Table 18.8, and the computed effects are displayed in Table 18.9.

To identify those effect estimates that are distinguishable from the noise in the experimental environment, a normal probability plot may be constructed. The normal plot associated with the effects of Table 18.9 is shown in Fig. 18.15. In the figure a straight line has been drawn through four of the effect estimates that have values close to zero. This line passes near the (0, 50%) coordinate, and the scatter of the four smaller effect estimates about this line appears fairly

TABLE 18.8	Calculation Matrix of the Surface Finish Experiment							
Test	I	1	2	3	12	13	23	123	y_i
1	+	−	−	−	+	+	+	−	52
2	+	+	−	−	−	−	+	+	63
3	+	−	+	−	−	+	−	+	213
4	+	+	+	−	+	−	−	−	206
5	+	−	−	+	+	−	−	+	31
6	+	+	−	+	−	+	−	−	28
7	+	−	+	+	−	−	+	−	110
8	+	+	+	+	+	+	+	+	105

TABLE 18.9	Effect Estimates for the Surface Finish Experiment (microinches)		
Average =	101	E_{12} =	−5
E_1 =	−1	E_{13} =	−3
E_2 =	115	E_{23} =	−37
E_3 =	−65	E_{123} =	4

random. It is seen that the other three effect estimates fall well off the line. We therefore conclude that the main effects 2, 3, and the interaction effect 23, are important.

Model Development and Checking

For a 2^3 factorial design the response is assumed to be described by a model of the following form:

$$y = b_0 + b_1 x_1 + b_2 x_2 + b_3 x_3 + b_{12} x_1 x_2 + b_{13} x_1 x_3$$
$$+ b_{23} x_2 x_3 + b_{123} x_1 x_2 x_3 + \epsilon. \tag{18.15}$$

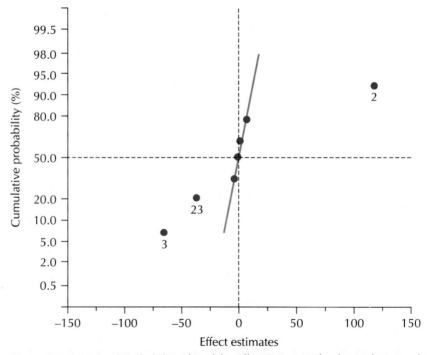

Figure 18.15 Normal Probability Plot of the Effect Estimates for the Surface Finish Experiment

TABLE 18.10		**Predicted Response and Model Residuals for Surface Finish Experiment**				
Test	x_1	x_2	x_3	y_i	$\hat{y}_i$	e_i
1	$-$	$-$	$-$	52	57.5	-5.5
2	$+$	$-$	$-$	63	57.5	5.5
3	$-$	$+$	$-$	213	209.5	3.5
4	$+$	$+$	$-$	206	209.5	-3.5
5	$-$	$-$	$+$	31	29.5	1.5
6	$+$	$-$	$+$	28	29.5	-1.5
7	$-$	$+$	$+$	110	107.5	2.5
8	$+$	$+$	$+$	105	107.5	-2.5

Since only the significant effects (model coefficients) need be included in the model, the fitted model for surface finish is given by

$$\hat{y} = \hat{b}_0 + \hat{b}_2 x_2 + \hat{b}_3 x_3 + \hat{b}_{23} x_2 x_3, \tag{18.16}$$

and substituting the values for $\hat{b}_i = E_i/2$ and $\hat{b}_0 = $ average response, we obtain.

$$\hat{y} = 101 + 57.5 x_2 - 32.5 x_3 - 18.5 x_2 x_3.$$

With a model for surface finish now developed, it must be checked for any potential inadequacies through study of the model residuals. Table 18.10 displays the measured and predicted responses along with the model residuals for each of the tests. The residuals of Table 18.10 should be randomly clustered about zero, and if plotted as a function of the test order, $\hat{y}$, x_1, x_2, and x_3 should show no trends/patterns, or in general, any structure whatsoever. Figures 18.16 to 18.21 display these residual plots. An examination of these plots shows no obvious structure remaining in the residuals, and thus we may conclude that our model, Eq. (18.16), has no apparent inadequacies.

Using the Model for Quality Improvement

With a prediction model now developed and checked for any potential inadequacies, the study team next turned its attention to using the model to find a solution to the problem at hand. In short, it was desired to find values for spindle speed, feedrate, and nose radius that center the turning process at a surface finish of 100 microinches, on the average. In terms of the fitted model, the team wanted to know values of x_1, x_2, and x_3 that produce a predicted surface finish of 100 microinches. Since the prediction model of Eq. (18.16) does not depend on the level of the spindle speed (x_1), values for x_2 and x_3 are sought that satisfy Eq. (18.16) when $\hat{y} = 100$.

Solving Eq. (18.16) for x_3 gives

$$x_3 = \frac{\hat{y} - 101 - 57.5 x_2}{-32.5 - 18.5 x_2}. \tag{18.17}$$

(Continued on page 643.)

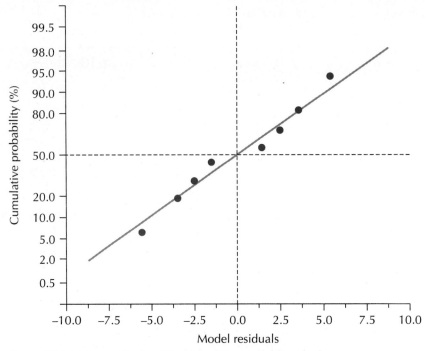

Figure 18.16 Normal Probability Plot of the Model Residuals

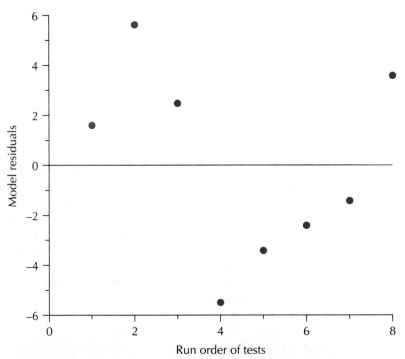

Figure 18.17 Residuals Plotted Against the Run Order of the Tests

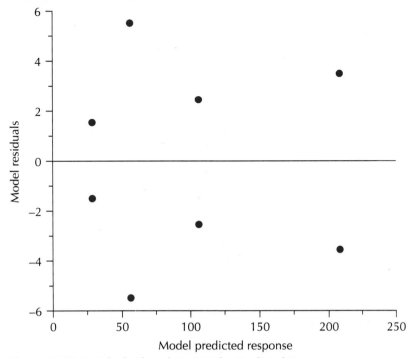

Figure 18.18 Residuals Plotted Against the Predicted Responses

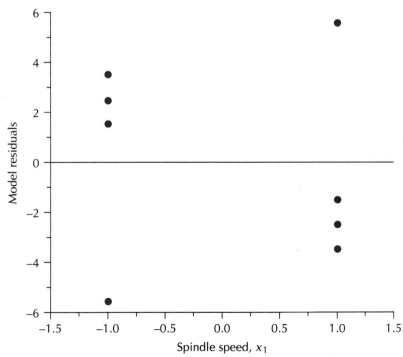

Figure 18.19 Residuals Versus Spindle Speed Level: 2400 Revolutions per Minute (-1), 2600 Revolutions per Minute $(+1)$

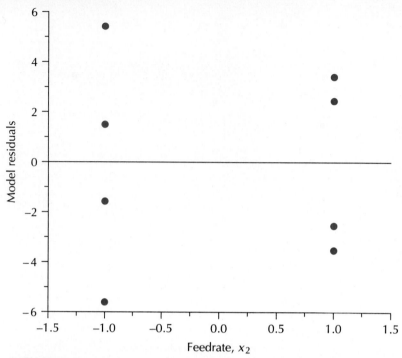

Figure 18.20 Residuals Versus Feed Rate Level: 0.005 Inch per Revolution (-1), 0.010 Inch per Revolution ($+1$)

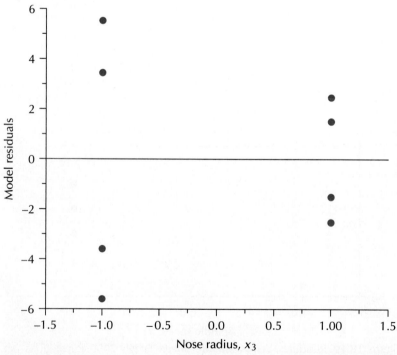

Figure 18.21 Residuals Versus Nose Radius Level: $\frac{1}{64}$ Inch (-1), $\frac{1}{32}$ Inch ($+1$)

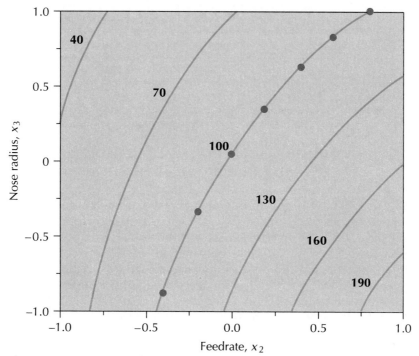

Figure 18.22 Contours of Constant Predicted Surface Finish in Microinches

Using Eq. (18.17), we find that contours of constant predicted surface finish may be constructed as a function of x_2 and x_3. Such a contour plot was developed by the study team and is shown in Fig. 18.22. Table 18.11 shows some of the contour construction calculations for a contour value of 100 microinches.

An examination of Fig. 18.22 shows that there are a number of combinations of variables x_2 and x_3 that produce a predicted surface finish of 100 microinches. Based on this contour plot, the study team decided to select the largest value for x_2 (feedrate), corresponding to an x_3 (nose radius) value of $+1$. The reasoning behind the selection of this particular x_2, x_3 combination was that in addition to it producing the predicted surface finish of 100 microinches, the larger value for x_2 would also lead to higher productivity. The same desire for higher productivity led to selection of the level for variable x_1 (i.e., $x_1 = +1$). In summary, therefore, the following levels were selected for x_1, x_2, and x_3:

$$x_1 = +1 \qquad \text{spindle speed} = 2600 \text{ rev/min}$$
$$x_2 = +0.8 \qquad \text{feedrate} = 0.0095 \text{ in./rev}$$
$$x_3 = +1 \qquad \text{nose radius} = \tfrac{1}{32} \text{ in.}$$

It may be noted that when these conditions were applied to the turning process in confirmatory tests, the process was indeed recentered at approximately 100 microinches. Furthermore, the selected conditions provided for a very substantial increase in the productivity from the original operating conditions.

TABLE 18.11	Contour Calculations for $\hat{y} = 100$ microinches									
x_2	-1.0	-0.8	-0.6	-0.4	-0.2	0.0	0.2	0.4	0.6	0.8	1.0
x_3	-4.036	-2.542	-1.565	-0.876	-0.365	-0.031	0.345	0.601	0.814	0.994	1.147

18.7
Simple Decomposition of the Variation in Data from an Experiment

We have emphasized throughout this book the importance of identifying sources of variation and attributing them to certain causal forces. *Analysis of variance* (ANOVA) is a widely used technique for decomposing the total variation in a set of data into sources of variation, each of which is considered to be important if it contributes a significant amount of variation to the experimental results. We will use this technique to judge effect importance and to identify the effects to be included in a model for the response, or simply to communicate to others the extent to which certain factors account for the structural variation in the data. The amount of variation accounted for by a fitted model, for example, can then be used as a measure of the model's performance or capability or the lack of good fit or adequacy. First, however, we must introduce some basic concepts that will be needed later for this analysis of variance procedure.

To introduce several of the concepts associated with an analysis of variance, we will use the data from a replicated 2^2 factorial design that is displayed in Table 18.12. From the table we may observe that there are $m = 2^2 = 4$ unique test conditions, with $n = 5$ runs at each test condition. The average response for each test condition, $\bar{y}_i$, is also displayed in the table, and is given by

$$\bar{y}_i = \frac{\sum_{j=1}^{n} y_{ij}}{n},$$

where i is the index associated with the test number and j is the index associated with the run number within a test. The overall or grand average of all $nm = 20$ responses may be calculated as

TABLE 18.12	Data for Replicated 2^2 Factorial Design							
Test	x_1	x_2	y_{i1}	y_{i2}	y_{i3}	y_{i4}	y_{i5}	$\bar{y}_i$
1	-1	-1	11	7	10	15	7	10
2	$+1$	-1	48	43	52	55	47	49
3	-1	$+1$	31	24	27	23	20	25
4	$+1$	$+1$	37	33	34	37	34	35

$$\bar{\bar{y}} = \frac{\sum_{i=1}^{m}\sum_{j=1}^{n} y_{ij}}{nm} = 29.75.$$

The total variation in the data set presented in Table 18.12 is given by

$$SS(total) = \sum_{i=1}^{m}\sum_{j=1}^{n} y_{ij}^2 = 21{,}969.$$

We will refer to this quantity as the total sum of squares of the data, SS(total).

As suggested at the outset, it may be desirable to decompose the total variation into several sources identified by the experimental design. For now we will consider decomposing the total variation into sources due to the mean, test-to-test (between test) variation, and variation within a test condition. For such a decomposition, the response may be represented as

$$y_{ij} = \bar{\bar{y}} + (\bar{y}_i - \bar{\bar{y}}) + (y_{ij} - \bar{y}_i). \qquad (18.18)$$

      ↑   ↑ Within-test conditions

       Between-test conditions

It should be noted that for this equation, the quantity on the right-hand side of the equation simplifies to the quantity on the left-hand side.

To see the decomposition in a variation sense, both sides of Eq. (18.18) are squared and summed over all runs (j) and tests (i) to obtain

$$\sum_{i=1}^{m}\sum_{j=1}^{n} y_{ij}^2 = \sum_{i=1}^{m}\sum_{j=1}^{n} \bar{\bar{y}}^2 + \sum_{i=1}^{m}\sum_{j=1}^{n} (\bar{y}_i - \bar{\bar{y}})^2 + \sum_{i=1}^{m}\sum_{j=1}^{n} (y_{ij} - \bar{y}_i)^2$$

$$+ 2\sum_{i=1}^{m}\sum_{j=1}^{n} \bar{\bar{y}}(\bar{y}_i - \bar{\bar{y}}) + 2\sum_{i=1}^{m}\sum_{j=1}^{n} \bar{\bar{y}}(y_{ij} - \bar{y}_i)$$

$$+ 2\sum_{i=1}^{m}\sum_{j=1}^{n} (\bar{y}_i - \bar{\bar{y}})(y_{ij} - \bar{y}_i).$$

The cross-product terms drop from this expression, since

$$\sum_{i=1}^{m}\sum_{j=1}^{n} \bar{\bar{y}}(\bar{y}_i - \bar{\bar{y}}) = n\bar{\bar{y}}\sum_{i=1}^{m}(\bar{y}_i - \bar{\bar{y}}) = n\bar{\bar{y}}\left(\sum_{i=1}^{m}(\bar{y}_i) - m\bar{\bar{y}}\right)$$

$$= n\bar{\bar{y}}(m\bar{\bar{y}} - m\bar{\bar{y}}) = 0$$

$$\sum_{i=1}^{m}\sum_{j=1}^{n} \bar{\bar{y}}(y_{ij} - \bar{y}_i) = \bar{\bar{y}}\sum_{i=1}^{m}\sum_{j=1}^{n}(y_{ij} - \bar{y}_i) = \bar{\bar{y}}\sum_{i=1}^{m}\left(\sum_{j=1}^{n}(y_{ij}) - n\bar{y}_i\right)$$

$$= \bar{\bar{y}}\sum_{i=1}^{m}(n\bar{y}_i - n\bar{y}_i) = 0$$

$$\sum_{i=1}^{m}\sum_{j=1}^{n}(\bar{y}_i - \bar{\bar{y}})(y_{ij} - \bar{y}_i) = \sum_{i=1}^{m}(\bar{y}_i - \bar{\bar{y}})\left(\sum_{j=1}^{n}(y_{ij}) - n\bar{y}_i\right)$$

$$= \sum_{i=1}^{m}(\bar{y}_i - \bar{\bar{y}})(n\bar{y}_i - n\bar{y}_i) = 0.$$

We therefore obtain the following equation, which expresses the total variation in the data, SS(total), as a function of the variation due to the mean, SS(mean); the variation between tests, SS(between tests); and the variation within tests, SS(within tests):

$$\sum_{i=1}^{m} \sum_{j=1}^{n} y_{ij}^2 = nm\bar{\bar{y}}^2 \qquad + n\sum_{i=1}^{m} (\bar{y}_i - \bar{\bar{y}})^2 \qquad + \sum_{i=1}^{m} \sum_{j=1}^{n} (y_{ij} - \bar{y}_i)^2 \qquad (18.19)$$

SS(total) = SS(mean) + SS(between tests) + SS(within tests).

Based on our data, we may numerically evaluate the three terms on the right-hand side of Eq. (18.19):

$$\text{SS(mean)} = nm\bar{\bar{y}}^2 = (5)(4)(29.75)^2 = 17{,}701.25$$

$$\text{SS(between tests)} = n\sum_{i=1}^{m} (\bar{y}_i - \bar{\bar{y}})^2$$
$$= 5[(10 - 29.75)^2 + (49 - 29.75)^2 + (25 - 29.75)^2$$
$$+ (35 - 29.75)^2]$$
$$= 4053.75$$

$$\text{SS(within tests)} = \sum_{i=1}^{m} \sum_{j=1}^{n} (y_{ij} - \bar{y}_i)^2$$
$$= (11 - 10)^2 + (7 - 10)^2 + (10 - 10)^2$$
$$+ (15 - 10)^2 + \cdots + (37 - 35)^2 + (34 - 35)^2$$
$$= 214.$$

As expected, the sum of the three sources of variation (17,701.25 + 4053.75 + 214 = 21,969) is equal to the total variation in the data. A simple but meaningful interpretation of all of this is that representing all of the data by a grand average $\bar{\bar{y}}$ accounts for 80% of the variation in the data. The fact that test conditions vary due to the changes in the two factors accounts for another 19% of the total variation. Within-test-condition variation (experimental error) accounts for only 1% of the total variation in the data.

18.8

Formalization of the Analysis of Variance Method

When ANOVA is used to examine the relative importance/significance of variable effects from a designed experiment, a more formalized methodology should be invoked. The following sets down that methodology.

Estimating the Variation of the Response Using Within- and Between-Test-Condition Variation

We have seen previously that an estimate of the true variance of a response, σ_y^2, may be obtained for a replicated two-level factorial experiment by pooling the sample variances for each test condition. When each of the m unique test conditions has n runs, this pooled variance may be expressed as

$$s_p^2 = \frac{s_1^2 + s_2^2 + \cdots + s_m^2}{m} = \frac{\sum\limits_{i=1}^{m} s_i^2}{m} = \frac{\sum\limits_{i=1}^{m} \left[\sum\limits_{j=1}^{n} (y_{ij} - \bar{y}_i)^2/(n-1) \right]}{m}$$

or

$$s_p^2 = \frac{\sum\limits_{i=1}^{m} \sum\limits_{j=1}^{n} (\bar{y}_{ij} - \bar{y}_i)^2}{(n-1)m} = \frac{\text{SS(within tests)}}{(n-1)m}.$$

Thus the pooled variance, an estimate of the true variance of the response, σ_y^2, may be obtained by dividing the within-tests sum of squares by the product of $(n-1)$ and m. This product, the denominator in the equation above, is often referred to as the *within-tests degrees of freedom*, or DOF(within tests). For our example the pooled variance is

$$s_p^2 = \frac{\text{SS(within tests)}}{\text{DOF(within tests)}} = \frac{214}{(5-1)\,4} = 13.375 = \text{MS(within tests)}.$$

This quantity is often referred to as the *mean square within tests* or MS(within tests). Here, this estimate of the variance is based on 16 degrees of freedom.

In summing the within-test sums of squares across the m unique test conditions to obtain the MS(within tests), we assume that the true variance of the response, σ_y^2, is the same for each test. Again, this assumption may be tested using Bartlett's test as described earlier in this chapter.

Let us, for now, also assume that the true mean for each test condition is the same. That is, we assume that changes in the factor levels do not lead to real changes in the response. Under this assumption, we may calculate another estimate of σ_y^2 using the deviation of the m test averages from the grand average, $\bar{\bar{y}}$.

We can consider the $\bar{y}_i$'s for a factorial experiment as a sample of size m (m unique test conditions) whose mean is $\bar{\bar{y}}$. The sample variance of these averages is then

$$s_{\bar{y}}^2 = \frac{\sum\limits_{i=1}^{m} (\bar{y}_i - \bar{\bar{y}})^2}{m-1}.$$

Since each $\bar{y}_i$ is the mean of n individual observations, we can obtain another estimate of σ_y^2 by multiplying both sides of the above equation by n:

$$ns_{\bar{y}}^2 = \frac{n \sum\limits_{i=1}^{m} (\bar{y}_i - \bar{\bar{y}})^2}{m-1}$$

$$= \frac{\text{SS(between tests)}}{\text{DOF(between tests)}} = \text{MS(between tests)}.$$

As can be seen, the foregoing estimate for the variance is equivalent to dividing the between-tests sum of squares by the between-tests degrees of freedom. This estimate of the variance is referred to as the mean square between tests. For our continuing example, the MS(between tests) is

$$MS(\text{between tests}) = \frac{SS(\text{between tests})}{DOF(\text{between tests})} = \frac{4053.75}{4 - 1} = 1351.25.$$

This estimate of the variance is based on 3 degrees of freedom.

So under the assumption that the true mean and variance of the response are the same for each test condition, we have two estimates of the variance, MS(within tests) = 13.375 having 16 degrees of freedom and MS(between tests) = 1351.25 having 3 degrees of freedom. If our assumption is true, these two estimates of the response variance should be close in magnitude. If only the assumption about the mean is not true (i.e., the true mean is different for at least one of the test conditions), the estimate of the variance based on differences between the test averages and the grand average, MS(between tests), will be inflated. For the example being considered here, it appears that the two sample variances are not very close in magnitude, and thus the MS(between tests) has been inflated by differences among the true means for the test conditions. To examine this more carefully, we must consider another statistical distribution.

Distribution of the Ratio of Two Sample Variances

Let s_1^2 and s_2^2 be the variances of two independent samples of size n_1 and n_2, respectively, taken from two normal populations having a common variance σ_y^2. Under this condition, the quantity s_1^2/s_2^2 follows the F distribution. The F distribution is generally a skewed distribution similar to the shape of a chi-square distribution. An F distribution is defined by the two parameters, v_1 and v_2, which are in this case the degrees of freedom associated with s_1^2 and s_2^2, respectively.

Figure 18.23 shows some typical F distributions where $F_{v_1, v_2, 1-\alpha}$ is the critical value of F such that probability is α that F is equal to or greater than $F_{v_1, v_2, 1-\alpha}$. Table A.6 gives the critical F values for various combinations of v_1 and v_2 and several values for α. In using these tables it is noted that v_1 always refers to the degrees of freedom associated with the sample variance in the numerator of the F ratio.

Under the null hypothesis that the true means for each test condition are the same, we have two estimates for the common variance σ_y^2. For our example, the ratio of these estimates may be taken and compared to the appropriate F distribution:

$$F_{\text{calc}} = \frac{MS(\text{between tests})}{MS(\text{within tests})} = \frac{1351.25}{13.375} = 101.03.$$

This calculated F value, under our null hypothesis, should follow an F distribution with 3 and 16 degrees of freedom, where $3 = (m - 1)$ is the degrees of freedom for the numerator, and $16 = m(n - 1)$ is the degrees of freedom for

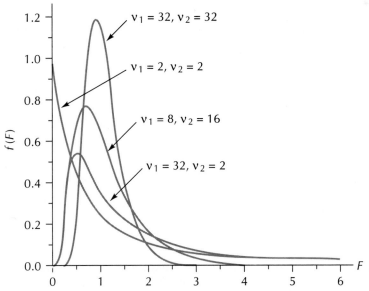

Figure 18.23 F Distributions for ν_1, $\nu_2 = (32, 32), (2, 2), (8, 16),$ and $(32, 2)$

the denominator. Referring to Table A.6, it is observed that $F_{3,16,0.99} = 5.29$. Therefore, since our calculated F value of 101.03 far exceeds the critical value of F, we must reject our null hypothesis that the true means are the same for each test condition. As suspected, differences between the true means have inflated the variance estimate obtained between tests, MS(between tests). In rejecting H_0, we say that the true means are not the same for at least some of the test conditions. It should be noted that the MS(within tests) is in general an unbiased estimate for the variance, since it is based on only the variation present within the test conditions due to replication. For this reason, the within-test variation is often termed the *pure error variation*.

ANOVA Table Construction

The information calculated during an analysis of variance procedure is often summarized in an analysis of variance (ANOVA) table. Such a table succinctly presents all pertinent information in a nicely structured format. The ANOVA table for our replicated 2^2 factorial experiment is shown in Table 18.13.

In constructing such an ANOVA table, the following points should be noted:

1. The sum of squares associated with the various sources must add up to the total sum of squares.
2. The degrees of freedom associated with the various sources must add up to the total degrees of freedom.

TABLE 18.13   ANOVA Table for Replicated 2^2 Factorial Design

Sources of Variation	Sum of Squares	Degrees of Freedom	Mean Square	Calculated F Value
Mean	17,701.25	1	17,701.250	1,323.46
Between tests	4,053.75	3	1,351.250	101.03
Pure error	214.00	16	13.375	
Total	21,969.00	20		

3. The value in the mean square column is obtained by dividing the sum of squares by the degrees of freedom. A mean square value for the "total" is typically not calculated.
4. The F value is calculated by dividing any other mean square by the "pure error" mean square.

Note that in Table 18.13, mean square and F values have been calculated for the mean or grand average. Under the hypothesis/assumption that the true mean is zero, this mean square value is also an estimate of the true variance of the response. To test this hypothesis, the ratio of the MS(mean) and the MS(pure error) values may be compared to an F distribution with 1 and 16 degrees of freedom. Since the calculated F value of 1323.46 is far greater than the critical F value from Table A.6 ($F_{1,16,0.99} = 8.53$), we conclude that the mean is different from zero (i.e., the null hypothesis is rejected).

The equations used to obtain an ANOVA table like that displayed above for a replicated two-level factorial design are summarized in Table 18.14. Clearly, the equations of Table 18.14 are quite general and apply to situations where more than two levels of each factor are present.

TABLE 18.14   ANOVA Table Equations

Sources of Variation	Sum of Squares (SS)	Degrees of Freedom (DOF)	Mean Square (MS)	F Ratio
Mean	$nm\bar{\bar{y}}^2$	1	SS (Mean)/ DOF (Mean)	MS (Mean)/ MS (Pure error)
Between tests	$n\sum_{i=1}^{m}(\bar{y}_i - \bar{\bar{y}})^2$	$m - 1$	SS (Between tests)/ DOF (Between tests)	MS (Between tests)/ MS (Pure error)
Pure error (Within tests)	$\sum_{i=1}^{m}\sum_{j=1}^{n}(y_{ij} - \bar{y}_i)^2$	$m(n - 1)$	SS (Pure error)/ DOF (Pure error)	
Total	$\sum_{i=1}^{m}\sum_{j=1}^{n}y_{ij}^2$	mn		

18.9

Use of Analysis of Variance for Two-Level Factorial Design Effect Evaluation

In Sections 18.7 and 18.8 the analysis of variance concept was introduced and used to decompose the total variation of a set of data into various sources. The F distribution was then introduced and used to judge if the variation accounted for by a specific source was more than would arise simply due to the replication or pure error variation. We will now apply this ANOVA procedure to study models developed from two-level factorial designs.

Replicated 2^2 Factorial Design Revisited

Let us once again consider the replicated 2^2 factorial design presented in Table 18.12.

Test	x_1	x_2	y_{i1}	y_{i2}	y_{i3}	y_{i4}	y_{i5}	$\bar{y}_i$
1	-1	-1	11	7	10	15	7	10
2	$+1$	-1	48	43	52	55	47	49
3	-1	$+1$	31	24	27	23	20	25
4	$+1$	$+1$	37	33	34	37	34	35

For this factorial experiment, the following effects may be calculated:

$$\text{average} = 29.75, \quad E_1 = 24.50, \quad E_2 = 0.50, \quad E_{12} = -14.50.$$

Based on these calculated effect estimates, we may develop a model to predict the response. For now, let us make no decision as to which of the model coefficients are important and which are not. Let us initially consider a model with all possible terms. Under this condition, the prediction equation becomes

$$\hat{y} = \hat{b}_0 + \hat{b}_1 x_1 + \hat{b}_2 x_2 + \hat{b}_{12} x_1 x_2$$

or substituting for $\hat{b}$ the effect estimate, E divided by 2,

$$\hat{y} = 29.75 + 12.25 x_1 + 0.25 x_2 + -7.25 x_1 x_2.$$

Recall that $\hat{b}_0$ is the sample mean or average response.

Let us now consider a decomposition of the variability in the data captured by the model into components due to each effect. As we know, the predicted response for the ith test may be expressed as

$$\hat{y}_i = \hat{b}_0 + \hat{b}_1 x_{i1} + \hat{b}_2 x_{i2} + \hat{b}_{12} x_{i1} x_{i2}.$$

Squaring both sides of this equation, and summing over all runs, j, and tests, i, gives the following decomposition for the model variability, SS(model):

$$\sum_{i=1}^{m}\sum_{j=1}^{n}\hat{y}_i^2 = \sum_{i=1}^{m}\sum_{j=1}^{n}\hat{b}_0^2 + \sum_{i=1}^{m}\sum_{j=1}^{n}\hat{b}_1^2 x_{i1}^2$$

$$+ \sum_{i=1}^{m}\sum_{j=1}^{n}\hat{b}_2^2 x_{i2}^2 + \sum_{i=1}^{m}\sum_{j=1}^{n}\hat{b}_{12}^2 x_{i1}^2 x_{i2}^2$$

or

$$n\sum_{i=1}^{m}\hat{y}_i^2 = n\left(\sum_{i=1}^{m}\hat{b}_0^2 + \sum_{i=1}^{m}\hat{b}_1^2 x_{i1}^2 + \sum_{i=1}^{m}\hat{b}_2^2 x_{i2}^2 + \sum_{i=1}^{m}\hat{b}_{12}^2 x_{i1}^2 x_{i2}^2\right). \qquad (18.20)$$

It is noted that for the case of a 2^2 design (in fact, any 2^k design), each of the terms in the parentheses on the right-hand side of Eq. (18.20) can be simplified to the form $\hat{b}_i^2 m$. Further, since $\hat{b}_i = E_i/2$ for 2^k designs, then for the example at hand,

$$\text{SS(model)} = (nm)(\text{average})^2 + \frac{(nm)E_1^2}{4} + \frac{(nm)E_2^2}{4} + \frac{(nm)E_{12}^2}{4}$$

$$= \text{SS(mean)} + \text{SS}(E_1) + \text{SS}(E_2) + \text{SS}(E_{12}).$$

In general, for a 2^k design where each of the $m = 2^k$ test conditions is performed n times, the sum of squares due to each effect, E, is given by

$$\text{SS}(E) = \frac{(nm)E^2}{4}, \text{ with one degree of freedom.}$$

To evaluate the appropriateness of a postulated model given the estimated effects from an experiment, we must determine the significance of each effect in terms of its contribution to the total sum of squares of the data. Only those effects that account for a large amount of variation will be included in the final prediction (fitted) model; the remaining effects will be ignored because their variability is not any larger than would be expected just due to experimental error. To evaluate the magnitude of the variability described by a single effect, we will compare it to an estimate of the experimental error (response) variance.

To illustrate the use of ANOVA table for significance tests of the estimated effects and the prediction model, let us consider the replicated 2^2 factorial experiment that had the following effect estimates:

average = 29.75, E_1 = 24.50, E_2 = 0.50, E_{12} = −14.50.

Based on these effect estimates the ANOVA table of Table 18.15 was constructed. We will now refer to that portion of the total variation in the data not explained by the model as the residual (i.e., that which is "left over"). In general, the residual variation is composed of two parts, that due to pure error (replication) and that due to what we will call *lack of fit*. For now, this residual is composed solely of the pure error (replication) variation. Each of the calculated F values in the ANOVA table may be compared to an F distribution with 1 and 16 degrees of freedom. Since $F_{1,16,0.95} = 4.49$, it may be concluded that

TABLE 18.15	ANOVA Table for the Model That Includes All Terms for Replicated 2^2 Factorial Design			
Source of Variation	Sum of Squares	Degrees of Freedom	Mean Square	Calculated F Value
Mean	17,701.25	1	17,701.250	1,323.46
E_1	3,001.25	1	3,001.250	224.39
E_2	1.25	1	1.250	0.09
E_{12}	1,051.25	1	1,051.250	78.60
Residual				
Pure error	214.00	16	13.375	
Total	21,969.00	20		

the average, E_1, and E_{12} are statistically significant since their calculated F values exceed 4.49.

Significance Tests for Model Lack of Fit

The fact that the initially postulated model is fit to the data is no guarantee that the model will provide good predictions of the process behavior. If the model postulation is misguided, the predictions will be poor even if the experiment is well-designed and carried out. Earlier, we said that in diagnostically checking the model, two basic questions are addressed:

1. Are all the terms initially included in the model necessary?
2. Are the terms initially included in the model sufficient?

We have seen several ways that we can address the first of these two questions, both in Chapter 17 and in Chapter 18 to this point. The analysis of the residual plots has also given us some insights into the second question. We now will examine the second of these questions in a more rigorous fashion through the technique of analysis of variance.

In the last section, we concluded that there were 2 effects (E_1 and E_{12}) plus the average response that were statistically significant, that is, their true effects differed from zero. One effect, E_2, was judged not to be important. Consequently, the term associated with E_2 should not be present in the prediction model. Thus the fitted prediction model for the replicated 2^2 factorial is

$$\hat{y} = 29.75 + 12.25x_1 - 7.25\, x_1 x_2.$$

The predicted responses for the four tests based on this model are listed in Table 18.16.

We have shown earlier that the sum of squares due to a model is equal to the total of the sums of squares associated with each of the effects in the model.

TABLE 18.16	Data for Replicated 2^2 Factorial Design with Predicted Responses Using the Fitted Model							
Test	x_1	x_2	y_{i1}	y_{i2}	y_{i3}	y_{i4}	y_{i5}	$\bar{y}_i$	$\hat{y}$
1	-1	-1	11	7	10	15	7	10	10.25
2	$+1$	-1	48	43	52	55	47	49	49.25
3	-1	$+1$	31	24	27	23	20	25	24.75
4	$+1$	$+1$	37	33	34	37	34	35	34.75

For our example,

$$SS(\text{model}) = SS(\text{mean}) + SS(E_1) + SS(E_{12})$$
$$= 17{,}701.25 + 3001.25 + 1051.25 = 21{,}753.75.$$

Since the total sums of squares has been previously calculated to be

$$SS(\text{total}) = 21{,}969.00,$$

the SS(residual) may be calculated by subtraction, as can the degrees of freedom associated with the residual sum of squares,

$$SS(\text{residual}) = SS(\text{total}) - SS(\text{model})$$
$$= 21{,}969.00 - 21{,}753.75$$
$$= 215.25$$
$$DOF(\text{residual}) = DOF(\text{total}) - DOF(\text{model})$$
$$= 20 - 3$$
$$= 17.$$

Alternatively, we could calculate the residual sum of squares, as before, as the sum of the squared deviations of the data from their model predictions,

$$SS(\text{residual}) = \sum_{i=1}^{m} \sum_{j=1}^{n} (y_{ij} - \hat{y}_i)^2$$
$$= (11 - 10.25)^2 + (7 - 10.25)^2 + \cdots + (37 - 34.75)^2$$
$$+ (34 - 34.75)^2$$
$$= 215.25.$$

Evaluating the residual sum of squares based on this equation, we find that the SS(residual) is, again, equal to 215.25.

We now note that the residual sum of squares is really composed of two parts, the pure error sum of squares due to replication and the remainder of all of the variation in the data that is (1) not accounted for by the model and (2) that is not a result of the replication in the data. This latter component we will refer to as the "lack-of-fit" sum of squares. Since the pure error sum of squares has already been determined to be 214.00, then, by subtraction,

$$SS(\text{lack of fit}) = SS(\text{residual}) - SS(\text{pure error})$$
$$= 215.25 - 214.00$$
$$= 1.25.$$

All of the above can be laid out in an ANOVA table, as shown in Table 18.17.

The lack-of-fit variation will be indistinguishable from the pure error variation if the model is an adequate one. Therefore, a model will be considered inadequate (suffer from lack of fit) if the calculated F value for the ratio of the $MS(LOF)$ to the MS(pure error) exceeds the tabulated value for F with the appropriate degrees of freedom,

$$F_{LOF} = \frac{MS(LOF)}{MS(\text{pure error})}$$

$$= \frac{1.25}{13.375}$$

$$= 0.09.$$

Since the calculated F value for the lack of fit, $F_{LOF} = 0.09$, is less than $F_{1,16,0.95} = 4.49$, we have no evidence to suggest any significant lack of fit.

It should be noted that for this example, the lack-of-fit sum of squares is the sum of squares associated with the estimated main effect of variable 2, the term in the original model that was found to be insignificant,

$$SS(E_2) = \frac{(nm)E_2^2}{4}$$

$$= \frac{(4)(5)(0.5)^2}{4}$$

$$= 1.25.$$

TABLE 18.17   ANOVA for Checking the Model for the Replicated 2^2 Factorial Experiment

Source of Variation	Sum of Squares	Degrees of Freedom	Mean Square	Calculated F Value
Model	21,753.75	3		
Mean	17,701.25	1	17,701.250	1,323.46
Model − (mean)	4,052.50	2	2,026.250	151.50
Residual	215.25	17		
LOF	1.25	1	1.250	0.09
Pure error	214.00	16	13.375	
Total	21,969.00	20		

Once dropped from the model, such terms become part of the residual, that is, they constitute variation sources that cannot be distinguished from the noise/error in the experiment.

The purpose of checking the model for lack of fit is to reveal problems with the model-building process. In dealing with 2^k experiments that involve a larger number of variables, say four or more, we might simply assume that higher-order interactions can be neglected (third and higher, perhaps) from the outset and not even estimate these effects; that is, we may not include them in the calculation matrix. Subsequently, examination of the model lack of fit might reveal inadequacies that could be overcome if one or more of these interactions were added to the model. It may also be possible that the phenomenon under study may contain structure—say, a quadratic effect—that cannot be captured through the data by the associated model. Such will not be revealed through the lack-of-fit analysis. In this case, additional experimentation will be required to build an adequate model.

An Overall Measure of Model Performance

The fact that a model is significant (contains one or more important terms) and that the model does not suffer from lack of fit does not necessarily mean that the model is a good one. If the experimental environment is quite noisy or some important variables have been left out of the experiment, then it is possible that the portion of the variability in the data not explained by the model (the residual) could be large. To quantify this, we introduce a measure of the model's overall performance, a quantity called the *coefficient of determination*, denoted by R^2. This R^2 (which is not the square of a range) is the percentage of the total variability in the data that is accounted for by the model. The percent variation accounted for by the model is

$$R^2 = \frac{SS(\text{model})}{SS(\text{total})}. \tag{18.21}$$

For our replicated 2^2 factorial we obtain

$$R^2 = \frac{21{,}753.75}{21{,}969} = 0.9902 \text{ or } 99.02\%.$$

The R^2 value is given by the equation above; however, often it is more meaningful to adjust or correct the SS(model) and SS(total) for the mean. Hence we define an R^2_{cor} value given by

$$R^2_{\text{cor}} = \frac{SS(\text{model}) - SS(\text{mean})}{SS(\text{total}) - SS(\text{mean})}. \tag{18.22}$$

For our example, R^2_{cor} is, therefore,

$$R^2_{\text{cor}} = \frac{21{,}753.75 - 17{,}701.25}{21{,}969 - 17{,}701.25} = 0.9496 \text{ or } 94.96\%.$$

The interpretation of R^2_{cor} is that the fitted model (E_1 and E_{12}) explains about 95% of the variability in the responses about the average of 29.75.

18.10

Use of Analysis of Variance on the Glove Box Door Alignment Study

Let us return to the glove box door parallelism experiment and examine how the ANOVA procedure described in the preceding sections may be applied to this problem. The designed experiment and responses are given again in Table 18.18. The calculated effect estimates for this experiment are also summarized below.

average $= -0.087$

$E_1 = -0.654$	$E_{13} = -0.117$	$E_{123} = 0.172$
$E_2 = 0.794$	$E_{14} = -0.031$	$E_{124} = 0.101$
$E_3 = 0.638$	$E_{23} = -0.191$	$E_{134} = -0.138$
$E_4 = 0.322$	$E_{24} = -0.154$	$E_{234} = -0.104$
$E_{12} = 0.147$	$E_{34} = 0.009$	$E_{1234} = 0.121$

TABLE 18.18 2^4 Design and Results of the Glove Box Door Experiment

Test	x_1	x_2	x_3	x_4	Run 1, y_{i1}	Run 2, y_{i2}	Average, $\bar{y}_i$
1	−	−	−	−	−1.44	−0.08	−0.760
2	+	−	−	−	−1.79	−1.01	−1.400
3	−	+	−	−	0.39	0.17	0.280
4	+	+	−	−	−0.50	−0.24	−0.370
5	−	−	+	−	−0.20	0.17	−0.015
6	+	−	+	−	−0.79	−0.64	−0.715
7	−	+	+	−	1.22	0.28	0.750
8	+	+	+	−	0.21	0.28	0.245
9	−	−	−	+	−0.40	−0.65	−0.525
10	+	−	−	+	−0.63	−1.19	−0.910
11	−	+	−	+	0.47	0.44	0.455
12	+	+	−	+	−0.01	−0.03	−0.020
13	−	−	+	+	1.29	0.64	0.965
14	+	−	+	+	−1.17	0.14	−0.515
15	−	+	+	+	0.48	1.06	0.770
16	+	+	+	+	0.40	0.34	0.370

As we saw in Section 18.9 for replicated 2^k designs, the formulas for calculating the various sum of squares values are

$$\text{SS(mean)} = nm\hat{b}_0^2$$

$$\text{SS(effect)} = \frac{nmE^2}{4} \quad \text{for each of the } (m-1) \text{ effects}$$

$$\text{SS(pure error)} = m(n-1)s_p^2$$

$$\text{SS(total)} = \sum_{i=1}^{m} \sum_{j=1}^{n} y_{ij}^2,$$

where n is the number of repeat trials (replications) for each unique test condition, and m is equal to 2^k. It is assumed here that each unique test condition is repeated an equal (n) number of times.

Alternatively, the pure error sum of squares may be calculated by subtraction.

$$\text{SS(pure error)} = \text{SS(total)} - \text{SS(mean)} - [\text{all } (m-1) \text{ SS}(E)\text{'s}].$$

ANOVA Table

Based on the equations above and the replicated 2^4 design given in Table 18.18, the ANOVA table in Table 18.19 was constructed. The calculated F values in the ANOVA table under the null hypothesis that the true variable effects are zero are distributed according to the F distribution with 1 and 16 degrees of freedom. Since the calculated F values associated with E_1, E_2, and E_3 exceed the critical value for F at a 5% significance level ($F_{1,16,0.95} = 4.49$), we may conclude that their true mean effects are not zero. In other words, the variability accounted for by E_1, E_2, and E_3 is larger than what can be explained solely by the forces of experimental error. It is noted that this is the same conclusion that was reached based on using the confidence interval approach. Note again that the mean response is not different from zero.

An alternative method to check the significance of the effects is to compare the MS(E)'s against a critical value of the mean square that can be obtained as the product of MS(pure error) and the critical F value. This approach requires fewer calculations, in that F values need not be calculated for each effect. Since the critical F value ($F_{1,16,0.95}$) is 4.49, the critical MS value is given by

$$\text{MS(critical)} = \text{MS(pure error)}F_{1,16,0.95} = (0.2024)(4.49) = 0.9088.$$

Comparing the MS(E)'s with MS(critical), only MS(E_1), MS(E_2), and MS(E_3) exceed the critical mean square. As before, we conclude that the main effects of variables 1, 2, and 3 are the only effects that are statistically different from zero.

Checking for Model Adequacy and Lack of Fit

We have concluded that there are three effects (E_1, E_2, and E_3) that differ significantly from zero. Consequently, a model to predict the parallelism response need only contain terms associated with these effects. The model for the predicted response is therefore given by

$$\hat{y} = -0.327x_1 + 0.397x_2 + 0.319x_3. \tag{18.23}$$

The sum of squares accounted for by the model may be obtained by adding the sums of squares due to E_1, E_2, and E_3:

$$SS(model) = SS(E_1) + SS(E_2) + SS(E_3) = 3.42565 + 5.04825 + 3.25763$$
$$= 11.73153.$$

Based on this model, predictions may be made for each of the $2^4 = 16$ test conditions. Since the model residuals are the differences between the model predictions and measured responses, the residual sum of squares is given by (refer to Table 18.3 for the model predictions):

$$SS(residual) = \sum_{i=1}^{m} \sum_{j=1}^{n} (y_{ij} - \hat{y}_i)^2$$
$$= [-1.44 - (-0.389)]^2$$
$$+ [-0.080 - (-0.389)]^2 + \cdots + (0.400 - 0.389)^2$$
$$+ (0.340 - 0.389)^2$$
$$= 5.75597.$$

TABLE 18.19   ANOVA for the Effects of the Glove Box Alignment Study

Source of Variation	Sum of Squares	Degrees of Freedom	Mean Square	Calculated F Value
Mean	0.24325	1	0.24325	1.202
E_1	3.42565	1	3.42565	16.922
E_2	5.04825	1	5.04825	24.938
E_3	3.25763	1	3.25763	16.092
E_4	0.82883	1	0.82883	4.094
E_{12}	0.17258	1	0.17258	0.853
E_{13}	0.10928	1	0.10928	0.540
E_{14}	0.00750	1	0.00750	0.037
E_{23}	0.29070	1	0.29070	1.436
E_{24}	0.19065	1	0.19065	0.942
E_{34}	0.00070	1	0.00070	0.003
E_{123}	0.23633	1	0.23633	1.167
E_{124}	0.08100	1	0.08100	0.400
E_{134}	0.15263	1	0.15263	0.754
E_{234}	0.08715	1	0.08715	0.431
E_{1234}	0.11640	1	0.11640	0.575
Pure error	3.23895	16	0.20243	
Total	17.48750	32		

Alternatively, since the SS(model) + SS(residual) is equal to the total sum of squares [SS(total) = 17.4875] the SS(residual) could have been obtained by subtraction.

The residual sum of squares is the variability in the data not described by the model. Part of this residual variability is due to the pure error sum of squares, SS(pure error) = 3.23895, with the remainder equal to the lack-of-fit sum of squares, SS(LOF). The lack-of-fit sum of squares, SS(LOF), may be calculated by subtracting SS(pure error) from SS(residual) [i.e., SS(LOF) = 5.75597 − 3.23895 = 2.51702]. The SS(LOF) is in reality the total of the sums of squares of the 12 insignificant effects plus the sum of squares of the sample mean response. When we add up all these sums of squares from Table 18.19, we likewise get SS(LOF) = 2.51702 with 13 degrees of freedom.

We may now evaluate the performance of the fitted prediction model by checking to see if it has a significant amount of lack of fit. This is most conveniently done via the ANOVA table shown in Table 18.20. The critical value of F with 3 and 16 degrees of freedom at a 5% level for checking the overall model significance is 3.239, which is smaller than the F calculated for the model, shown in the last column of Table 18.20. This suggests that the fitted model is statistically significant in its explanation of the variation in the responses. To check for potential model lack of fit, we compare the calculated F for the LOF/pure error ratio against the critical F value with 13 and 16 degrees of freedom, at a 5% level. Since this critical F of 2.397 is greater than the calculated F of 0.956, we have no evidence to suggest any significant lack of fit.

We may also examine the performance of the fitted model by calculating the coefficient of determination, R^2,

$$R^2 = \frac{SS(model)}{SS(total)} = \frac{11.73153}{17.4875} = 0.6709 \text{ or } 67.09\%.$$

The interpretation of R^2 is that the fitted model explains about 67% of the variability in the responses. Approximately one-third of the response variability

TABLE 18.20 ANOVA for Checking the Model of Glove Box Door Parallelism				
Source of Variation	Sum of Squares	Degrees of Freedom	Mean Square	Calculated F Value
Model	11.73153	3	3.91051	19.318
Residual				
Lack of fit	2.51702	13	0.19362	0.956
Pure error	3.23893	16	0.20243	
Total	17.48750	32		

is not explained by the fitted model. This may be indicating that variables not considered in the original design (i.e., variables that are currently part of the noise structure) are important. Potentially, if these variables were added to the model, the model would describe more of the variability in the alignment response. Of course, it is also possible (and perhaps likely) that noise factors operating in the experimental environment are responsible for much of the unexplained variation. The use of SPC prior to the experiment or techniques such as blocking during the experiment could alleviate this problem.

Exercises

18.1. Describe the four stages associated with the model building process. For each stage discuss the specific task(s) and associated methods that may be used.

18.2. What two specific problems can arise with a model developed from any designed experiment? That is, given a fitted model, what specific types of inadequacies can this model have in terms of its ability to describe the phenomenon of interest?

18.3. A replicated 2^4 factorial design is performed, and a model developed based on the experimental results. List the types of residual plots that should be prepared to check the model. In each case, describe the patterns in the residual plot that would lead you to believe that the model is adequate.

18.4. A 2^3 factorial design was conducted to study the influence of temperature (x_1), pressure (x_2), and cycle time (x_3) on the occurrence of splay in an injection molding process. For each of the eight unique trials, 50 parts were made and the response observed was the number of incidences of the occurrence of splay on the part surface across all 50 parts. The design matrix and test responses are given in the table.

Test	1	2	3	y
1	−	−	−	12
2	+	−	−	15
3	−	+	−	24
4	+	+	−	17
5	−	−	+	24
6	+	−	+	16
7	−	+	+	24
8	+	+	+	28

a. Calculate all variable effects.
b. Construct a normal probability plot for the effect estimates. Which of the effects appear to be statistically significant?

   c. Write down the appropriate mathematical model, including only those terms that appear to be significant.

   d. Calculate the model residuals.

   e. Prepare all the residual plots necessary to check the model diagnostically. Comment.

18.5. A 2^3 factorial design was used to study the possible effects of temperature, time, and pressure on the yield of a chemical process. The data, including experimental results, are shown in the tables.

Variable	Low Level	High Level
1. Temperature (°F)	80	100
2. Time (min)	5	7
3. Pressure (psi)	120	140

Test	1	2	3	y
1	−	−	−	10
2	+	−	−	20
3	−	+	−	4
4	+	+	−	10
5	−	−	+	8
6	+	−	+	18
7	−	+	+	6
8	+	+	+	12

   a. Calculate the effects and construct a normal probability plot for the effect estimates. Which of the effects appear to be statistically significant?

   b. Write down the appropriate mathematical model, including only those terms that appear to be significant. Calculate the model residuals.

   c. Prepare all the residual plots necessary to check the model diagnostically.

   d. Prepare a contour plot for predicted yield that displays temperature and time on the horizontal and vertical axes, respectively, for a pressure of 130 psi.

   e. Within the range of the variables studied, where is the yield a maximum? What is the predicted yield at this point?

18.6. Consider again the 2^3 factorial design of Exercise 17.13.

   a. Calculate the variable effect estimates.

   b. Construct an ANOVA table to identify which of the effect estimates are statistically significant.

   c. Develop a model for the response based on the results of part (b). Construct an ANOVA table to see if the developed model suffers from lack of fit.

18.7. A chemical engineer performed the experiments shown in the first table, randomizing the order of the tests within each week. The response of interest is the chemical yield, and the run order is given in parentheses.

Test	Temperature	Catalyst	pH	Week 1	Week 2	Week 3
1	−	−	−	60.4 (5)	62.1 (10)	63.4 (19)
2	+	−	−	64.1 (8)	79.4 (11)	74.0 (23)
3	−	+	−	59.6 (1)	61.2 (13)	57.5 (17)
4	+	+	−	66.7 (3)	67.3 (16)	68.9 (22)
5	−	−	+	63.3 (7)	66.0 (9)	65.3 (24)
6	+	−	+	91.2 (4)	77.4 (14)	84.9 (18)
7	−	+	+	68.1 (6)	71.3 (12)	68.6 (21)
8	+	+	+	75.3 (2)	77.1 (15)	76.1 (20)

The variable levels are given in the second table.

Level	Temperature (°F)	Catalyst	pH
−	170	A	6
+	180	B	7

a. Calculate the effects.
b. Calculate the sample pooled variance, s_p^2, based on the replication within each test run, and use this estimate to determine the sample variance of an effect. Develop 95% confidence intervals for each effect estimate. Find the statistically significant effects.
c. Write down the appropriate mathematical model, including only those terms found to be significant. Calculate the model residuals, and prepare all the residual plots necessary to check the model diagnostically.
d. Prepare a contour plot for predicted yield that displays temperature and pH on the horizontal and vertical axes, respectively, for catalyst type B.
e. Within the range of the variables studied, where is the yield a maximum? What is the predicted yield at this point?

18.8. Consider again the data presented in Exercise 18.7, and use Bartlett's test to check the homogeneity of the variance of the response. Prepare a graphical representation of the experimental design, and graphically depict the result of Bartlett's test.

18.9. Recall the injection molding study of Chapter 16.

Variable	Unit	Low Level	High Level
Orifice size, O	mm	1.30	1.50
Pump setting, P		4.00	4.50
Isocynate temperature, T	°C	22	30

Test	x_1	x_2	x_3	Test Order	y
1	−1	−1	−1	6	1550
2	+1	−1	−1	8	1925
3	−1	+1	−1	1	2150
4	+1	+1	−1	2	2350
5	−1	−1	+1	5	1525
6	+1	−1	+1	3	1800
7	−1	+1	+1	4	2175
8	+1	+1	+1	7	2200

The calculated effect estimates for the impingement pressure (psi) response were

$$\text{average} = 1959.375 \quad E_{12} = -106.25$$
$$E_1 = 218.75 \quad E_{13} = -68.75$$
$$E_2 = 518.75 \quad E_{23} = 6.25$$
$$E_3 = -68.75 \quad E_{123} = -18.75.$$

a. Plot these effect estimates on normal probability paper, and postulate a model for the impingement pressure.
b. Construct a contour plot for the response surface over the range of variables tested.
c. If it is desired to control the isocynate impingement pressure to values of 1750 and 1950 psi for the next phase of the experiment (refer to Fig. 16.2), identify appropriate settings for the orifice size, pump setting, and isocynate temperature.

18.10. Referring to Exercise 18.9, suppose that the model proposed to describe the impingement pressure response was

$$\hat{y} = 1959.375 + 259.375x_2.$$

a. Calculate all the model residuals.
b. Construct all appropriate residual plots.
c. If the proposed model is found to be inadequate, propose a new model, and conduct the necessary model diagnosis procedures.

18.11. A 2^3 factorial experiment was designed to test the effect of three different variables on the ultimate tensile strength of a weld. The three variables and the levels used are shown along with the test results in the tables.

Variable	Low Level	High Level
1. Ambient temperature (°F)	20	40
2. Wind velocity (mph)	0	20
3. Bar size (in.)	1/2	3/4

Test	1	2	3	y_1	y_2
1	−	−	−	88.6	90.2
2	+	−	−	92.6	83.4
3	−	+	−	70.3	74.9
4	+	+	−	82.6	99.2
5	−	−	+	78.7	80.1
6	+	−	+	99.7	95.5
7	−	+	+	80.7	76.3
8	+	+	+	93.7	91.7

Use Bartlett's test to check the homogeneity of the variance of the response. Prepare a graphical representation of the experimental design, and graphically depict the result of Bartlett's test.

18.12. Consider again the 2^3 factorial design of Exercise 18.11. Assume that the following model form is postulated to describe the data given:

$$y = b_0 + b_1x_1 + b_2x_2 + b_3x_3 + \epsilon.$$

a. Estimate the variable effects and use them to estimate the coefficients in the model above. Construct an ANOVA table to determine which of the model coefficients are significant.

b. Construct an ANOVA table to see if the model developed suffers from lack of fit.

c. If the model does suffer from lack of fit, postulate a new model and perform a complete analysis of the new model.

18.13. A 2^3 factorial design was performed. Each test, or unique combination of the three variables, was performed three times. The variable levels and results of the tests are shown in the tables.

Variable	Low Level	High Level
1. Temperature (°C)	80	120
2. Pressure (psi)	50	70
3. Reaction time (min)	5	15

Test	x_1	x_2	x_3	Trial 1	Trial 2	Trial 3	Average
1	80	50	5	61.43	58.58	57.07	59.03
2	120	50	5	75.62	77.57	75.75	76.31
3	80	70	5	27.51	34.03	25.07	28.87
4	120	70	5	51.37	48.49	54.37	51.41
5	80	50	15	24.80	20.69	15.41	20.30
6	120	50	15	43.58	44.31	36.99	41.63
7	80	70	15	45.20	49.53	50.29	48.34
8	120	70	15	70.51	74.00	74.68	73.07

Use Bartlett's test to check the homogeneity of the variance of the response. Prepare a graphical representation of the experimental design, and graphically depict the result of Bartlett's test.

18.14. An in-depth examination of the experiment described in Exercise 18.13 revealed that the eight tests (24 trials) were actually conducted over a 2-week period. During the first week, tests 1, 4, 6, and 7 were performed, and during the second week tests 2, 3, 5, and 8 were performed. Concern then arose as to whether the experimental error variation was the same during weeks 1 and 2. Perform a statistical test to see if the pure error variation, as characterized by the replicated responses, is different from one week to another.

18.15. Consider again the 2^3 factorial design presented in Exercise 18.13.
   a. Calculate the between test and within test (pure error) sums of squares for the data.
   b. Using the results of part (a), calculate the associated mean squares and conduct a test to determine if the variation between tests is larger than one might expect given the variation within tests.

18.16. Consider again the factorial design presented in Exercise 18.13.
   a. Calculate the variable effect estimates.
   b. Construct an ANOVA table to identify which effects appear to be important.
   c. Develop a model for the response based on the results of part (b). Construct an ANOVA table to see if the model developed suffers from lack of fit.

18.17. This problem is adapted from Example 9.2 in Montgomery.[2] A factorial experiment was performed in a pilot plant to study the effects of four factors: temperature (x_1), pressure (x_2), concentration of reactant (x_3), and stirring rate (x_4) on the filtration rate of a product. A complete analysis of the experiment led to the following model, which predicts filtration rate as a function of these four factors:

$$\hat{y} = 70.06 + 10.82x_1 + 4.94x_3 + 7.32x_4 - 9.07x_1x_3 + 8.32x_1x_4.$$

The filtration rate is measured in units of percent of theoretical value, and the levels of the variables over which the experiment was conducted were as given in the table.

Variable	Low Level $(-)$	High Level $(+)$
Temperature (°C)	140	180
Pressure (psi)	100	120
Concentration of reactant (%)	4	8
Stirring rate (rpm)	50	70

[2] D. C. Montgomery, *Design and Analysis of Experiments*, 3rd ed., Wiley, New York, 1990.

  a. Construct a contour plot for the following stirring rates: 50, 60, and 70 rpm.
  b. Make recommendations for the process operation so that the filtration rate is closest to the predicted theoretical value (i.e., the response is closest to 100%).

18.18. A 2^2 factorial design was run to test the relative importance of fixture type (x_1) and workplace layout (x_2) on the time to complete an assembly task. Each of the four test conditions was run four times. Each trial observed the total time to complete 20 consecutive assemblies. The data are given in the following table.

| | Workplace Layout | | | |
	A		B	
Fixture Type 1	10.24	9.78	14.30	13.48
	9.96	10.23	12.10	11.98
Fixture Type 2	9.90	8.50	9.82	10.56
	8.86	8.54	9.42	9.50

  a. Calculate all the variable effects. Use the four replicates for each trial to estimate the variance of an effect. Develop 95% confidence intervals to determine which effect estimates are significant.
  b. Calculate the model residuals. Construct all appropriate residual plots. Comment on the appearance of the plots.

18.19. Consider again the replicated 2^2 factorial design of Exercise 18.18.
  a. Construct an ANOVA table to see if the variation between the various fixture, layout combinations is statistically significant.
  b. Calculate the variable effect estimates.
  c. Construct an ANOVA table to identify which of the effect estimates are statistically significant.

18.20. Consider the following residual plots developed from the results of two-level factorial designs.

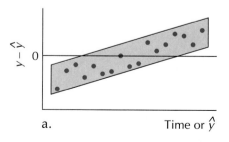

a.   Time or $\hat{y}$

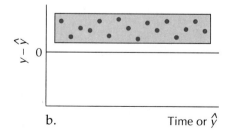

b.   Time or $\hat{y}$

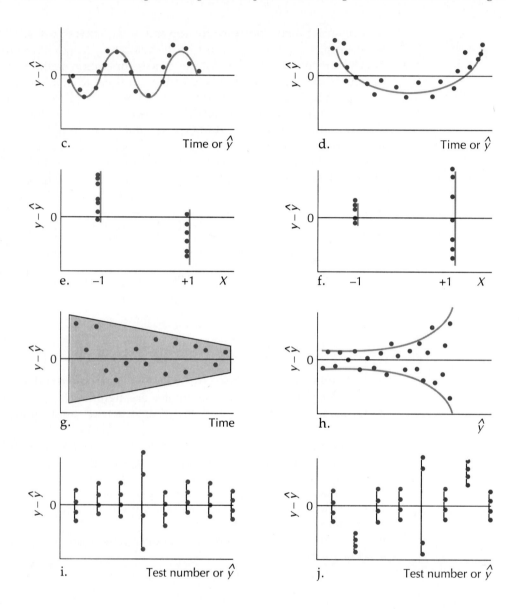

The residuals show the discrepancies between the model and the actual data. For each residual plot, describe in words what could be causing the residual patterns displayed. Be very specific. Where appropriate, present a model form that might describe the response of interest, and then propose an improved model.

18.21. A 2^4 factorial design was conducted to study the effect of four variables [tool rake angle (degrees), cutting speed (surface feet per minute), feed-rate (inches per revolution), and tool nose radius (inches)] on the finish of a machined surface. The table describes the conditions for the 16 tests that were performed as well as the measured surface finish (R_a) value that was observed.

Test	Side Rake Angle (deg)	Cutting Speed (sf/min)	Feedrate (in./rev)	Nose Radius (in.)	Surface Finish (μin.)
1	0	600	0.005	1/64	74
2	10	600	0.005	1/64	85
3	0	900	0.005	1/64	56
4	10	900	0.005	1/64	37
5	0	600	0.010	1/64	132
6	10	600	0.010	1/64	284
7	0	900	0.010	1/64	170
8	10	900	0.010	1/64	202
9	0	600	0.005	1/32	22
10	10	600	0.005	1/32	15
11	0	900	0.005	1/32	25
12	10	900	0.005	1/32	13
13	0	600	0.010	1/32	156
14	10	600	0.010	1/32	76
15	0	900	0.010	1/32	98
16	10	900	0.010	1/32	136

a. Calculate the effects and construct a normal probability plot for the effect estimates. Which of the effects appear to be statistically significant?

b. Write down the appropriate mathematical model, including only those terms that appear to be significant. Calculate the model residuals.

c. Prepare all the residual plots necessary to check the model diagnostically.

d. If the proposed model is found to be inadequate, propose a new model, and conduct the steps necessary to develop and check the new model.

18.22. A replicated 2^2 factorial experiment was performed, the results of which are shown in the table.

x_1	x_2	y_{i1}	y_{i2}
−	−	0.50	0.46
+	−	0.42	0.37
−	+	0.33	0.30
+	+	0.11	0.16

a. Calculate the effect estimates (i.e., the average, E_1, E_2, and E_{12}).

b. Construct an ANOVA table to identify which of the effect estimates are significant.

18.23. A 2^2 factorial design was conducted to study the effect of welding temperature (°F) and flux quantity (in./min) on the resulting weld strength (psi). The table describes the conditions that were examined in the experiment and the associated measured weld strengths (note that each unique combination of temperature and flux quantity was performed three times). The table also shows the average and sample variance for the weld strength calculated for each unique combination of the temperature (variable 1) and flux quantity (variable 2).

Test	x_1, Temp.	x_2, Flux Quantity	Trial 1	Trial 2	Trial 3	Average	Sample Variance
1	300	10	3088	2724	3586	3132.67	187,257.33
2	500	10	799	1202	1079	1026.67	42,656.33
3	300	20	813	1124	1101	1012.67	30,032.33
4	500	20	2806	2946	3336	3029.33	75,433.33

a. It is assumed that a model of the form $y = b_0 + b_1x_1 + b_2x_2 + \epsilon$ may be used to describe the response. Estimate the coefficients in the postulated model using calculated variable effect estimates.

b. Construct an ANOVA table to see if the model developed suffers from lack of fit.

c. If the model examined in part (b) suffers from lack of fit, propose a new model, and construct an ANOVA table to check it for lack of fit.

18.24. A 2^4 factorial design was conducted to study the effect of four variables on a response of interest. The table shows the conditions that were examined in the experiment and the associated response (note that each unique combination of the variables was performed twice).

x_1	x_2	x_3	x_4	Run 1	Run 2
−	−	−	−	13	28
−	+	−	−	9	18
+	+	+	−	38	33
+	+	−	−	57	55
−	−	+	−	52	34
−	+	+	−	30	42
+	−	−	−	41	53
+	−	+	−	38	41
−	−	−	+	12	31
−	+	−	+	30	23
+	+	+	+	53	32
+	+	−	+	39	45
−	−	+	+	49	41
−	+	+	+	43	33
+	−	−	+	60	62
+	−	+	+	50	27

      a. Construct an ANOVA table to see if the variation between the tests is statistically significant.

      b. Calculate the variable effect estimates.

      c. Construct an ANOVA table to identify which of the effect estimates are statistically significant.

      d. Develop a model for the response based on the results of part (b). Construct an ANOVA table to see if the model developed suffers from lack of fit.

18.25. A convenience store owner is trying to determine if the sales are different for four types of soda stocked in the cooler. The sales (in dollars) for each type for three randomly selected weeks are to be used for the analysis.

Soda Brand	Regular	Diet
Caffeine-Free Cola	65	79
	68	71
	73	85
Cola-Caf	80	71
	74	83
	77	75

      a. Construct a simple ANOVA table to determine whether or not there exists a significant difference between the various soda types.

      b. Summarize and comment on the results.

18.26. A dog food company is interested in how dogs respond to the taste of one of its new products. Two males and two females were selected from each of two different breeds of dogs for the experiment. To assess the taste of the new product, the number of ounces of the product consumed by each dog during one meal was recorded. The amount of the new product consumed by each dog is shown in the table.

Dog Breed	Male	Female
Collie	15, 14	12, 14
Dachshund	7, 9	8, 6

      a. Construct an ANOVA table to see if the variation between the various breed/sex combinations is statistically significant.

      b. Calculate the variable effect estimates.

      c. Construct an ANOVA table to identify which of the effect estimates are statistically significant.

18.27. An experimenter is studying the number of defectives observed during a 1-hour time period for four different inspectors and three different fixturing devices. The data were collected by randomly selecting two observations (1-hour inspection results) for each inspector/fixture combination from the scrap report record for a given week.

Fixture	Inspector A	B	C	D
I	7	5	2	6
	6	6	4	6
II	4	2	0	0
	4	4	2	2
III	5	4	4	3
	4	5	4	4

The data displayed in the table may be described with a model of the following form:

$$y_{ijk} = \bar{\bar{\bar{y}}} + (\bar{\bar{y}}_i - \bar{\bar{\bar{y}}}) + (\bar{\bar{y}}_j - \bar{\bar{\bar{y}}}) + \text{interaction} + (y_{ijk} - \bar{y}_{ij}),$$

where y_{ijk} = response for the ith fixture, jth inspector, and kth replicate
$\bar{\bar{\bar{y}}}$ = grand average over all 24 trials
$\bar{\bar{y}}_i$ = average response for the ith fixture
$\bar{\bar{y}}_j$ = average response for the jth inspector
$\bar{y}_{ij}$ = average response for the ith fixture and jth inspector.

a. Find the expression for the interaction term in the model.
b. Derive expressions for the sum of squares due to the grand average, fixtures, inspectors, fixture/inspector interaction, and pure error.
c. Construct an ANOVA table for the data. Perform the tests of significance and comment on the results.

18.28. A fisherman has conducted a 2^5 factorial design to determine the effects of five factors on the number of fish he catches in a four-hour period. The variables studied, and their levels for experiment are shown in the table.

Variable	Low Level	High Level
1. Fishing location	Pier	Boat
2. Bait type	Worms	Night crawlers
3. Time of day	Day	Night
4. Weather	No rain	Rain
5. Hook size	Small	Large

The experimental trials were all performed at Jack's Fish Farm in northern Wisconsin—stocked with bluegills. The following table gives the results of the experiment.

Test	x_1	x_2	x_3	x_4	x_5	y
1	−	−	−	−	−	25
2	+	−	−	−	−	26
3	−	+	−	−	−	31
4	+	+	−	−	−	34
5	−	−	+	−	−	24
6	+	−	+	−	−	26
7	−	+	+	−	−	32
8	+	+	+	−	−	36
9	−	−	−	+	−	30
10	+	−	−	+	−	36
11	−	+	−	+	−	40
12	+	+	−	+	−	43
13	−	−	+	+	−	30
14	+	−	+	+	−	34
15	−	+	+	+	−	40
16	+	+	+	+	−	42
17	−	−	−	−	+	2
18	+	−	−	−	+	21
19	−	+	−	−	+	26
20	+	+	−	−	+	43
21	−	−	+	−	+	4
22	+	−	+	−	+	20
23	−	+	+	−	+	25
24	+	+	+	−	+	44
25	−	−	−	+	+	10
26	+	−	−	+	+	27
27	−	+	−	+	+	33
28	+	+	−	+	+	51
29	−	−	+	+	+	9
30	+	−	+	+	+	27
31	−	+	+	+	+	32
32	+	+	+	+	+	53

a. Calculate all effect estimates.
b. Calculate an estimate of the error variation using the sums of squares associated with three-factor and higher-order interactions.
c. Construct an ANOVA table to identify which of the effect estimates are statistically significant.

19

Two-Level Fractional Factorial Designs

19.1

Rationale for, and Consequences of, Fractionation of Two-Level Factorials

In previous chapters we have studied a most useful class of experimental designs—two-level factorial designs. These designs constitute a powerful way to experiment with several factors simultaneously, revealing the extent and nature of variable interactions in a straightforward manner. These designs have great attraction because they are easy to set up, require relatively simple calculations to analyze the data, and have an appealing geometric representation that greatly enhances the interpretation of the results. Two-level factorial designs can deal simultaneously with variables whose levels are both continuous and discrete (i.e., quantitative or qualitative in nature).

Redundancy in Two-Level Factorials

Although the class of two-level factorial designs appears to be an efficient way to deal with several factors simultaneously, this efficiency seems to disappear as the number of variables to be studied grows. Since the two-level factorial requires the consideration of all possible combinations of k variables at two levels each, a 10-variable experiment would require $2^{10} = 1024$ tests. Such a test plan is simply prohibitive in size. Using the data from a 2^{10} factorial experiment the investigator could, in theory, estimate the following variable effects:

1	Mean response
10	Main effects
45	Two-factor interaction effects
120	Three-factor interaction effects
210	Four-factor interaction effects
252	Five-factor interaction effects
210	Six-factor interaction effects
120	Seven-factor interaction effects
45	Eight-factor interaction effects
10	Nine-factor interaction effects
1	Ten-factor interaction effect
1024	Variable effects.

Although the volume of information from the experiment may seem impressive at first glance, one may wish to question the practical significance of the vast majority of it.

In dealing with phenomena involving continuous variables, it is not unreasonable to expect that the relationship which describes the influence of the variables on a response of interest constitutes a relatively smooth response surface. When dealing with qualitative (discrete) variables, it is usually the case that the responses are similar at different levels of such variables. It should therefore not be too surprising to find that higher-order variable interactions are often very small—negligible—in magnitude. In fact, usually interaction effects involving three factors or more can simply be ignored without causing problems in interpretation of results.

Since the vast majority of the estimatable variable effects in a 2^{10} factorial are three-factor and above interactions (968 of the 1024 variable effects), it would appear that the more relevant information—main and two-factor interaction effects—could be obtained through an experiment requiring many fewer tests than the full factorial. Experimentation with large numbers of variables generally arises out of uncertainty as to which variables have the dominant influence on the response of interest. In the early stages of an investigation the combined opinions of those involved with the situation under study can lead to an often long list of potentially important variables. In the end, however, only a subset of all of these variables will be proven to be important. Therefore, when considering 10 variables that together constitute 10 average/main effects and 45 two-factor interactions, perhaps as few as four to eight of these effects will really prove to be important.

The negligible magnitude of the many higher-order interactions, together with the fact that only a few variables will have significant influence on the response of interest, constitutes a tremendous amount of redundancy in two-level full factorials dealing with large numbers of variables; that is, many of the variable effects will be shown to be essentially zero. G. E. P. Box has referred to this phenomenon as the *sparsity of variable effects*.[1] If this is the case, one should be able to conduct fewer than the full factorial number of tests without loss of relevant information or confusion.

Consequences of Fractionating a Full Factorial

Suppose that an investigator wished to study three variables, ambient temperature, wind velocity, and bar size, to determine their possible effect on the ultimate strength of welded rail steel bars. A 2^3 full factorial was designed, providing for the calculation of three main effects, three two-factor interaction effects, and one three-factor interaction effect.

Suppose that the investigator then wished to consider a fourth variable, type of welding flux, but that the full factorial, $2^4 = 16$ tests, could not be considered. Rather, only eight tests could be run. Is it possible to conduct an experiment on four factors with only eight tests, and if so, what useful information can be obtained from the results?

Based on our assumption about three-factor and higher-order interaction effects being negligible, one could consider assigning the column of plus and minus signs associated with the 123 interaction to the fourth variable. This should be acceptable since we believe that the 123 interaction is very small in magnitude. Therefore, we can use this column to define the levels of variable 4 for the eight tests, and also use it when we estimate the main effect of variable 4.

In using the 123 column to introduce a fourth variable into the experiment, the new design matrix for the eight tests to be conducted becomes

Design Matrix

Test	1	2	3	4 = 123
1	−	−	−	−
2	+	−	−	+
3	−	+	−	+
4	+	+	−	−
5	−	−	+	+
6	+	−	+	−
7	−	+	+	−
8	+	+	+	+

Once the tests are conducted in accordance with the test recipes defined by the design matrix, the calculation matrix is determined to provide for the estimation of the interaction effects. Expanding the design matrix above, we obtain the following calculation matrix by forming all possible products of columns 1 through 4.

---

[1] G. E. P. Box and R. D. Meyer, "An Analysis for Unreplicated Fractional Factorials," *Technometrics*, Vol. 28, 1986, pp. 11–18.

Calculation Matrix

Test	I	1	2	3	4	12	13	14	23	24	34	123	124	134	234	1234
1	+	−	−	−	−	+	+	+	+	+	+	−	−	−	−	+
2	+	+	−	−	+	−	−	+	+	−	−	+	−	−	+	+
3	+	−	+	−	+	−	+	−	−	+	−	+	−	+	−	+
4	+	+	+	−	−	+	−	−	−	−	+	−	−	+	+	+
5	+	−	−	+	+	+	−	−	−	−	+	+	+	−	−	+
6	+	+	−	+	−	−	+	−	−	+	−	−	+	−	+	+
7	+	−	+	+	−	−	−	+	+	−	−	−	+	+	−	+
8	+	+	+	+	+	+	+	+	+	+	+	+	+	+	+	+

Examination of the calculation matrix above reveals that many of the columns are identical. In particular, of the 16 columns, only eight are unique; each unique column appears twice. The following pairs of variable effects are represented in the calculation matrix by the same column of plus and minus signs:

1 and 234	12 and 34
2 and 134	13 and 24
3 and 124	23 and 14
4 and 123	average (I) and 1234.

What does all this mean? When you multiply, for example, the 12 column by the data, sum and divide by 4 do you get an estimate of the two-factor interaction 12? Or the two-factor interaction 34? Or both? The interactions 12 and 34 are said to be *confounded* or *confused*.

The interactions 12 and 34 are said to be *aliases* of the unique column of plus and minus signs defined by $(+ - - + + - - +)$. Use of this column for effect estimation produces a number (estimate) that is actually the sum of the two-factor interaction effects 12 and 34. Similarly, 1 and 234 are confounded effects, 2 and 134 are confounded effects, and so on. It seems that the *innocent* act of using the 123 column to introduce a fourth variable into a 2^3 full factorial scheme has created a lot of confounding among the variable effects.

The eight unique columns in the calculation matrix are used to obtain the linear combinations $l_0, l_1, \ldots, l_{123}$ of confounded effects when their signs are applied to the data, and the result is summed and then divided by 4 (divide by 8 for l_0):

l_0 estimates mean $+ (\tfrac{1}{2})(1234)$	l_{12} estimates 12 $+$ 34
l_1 estimates 1 $+$ 234	l_{13} estimates 13 $+$ 24
l_2 estimates 2 $+$ 134	l_{23} estimates 23 $+$ 14
l_3 estimates 3 $+$ 124	l_{123} estimates 4 $+$ 123.

Some of this confounding can be eliminated by invoking the assumption that third- and higher-order effects are negligible, leading to clear estimates of all main effects. But the six two-factor interactions are still hopelessly confounded.

If we assume three- and four-factor interactions can be neglected, the experiment produces the following linear combinations:

l_0 estimates mean l_{12} estimates 12 + 34
l_1 estimates 1 l_{13} estimates 13 + 24
l_2 estimates 2 l_{23} estimates 23 + 14
l_3 estimates 3 l_{123} estimates 4.

The four-variable, eight-test, two-level experiment discussed thus far is referred to as a two-level fractional factorial design since it considers only a fraction of the tests defined by the full factorial. In this case we have created a one-half fraction design. It is commonly referred to as a 2^{4-1} fractional factorial design. It is a member of the general class of 2^{k-p} fractional factorial designs. For these designs

1. k variables are examined
2. in 2^{k-p} tests
3. requiring that the p of the variables be introduced into the full factorial in $k - p$ variables
4. by assigning them to interaction effects in the first $k - p$ variables.

Example. 2^{4-1} Fractional Factorial

1. 4 variables are studied
2. in $2^{4-1} = 8$ tests
3. $p = 1$ of the variables is introduced into a 2^3 full factorial
4. by assigning it to the interaction 123 (i.e., let 4 = 123).

Many other useful fractional factorials can be developed, some dealing with rather large numbers of variables in relatively few tests. The 2^{4-1} fractional factorial design just examined is one of the more *simple* fractional factorial designs, although the extensive confounding in this design seems a little *confusing*. It can get much worse. Therefore, we need a system to set up such designs easily and to determine quickly the precise nature/pattern of the confounding of the variable effects.

19.2

System to Define the Confounding Pattern of a Two-Level Fractional Factorial

Suppose that an investigator wishes to study the potential effects that five variables may have on the output of a certain process using some type of two-level factorial experiment. If all possible combinations of five variables at two levels each are to be considered, then $2^5 = 32$ tests must be conducted. His boss informs him that due to time and budget limitations he will only be able to run 8 tests, not 32. How might the investigator reconsider his original test plan so as to accommodate his boss' constraints and still gain some useful information about the five variables?

Base Design[2] and Introduction of Additional Variables

If only eight tests are to be conducted using a two-level scheme, only three variables can be examined in a full two-level factorial test plan. The eight unique test settings/conditions and the effect calculation matrix for a 2^3 full factorial design are as follows:

Test	Average I	1	2	3	12	13	23	123	y
1	+	−	−	−	+	+	+	−	y_1
2	+	+	−	−	−	−	+	+	y_2
3	+	−	+	−	−	+	−	+	y_3
4	+	+	+	−	+	−	−	−	y_4
5	+	−	−	+	+	−	−	+	y_5
6	+	+	−	+	−	+	−	−	y_6
7	+	−	+	+	−	−	+	−	y_7
8	+	+	+	+	+	+	+	+	y_8

We recall that:

- The columns with headings 1, 2, and 3 constitute the definition of the eight unique test settings for the three variables denoted heretofore simply as 1, 2, and 3 [e.g., test 1 is conducted with all three variables at their low (−) levels].
- All eight columns of + and − signs constitute the calculation matrix and provide for the independent estimation of the mean (column I), main effects of variables 1, 2, and 3 (columns 1, 2, and 3), the two-factor interaction effects among the three variables (columns 12, 13, and 23), and the three-factor interaction among all three variables (column 123).

The calculation matrix is obtained directly from the design matrix (columns 1, 2, and 3) and is used to estimate the main effects and interaction effects for the variables under study. The columns with headings 12, 13, 23, and 123 are each obtained by multiplying together for each row (eight rows) the signs of the columns in the design matrix.

Now the dilemma of our investigator is that he would like to study five variables, not three. How might this be accomplished within the structure of

[2] This element of the fractional factorial design construction procedure was introduced by these authors some time ago in their classroom lectures on the subject to help avoid the problem that often arises of not properly identifying the calculation matrix for a given fractional factorial.

the 2^3 factorial design presented above? To define test conditions for five variables he will need five columns of $+$ and $-$ signs. In the 2^3 design he has only three such columns; 1, 2, and 3. Where will he find the other two?

In designing fractional factorial experiments, we introduce additional variables into the base design by borrowing columns initially assigned to interaction effects in the base design variables. *The base design is the full factorial design associated with the number of tests we wish to run.* For the case under consideration:

1. Five variables will be studied using only eight tests.
2. Only three variables can be accommodated in a complete or full factorial scheme with eight tests.
3. Therefore, a 2^3 design is the base design.
4. Two variables must be further introduced in the 2^3 base design.
5. Columns 12, 13, 23, and 123 are available to introduce these two additional variables.
6. The new test plan will be called a 2^{5-2} fractional factorial design:
   - Two levels of each variable.
   - Five variables under study.
   - $2^{5-2} = 8$ tests to be run.
   - Two variables introduced into the 2^3 base design.

For the five-variable, eight-test fractional factorial under study, let us introduce variables 4 and 5 into the 2^3 base design by assigning them to the 12 and 13 columns, respectively.

Test	I	1	2	3	4 / 12	5 / 13	23	123	y
1	+	−	−	−	+	+	+	−	y_1
2	+	+	−	−	−	−	+	+	y_2
3	+	−	+	−	−	+	−	+	y_3
4	+	+	+	−	+	−	−	−	y_4
5	+	−	−	+	+	−	−	+	y_5
6	+	+	−	+	−	+	−	−	y_6
7	+	−	+	+	−	−	+	−	y_7
8	+	+	+	+	+	+	+	+	y_8

In the matrix of $+$ and $-$ signs, the first five columns (excluding I) represent the eight unique test conditions for the five variables [e.g., for test 1, fix variables 1, 2, and 3 at their low ($-$) levels and variables 4 and 5 at their high ($+$) levels]. These five columns constitute the design (recipe) matrix for this 2^{5-2} fractional factorial design.

All seven columns plus a column of all plus signs (the I column) constitute the calculation matrix for this 2^{5-2} design. This means that all average/main and interaction effects among the five variables will be contained (confounded) within eight linear combinations of the data. The question that remains is to determine exactly which effects are confounded with each other.

Design Generators and the Defining Relation

From now on, when we refer to a column heading (e.g., 1 or 23 or 123) we should imagine a column of + and − signs directly under it. Our 2^{5-2} fractional factorial design was generated by setting the 4-column equal to the 12-column and the 5-column equal to the 13-column. In the interest of convenience we will denote these as 4 = 12 and 5 = 13, where the = sign really implies an identity between columns of + and − signs, for example

4	=	12
+		+
−		−
−		−
+	=	+
+		+
−		−
−		−
+		+.

Now, if any column of + and − signs is multiplied by itself, a column of all + signs is produced. We will denote such a column by the heading *I:* for example,

4	×	4	=	*I*
+		+		+
−		−		+
−		−		+
+	×	+	=	+
+		+		+
−		−		+
−		−		+
+		+		+.

This simple operation will prove to be very useful. Since the 2^{5-2} fractional factorial design was generated by setting

$$4 = 12$$
$$5 = 13$$

and given the definition of *I* above, then if we multiply both sides of the two "equations" above by 4 and 5, respectively, then

$$4 \times 4 = 12 \times 4, \qquad 5 \times 5 = 13 \times 5,$$

which reduces to

$$I = 124, \qquad I = 135.$$

These two identities are referred to as our design *generators*. While both the left- and right-hand sides of the equation above represent columns of all + signs, the right-hand side retains the individual column headings that produced the column of all + signs by their product. That is,

I column = 1 column × 2 column × 4 column.

Now since both the 124 and 135 columns equal I, their product must also equal I:

(124) × (135) = I,

or, rearranging numbers (columns),

(1)(1)2345 = I.

But (1)(1) = I and any column multiplied by a column of plus signs (I) remains unchanged. Therefore, I is also equal to 2345:

I = 2345.

Hence we have

I = 124 = 135 = 2345,

an identity comprised of the design generators and their products in all possible combinations (in this case only one product). The identity

I = 124 = 135 = 2345

is referred to as the *defining relation* of this 2^{5-2} fractional factorial design, and through it we can reveal the complete aliasing/confounding structure of this fractional factorial design.

Revelation of the Complete Confounding Pattern

Returning to the original base design calculation matrix, we recall that we have seven independent columns of + and − signs, and an eighth column for I (a column of all + signs):

Test	I	1	2	3	12	13	23	123
1	+	−	−	−	+	+	+	−
2	+	+	−	−	−	−	+	+
3	+	−	+	−	−	+	−	+
4	+	+	+	−	+	−	−	−
5	+	−	−	+	+	−	−	+
6	+	+	−	+	−	+	−	−
7	+	−	+	+	−	−	+	−
8	+	+	+	+	+	+	+	+
	l_0	l_1	l_2	l_3	l_{12}	l_{13}	l_{23}	l_{123}

By letting $4 = 12$ and $5 = 13$, we have created many aliased effects. To find the aliases of the column headings above $(1, 2, 3, \ldots, 123)$, we multiply each by every term (including I) in the defining relation:

defining relation: $\quad I = 124 = 135 = 2345.$

For column heading 1:

$$(1)I = (1)124 = (1)135 = (1)2345.$$

Removing all I's [recall that $(1)(1) = I$], we have

$$1 = 24 = 35 = 12345.$$

That is, the aliases of 1 are 24, 35, and 12345. Therefore, when we multiply the 1 column by the y column, sum, and divide by 4, we obtain an estimate of the sum (linear combination) of 1, 24, 35, and 12345.[3] We conveniently denote this sum of confounded variable effects as l_1 (l for linear combination of the effects). That is,

$$l_1 \text{ estimates } 1 + 24 + 35 + 12345.$$

Similarly, moving to column headings 2, 3, and so on, we find that

$$
\begin{array}{rcrcrcr}
2 &=& 14 &=& 1235 &=& 345 \\
3 &=& 1234 &=& 15 &=& 245 \\
12 &=& 4 &=& 235 &=& 1345 \\
13 &=& 234 &=& 5 &=& 1245 \\
23 &=& 134 &=& 125 &=& 45 \\
123 &=& 34 &=& 25 &=& 145.
\end{array}
$$

Hence

$$
\begin{array}{l}
l_2 \text{ estimates } \quad 2 + \quad 14 + 1235 + 345 \\
l_3 \text{ estimates } \quad 3 + 1234 + \quad 15 + 245
\end{array}
$$

$$\vdots$$

$$l_{123} \text{ estimates } 123 + \quad 34 + \quad 25 + 145.$$

We have now defined the complete confounding pattern of this 2^{5-2} fractional factorial design and we know precisely what effect combinations we can obtain from the data.

Summary

1. We want to study five variables in only eight tests in a two-level factorial scheme.

---

[3] Actually, l_1 estimates the sum of the expected value of effects, E_1, E_{24}, E_{35}, and E_{12345}. That is, l_1 estimates the sum of the mean values of these effects. From this point on, when we use the notation l_1 estimates $1 + 24 + 35 + 12345$, it should be understood that 1, 24, 35, 12345 are mean values, while E_1, E_{24}, E_{35}, E_{12345} continue to be considered estimates of these mean values.

2. The base design is therefore a $2^3 = 8$ test two-level full factorial in the base design variables 1, 2, and 3.

3. To the base design we introduce variables 4 and 5 by assigning them to interaction columns in the base design: for example,

$$4 = 12$$
$$5 = 13.$$

4. The design generators are therefore

$$I = 124$$
$$I = 135,$$

and hence the design defining relation is

$$I = 124 = 135 = 2345.$$

5. The defining relation produces the confounding pattern or alias structure:

$$
\begin{aligned}
I &= & 124 &= & 135 &= & 2345 \\
1 &= & 24 &= & 35 &= & 12345 \\
2 &= & 14 &= & 1235 &= & 345 \\
3 &= & 1234 &= & 15 &= & 245 \\
12 &= & 4 &= & 235 &= & 1345 \\
13 &= & 234 &= & 5 &= & 1245 \\
23 &= & 134 &= & 125 &= & 45 \\
123 &= & 34 &= & 25 &= & 145.
\end{aligned}
$$

For the confounding structure shown above, it should be noted that each column heading (I, 1, 2, 3, . . . , 123) is confounded with three effects (e.g., I is confounded with 124, 135, and 2345). Thus, each of the eight rows in the confounding structure contains four effects. For a full factorial in five variables ($2^5 = 32$ tests) there are 31 column headings plus I in the calculation matrix. In particular, for a 2^5 full factorial calculation matrix, there are columns for

1	Mean
5	Main effects
10	Two-factor interaction effects
10	Three-factor interaction effects
5	Four-factor interaction effects
1	Five-factor interaction effect
32	Variable effects.

Examining the aliasing structure for the 2^{5-2} design, we observed that all 32 variable effects (including the mean, I) are accounted for (8 rows × 4 effects/row). It is a good idea to verify that this is the case.

6. The eight columns (including I) in the 2^{5-2} fractional factorial design

produce the following linear combinations of effects which can be estimated:[4]

l_0 estimates mean $+ \frac{1}{2}(124 + 135 + 2345)$
l_1 estimates 1 $+$ 24 $+$ 35 $+$ 12345
l_2 estimates 2 $+$ 14 $+$ 1235 $+$ 345
l_3 estimates 3 $+$ 1234 $+$ 15 $+$ 245
l_{12} estimates 12 $+$ 4 $+$ 235 $+$ 1345
l_{13} estimates 13 $+$ 234 $+$ 5 $+$ 1245
l_{23} estimates 23 $+$ 134 $+$ 125 $+$ 45
l_{123} estimates 123 $+$ 34 $+$ 25 $+$ 145.

7. Often, we assume that the majority of the variability in the data can be explained by the presence of main effects and two-factor interaction effects among the variables. Under such an assumption, the linear combinations of effects that can be estimated from the 2^{5-2} fractional factorial under study are

l_0 estimates mean
l_1 estimates 1 $+$ 24 $+$ 35
l_2 estimates 2 $+$ 14
l_3 estimates 3 $+$ 15
l_{12} estimates 12 $+$ 4
l_{13} estimates 13 $+$ 5
l_{23} estimates 23 $+$ 45
l_{123} estimates 34 $+$ 25.

19.3

Procedure for Design Characterization: Another Example

Suppose that six variables are to be studied in a two-level experiment but only eight tests can be conducted. How is such an experiment set up, and what is the precise nature of the confounding among all variable effects?

Step 1: Defining the Base Design

Since only eight tests are to be conducted, the base design for this fractional factorial is a 2^3 full factorial. The complete calculation matrix for the 2^3 base design is

---

[4] Recall that the divisor for the I (mean) column is *eight*, not *four*.

Test	I	1	2	3	12	13	23	123	y
1	+	−	−	−	+	+	+	−	y_1
2	+	+	−	−	−	−	+	+	y_2
3	+	−	+	−	−	+	−	+	y_3
4	+	+	+	−	+	−	−	−	y_4
5	+	−	−	+	+	−	−	+	y_5
6	+	+	−	+	−	+	−	−	y_6
7	+	−	+	+	−	−	+	−	y_7
8	+	+	+	+	+	+	+	+	y_8
Divisor	8	4	4	4	4	4	4	4	

Step 2: Introduction of Additional Variables

The interaction columns 12, 13, 23, and 123 are available for the introduction of the variables 4, 5, and 6. In some cases the choice of which interactions to use is critical. We discuss this choice later. In this case we will use the interactions 12, 13, and 23 to establish the design generators.

$$4 = 12$$
$$5 = 13$$
$$6 = 23.$$

The design matrix for this 2^{6-3} fractional factorial design is therefore

Test	1	2	3	4	5	6
1	−	−	−	+	+	+
2	+	−	−	−	−	+
3	−	+	−	−	+	−
4	+	+	−	+	−	−
5	−	−	+	+	−	−
6	+	−	+	−	+	−
7	−	+	+	−	−	+
8	+	+	+	+	+	+

The eight tests may now be conducted in accordance with these test recipes.

Step 3: Obtaining the Defining Relation for the Design

The design was set up by introducing variables 4, 5, and 6 using the interactions 12, 13, and 23, respectively. Therefore, the design generators are

$$I = 124, \qquad I = 135, \qquad I = 236.$$

Remembering that

$$4 = 12$$
$$4 \times 4 = 12 \times 4,$$

but since

$4 \times 4 = I$, a column of $(+)$ signs, we have
$$I = 124.$$

Next, we obtain the defining relation. The defining relation consists of the generators plus all of their products taken two at a time, three at a time, . . . , p at a time ($p = 3$ for this case). The defining relation is given by

$$I = 124, \qquad I = 135, \qquad I = 236 \qquad \text{(the generators)}$$

plus two-at-a-time products:

$$(124)(135) = 2345$$
$$(124)(236) = 1346$$
$$(135)(236) = 1256$$

plus the three-at-a-time product:

$$(124)(135)(236) = 456.$$

The complete defining relation is therefore

$$I = 124 = 135 = 236 = 2345 = 1346 = 1256 = 456.$$

Step 4: Revealing the Complete Confounding Structure of the Design

Once the defining relation has been obtained, the complete confounding pattern may be determined. Since the experiment involves eight tests, there are eight unique linear combinations of effects that can be estimated. These are defined by the eight columns of the calculation matrix including I. Moving from left to right, these columns are identified as $l_0, l_1, l_2, \ldots, l_{123}$:

l_0	l_1	l_2	l_3	l_{12}	l_{13}	l_{23}	l_{123}
I	1	2	3	12	13	23	123.

That is, the use of each column for estimation produces an estimate of a certain linear combination of effects. To determine each set of confounded effects, we multiply, in turn, each column heading by every term in the defining relation, including I. For example, for the column denoted 1,

$$(1)I = (1)124 = (1)135 = (1)236 = (1)2345 = (1)1346 = (1)1256 = (1)456,$$

or

$$1 = 24 = 35 = 1236 = 12345 = 346 = 256 = 1456.$$

Assuming that third- and higher-order interactions can be neglected, the main effect of variable 1 is therefore confounded with the two-factor interactions 24 and 35. That is,

l_1 estimates $1 + 24 + 35$.

Summary: Relevant Confounding Structure

For this 2^{6-3} design, assuming third- and higher-order interactions can be neglected, we obtain

l_0 estimates mean
l_1 estimates $1 + 24 + 35$
l_2 estimates $2 + 14 + 36$
l_3 estimates $3 + 15 + 26$
l_{12} estimates $12 +   4 + 56$
l_{13} estimates $13 +   5 + 46$
l_{23} estimates $23 +   6 + 45$
l_{123} estimates $34 + 25 + 16.$

It is important to check to make sure that all six main effects and all 15 two-factor interactions have been accounted for. A quick check above shows that they have been. In general, there are $k!/2!(k - 2)!$ two-factor interactions.

19.4

Concept of Resolution of Two-Level Fractional Factorial Designs

We have seen previously that the introduction of additional variables into two-level full factorials gives rise to confounding or aliasing of variable effects. It would be desirable to make this introduction in such a way as to confound low-order effects (main effects and two-factor interactions) not with each other but with higher-order interactions. Then, under the assumption that third- and higher-order interactions can be neglected, the low-order effects become, in a sense, unconfounded by this assumption.

Selecting the Preferred Generators

To illustrate, consider the study of five variables in just 16 tests (the full factorial would require $2^5 = 32$ tests). One additional variable—the fifth variable—must be introduced into a $2^4 = 16$ run base design. Any of the interactions in the first four variables could be used for this purpose:

$$12, \quad 13, \quad 14, \quad 23, \quad 24, \quad 34$$
$$123, \quad 124, \quad 134, \quad 234$$
$$1234.$$

If any one of the two-factor interactions are used, say, $5 = 12$, the design generator becomes

$$I = 125,$$

which is also the defining relation. Therefore, at least some of the main effects will be confounded with two-factor interactions:

$$1 = 25, \quad 2 = 15, \quad 5 = 12.$$

If any one of the three-factor interactions are used to introduce the fifth variable, the situation is greatly improved, at least for the estimation of main effects. For example, if we let $5 = 123$, then

$$I = 1235$$

is the generator and defining relation. So some main effects are confounded with, at worst, three-factor interactions, while two-factor interactions are confounded with each other: for example,

$$1 = 235, \quad 2 = 135, \quad 3 = 125, \quad 5 = 123, \quad 12 = 35,$$
$$13 = 25, \quad 23 = 15.$$

If the four-factor interaction in the first four variables is used to introduce the fifth variable (i.e., $5 = 1234$), an even more desirable result is obtained (the best under these circumstances). The generator and defining relation for this situation is

$$I = 12345.$$

Therefore,

$$1 = 2345, \quad 2 = 1345, \quad 3 = 1245, \quad 4 = 1235, \quad 5 = 1234,$$
$$12 = 345, \quad 13 = 245, \quad 14 = 235, \quad 15 = 234, \quad 23 = 145,$$
$$24 = 135, \quad 25 = 134, \quad 34 = 125, \quad 35 = 124, \quad 45 = 123.$$

In this last case:

1. All main effects are confounded with four-factor interactions.
2. All two-factor interactions are confounded with three-factor interactions.

The varying confounding structures produced by using different orders of variable interactions to introduce the fifth variable in the example above are described by the concept of the resolution of fractional factorial designs: "The *resolution* of a two-level fractional factorial design is defined to be equal to the number of letters (numbers) in the shortest length word (term) in the defining relation, excluding *I*." If the defining relation of a certain design is

$$I = 124 = 135 = 2345,$$

the design is of resolution three, denoted as resolution III, since the words "124" and "135" have three letters each. If the defining relation of a certain design is

$$I = 1235 = 2346 = 1456,$$

then the design is of resolution IV (1235, 2346, and 1456 each have four letters). The last design we examined above, which had the defining relation ($I = 12345$), is a resolution V design.

- If a design is of resolution III, this means that at least some main effects are confounded with two-factor interactions.
- If a design is of resolution IV, this means that at least some main effects are confounded with three-factor interactions, while at least some two-factor interactions are confounded with other two-factor interactions.
- If a design is of resolution V, this means that at least some main effects are confounded with four-factor interactions and some two-factor interactions are confounded with three-factor interactions.

It may be noted at this point that the number of words in the defining relation for a 2^{k-p} fractional factorial design is equal to 2^p. Thus for 2^{6-3} fractional factorial ($k = 6$ and $p = 3$), there are $2^3 = 8$ words in the defining relation.

Example. Design Resolution/Selection of Generators

A 2^{6-2} fractional factorial design is set up by introducing variables 5 and 6 via

$$5 = 123, \qquad 6 = 1234.$$

What is the resolution of this design? The design generators are

$$I = 1235, \qquad I = 12346.$$

The defining relation is

$$I = 1235 = 12346 = 456.$$

Therefore, the design is of resolution III. What would the resolution be if the generators were

$$5 = 123, \qquad 6 = 124?$$

The defining relation is

$$I = 1235 = 1246 = 3456.$$

Now, the design is of resolution IV. It is clear that the selection of the proper design generators is very important.

Summary: The Concept of Design Resolution

From the discussion of design resolution above, several observations can be made:

1. Higher-resolution designs seem more desirable since they provide the opportunity for low-order effect estimates to be determined in an unconfounded state, assuming that higher-order interaction effects can be neglected.
2. There is a limit to the number of variables that can be considered in a fixed number of tests while maintaining a prespecified resolution requirement.
3. No more than $(n - 1)$ variables can be examined in n tests (n is a power of 2, e.g., 4, 8, 16, 32, . . .) to maintain a design resolution of at least III. Such designs are commonly referred to as saturated designs. Examples are

$$2^{3-1}, \quad 2^{7-4}, \quad 2^{15-11}, \quad 2^{31-26}.$$

For saturated designs all interactions in the base design variables are used to introduce additional variables.

19.5

Case Study Application: The Sandmill Experiment Revisited

Previously, we described an investigation conducted at an automotive plant that was concerned with a process used to manufacture vinyl film. At the beginning of the study, the plant employed vertical sandmills to disperse pigments for the production of colorants for the vinyl film. As has been noted, the plant was considering the acquisition of a new type of sandmill, a horizontal sandmill. The manufacturer of the new machine claimed that the horizontal sandmill would improve both productivity and quality. It was decided to perform some tests to assess the superiority of the horizontal sandmill with respect to the vertical sandmill, since the sandmill replacement cost would be approximately $180,000. Arrangements were made with the manufacturer for some trial runs on the horizontal sandmill.

As the planning of the sandmill comparison experiment continued to evolve, questions were raised concerning the operating conditions of the machines. Past experience with the vertical sandmill had shown that grind fineness (measured using the Hegman scale from 0/coarse to 10/fine) was affected by both the temperature setting of the process and the flow rate. While targeted values had been established some time in the past for these variables (as a function of pigment type and pigment concentration), it was not at all clear which values to use for the horizontal machine. In addition, there was some concern about the "goodness" of the target values actually being used for the vertical machine. Therefore, it was decided to consider the two process parameters (temperature and flow rate) in addition to the machine type, pigment type, and pigment concentration in an experiment. Table 19.1 lists the variables and variable levels that were chosen for the experiment.

To consider this group of five factors in accordance with a two-level full factorial design scheme seemed prohibitive. A total of $2^5 = 32$ unique conditions would need to be tested even without any provision for replication. Given the time constraints and the expense of so many tests, this approach did not seem feasible. It was decided, therefore, to conduct a 2^{5-1} fractional factorial design (i.e., a one-half fraction design).

Design and Conduct of the Experiment

As stated, it was decided to use a 2^{5-1} fractional factorial design to study the five factors listed in Table 19.1. For a 2^{5-1} fractional factorial, the base design calculation matrix is based on a 2^4 full factorial design:

Test	I	1	2	3	4	12	13	14	23	24	34	123	124	134	234	1234
1	+	−	−	−	−	+	+	+	+	+	+	−	−	−	−	+
2	+	+	−	−	−	−	−	−	+	+	+	+	+	+	−	−
3	+	−	+	−	−	−	+	+	−	−	+	+	+	−	+	−
4	+	+	+	−	−	+	−	−	−	−	+	−	−	+	+	+
5	+	−	−	+	−	+	−	+	−	+	−	+	−	+	+	−
6	+	+	−	+	−	−	+	−	−	+	−	−	+	−	+	+
7	+	−	+	+	−	−	−	+	+	−	−	−	+	+	−	+
8	+	+	+	+	−	+	+	−	+	−	−	+	−	−	−	−
9	+	−	−	−	+	+	+	−	+	−	−	−	+	+	+	−
10	+	+	−	−	+	−	−	+	+	−	−	+	−	−	+	+
11	+	−	+	−	+	−	+	−	−	+	−	+	−	+	−	+
12	+	+	+	−	+	+	−	+	−	+	−	−	+	−	−	−
13	+	−	−	+	+	+	−	−	−	−	+	+	+	−	−	+
14	+	+	−	+	+	−	+	+	−	−	+	−	−	+	−	−
15	+	−	+	+	+	−	−	−	+	+	+	−	−	−	+	−
16	+	+	+	+	+	+	+	+	+	+	+	+	+	+	+	+

A fifth variable may be introduced into the structure by assigning it to one of the interaction columns. To obtain the highest design resolution, variable 5 was

TABLE 19.1	Variables and Their Levels for the Sandmill Experiment		

| | | Level | |
Variable	Name	Low (−)	High (+)
1	Sandmill/machine type	Vertical	Horizontal
2	Pigment type	Blue	Red
3	Pigment concentration (%)	10	15
4	Processing temperature (°F)	140	160
5	Flow rate (gal/hr)	20	30

introduced using the 1234 interaction column. Using this assignment, the design matrix becomes

Test	1	2	3	4	5
1	−	−	−	−	+
2	+	−	−	−	−
3	−	+	−	−	−
4	+	+	−	−	+
5	−	−	+	−	−
6	+	−	+	−	+
7	−	+	+	−	+
8	+	+	+	−	−
9	−	−	−	+	−
10	+	−	−	+	+
11	−	+	−	+	+
12	+	+	−	+	−
13	−	−	+	+	+
14	+	−	+	+	−
15	−	+	+	+	−
16	+	+	+	+	+

If the $+/-$ signs in the design matrix are replaced by the actual levels for the variables, we obtain the recipes for each test condition (Table 19.2). The test conditions in the design matrix were performed in a random order (using the indicated run order) and the Hegman values displayed in Table 19.2 were obtained.

Confounding Structure and Linear Combinations of Effects

As has been noted, the 2^{5-1} fractional factorial design described above was obtained by introducing variable 5 through the 1234 interaction column. Hence the generator and defining relation for this design is

$$I = 12345.$$

	Machine Type	Pigment Type	Pigment Conc. (%)	Processing Temp. (°F)	Flow Rate (gal/hr)	Hegman Value,	Run
Test	1	2	3	4	5	y	Order
1	Vertical	Blue	10	140	30	6.25	5
2	Horizontal	Blue	10	140	20	6.25	11
3	Vertical	Red	10	140	20	7.75	8
4	Horizontal	Red	10	140	30	6.75	2
5	Vertical	Blue	15	140	20	6.25	13
6	Horizontal	Blue	15	140	30	5.25	10
7	Vertical	Red	15	140	30	6.75	4
8	Horizontal	Red	15	140	20	6.75	14
9	Vertical	Blue	10	160	20	7.00	6
10	Horizontal	Blue	10	160	30	5.25	12
11	Vertical	Red	10	160	30	8.00	16
12	Horizontal	Red	10	160	20	7.00	1
13	Vertical	Blue	15	160	30	5.50	15
14	Horizontal	Blue	15	160	20	5.25	7
15	Vertical	Red	15	160	20	7.75	3
16	Horizontal	Red	15	160	30	6.50	9

TABLE 19.2   Test Condition Recipes

As can be seen, this design is of resolution V since the length of the smallest word in the defining relation is five (excluding I). Therefore, at least some of the main effects (in this case all) will be confounded with four-factor interactions, and at least some of the two-factor interactions (in this case all) will be confounded with three-factor interactions.

Given the defining relation, the patterns of confounded effects may be obtained by multiplying the column headings in the base design calculation matrix (I, 1, 2, 3, 4, 12, 13, . . . , 1234) by every term in the defining relation. For example, for the column denoted 1, we have

$$(1)I = (1)12345 \qquad \text{or} \qquad 1 = 2345.$$

As another example, for the column denoted 123, we have

$$(123)I = (123)12345 \qquad \text{or} \qquad 123 = 45.$$

The complete set of linear combinations of confounded effects is then

l_0 estimates mean + $\frac{1}{2}$(12345) l_{23} estimates   23 + 145
l_1 estimates 1 + 2345 l_{24} estimates   24 + 135
l_2 estimates 2 + 1345 l_{34} estimates   34 + 125
l_3 estimates 3 + 1245 l_{123} estimates 123 + 45
l_4 estimates 4 + 1235 l_{124} estimates 124 + 35
l_{12} estimates 12 + 345 l_{134} estimates 134 + 25
l_{13} estimates 13 + 245 l_{234} estimates 234 + 15
l_{14} estimates 14 + 235 l_{1234} estimates 1234 +   5.

Under the assumption that three-factor interactions and higher are negligible, this set of linear combinations reduces to:

l_0 estimates mean		l_{23}	estimates 23
l_1 estimates 1		l_{24}	estimates 24
l_2 estimates 2		l_{34}	estimates 34
l_3 estimates 3		l_{123}	estimates 45
l_4 estimates 4		l_{124}	estimates 35
l_{12} estimates 12		l_{134}	estimates 25
l_{13} estimates 13		l_{234}	estimates 15
l_{14} estimates 14		l_{1234}	estimates 5.

An examination of these linear combinations demonstrates the advantage of designs having resolution V or higher. Under the assumption that three-factor interactions and higher are negligible, such designs provide unaliased estimates of all main and two-factor interaction effects.

Estimation of Effects

The estimates (l_i's) associated with the linear combinations of confounded effects are obtained by multiplying each column of plus and minus signs in the base design calculation matrix by the column of responses (Hegman values), summing, and then dividing by 8 (for the average, divide by 16). As an example, consider the calculation of the estimate l_1:

Column 1		Column y	
−		6.25	
+		6.25	
−		7.75	
+		6.75	
−		6.25	
+		5.25	
−	×	6.75	$= (-6.25 + 6.25 - 7.75 + 6.75 - 6.25 + 5.25$
+		6.75	$\quad - 6.75 + 6.75 - 7.00 + 5.25 - 8.00 + 7.00$
−		7.00	$\quad - 5.50 + 5.25 - 7.75 + 6.50)/8$
+		5.25	$= -0.78125 = l_1$
−		8.00	
+		7.00	
−		5.50	
+		5.25	
−		7.75	
+		6.50	

The estimates are summarized as follows:

$$l_0 = 6.51563 \rightarrow \text{estimates mean}$$
$$l_1 = -0.78125 \rightarrow \text{estimates}\ \ 1$$
$$l_2 = 1.28125 \rightarrow \text{estimates}\ \ 2$$
$$l_3 = -0.53125 \rightarrow \text{estimates}\ \ 3$$
$$l_4 = 0.03125 \rightarrow \text{estimates}\ \ 4$$
$$l_{12} = -0.03125 \rightarrow \text{estimates}\ 12$$
$$l_{13} = 0.15625 \rightarrow \text{estimates}\ 13$$
$$l_{14} = -0.28125 \rightarrow \text{estimates}\ 14$$
$$l_{23} = 0.09375 \rightarrow \text{estimates}\ 23$$
$$l_{24} = 0.28125 \rightarrow \text{estimates}\ 24$$
$$l_{34} = -0.03125 \rightarrow \text{estimates}\ 34$$
$$l_{123} = 0.03125 \rightarrow \text{estimates}\ 45$$
$$l_{124} = -0.03125 \rightarrow \text{estimates}\ 35$$
$$l_{134} = 0.15625 \rightarrow \text{estimates}\ 25$$
$$l_{234} = 0.09375 \rightarrow \text{estimates}\ 15$$
$$l_{1234} = -0.46875 \rightarrow \text{estimates}\ \ 5.$$

Previously, in our analysis of two-level full factorial designs, we employed normal probability plots to judge the importance of variable effects. This same graphical technique may be employed usefully in some cases in analyzing two-level fractional factorial designs. Figure 19.1 displays a normal probability plot of the estimates (l_i's) given above.

Summary

An examination of the normal probability plot of Fig. 19.1 reveals that the following effects are important (under the assumption that three-factor and higher-order interactions are negligible):

1, 2, 3, 5, 14, 24.

These main and two-factor interaction effects may be interpreted as follows:

1. **Main Effect of Machine Type (Variable 1).** The vertical machine, on average, yields Hegman values 0.78125 higher than the horizontal machine. This effect was large relative to the other effects.
2. **Main Effect of Pigment Type (Variable 2).** Blue pigment, on average, yields Hegman values 1.28125 lower than red pigment. Blue pigment is known to be generally more difficult to grind.
3. **Main Effect of Pigment Concentration (Variable 3).** A 10% pigment concentration, on average, yields Hegman values 0.53125 higher than a concentration of 15%.

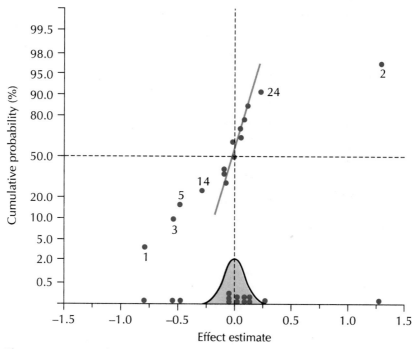

Figure 19.1 Normal Probability Plot of Effect Estimates

4. **Main Effect of Flow Rate (Variable 5).**   A 20 gal/hr flow rate, on average, gives Hegman values 0.46875 higher than a flow rate of 30 gal/hr.
5. **Two-Factor Interaction: Machine by Temperature (14 Interaction).**   This interaction (Fig. 19.2) indicates that the vertical machine performs somewhat better at higher temperatures, while the horizontal machine performs somewhat better at lower temperatures. However, the low-temperature horizontal mill results were poorer than either high- or low-temperature vertical mill results.
6. **Two-Factor Interaction: Pigment Type by Temperature (24 Interaction).** This interaction (Fig. 19.3) indicates that for the blue pigment, better Hegman values are obtained at the low level of temperature, while for the red pigment, the high level of temperature is better. This result is important in that it begins to provide information of value with respect to the optimization of the process.

Based on the results of this experiment, there appears to be no advantage in using the horizontal sandmill in place of the vertical sandmill. In fact, the vertical sandmill appears to be superior. Since a capital expenditure was being considered for the purchase of horizontal sandmills in the plant, this experiment effectively produced a savings of $180,000. Furthermore, it was learned that a lower pigment concentration actually produces higher/better Hegman values.

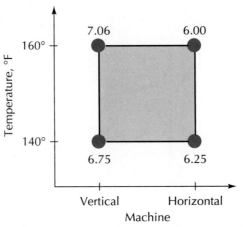

Figure 19.2 Machine by Temperature Interaction

This result is important because a reduction in the pigment concentration lowers the raw material costs. Finally, both processing variables temperature and flow rate greatly influence the grind fineness (the former through the temperature by pigment interaction). Further study of these two processing variables is needed from a process optimization standpoint.

19.6

Orthogonal Arrays and Two-Level Fractional Factorial Designs

The purpose of this section is to present some examples of the use of orthogonal arrays as recommended by Taguchi and to discuss the relationship between these designs and the more traditional treatment of fractional factorial designs. We do not attempt to contrast the overarching philosophical approaches encompassing these methods but rather, attempt only to compare experimental design structures from a more mechanical point of view. However, it is impossible to separate the mechanics of the use of these experimental design structures completely from the frameworks in which they are implemented. We discuss only experimental designs that consider each variable at two levels. It should be noted that both in the context of Taguchi methods and in the more traditional design of experiments framework, designs that vary factors over more than two levels are commonly used.

The use of orthogonal arrays has been widespread for almost 60 years. As pointed out by Box and Meyer,[5] Tippett[6] employed a 125th fraction of a 5^5

[5] G. E. P. Box and R. D. Meyer, "An Analysis for Unreplicated Fractional Factorials," *Technometrics*, Vol. 28, 1986, pp. 11–18.

[6] L. H. C. Tippett, *Applications of Statistical Methods to the Control of Quality in Industrial Production*, Manchester Statistical Society, Manchester, England, 1934.

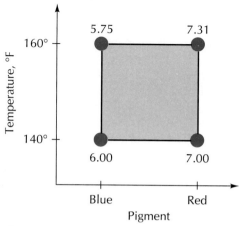

Figure 19.3 Pigment Type by Temperature Interaction

factorial for screening experiments on a cotton spinning machine. The theory of fractional factorial designs was first worked out by Finney[7] and Rao.[8] Other orthogonal arrays were introduced by Plackett and Burman.[9] Two- and three-level fractional factorial designs gained widespread attention and industrial application beginning in the 1950s with the publication of numerous papers on theory and applications. Papers such as "The 2^{k-p} Fractional Factorial Designs," Parts I and II, *Technometrics*, Vol. 3, p. 311, by G. E. P. Box and J. S. Hunter (1961) provided much useful guidance to the practitioner in the adroit use of these experimental design structures. Beginning around 1950, Taguchi and his colleagues began extensive use of this class of experimental designs. Since the early 1980s, the use of Taguchi methods and orthogonal arrays has been widespread in the United States.

Numerous examples of the use of orthogonal arrays can be found in the industrial experimental design literature of the Western world in the post–World War II era. In fact, these authors recently browsed through a 1950 publication of the Shell Oil Company that advocated the use of "magic squares" in industrial experimentation in the chemical processing industry.

Two-Level Fractional Factorial Design Structures: The Issue of Confounding

To many, the most significant property of the designs we are examining here is orthogonality—the ability to separate out the individual effects of several

[7] D. J. Finney, "The Fractional Replication of Factorial Arrangements," *Annals of Eugenics*, Vol. 12, 1945, pp. 291–301.
[8] C. R. Rao, "Factorial Experiments Derivable from Combinatorial Arrangements of Arrays," *Journal of the Royal Statistical Society, Ser. B*, Vol. 9, 1947, pp. 128–140.
[9] R. L. Plackett and J. P. Burman, "Design of Optimal Multifactorial Experiments," *Biometrika*, Vol. 23, 1946, pp. 305–325.

variables on a response of interest. The words "not mixed or confounded" have been used to describe the property of orthogonality in orthogonal arrays. Some caution might be given to the use of these descriptions, especially the words "not confounded," since it is not so much the design structure itself, but rather the assumptions used in its application, that justify the terminology. When anything less than a full factorial is under consideration, confounding or mixing of effects is present. Through thoughtful design, attractive confounding structures can be found, as we show in this chapter and in Chapter 20. Even so, confounding can be removed only by assumptions made about the physical system, not about the experimental design.

We have seen that an important feature of two-level factorial experiments is the potential for the estimation of variable interactions. A two-factor interaction measures the manner in which the effect of one factor changes/is influenced by the level of a second factor. When two-level factorials are fractionated, the ability to determine all of the interactions among a set of factors is lost. However, as we have seen, judicious fractionation leads to the ability to obtain knowledge on low-order interactions under the assumption that higher-order interactions are of negligible importance. We will soon see that orthogonal arrays are highly fractionated factorial designs. The information they produce is a function of two things: the nature of their confounding patterns and the assumptions made about the physical system to which they are applied.

It is well known that highly fractionated factorials suffer from the problem of having main effects of factors confounded/aliased with lower-order interactions effects, often with two-factor interactions. This was well documented earlier in this chapter and is discussed again in Chapter 20. Were it not for the *principle of redundancy*, which has been referred to as the *sparsity of variable effects*, and the "sequential assembly" (to be discussed in Chapter 20) of experimental designs, such highly fractionated designs might be of limited utility.

Analysis of the $L_8(2^7)$ Orthogonal Array Design Structure

Taguchi refers to the experimental design structure in Table 19.3 as an "$L_8(2^7)$ orthogonal array."[10] For the case illustrated, each of the seven columns is assigned to a different factor (A to G, 1 to 7). The two levels of each factor are designated 1 and 2.

We recognize this structure as representing the design matrix of a certain 2^{7-4} fractional factorial design. The alternative way that we are familiar with viewing this structure is shown in Table 19.4. The order of the columns, left to right, has been rearranged (note the cross-referenced column designations at the bottom), and we recognize this $L_8(2^7)$ orthogonal array as a member of the general family of 2^{7-4} fractional factorial designs. The concept of families of

---

[10] The next three examples in the chapter are based on experimental designs in G. Taguchi and Yu-In Wu, *Introduction to Off-Line Quality Control*, Central Japan Quality Control Association, Meieki Nakamura-Ku Magaya, Japan, 1979.

TABLE 19.3 $L_8(2^7)$ Orthogonal Array

Test	Factor							Result
	A 1	B 2	C 3	D 4	E 5	F 6	G 7	
1	1	1	1	1	1	1	1	y_1
2	1	1	1	2	2	2	2	y_2
3	1	2	2	1	1	2	2	y_3
4	1	2	2	2	2	1	1	y_4
5	2	1	2	1	2	1	2	y_5
6	2	1	2	2	1	2	1	y_6
7	2	2	1	1	2	2	1	y_7
8	2	2	1	2	1	1	2	y_8

two-level fractional factorials will be discussed in Chapter 20. The particular member of this family shown above is defined by the generators

$$I = -124, \qquad I = -135, \qquad I = -236, \qquad I = +1237.$$

It is noted that some of the generators carry a minus sign. The significance of this will be discussed in Chapter 20.

We have seen that when used to assign or consider seven variables, the $L_8(2^7)$ orthogonal array shown previously is equivalent to a particular 2_{III}^{7-4} fractional factorial design with generators: $I = -124$, $I = -135$, $I = -236$, and $I = 1237$. The "III" in the designation "2_{III}^{7-4}" refers to the resolution of the design, which is three, meaning that at least some (in this case, all) main effects are confounded or confused with two-factor interactions.

In the general context of the Taguchi methods, assignment of seven vari-

TABLE 19.4 2_{III}^{7-4} Fractional Factorial Design

Test	1	2	3	−12 4	−13 5	−23 6	+123 7
1	−	−	−	−	−	−	−
2	+	−	−	+	+	−	+
3	−	+	−	+	−	+	+
4	+	+	−	−	+	+	−
5	−	−	+	−	+	+	+
6	+	−	+	+	−	+	−
7	−	+	+	+	+	−	−
8	+	+	+	−	−	−	+
	D (4)	B (2)	A (1)	F (6)	E (5)	C (3)	G (7)

ables to the $L_8(2^7)$ orthogonal array and the subsequent interpretation of the experimental results assumes that variable interactions at all levels may be neglected unless specifically identified by a column in the array. In the context of the approach taken earlier in this chapter, the 2_{III}^{7-4} fractional factorial will be interpreted with more caution due to the explicit recognition of the potential importance of two-factor interaction effects, and the fact that they are confounded with the main effects of the seven variables under study.

Examination of the Use of an $L_{16}(2^{15})$ Orthogonal Array

The $L_{16}(2^{15})$ orthogonal array of Table 19.5 appears in *Introduction to Off-Line Quality Control* by Taguchi and Wu as Exercise 2 at the end of Chapter 6.

In this application of the $L_{16}(2^{15})$ orthogonal array, we see that the following assignments have been made:

- Eight columns have been assigned to variable main effects: A, B, C, D, E, F, G, H.
- Three columns have been assigned to two-factor interactions: $A \times B$, $B \times D$, $A \times D$.
- Four columns all designated as e have been assigned to the estimation of error.

TABLE 19.5 $L_{16}(2^{15})$ Orthogonal Array

Test	F 1	A 2	e 3	B 4	e 5	A×B 6	E 7	C* 8	H 9	e 10	B×D 11	e 12	A×D 13	G 14	D 15	Response
1	1	1	1	1	1	1	1	1	1	1	1	1	1	1	1	
2	1	1	1	1	1	1	1	2	2	2	2	2	2	2	2	
3	1	1	1	2	2	2	2	1	1	1	1	2	2	2	2	
4	1	1	1	2	2	2	2	2	2	2	2	1	1	1	1	
5	1	2	2	1	1	2	2	1	1	2	2	1	1	2	2	
6	1	2	2	1	1	2	2	2	2	1	1	2	2	1	1	
7	1	2	2	2	2	1	1	1	1	2	2	2	2	1	1	
8	1	2	2	2	2	1	1	2	2	1	1	1	1	2	2	
9	2	1	2	1	2	1	2	1	2	1	2	1	2	1	2	
10	2	1	2	1	2	1	2	2	1	2	1	2	1	2	1	
11	2	1	2	2	1	2	1	1	2	1	2	2	1	2	1	
12	2	1	2	2	1	2	1	2	1	2	1	1	2	1	2	
13	2	2	1	1	2	2	1	1	2	2	1	1	2	2	1	
14	2	2	1	1	2	2	1	2	1	1	2	2	1	1	2	
15	2	2	1	2	1	1	2	1	2	2	1	2	1	1	2	
16	2	2	1	2	1	1	2	2	1	1	2	1	2	2	1	

Source: Adapted from G. Taguchi and Y. Wu, *Introduction to Off-Line Quality Control*, Central Japan Quality Control Association, Meieki Nakamura-Ku Magaya, Japan, 1979.
* Column 8 was labeled as variable D in Taguchi and Wu, but has been corrected here and labeled as variable C.

The $L_{16}(2^{15})$ orthogonal array in Table 19.5 can be examined through the eyes of the class of 2^{k-p} fractional factorial designs. Table 19.6 shows the design matrix for a 2^{8-4} fractional factorial design. This design matrix was constructed in the following way:

1. Columns 1 to 4 are the columns in the $L_{16}(2^{15})$ orthogonal array assigned to variables C, B, A, and F, respectively.
2. Columns 5 to 8 are the columns in the $L_{16}(2^{15})$ orthogonal array assigned to variables E, H, G, and D, respectively.

It is interesting to note what we discover when we examine how columns 1 to 4 were used to define columns 5 to 8. Close examination of the $L_{16}(2^{15})$ orthogonal array of Table 19.5 (and Table 19.6) shows that the columns assigned to E, G, H, and D are the following products of the columns assigned to C, B, A, and F:

$$5 = 234 \qquad\qquad 6 = -14$$
$$E = B \times A \times F \qquad H = -(C \times F)$$
$$7 = 123 \qquad\qquad 8 = -1234$$
$$G = C \times B \times A \qquad D = -(C \times B \times A \times F).$$

TABLE 19.6 2^{8-4} Fractional Factorial Design

Test	1	2	3	4	234 5	−14 6	123 7	−1234 8
1	−	−	−	−	−	−	−	−
2	+	−	−	−	−	+	+	+
3	−	+	−	−	+	−	+	+
4	+	+	−	−	+	+	−	−
5	−	−	+	−	+	−	+	+
6	+	−	+	−	+	+	−	−
7	−	+	+	−	−	−	−	−
8	+	+	+	−	−	+	+	+
9	−	−	−	+	+	+	−	+
10	+	−	−	+	+	−	+	−
11	−	+	−	+	−	+	+	−
12	+	+	−	+	−	−	−	+
13	−	−	+	+	−	+	+	−
14	+	−	+	+	−	−	−	+
15	−	+	+	+	+	+	−	+
16	+	+	+	+	+	−	+	−
	C	B	A	F	E	H	G	D

[Variable "names" given by Taguchi and Wu
in the $L_{16}(2^{15})$ orthogonal array]

Hence the design generators of this design are

$$I = 2345, \quad I = 1237, \quad I = -146, \quad I = -12348.$$

It follows, then, that the associated defining relation is

$$
\begin{aligned}
I &= 2345 = -146 = 1237 = -12348 \\
&= -2356 = 1457 = -158 = -23467 \\
&= 2368 = -478 = 1678 = -23578 \\
&= 4568 = -567 = 12345678.
\end{aligned}
$$

The results here may be a little surprising. They show not only that the $L_{16}(2^{15})$ orthogonal array is a 2^{8-4} fractional factorial design, but that in particular, this one is a resolution III 2^{8-4} fractional factorial. Note the terms in the defining relation: -146, -258, -478, -567. This is surprising because we know that it is possible to design a 16-run two-level fractional factorial to examine eight variables that is resolution IV. In fact, Box, Hunter, and Hunter, in *Statistics for Experimenters*,[11] recommend the 2^{8-4} design with generators

$$
\begin{array}{cccc}
5 = 234 & 6 = 134 & 7 = 123 & 8 = 124 \\
(I = 2345 & I = 1346 & I = 1237 & I = 1248).
\end{array}
$$

This produces a resolution IV design. Recall that a resolution III design confounds at least some main effects with two-factor (and higher-order) interactions, while a resolution IV design confounds at least some main effects with three-factor (and higher-order) interactions (but not two-factor interactions).

Under the Taguchi methods approach to the design of experiments, the philosophy toward interaction effects is somewhat different. As a result, the bottom line is that unless expressly identified as present/real and assigned to the orthogonal array, interactions at *all* levels are assumed to be negligible. We must remember, however, that although different philosophies may be used to interpret the workings of the real world, those "workings" are unchanged by the assumption. Hence, it is probably always wise to select orthogonal arrays of the highest possible resolution so as to minimize the effect that the assumptions we do make about interaction effects has on the interpretation of the results.

Another Interesting Orthogonal Array Application

Another interesting example of the use of orthogonal arrays by Taguchi concerns an arc welding experiment performed by the National Railway Cooperation of Japan in 1959.[12] Nine variable main effects and four specific two-factor interactions were studied using the $L_{16}(2^{15})$ orthogonal array shown in Table 19.7.

---

[11] G. E. P. Box, W. G. Hunter, and J. S. Hunter, *Statistics for Experimenters*, Wiley, New York, 1978.
[12] See footnote 10.

TABLE 19.7 $L_{16}(2^{15})$ Orthogonal Array Example

Factor	Number of Levels	First Level	Second Level
A: Kind of welding rods	2	A_1 = J100	A_2 = B17
B: Drying methods of rods	2	B_1 = no drying	B_2 = one-day drying
C: Welded materials	2	C_1 = SS41	C_2 = SB35
D: Thickness of welded material	2	D_1 = 8 mm	D_2 = 12 mm
E: Angle of welding device	2	E_1 = 70 degrees	E_2 = 60 degrees
F: Opening of welding device	2	F_1 = 1.5 mm	F_2 = 3 mm
G: Current	2	G_1 = 150 A	G_2 = 130 A
H: Welding methods	2	H_1 = weaving	H_2 = single
I: Preheating	2	I_1 = no preheating	I_2 = 150 degrees preheating

Test	A	G	A × G	H	A × H	G × H	B	D	E	F	I	e	e	C	A × C
1	1	1	1	1	1	1	1	1	1	1	1	1	1	1	1
2	1	1	1	1	1	1	1	2	2	2	2	2	2	2	2
3	1	1	1	2	2	2	2	1	1	1	1	2	2	2	2
4	1	1	1	2	2	2	2	2	2	2	2	1	1	1	1
5	1	2	2	1	1	2	2	1	1	2	2	1	1	2	2
6	1	2	2	1	1	2	2	2	2	1	1	2	2	1	1
7	1	2	2	2	2	1	1	1	1	2	2	2	2	1	1
8	1	2	2	2	2	1	1	2	2	1	1	1	1	2	2
9	2	1	2	1	2	1	2	1	2	1	2	1	2	1	2
10	2	1	2	1	2	1	2	2	1	2	1	2	1	2	1
11	2	1	2	2	1	2	1	1	2	1	2	2	1	2	1
12	2	1	2	2	1	2	1	2	1	2	1	1	2	1	2
13	2	2	1	1	2	2	1	1	2	2	1	1	2	2	1
14	2	2	1	1	2	2	1	2	1	1	2	2	1	1	2
15	2	2	1	2	1	1	2	1	2	2	1	2	1	1	2
16	2	2	1	2	1	1	2	2	1	1	2	1	2	2	1

Nine columns are assigned to variables: A, G, H, B, D, E, F, I, C
Four columns are assigned to interactions: A × G, A × H, G × H, A × C
Two columns are assigned to error (e) estimation

This orthogonal array may be recognized as a 2^{9-5} fractional factorial as follows:

Test	D 1	H 2	G 3	A 4	B 5	E 6	F 7	I 8	C 9
1	−	−	−	−	−	−	−	−	−
2	+	−	−	−	−	+	+	+	+
3	−	+	−	−	+	−	−	−	+

(continued)

Test	D 1	H 2	G 3	A 4	B 5	E 6	F 7	I 8	C 9
4	+	+	−	−	+	+	+	+	−
5	−	−	+	−	+	−	+	+	+
6	+	−	+	−	+	+	−	−	−
7	−	+	+	−	−	−	+	+	−
8	+	+	+	−	−	+	−	−	+
9	−	−	−	+	+	+	−	+	+
10	+	−	−	+	+	−	+	−	−
11	−	+	−	+	−	+	−	+	−
12	+	+	−	+	−	−	+	−	+
13	−	−	+	+	−	+	+	−	−
14	+	−	+	+	−	−	−	+	+
15	−	+	+	+	+	+	+	−	+
16	+	+	+	+	+	+	−	+	−

This design is a 2_{III}^{9-5} fractional factorial with generators

$$I = 2345, \quad I = -146, \quad I = -137, \quad I = 1348, \quad I = -12349.$$

The defining relation for this design is

$$I = 2345 = -146 = -137 = 1348 = -12349$$
$$= -12356 = 2567 = 12345678 = -156789 = -1345679$$
$$= -24568 = 135689 = 4569 = -12457 = -23578 = 145789$$
$$= 3579 = 1258 = -34589 = -159 = 3467 = 1678 = -2346789$$
$$= -12679 = -368 = 124689 = 2369 = -478 = 123789 = 2479 = -289.$$

The linear combinations of aliased effects that can be estimated, assuming third- and higher-order interactions can be neglected, are as follows (in the regular calculation matrix order):

l_0 estimates I (mean) l_8 estimates $23 + 45 + 69$
l_1 estimates $1 - 46 - 37 - 59$ l_9 estimates $24 + 35 + 79$
l_2 estimates $2 - 89$ l_{10} estimates $34 + 25 + 18 + 67$
l_3 estimates $3 - 17 - 68$ l_{11} estimates $-27 - 49 - 56$
l_4 estimates $4 - 16 - 78$ l_{12} estimates $-26 - 39 - 57$ (error)
l_5 estimates $12 + 58$ (error) l_{13} estimates $8 - 36 - 47 - 29$
l_6 estimates $-7 + 13 + 48$ l_{14} estimates $5 - 19$
l_7 estimates $-6 + 14 + 38$ l_{15} estimates $-9 + 15 + 28.$

Now the $L_{16}(2^{15})$ or 2_{III}^{9-5} fractional factorial design above is of resolution III, that is, at least some (in this case, all) of the main effects are confounded with two-factor interactions. In fact, it is not possible to examine the effects of nine variables in a 16-run two-level fractional factorial and have a design resolution better than III. However, it is possible to find another resolution III 2^{9-5} design/orthogonal array that has potentially a more attractive alias structure.

As an alternative to the design above, Box, Hunter, and Hunter recommend a 2_{III}^{9-5} design with the following generators:

$$I = 1235, \quad I = 2346, \quad I = 1347, \quad I = 1248, \quad I = 12349.$$

The reason for this choice over the one above [the $L_{16}(2^{15})$] becomes apparent when we examine the complete alias structure of the design. For this 2_{III}^{9-5} design, the linear combinations of aliased effects are (again, assuming third- and higher-order interactions can be neglected)

l_0 estimates mean	l_8 estimates $23 + 15 + 46 + 78$
l_1 estimates $1 + 69$	l_9 estimates $24 + 36 + 18 + 57$
l_2 estimates $2 + 79$	l_{10} estimates $34 + 26 + 17 + 58$
l_3 estimates $3 + 89$	l_{11} estimates $5 + 49$
l_4 estimates $4 + 59$	l_{12} estimates $8 + 39$
l_5 estimates $12 + 35 + 48 + 67$	l_{13} estimates $7 + 29$
l_6 estimates $13 + 25 + 47 + 68$	l_{14} estimates $6 + 19$
l_7 estimates $14 + 37 + 28 + 56$	l_{15} estimates $9 + 45 + 16 + 27 + 38.$

When we look closely at the above we notice something quite interesting. Although main effects are still confounded with two-factor interactions, eight of the nine main effects are confounded with only one two-factor interaction. Furthermore, in each case that two-factor interaction involves variable 9 (i.e., 69, 79, 89, 59, 49, 39, 29, 19). The one factor that is confounded with four two-factor interactions is variable 9.

Essentially, we might use the following design strategy: Among the nine factors to be studied select the one that is least likely to be important and/or to interact with other factors. Assign this factor as variable 9. If your thoughts about this factor's unimportance are correct, the design becomes effectively a resolution IV design (i.e., it is a 2_{IV}^{8-4} design with main effects "clear" of two-factor interaction aliases). Again, what we have witnessed in this example is the importance of design resolution and the judicious choice of design generators.

Exercises

19.1. Suppose that it is desired to study eight variables at two levels each in a full factorial design scheme.
   a. How many tests are required?
   b. How many effects can be estimated? How many two-factor interactions can be estimated? How many three-factor interactions? How many four-factor interactions and above can be estimated?
   c. Comment on the general advisability of running such an experiment.
   d. If this experiment were performed, of what use might the higher-order interaction effect estimates be in the analysis?

19.2. It is desired to study the effect of five factors on the burr height in a metal stamping process. Only enough funds are available to conduct 16 tests; therefore, a 2^{5-1} fractional factorial design will be performed.

    a. What generator should be used to get the highest resolution fractional factorial? What is the defining relation?

    b. Show the complete set of linear combinations of confounded effects that result from the generator that you chose.

    c. How does this confounding structure simplify if third- and higher-order interactions are negligible?

19.3. Consider a five-variable, eight-run, two-level fractional factorial design.

    a. Develop the confounding structure based on the following sets of generators:

       (i) $4 = 12$, $5 = 13$.

       (ii) $4 = 12$, $5 = 123$.

    b. Does there appear to be any reason to choose one set of generators over another in part (a)? Indicate the reasons for your choice of generators clearly.

19.4. Consider a 2^{8-4} fractional factorial design.

    a. Write down the defining relation for each of the following candidate designs:

       (i) $5 = 234$, $6 = 134$, $7 = 123$, $8 = 124$.

       (ii) $5 = 234$, $6 = 134$, $7 = 123$, $8 = 1234$.

    b. Which of these designs is preferred? Why?

19.5. Consider a 2^{7-4} fractional factorial design with the generators $4 = 23$, $5 = 12$, $6 = 123$, and $7 = 13$.

    a. Write down the defining relation for this design.

    b. Write down the linear combinations of effects that may be estimated from such a design.

    c. Under the assumption that three-factor interactions and higher are negligible, how do the linear combinations of effects that can be estimated simplify?

19.6. Consider a 2^{7-2} fractional factorial design with the generators $6 = 123$ and $7 = 345$.

    a. Write down the defining relation for this design.

    b. Write down the linear combinations of effects that may be estimated from such a design.

    c. Under the assumption that three-factor interactions and higher are negligible, how do the linear combinations of effects simplify?

19.7. A 2^{6-1} fractional factorial design was conducted. The linear combinations of effects that were estimated based on the measured responses from the experiment are as follows.

$$l_0 = \quad 98.875 \text{ estimates } I + 123456/2 \qquad l_{123} = \quad -2.875 \text{ estimates } 123 + 456$$
$$l_1 = \quad 36.750 \text{ estimates } 1 + 23456 \qquad l_{124} = \quad 4.625 \text{ estimates } 124 + 356$$
$$l_2 = \quad -5.375 \text{ estimates } 2 + 13456 \qquad l_{125} = \quad -0.625 \text{ estimates } 125 + 346$$
$$l_3 = \quad 1.750 \text{ estimates } 3 + 12456 \qquad l_{134} = \quad -0.500 \text{ estimates } 134 + 256$$

l_4 = 0.750 estimates 4 + 12356

l_5 = 1.750 estimates 5 + 12346

l_{12} = −23.875 estimates 12 + 3456

l_{13} = 2.750 estimates 13 + 2456

l_{14} = −2.250 estimates 14 + 2356

l_{15} = −0.500 estimates 15 + 2346

l_{23} = 0.125 estimates 23 + 1456

l_{24} = −0.375 estimates 24 + 1356

l_{25} = 0.625 estimates 25 + 1346

l_{34} = −3.000 estimates 34 + 1256

l_{35} = 5.750 estimates 35 + 1246

l_{45} = −0.750 estimates 45 + 1236

l_{135} = 3.500 estimates 135 + 246

l_{145} = 3.500 estimates 145 + 236

l_{234} = 2.875 estimates 234 + 156

l_{235} = 2.125 estimates 235 + 146

l_{245} = −2.625 estimates 245 + 456

l_{345} = 2.000 estimates 345 + 126

l_{1234} = 17.875 estimates 1234 + 56

l_{1235} = −1.625 estimates 1235 + 46

l_{1245} = −1.625 estimates 1245 + 36

l_{1345} = −2.250 estimates 1345 + 26

l_{2345} = −1.625 estimates 2345 + 16

l_{12345} = −39.875 estimates 12345 + 6

a. Write down the generator and the defining relation for this design.

b. Based on the assumption that third- and higher-order interactions are negligible, several of the linear combinations shown above may be used to estimate the standard error of an effect estimate. Under this assumption, develop 95% confidence intervals for the true mean effect values to determine which effect estimates/linear combinations of effects are significant. Comment on the results.

19.8. Consider a 2^{5-2} fractional factorial design with the generators $I = 124$ and $I = 135$.

a. What is the defining relation for this design?

b. For this design, complete the recipe matrix shown in the table.

Test	Variable				
	1	2	3	4	5
1	−	−	−		
2	+	−	−		
3	−	+	−		
4	+	+	−		
5	−	−	+		
6	+	−	+		
7	−	+	+		
8	+	+	+		

c. Write down the complete confounding pattern for this design; that is, what linear combinations of effects can be estimated from this design?

d. What is the resolution of this design? What does the resolution tell us about our ability to interpret the results of such an experiment?

19.9. The table provides the design matrix for a certain two-level fractional factorial experimental design.

Test	Variable 1	2	3	4	5
1	−	−	−	+	−
2	+	−	−	−	+
3	−	+	−	−	+
4	+	+	−	+	−
5	−	−	+	+	+
6	+	−	+	−	−
7	−	+	+	−	−
8	+	+	+	+	+

a. Write down the generators and the defining relation for this design.
b. What is the resolution of this design?
c. Write down all of the linear combinations of confounded effects that can be estimated from the result of this experiment. Assume that all interaction effects may be important.
d. Suppose that it was decided to run only a four-variable experiment using eight tests (i.e., a 2^{4-1} fractional factorial design). What generator would you propose for this design? How would the design resolution differ, if at all, for this new 2^{4-1} design as compared with the original 2^{5-2} design?

19.10. Consider a 2^{7-3} fractional factorial design with the generators $I = 1235$, $I = 1246$, and $I = 1347$.
a. Write down the defining relation for this design.
b. What is the resolution of this design? What does this tell us?
c. Write down all linear combinations of effects that can be estimated from this experiment. Assume that third- and higher-order interaction effects can be neglected.

19.11. Consider a 2^{8-4} fractional factorial design.
a. How many variables does this design have?
b. How many runs are involved in this design?
c. How many levels are used for each of the variables?
d. How many generators are there for this design?
e. How many words are there in the defining relation (including I)?

19.12. Construct a 2^{6-2} fractional factorial design with as high a resolution as possible.
a. Write down the design matrix and the calculation matrix for this design.
b. What are the generators for your design?
c. What is the defining relation for your design?
d. What is confounded with the main effect of variable 3 in your design?
e. What is confounded with the two-factor interaction 12 in your design?

19.13. It is proposed to conduct a 2^{8-3} fractional factorial design to study a certain process. After some discussion, it is decided that only 16 tests can be conducted, and so a 2^{8-4} design is proposed. Will the increased level of fractionation seriously affect the nature of the effect estimates, or the density of the confounding structure, that can be obtained from the experiment? Demonstrate your answer clearly.

19.14. A consulting firm engaged in road-building work is asked by one of its clients to carry out an experimental study to determine the effects of six variables on the physical properties of a certain kind of asphalt. Call these variables A, B, C, D, E, and F.

   a. If a two-level fractional factorial design of at least resolution IV is required for this situation, what is the fewest number of tests in which this can be accomplished?

   b. What would be the generator for the best 32-run design you would propose for this situation? What is the design resolution for this situation?

   c. If a seventh variable were to be added, what is the highest-resolution 32-run fractional factorial design that can be obtained? What generators would you propose for this design?

19.15. A welding experiment has been conducted to study the effect of five variables on the heat input in watts.[13] The variables and their settings are shown in the first table.

Variable	Low Level	High Level
x_1: Open-circuit voltage (V)	31	34
x_2: Slope	11	6
x_3: Electrode melt-off rate (in./min)	162	137
x_4: Electrode diameter (in.)	0.045	0.035
x_5: Electrode extension (in.)	0.375	0.625

The second table summarizes the levels of the five variables that were used for each of the tests (note that the tests are shown in the randomized order in which they were run), along with the measured heat input that was produced for that run.

Run	x_1	x_2	x_3	x_4	x_5	y
1	34	11	162	0.045	0.375	4141
2	31	6	162	0.045	0.375	3790
3	31	6	162	0.035	0.625	2319
4	34	6	137	0.045	0.375	3765
5	31	6	137	0.035	0.375	2485
6	34	6	162	0.045	0.625	4061

(*continued*)

[13] D. A. J. Stegner, S. M. Wu, and N. R. Braton, "Prediction of Heat Input for Welding," *Welding Journal Research Supplement*, Vol. 1, March 1967.

Run	x_1	x_2	x_3	x_4	x_5	y
7	31	6	137	0.045	0.625	3507
8	34	11	137	0.045	0.625	3425
9	34	6	137	0.035	0.625	2450
10	34	11	137	0.035	0.375	2466
11	31	11	162	0.035	0.375	2580
12	31	11	162	0.045	0.625	3318
13	31	11	137	0.035	0.625	1925
14	34	11	162	0.035	0.625	2450
15	31	11	137	0.045	0.375	3431
16	34	6	162	0.035	0.375	3067

   a. The kind of experiment shown in the second table is a 2^{k-p} fractional factorial design. What are the values for k and p? Write out the complete calculation matrix for this design.
   b. What is the generator for this fractional factorial? What is the defining relation?
   c. What effects are associated with each linear combination that may be calculated?
   d. What are the numerical values for each linear combination of effects?
   e. Under the assumption that three-factor and higher-order interactions are negligible, interpret the results of this experiment.

19.16. An experimenter wishes to study seven variables at two levels each and chooses to do so with a 32-run resolution IV design with the generators $I = 12346$ and $I = 12457$. A colleague argues that he can achieve the same resolution with a design that employs only 16 runs. Is the colleague correct? If so, write down generators and the defining relation for such a design. Write out all of the linear combinations of effects that can be estimated for each design. Comment on the relative merits of each design.

19.17. A 2^{4-1} fractional factorial design was conducted on a chemical process by assigning variable 4 to the 123 interaction column.

Variable	Low Level	High Level
1. Feedrate (liters/min)	5	20
2. Catalyst (%)	A	B
3. Temperature (°C)	200	220
4. Concentration (%)	5	7

The second table summarizes the 8 tests that were run, including the levels of each of the four variables and the yield (% reacted) for each test.

Test	1	2	3	4	y (% reacted)
1	−	−	−	−	33
2	+	−	−	+	51
3	−	+	−	+	44
4	+	+	−	−	40
5	−	−	+	+	35
6	+	−	+	−	82
7	−	+	+	−	46
8	+	+	+	+	69

    a. Write down all of the linear combinations of effects that can be estimated from this experiment (assume that third- and higher-order interactions are negligible).

    b. Calculate numerical values for the effect estimates and determine which effects are significant using a normal probability plot of the effect estimates.

    c. Propose a tentative mathematical model for the response.

19.18. A newly hired engineer has been asked to analyze the experimental results of the fractional factorial design shown in the table. It seems that this experiment was designed and performed by an employee who is now working for a competitor of the company. The response of interest is the number of particles per million of a certain contaminant. The response was available for all of the tests, but as is evident, some of the test settings were not readable. The design matrix is given in the order in which the tests were performed.

Run	1	2	3	4	5	6	y
1	?	?	?	?	?	?	16
2	−	+	−	−	+	−	23
3	+	+	−	−	−	+	15
4	−	−	−	+	−	+	19
5	?	?	?	?	?	?	21
6	+	−	−	+	+	−	14
7	−	−	+	−	+	+	27
8	+	−	+	−	−	−	12

    a. What kind of a two-level design is this? Be as specific as you can in defining this experiment.

    b. Provide the missing information concerning the variable settings.

    c. Provide estimates of what variable effects (or linear combination of effects) you can. Can you say anything about the results of this experiment?

19.19. Some tests were carried out on a newly designed automobile. Five variables were studied, as shown in the first table.

Variable	Low Level	High Level
1. Engine size (L)	4.0	4.5
2. Fuel octane	87	93
3. Tire pressure (psi)	22	28
4. Driving speed (mph)	45	55
5. Air conditioning	Off	On

The objective of the experiment was to determine the effects of these variables on the fuel economy. The experimental results are shown in the following table.

Test	x_1	x_2	x_3	x_4	x_5	y(mpg)
1	−1	−1	−1	−1	1	25
2	1	−1	−1	−1	−1	21
3	−1	1	−1	−1	−1	26
4	1	1	−1	−1	1	22
5	−1	−1	1	−1	−1	29
6	1	−1	1	−1	1	20
7	−1	1	1	−1	1	29
8	1	1	1	−1	−1	29
9	−1	−1	−1	1	−1	27
10	1	−1	−1	1	1	19
11	−1	1	−1	1	1	27
12	1	1	−1	1	−1	28
13	−1	−1	1	1	1	26
14	1	−1	1	1	−1	21
15	−1	1	1	1	−1	32
16	1	1	1	1	1	25

a. Completely define the linear combinations of effects that can be obtained from this experiment.
b. Analyze the data. Propose a tentative mathematical model and do a complete analysis of the model residuals. Revise the model, if necessary, and repeat the residual analysis.

19.20. A two-level fractional factorial design was run with 16 unique test conditions. One of the linear combinations of confounded effects that can be estimated from this particular design is 13 + 2345 + 1246 + 56. Determine the design generators. Show all work. What is the resolution of this design? What does *resolution* mean?

19.21. It is desired to study seven variables in a two-level factorial or fractional factorial scheme. Write down and discuss all the options available to the experimenter from the most highly fractionated design to the full factorial. Discuss the advantages and disadvantages of each design. Which design would be best if:

a. Only main effects were felt to be important?

b. Two or three of the variables are likely to have two-factor interactions with each other and/or some of the other factors?

c. Main effects and two-factor interactions may be important but the estimation of main effects is of particular importance?

19.22. Consider the following two 2^{9-4} fractional factorial designs:

     (i) $I = 23456$, $I = 13457$, $I = 12458$, $I = 12359$.

    (ii) $I = 1236$, $I = 1247$, $I = 1358$, $I = 3459$.

a. Determine the defining relation for each of these designs.

b. Which of these designs is preferred? Why?

c. If it is known that variables 6, 7, 8, and 9 cannot physically interact with each other or any of the other five variables, which design is preferred? Why?

19.23. Consider a 2^{9-3} fractional factorial design.

a. How many variables does the design consider?

b. How many tests are required, assuming no replication?

c. How many generators are there for this design?

d. How many words are there in the defining relation (including I)?

19.24. Is it possible to develop a resolution V two-level fractional factorial for six variables that requires only 16 tests? Explain your answer.

19.25. What is the maximum number of variables that could be examined in a 64-run two-level fractional factorial design and maintain at least a resolution V design? If it is desired to consider two additional variables in the scheme you just proposed, will the number of tests have to increase to maintain a resolution V design? How many tests are now required?

19.26. Consider the following defining relation for a two-level fractional factorial: $I = 1236 = 1247 = 23458 = 3467 = 1568 = 13578 = 25678$.

a. How many generators are there for this design?

b. What is the resolution of this design?

c. How many tests are required, assuming no replication?

d. If the number of tests were doubled, what would happen to the resolution of the design? (Assume the best-case situation.)

e. If one more variable is added to the design but the number of tests is held constant, will this affect the resolution of the design? What if two more variables are added?

19.27. Following are the generators for a 2^{9-5} resolution III design: $I = 1235$, $I = 2346$, $I = 1347$, $I = 1248$, $I = 12349$. This design has a somewhat interesting characteristic that could be particularly useful if a high premium is placed on keeping the number of tests in the experiment to a minimum. What is this characteristic? You will probably have to write down all the linear combinations of effects that can be estimated from this design to discover what is unique about this nine-variable, 16-run, two-level fractional factorial.

19.28. It is desired to study nine variables in a fractional factorial scheme requiring 128 tests or less. Write down and discuss all of the options available to the experimenter from the most highly fractionated design to the least fractionated design (remember, no more than 128 tests). Discuss the advantages and disadvantages of each design.

19.29. Following are the generators for three 16-run fractional factorial designs:

  (i) 5 = 123, 6 = 234.
  (ii) 5 = 123, 6 = 234, 7 = 134.
  (iii) 5 = 123, 6 = 234, 7 = 134, 8 = 124.

   a. Determine the defining relation for each design. What is the resolution for each design?
   b. Write out the linear combinations of effects that may be determined for each design assuming that three-factor and higher-order interactions are negligible.
   c. Explain what happens in this case to the confounding structure as the number of variables studied in 16 tests is increased from 6 to 7 to 8?

19.30. A 2^{6-2} fractional factorial design was performed to assess the effect of six variables on the cutting force produced during the machining of an aluminum alloy. The variables studied, and their levels for the experiment, are shown in the first table. The design matrix for the experiment and the response (cutting force, in pounds) are shown in the second table.

Variable	Low Level	High Level
1. Feedrate (in./rev)	0.005	0.010
2. Depth of cut (in.)	0.050	0.100
3. Cutting speed (sf/min)	2000	2500
4. Tool geometry	Positive	Negative
5. Machine	American	Monarch
6. Cutting fluid	Absent	Present

Test	1	2	3	4	5	6	y
1	−	−	−	−	−	−	21
2	+	−	−	−	+	−	59
3	−	+	−	−	+	+	57
4	+	+	−	−	−	+	122
5	−	−	+	−	+	+	29
6	+	−	+	−	−	+	52
7	−	+	+	−	−	−	45
8	+	+	+	−	+	−	110
9	−	−	−	+	−	+	31
10	+	−	−	+	+	+	70
11	−	+	−	+	+	−	65
12	+	+	−	+	−	−	119
13	−	−	+	+	+	−	35
14	+	−	+	+	−	−	61
15	−	+	+	+	−	+	57
16	+	+	+	+	+	+	121

    a. Completely define the linear combinations of effect that can be obtained from this experiment assuming that three-factor and higher-order interactions are negligible.

    b. Calculate numerical values for the linear combinations of effect and determine which of these are significant using a normal probability plot.

    c. Under the assumption that three-factor interactions and above are negligible, several of the linear combinations may be used to estimate the standard error of an effect estimate. Use this assumption to develop 95% confidence intervals for the true mean effect values. Which effects/linear combinations are significant? Comment on the results of the experiment.

19.31. An import/export company is working with a construction engineering firm to study the tensile strength of various concrete formulations. Six variables of interest have been identified, and are given in the first table.

Variable	Low Level	High Level
1. Specimen size (ft^3)	2	4
2. Amount of water	Low	High
3. Curing time (hr)	24	48
4. Mixing technique	By hand	Machine
5. Aggregate size	Fine	Coarse
6. Cement concentration	Low	High

A replicated $L_8(2^7)$ orthogonal array was performed to study the effects of these factors, and the results of this experiment are summarized in the second table.

Test	1	2	3	4	5	e	6	Run 1	Run 2
1	1	1	1	2	2	2	1	2.3	2.2
2	2	1	1	1	1	2	2	3.5	3.3
3	1	2	1	1	2	1	2	3.0	2.9
4	2	2	1	2	1	1	1	2.1	1.9
5	1	1	2	2	1	1	2	3.5	3.6
6	2	1	2	1	2	1	1	2.6	2.6
7	1	2	2	1	1	2	1	2.9	2.9
8	2	2	2	2	2	2	2	3.9	4.0

    a. Using the assumptions typically made when analyzing orthogonal arrays, calculate estimates of all the variable effects.

    b. Construct an ANOVA table to identify which of the variable effects determined in part (a) are significant.

    c. Develop a model for the response based on the results of part (b). Construct an ANOVA table to see if the model developed suffers from lack of fit.

    d. If the model developed in part (c) demonstrates a lack of fit, reanalyze the data using the fractional factorial design techniques described in this chapter, and comment on what might be the problem here in terms of how the study was initially formulated and analyzed.

19.32. A $L_8(2^7)$ orthogonal array was conducted to study the effect of seven variables on the finish of a machined surface (roughness average value, in microinches). The first table describes the levels for the seven variables that were studied in the experiment.

Variable	Low Level	High Level
1. Cutting speed (sfpm)	600	900
2. Feedrate (in./rev)	0.005	0.010
3. Nose radius (in.)	1/64	1/32
4. Depth of cut (in.)	0.050	0.100
5. Side rake (deg)	0	10
6. Lead angle (deg)	0	30
7. Back rake (deg)	0	10

The second table summarizes the conditions for the eight tests that were performed as well as the measured surface finish (R_a) value that was observed for each of the conditions.

Test	x_1	x_2	x_3	x_4	x_5	x_6	x_7	y
1	1	1	1	2	2	2	1	71
2	2	1	1	1	1	2	2	57
3	1	2	1	1	2	1	2	130
4	2	2	1	2	1	1	1	171
5	1	1	2	2	1	1	2	21
6	2	1	2	1	2	1	1	24
7	1	2	2	1	1	2	1	148
8	2	2	2	2	2	2	2	100

a. What are the generators and the defining relation for this design?

b. Using the assumptions typically made when analyzing orthogonal arrays, calculate estimates of all the variable effects.

c. Construct an ANOVA table to identify which of the variable effects determined in part (a) are significant.

d. Develop a model for the response based on the results of part (c). Construct an ANOVA table to see if the model developed suffers from lack of fit.

e. Prepare all the residual plots necessary to check the model diagnostically.

f. If the model developed in part (d) demonstrates a lack of fit or, based on the residual plots, is found to be inadequate, postulate a new model. Conduct all the steps necessary to develop and check the new model.

19.33. What is the maximum number of variables that can be examined using an $L_8(2^7)$ orthogonal array? What are the assumptions that generally govern the construction and interpretation of orthogonal arrays as promoted by Taguchi?

19.34. In the selection, implementation, and interpretation of the results of orthogonal arrays, as advocated by Taguchi, why is the concept of design resolution so important?

19.35. Suppose that an investigator decides to use an $L_{16}(2^{15})$ orthogonal array to study the effect of 18 factors on the yield of a chemical etching process. What is the most basic problem that arises in this situation? How would you propose to solve the problem? Propose at least two alternatives.

20

Sequential and Iterative Nature of Experimentation

20.1

Value of Sequential Experimentation: A Case Study

In the early stages of an investigation it is often the case that many variables seem to be of potential importance. It would not be uncommon for a project group or task force to draw up a list that could range from as few as 5 up to 15 or more variables. In the final analysis, perhaps only two or three of these variables may prove to be important. The problem is: Which ones are important? It therefore seems that the first task at hand is to conduct some experiments that will quickly reduce the number of variables under study to a few seemingly important ones which will then be the focus of further experimentation. For this screening task, two-level fractional factorial designs constitute a powerful and efficient tool.

Sometimes, situations are presented that require troubleshooting. Perhaps, problems have arisen with a particular manufacturing process, a procedure such as assembly of a component, or the design of a component seems to be giving trouble in the field. The initial task is to identify quickly the few key factors that seem to govern the situation at hand. If investigators attempt to get their arms around the entire problem by designing one comprehensive experiment, the resource requirements will probably be extensive and the final results of the experiment may be inconclusive due to poor selection of variable levels

		Level (weight %)	
Variable	Ingredient	Low (−)	High (+)
1	Soybean emulsion: 9.3% soybean solids	1.67	5.00
2	Vegetable fat: hydrogenated coconut oil	10.00	20.00
3	Carbohydrates: corn syrup solids	0.00	5.00
4	Emulsifiers: mono- and diglycerides	0.17	0.50
5	Primary stabilizer: hydroxypropyl methyl cellulose	0.00	0.50
6	Secondary stabilizer: microcrystalline cellulose	0.00	0.25
7	Salt: sodium chloride	0.00	0.10

TABLE 20.1   Variables Under Study Together with Their Low and High Settings*

* Amount of sucrose kept constant at 7%; water added to balance to 100%.

and/or a poorly controlled experimental environment precipitated by a large experiment.

In an experimental investigation, it is usually wise to take a sequential approach, building up knowledge gradually through a series of related experiments. In this regard, two-level fractional factorial designs serve as useful building blocks. To this end, an application of fractional factorial designs will be illustrated via an example that is adapted from a study described in the *Journal of Food Science*.[1]

Description of the Problem and Initial Experiment

Nondairy whipped topping is a fabricated food product that serves as a substitute for whipped cream dessert topping. It is generally formulated with sodium caseinate, vegetable fat, carbohydrates, and emulsifiers. Additives such as stabilizers, corn syrup solids, and salts are also widely incorporated into various formulations. Owing to the relatively high price of sodium caseinate, an investigation was conducted into using an aqueous soybean emulsion in place of the sodium caseinate. For an initial investigation to study the effects of the various ingredients and their concentrations on the resulting whipped topping, the seven variables and variable levels in Table 20.1 were selected.

In the evaluation of a whipped topping, the following characteristics are typically of interest: homogeneity of air cell distribution, stability, stiffness, stand-up, richness, sweetness, gumminess, greasy aftertaste, and texture

[1] E. T. S. Chow, L. S. Wei, R. E. DeVor, and M. P. Steinberg, "Application of Two-Level Fractional Factorial Designs in Development of a Soybean Whipped Topping," *Journal of Food Science*, Vol. 48, 1983, p. 230.

TABLE 20.2   Base Design Calculation Matrix

Test	I	1	2	3	12	13	23	123
1	+	−	−	−	+	+	+	−
2	+	+	−	−	−	−	+	+
3	+	−	+	−	−	+	−	+
4	+	+	+	−	+	−	−	−
5	+	−	−	+	+	−	−	+
6	+	+	−	+	−	+	−	−
7	+	−	+	+	−	−	+	−
8	+	+	+	+	+	+	+	+

mouthfeel. The overrun, or whipability, of a given formulation is given by the equation

$$\% \text{ overrun} = 100 \times \frac{W_1 - W_2}{W_2},$$

where W_1 is the weight of a given volume of liquid and W_2 is the weight of the same volume after whipping. Although each of the characteristics described above was studied during the original investigation, the analysis that follows focuses only on the overrun performance measure.

The design that was used for the initial experiment was a 2^{7-4} fractional factorial design. The *base design calculation matrix* for a 2^{7-4} design is shown in Table 20.2. Since four additional variables are to be introduced here, all of the interaction columns will be used and hence the design is a saturated fractional factorial. The recipe (design) matrix that was selected for this 2^{7-4} design is displayed in Table 20.3. Each trial was performed only once, in random order. Table 20.3 also displays the overrun associated with each of the experimental conditions.

TABLE 20.3   Design (Recipe) Matrix

Test	1	2	3	4	5	6	7	Overrun (%)
1	−	−	−	+	+	+	−	115
2	+	−	−	−	−	+	+	81
3	−	+	−	−	+	−	+	110
4	+	+	−	+	−	−	−	69
5	−	−	+	+	−	−	+	174
6	+	−	+	−	+	−	−	99
7	−	+	+	−	−	+	−	80
8	+	+	+	+	+	+	+	63

Generators, Defining Relation, and Confounding Structure

The 2^{7-4} fractional factorial design illustrated in the design matrix above was obtained by introducing variables 4, 5, 6, and 7 through the following interaction columns in the 2^3 base design calculation matrix:

$$4 = 12, \quad 5 = 13, \quad 6 = 23, \quad 7 = 123.$$

Hence the generators for this design are

$$I = 124, \quad I = 135, \quad I = 236, \quad I = 1237.$$

The defining relation for this design is obtained by adding to the generators all of their products taken two, three, and four at a time. The complete defining relation is

$$I = 124 = 135 = 236 = 1237$$
$$= 2345 = 1346 = 347 = 1256 = 257 = 167$$
$$= 456 = 1457 = 2467 = 3567$$
$$= 1234567.$$

This relationship completely defines the confounding structure of this design. This design is clearly of resolution III, since the length of the smallest word in the defining relation is 3 (excluding I). Therefore, at least some of the main effects are confounded with two-factor interactions.

Obtaining the Linear Combinations of Confounded Effects

Given the defining relation, the sets of confounded effects can be determined by multiplying the column headings in the base design calculation matrix by every term in the defining relation, including I. For example, for the column denoted 1, we have

$$(1)I = (1)124 \quad = (1)135 \quad = (1)236 \quad = (1)1237$$
$$= (1)2345 \quad = (1)1346 = (1)347 \quad = (1)1256 = (1)257 = (1)167$$
$$= (1)456 \quad = (1)1457 = (1)2467 = (1)3567$$
$$= (1)1234567,$$

which simplifies to

$$1 \quad = 24 \quad = 35 \quad = 1236 \quad = 237$$
$$= 12345 \quad = 346 = 1347 \quad = 256 \quad = 1257 = 67$$
$$= 1456 \quad = 457 = 12467 = 13567$$
$$= 234567.$$

Assuming that third- and higher-order interactions are negligible, we arrive at

$$1 = 24 = 35 = 67.$$

Hence the main effect of variable 1 is confounded with the two-factor interactions 24, 35, and 67.

As another example, for the column denoted 12, we have

$$
\begin{aligned}
(12)I &= (12)124 & = (12)135 & = (12)236 & = (12)1237 \\
&= (12)2345 & = (12)1346 & = (12)347 & = (12)1256 & = (12)257 & = (12)167 \\
&= (12)456 & = (12)1457 & = (12)2467 & = (12)3567 \\
&= (12)1234567,
\end{aligned}
$$

which simplifies to

$$
\begin{aligned}
12 &= 4 & = 235 & = 136 & = 37 \\
&= 1345 & = 2346 & = 12347 & = 56 & = 157 & = 267 \\
&= 12456 & = 2457 & = 1467 & = 123567 \\
&= 34567.
\end{aligned}
$$

Assuming that third- and higher-order interactions are negligible, we arrive at

$$12 = 4 = 37 = 56.$$

Hence the main effect of variable 4 is confounded with the two-factor interactions 12, 37, and 56.

The complete set of linear combinations of confounded effects to be estimated are (assuming third- and higher-order interactions can be neglected)

l_0 estimates mean
l_1 estimates $1 + 24 + 35 + 67$
l_2 estimates $2 + 14 + 36 + 57$
l_3 estimates $3 + 15 + 26 + 47$
l_{12} estimates $12 + 4 + 37 + 56$
l_{13} estimates $13 + 5 + 27 + 46$
l_{23} estimates $23 + 6 + 17 + 45$
l_{123} estimates $34 + 25 + 16 + 7.$

As noted previously, this is a resolution III design; at least some of the main effects are confounded with a two-factor interaction. In this case, each of the main effects is confounded with three two-factor interactions.

Estimation of Effects

The estimates for the linear combinations of confounded effects are obtained by multiplying each column of plus and minus signs in the calculation matrix by the overrun response column of Table 20.3, summing, and then dividing by 4 (for the sample mean, divide by 8). The estimates are shown in Table 20.4. Figure 20.1 is a normal probability plot of the effect estimates of Table 20.4. Figure 20.1 reveals that three of the linear combinations have estimates asso-

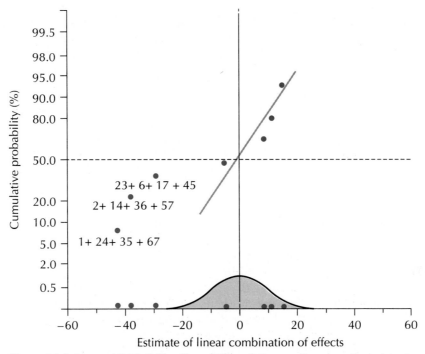

Figure 20.1 Normal Probability Plot of Effect Estimates Based on Tests 1 to 8

ciated with them that appear to be large relative to the others; that is, estimates that are not likely to have true mean values of zero.

Interpretation of Results

In attempting to explain the experimental results obtained up to this point, we might conclude that the three linear combinations of confounded effects determined above to be large, are so because the main effects of variables 1, 2, and 6 (the concentrations of soybean emulsion, vegetable fat, and secondary stabilizer) are the important variables. But we should bear in mind that there are many possible interpretations of these results. Looking at the confounded structure, we could say that:

1. The main effects of variables 1, 2, and 6 alone are important.
2. The main effects of variables 1 and 6, as well as the interactions 14 and/or 36, which involve variables with large main effects, could be responsible for the observed results.
3. Similarly, we might conclude that l_{23} is large not because the main effect of variable 6 is large but rather because the two-factor interactions 17 and/ or 23 are important.

TABLE 20.4	Linear Combinations and Estimates for the 2_{III}^{7-4} Fractional Factorial Design for Overrun (%)

$$l_0 = \quad 98.875 \text{ and estimates mean}$$
$$l_1 = -41.750 \text{ and estimates } 1 + 24 + 35 + 67$$
$$l_2 = -36.750 \text{ and estimates } 2 + 14 + 36 + 57$$
$$l_3 = \quad 10.250 \text{ and estimates } 3 + 15 + 26 + 47$$
$$l_{12} = \quad 12.750 \text{ and estimates } 12 + \quad 4 + 37 + 56$$
$$l_{13} = \quad -4.250 \text{ and estimates } 13 + \quad 5 + 27 + 46$$
$$l_{23} = -28.250 \text{ and estimates } 23 + \quad 6 + 17 + 45$$
$$l_{123} = \quad 16.250 \text{ and estimates } 34 + 25 + 16 + \quad 7$$

4. A similar interpretation can be made in terms of variables 2 and 6 by hypothesizing that l_1 is large in magnitude only because the two-factor interactions 24 and/or 67 are large.

It is possible that one might wish to adopt interpretation 1 above and run some additional tests, manipulating these factors to attempt to increase overrun. However, it is often the case that chemical processes of this nature are subject to nonlinear behavior, and hence a more conservative but revealing approach may be to run some additional tests to remove the ambiguity surrounding the results at this stage.

Mirror Image Design

If we wish to separate the main effects of the seven variables from the two-factor interactions, we can do this by using something called the *mirror image design*. This second 2^{7-4} fractional factorial is obtained by switching the signs (from $-$ to $+$, and vice versa) for **all** of the columns in the original *design (recipe) matrix* (Table 20.3). The mirror image design is shown in Table 20.5. Comparing the design matrix of Table 20.5 to the initial 2^{7-4} design (Table 20.3)

TABLE 20.5	Design Matrix for the Mirror Image Design							
Test	1	2	3	4	5	6	7	Overrun (%)
9	+	+	+	−	−	−	+	84
10	−	+	+	+	+	−	−	69
11	+	−	+	+	−	+	−	56
12	−	−	+	−	+	+	+	161
13	+	+	−	−	+	+	−	56
14	−	+	−	+	−	+	+	40
15	+	−	−	+	+	−	+	92
16	−	−	−	−	−	−	−	208

we see that the signs for all seven columns (main variables) have been switched. These new eight tests were run and the results are given in the rightmost column.

Generators and Defining Relation for the Mirror Image Design

As we examine this new design matrix for the second 2^{7-4} fractional factorial design, we find that the generators of this new design are different from those of the first. For example, the signs in the column associated with the fourth variable are no longer the product of columns 1 and 2; rather, column 4 is the negative of this product (i.e., $4 = -12$). Columns for the fifth and sixth variables are of a similar nature, while for the seventh variable we have the relationship $7 = 123$, as before. Therefore, the generators for the mirror image design are

$$I = -124, \quad I = -135, \quad I = -236, \quad I = 1237.$$

Note that only some sign changes have occurred, but the letters in the words of the generators have remained the same. The defining relation for the mirror image design is

$$\begin{aligned} I = &-124 = -135 = -236 = 1237 \\ &= 2345 = 1346 = -347 = 1256 = -257 = -167 \\ &= -456 = 1457 = 2467 = 3567 \\ &= -1234567. \end{aligned}$$

Again, while the effects in the linear combinations remain the same, some of the signs have changed. They have changed, however, in a way that will prove most useful.

When we use the defining relation to obtain the linear combinations of confounded effects, we find, for example, that (once again assuming that third- and higher-order interactions can be neglected)

l_1' estimates $1 - 24 - 35 - 67$.

Note that we write this linear combination with a prime (l_1') to distinguish it from l_1 from the first 2^{7-4} experiment. Note also the difference between l_1 and l_1':

l_1 estimates $1 + 24 + 35 + 67$
l_1' estimates $1 - 24 - 35 - 67$.

Hence it appears that when the results of the first design are combined with those of the second design, some simplification will occur that will greatly reduce the ambiguity of the initial results.

Using the mirror image design, eight additional tests (test numbers 9 through 16) were performed accordingly. The results were shown in the Table 20.5. For these eight tests, we again calculate the estimate for each linear combination of confounded effects. These are compared in Table 20.6 with the results of the original eight tests:

TABLE 20.6 Comparison of Original and Mirror Image Tests		

	Original Eight Tests	Mirror Image Design Additional Eight Tests
l_0 =	98.875 and estimates mean	l'_0 = 95.750 and estimates mean
l_1 =	-41.750 and estimates $1 + 24 + 35 + 67$	l'_1 = -47.500 and estimates $1 - 24 - 35 - 67$
l_2 =	-36.750 and estimates $2 + 14 + 36 + 57$	l'_2 = -67.000 and estimates $2 - 14 - 36 - 57$
l_3 =	10.250 and estimates $3 + 15 + 26 + 47$	l'_3 = -6.500 and estimates $3 - 15 - 26 - 47$
l_{12} =	12.750 and estimates $12 + 4 + 37 + 56$	l'_{12} = 63.000 and estimates $12 - 4 + 37 + 56$
l_{13} =	-4.250 and estimates $13 + 5 + 27 + 46$	l'_{13} = 2.500 and estimates $13 - 5 + 27 + 46$
l_{23} =	-28.250 and estimates $23 + 6 + 17 + 45$	l'_{23} = 35.000 and estimates $23 - 6 + 17 + 45$
l_{123} =	16.250 and estimates $34 + 25 + 16 + 7$	l'_{123} = -3.000 and estimates $-34 - 25 - 16 + 7$

Unconfounding the Main Effects

Based on the confounding structure of the original eight tests and that of the additional eight tests from the mirror image design and assuming that third- and higher-order interactions are negligible we can break some of the main effects away from the two-factor interactions by adding l_i and l'_i. That is,

$$\frac{l_1 + l'_1}{2} = (\tfrac{1}{2})[(-41.750) + (-47.500)] \text{ estimates } (\tfrac{1}{2})[(1 + 24 + 35 + 67)$$

$$+ (1 - 24 - 35 - 67)]$$

$$= -44.625 \text{ estimates } 1.$$

Similarly,

$$\frac{l_2 + l'_2}{2} = -51.875 \text{ estimates } 2$$

$$\frac{l_3 + l'_3}{2} = 1.875 \text{ estimates } 3$$

$$\frac{l_{12} + l'_{12}}{2} = 37.875 \text{ estimates } 12 + 37 + 56$$

$$\frac{l_{13} + l'_{13}}{2} = -0.875 \text{ estimates } 13 + 27 + 46$$

$$\frac{l_{23} + l'_{23}}{2} = 3.375 \text{ estimates } 23 + 45 + 17$$

$$\frac{l_{123} + l'_{123}}{2} = 6.625 \text{ estimates } 7.$$

We may also subtract l'_i from l_i to break some of the main effects away from the two-factor interactions. That is,

$$\frac{l_1 - l'_1}{2} = (\tfrac{1}{2})[(-41.750) - (-47.500)]$$

estimates $(\frac{1}{2})[1 + 24 + 35 + 67) - (1 - 24 - 35 - 67)]$
= 2.875 estimates $24 + 35 + 67$

$$\frac{l_2 - l'_2}{2} = \quad 15.125 \text{ estimates } 14 + 36 + 57$$

$$\frac{l_3 - l'_3}{2} = \quad 8.375 \text{ estimates } 15 + 26 + 47$$

$$\frac{l_{12} - l'_{12}}{2} = -25.125 \text{ estimates } 4$$

$$\frac{l_{13} - l'_{13}}{2} = \quad -3.375 \text{ estimates } 5$$

$$\frac{l_{23} - l'_{23}}{2} = -31.625 \text{ estimates } 6$$

$$\frac{l_{123} - l'_{123}}{2} = \quad 9.625 \text{ estimates } 34 + 25 + 16.$$

An estimate of the mean based on the information from both the initial and the mirror image design may be obtained by averaging the linear combinations l_0 and l'_0. This calculation is shown below, where the complete forms for l_0 and l'_0 have been displayed; i.e., no assumption about the negligibility of higher-order effects has been made:

$$l_0 = 98.875 \text{ estimates } I + (\tfrac{1}{2})(124 + 135 + 236 + 1237 + 2345 + 1346$$
$$+ 347 + 1256 + 257 + 167 + 456 + 1457 + 2467 + 3567$$
$$+ 1234567)$$

$$l'_0 = 95.570 \text{ estimates } I - (\tfrac{1}{2})(124 + 135 + 236 - 1237 - 2345 - 1346$$
$$+ 347 - 1256 + 257 + 167 + 456 - 1457 - 2467 - 3567$$
$$+ 1234567)$$

$$\frac{l_0 + l'_0}{2} = 97.313 \text{ estimates } I + (\tfrac{1}{2})(1237 + 2345 + 1346 + 1256 + 1457$$
$$+ 2467 + 3567).$$

Under the assumption that third- and higher-order interactions are negligible, this reduces to

$$\frac{l_0 + l'_0}{2} = 97.313 \text{ estimates mean.}$$

By taking the difference between l_0 and l'_0, the following result is obtained:

$$l_0 - l'_0 = 3.125 \text{ estimates } 124 + 135 + 236 + 347 + 257 + 167 + 456$$
$$+ 1234567.$$

If the assumption that third- and higher-order interactions are negligible is then applied to this result, we find that the difference $(l_0 - l'_0)$ has no effects associated with it. Thus the numerical value, 3.125, may be assumed to be due to the experimental error in the system. In general, if several of the linear combinations (or averages of linear combinations) have no effects associated with them, they may be assumed to be due to experimental error, and used to develop a confidence interval to assess the statistical significance of linear combinations.

Nature of the Two Combined 2_{III}^{7-4} Fractional Factorial Designs

When the mirror image design is combined with the original design, all of the main effects are uncoupled from the sets of three two-factor interactions they were confounded with: for example,

First design: l_1 estimates $1 + (24 + 35 + 67)$

Mirror image: l_1' estimates $1 - (24 + 35 + 67)$

Combined design: $\dfrac{l_1 + l_1'}{2}$ estimates 1

$\dfrac{l_1 - l_1'}{2}$ estimates $24 + 35 + 67$.

The combined design includes all 16 tests in the initial and mirror image designs. In the combined design, two-factor interactions are confounded with one another and the main effects are confounded with three-factor interactions. The combined (2^{7-3}) design is therefore of resolution IV. Earlier, it was seen that the average of l_0 and l_0' confounded the mean, I, with a set of four-factor interactions and that under the assumption of three-factor interactions and higher being negligible produced an unaliased estimate of the mean. The four-factor interactions confounded with the mean were

$$I = 1237 = 2345 = 1346 = 1256 = 1457 = 2467 = 3567.$$

This expression specifies the defining relation for the combined design. Such a defining relation for a 2_{IV}^{7-3} design is produced with the generators

$$I = 1237, \qquad I = 2345, \qquad I = 1346.$$

Thus for the combined design, variables 5, 6, and 7 are introduced through the interactions 234, 134, and 123 in a 2^{7-3} (2^4 base design) calculation matrix.

Interpretation of Results

Let us now look at the relative sizes of the main effects and sets of confounded interactions to see what physical interpretation of the data is possible at this stage. The estimates of the main effects and sets of three two-factor interactions are summarized in Table 20.7 and have been used to construct the normal probability plot displayed in Fig. 20.2. An examination of Table 20.7 and Fig. 20.2 shows five values that are much larger (in absolute value) than the others. Increasing the concentration of variables 1 (soybean emulsion), 2 (vegetable fat), 4 (emulsifiers), and 6 (secondary stabilizer) all appear to reduce the overrun

TABLE 20.7 Results of the Combined Designs		
Estimate of 1	=	−44.625
Estimate of 2	=	−51.875
Estimate of 3	=	1.875
Estimate of 12 + 37 + 56	=	37.875
Estimate of 13 + 27 + 46	=	−0.875
Estimate of 23 + 45 + 17	=	3.375
Estimate of 7	=	6.625
Estimate of error	=	3.125
Estimate of 24 + 35 + 67	=	2.875
Estimate of 14 + 36 + 57	=	15.125
Estimate of 15 + 26 + 47	=	8.375
Estimate of 4	=	−25.125
Estimate of 5	=	−3.375
Estimate of 6	=	−31.625
Estimate of 34 + 25 + 16	=	9.625

significantly. Additionally, the linear combination (12 + 37 + 56) also appears to be large.

Although the linear combination of effects 12, 37, and 56 appears to be important, we do not know which interaction (or combination of interactions) is responsible. Since the largest effects in Table 20.7 are associated with variables 1 and 2, it is quite possible that the 12 interaction is the important one. Furthermore, since variables 3, 5, and 7 do not appear to have main effects that exert appreciable influence on the overrun, one interpretation forwarded might be that the interactions 37 and 56 do not influence the overrun either. One tentative conclusion that may therefore be drawn based on the results of the 16 tests performed is that variables 1, 2, 4, and 6 and the interaction 12 are important with respect to the overrun.

Two-Way Diagram Analysis

Based on the assumptions that we have made above about the possible importance of variables 1, 2, 4, 6 and interaction 12, a two-way diagram may be constructed to interpret the effects of variables 1 and 2. This diagram is shown in Fig. 20.3. It is interesting to note the following from the two-way diagram.

1. Using a high level of vegetable fat (20%) and changing the concentration of soybean emulsion from 5% to 1.67% increases the overrun by 6.75% (68% to 74.75%). This increase is negligible for all practical purposes.
2. Using a low level of vegetable fat (10%) and changing the concentration of soybean emulsion from 5% to 1.67% increases the overrun by 82.5% (82% to 164.5%).

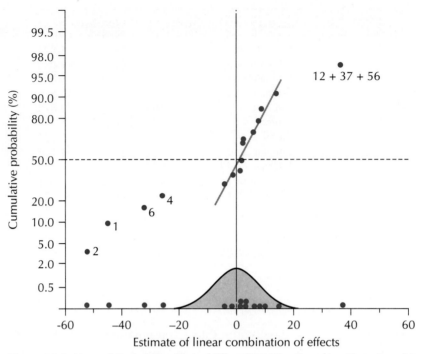

Figure 20.2 Normal Probability Plot of Effect Estimates Based on Tests 1 to 16

Clearly, the two-factor interaction between soybean emulsion and vegetable fat appears to be very important.

Summary

The effect of changing the concentration of the emulsifiers (variable 4) and secondary stabilizer (variable 6) is fairly clear. Reducing the levels for these variables appears to increase the amount of overrun. The concentrations of carbohydrates (3), primary stabilizer (5), and salt (7) appear to have no real effect on the response. Reducing the concentration for the soybean emulsion has little effect on the overrun when the vegetable fat concentration is at its high level. The key to a desirable (large) overrun response is the discovery that low concentrations for both the soybean emulsion and vegetable fat must be used simultaneously. "Hence main effects can be misleading without proper consideration of interaction effects."

This example illustrates how fractional factorial designs can be used for screening purposes. Only a relatively small number of tests were used to study the effects of many variables. If statistically designed experiments had not been used, chances are good that many more tests would have been required before similar conclusions could be reached. This example further reinforces the concept of *sequential assembly*, that is, combining the results of similarly structured

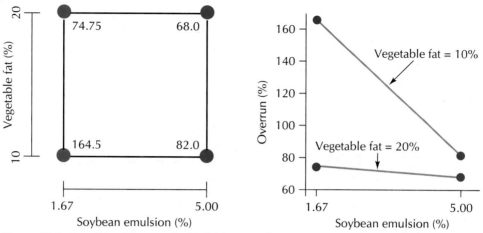

Figure 20.3 Two-Way Diagram for Variables 1 and 2

experimental designs to increase our knowledge and reduce the ambiguity that may surround the results of a single experiment.

20.2

Sequential Assembly of Fractional Factorials

We have emphasized previously the importance of running experiments in a sequential manner, gradually improving on our knowledge base. This notion is important to ensure that a lack of knowledge early in an investigation will not lead to wasting experimental resources (e.g., inappropriate variables or variable levels chosen in the context of one large experiment). To avoid such problems, it is wise to run a series of smaller but related experiments. Ambiguities that may arise in an early experiment can be removed by making additional information available from a follow-up experiment. It is possible to identify related experimental designs which, when combined, produce a clearer picture of the relative importance of the factors under study. This was evidenced in the preceding section through the use of the mirror image design.

In the soybean whipped topping example a 2_{III}^{7-4} fractional factorial design was run initially and the results were somewhat ambiguous. In particular, three of the linear combinations of effects seemed large, but the high degree of confounding among low-order effects (main effects and two-factor interactions) made several interpretations possible. Specifically, it was found from the first experiment that the seemingly large linear combinations of effects were

$$l_1 \text{ estimates } 1 + 24 + 35 + 67 = -41.750$$
$$l_2 \text{ estimates } 2 + 14 + 36 + 57 = -36.750$$
$$l_{23} \text{ estimates } 23 + 6 + 17 + 45 = -28.250.$$

The question is: Which effects are making these linear combinations large? One interpretation of the above is that the three main effects (1, 2, and 6) are important. Other interpretations are possible, of course.

Alternative Experimental Strategies: The Notion of Families of Fractional Factorial Designs

Since the initial 2^{7-4} fractional factorial design results for the soybean whipped topping example were so ambiguous, we performed the mirror image design to clarify things. One might ask: Why not perform the 2^{7-3}_{IV} 16-run experiment in the first place? This question is an important one and gets to the very heart of the sequential experimentation issue.

The 2^{7-4} design initially run, with generators

$$I = 124, \quad I = 135, \quad I = 236, \quad I = 1237,$$

is a $\frac{1}{16}$ fraction of the full 2^7 factorial (8 tests versus 128 tests). Among the 128 combinations of the seven variables, each over two levels, we selected a certain 8 tests based on the generators above. We should think of the 2^7 design as being comprised of 16 one-sixteenth fractions, of which the first design was one of the fractions and the mirror image design was another. This still leaves 14 other $\frac{1}{16}$ fractions that could have been chosen for the second design.

The two $\frac{1}{16}$ fractions of the 2^7 factorial that we ran were related through the following generators:

Original 2^{7-4}: $I = \quad 124, \quad I = \quad 135, \quad I = \quad 236, \quad I = 1237$
Mirror image: $I = -124, \quad I = -135, \quad I = -236, \quad I = 1237.$

The terms remained the same; only the signs changed. The remaining 14 members of this *family* of fractional factorials are defined by the other possible sign combinations:

$$I = \pm124, \quad I = \pm135, \quad I = \pm236, \quad I = \pm1237.$$

The generalized generators above define a *family* of fractional factorials. The design with all + signs on the generators is called the *principal fraction*. The remaining, in this case, 15 designs are the *alternate fractions*. The key point here is the following: Depending on the interpretation of the results of the principal fraction, we may choose any one of several other alternate fractions to achieve a particular result when the two fractions are combined.

As an alternative to the mirror image design, suppose that the larger linear combinations from the first experiment,

l_1 estimates $1 + 24 + 67 = -41.750$
l_2 estimates $2 + 14 + 36 = -36.750$
l_{23} estimates $23 + 6 + 17 = -28.250.$

TABLE 20.8			Principal Fraction Design Matrix					
Test	1	2	3	4	5	6	7	Overrun (%)
1	−	−	−	+	+	+	−	115
2	+	−	−	−	−	+	+	81
3	−	+	−	−	+	−	+	110
4	+	+	−	+	−	−	−	69
5	−	−	+	+	−	−	+	174
6	+	−	+	−	+	−	−	99
7	−	+	+	−	−	+	−	80
8	+	+	+	+	+	+	+	63
Generators: $I = 124$, $I = 135$, $I = 236$, $I = 1237$								

were interpreted somewhat differently. Suppose that our knowledge of the process led us to suspect a priori that variable 1 was a very important variable, and now the results seem to bear this out. In particular, we feel that the linear combinations above are large primarily because of the importance of the main effect of variable 1 and its two-factor interactions 14 and 17.

This interpretation could be clarified if we could run an additional 2^{7-4} fractional factorial that when combined with the principal fraction produced cleared or unconfounded estimates of the main effect of 1 and all of its two-factor interactions. Such a fraction exists; it is one of the 15 *alternate fractions* of the family we are working with. It is generated by folding (switching the signs of) only the variable 1 column in the original design (recipe) matrix.

For the soybean whipped topping experiment, this interpretation was in fact made, and the alternative fraction obtained by folding variable 1 was also performed. The principal and alternative fraction recipe matrices are shown in Tables 20.8 and 20.9 along with the overruns that were obtained for the tests.

TABLE 20.9			Alternative Fraction Design Matrix					
Test	1	2	3	4	5	6	7	Overrun (%)
17	+	−	−	+	+	+	−	66
18	−	−	−	−	−	+	+	171
19	+	+	−	−	+	−	+	147
20	−	+	−	+	−	−	−	122
21	+	−	+	+	−	−	+	51
22	−	−	+	−	+	−	−	148
23	+	+	+	−	−	+	−	49
24	−	+	+	+	+	+	+	14
Generators: $I = -124$, $I = -135$, $I = 236$, $I = -1237$								

The implications of folding the 1 column only can be seen clearly through an examination of the defining relation for the alternative fraction. The defining relation for this new second design is

$$
\begin{aligned}
I = &\ -124 &&= -135 &&= 236 &&= -1237 \\
= &\ 2345 &&= -1346 &&= 347 &&= -1256 &&= 257 &&= -167 \\
= &\ 456 &&= -1457 &&= 2467 &&= 3567 \\
= &\ -1234567.
\end{aligned}
$$

The linear combinations of effects obtained using this defining relation are

l_0'' estimates mean
l_1'' estimates $1 - 24 - 35 - 67$
l_2'' estimates $2 - 14 + 36 + 57$
l_3'' estimates $3 - 15 + 26 + 47$
l_{12}'' estimates $12 - 4 - 37 - 56$
l_{13}'' estimates $13 - 5 - 27 - 46$
l_{23}'' estimates $23 + 6 - 17 + 45$
l_{123}'' estimates $- 34 - 25 + 16 - 7.$

The l_i'' differ from the original l_i only in that in the new set of linear combinations, the signs of terms not including a 1 differ from the signs of terms that do have a 1 (i.e., 1, 12, 13, 14, 15, 16, 17). Therefore, if the l_i and l_i'' are added/subtracted, we obtain clear estimates of the main effect of variable 1 and all of its two factor interactions, under the assumption that third- and higher-order interactions can be neglected.

Using the alternative fraction design, tests 17 through 24 were used to calculate the estimate for each linear combination of confounded effects. These are compared in Table 20.10 with the linear combinations based on the original eight tests. When the confounding structures and effect estimates of the original eight tests are combined with those of the eight tests from the alternative fraction, the following structure is obtained:

$$
\frac{l_0 + l_0''}{2} = \quad 97.438 \text{ estimates mean}
$$

$$
\frac{l_1 + l_1''}{2} = -38.625 \text{ estimates } 1
$$

$$
\frac{l_2 + l_2''}{2} = -31.375 \text{ estimates } 2 + 36 + 57
$$

$$
\frac{l_3 + l_3''}{2} = -25.375 \text{ estimates } 3 + 26 + 47
$$

$$
\frac{l_{12} + l_{12}''}{2} = \quad 39.125 \text{ estimates } 12
$$

$$\frac{l_{13} + l''_{13}}{2} = 0.125 \text{ estimates } 13$$

$$\frac{l_{23} + l''_{23}}{2} = -35.125 \text{ estimates } 23 + 6 + 45$$

$$\frac{l_{123} + l''_{123}}{2} = 8.375 \text{ estimates } 16$$

$$l_0 - l''_0 = 2.875 \text{ estimates error}$$

$$\frac{l_1 - l''_1}{2} = -3.125 \text{ estimates } 24 + 35 + 67$$

$$\frac{l_2 - l''_2}{2} = -5.375 \text{ estimates } 14$$

$$\frac{l_3 - l''_3}{2} = 35.625 \text{ estimates } 15$$

$$\frac{l_{12} - l''_{12}}{2} = -26.375 \text{ estimates } 4 + 37 + 56$$

$$\frac{l_{13} - l''_{13}}{2} = -4.375 \text{ estimates } 5 + 27 + 46$$

$$\frac{l_{23} - l''_{23}}{2} = 6.875 \text{ estimates } 17$$

$$\frac{l_{123} - l''_{123}}{2} = 7.875 \text{ estimates } 34 + 25 + 7.$$

Interpretation of Results

Let us now look at results from the combined designs to see what physical interpretation of the data is possible based on the principal and alternative

TABLE 20.10   Comparison of Test Results

Principal Fraction Design Tests 1–8	Alternative Fraction Design Tests 17–24
$l_0 = 98.875$ estimates mean)	$l''_0 = 96.000$ estimates mean)
$l_1 = -41.750$ estimates $1 + 24 + 35 + 67$	$l''_1 = -35.500$ estimates $1 - 24 - 35 - 67$
$l_2 = -36.750$ estimates $2 + 14 + 36 + 57$	$l''_2 = -26.000$ estimates $2 - 14 + 36 + 57$
$l_3 = 10.250$ estimates $3 + 15 + 26 + 47$	$l''_3 = -61.000$ estimates $3 - 15 + 26 + 47$
$l_{12} = 12.750$ estimates $12 + 4 + 37 + 56$	$l''_{12} = -65.500$ estimates $12 - 4 - 37 - 56$
$l_{13} = -4.250$ estimates $13 + 5 + 27 + 46$	$l''_{13} = 4.500$ estimates $13 - 5 - 27 - 46$
$l_{23} = -28.250$ estimates $23 + 6 + 17 + 45$	$l''_{23} = -42.000$ estimates $23 + 6 - 17 + 45$
$l_{123} = 16.250$ estimates $34 + 25 + 16 + 7$	$l''_{123} = 0.500$ estimates $-34 - 25 + 16 - 7$

fractions. The linear combinations of effects and associated estimates are summarized below. Figure 20.4 displays a normal probability plot of the estimates.

Estimate of 1 = -38.625

Estimate of 2 + 36 + 57 = -31.375

Estimate of 3 + 26 + 47 = -25.375

Estimate of 12 = 39.125

Estimate of 13 = 0.125

Estimate of 23 + 6 + 45 = -35.125

Estimate of 16 = 8.375

Estimate of error = 2.875

Estimate of 24 + 35 + 67 = -3.125

Estimate of 14 = -5.375

Estimate of 15 = 35.625

Estimate of 4 + 37 + 56 = -26.375

Estimate of 5 + 27 + 46 = -4.375

Estimate of 17 = 6.875

Estimate of 34 + 25 + 7 = 7.875.

In examining the effects, our a priori suspicion concerning the importance of variable 1 seems to be borne out, as the main effect of variable 1 as well as the two factor interactions 12 and 15 appear to be important.

The estimates of the sets of confounded effects (2 + 36 + 57), (3 + 26 + 47), (23 + 6 + 45), and (4 + 37 + 56) are also relatively large. We do not know, however, which terms (or combination of terms) within the linear combinations are responsible. One conclusion that might be drawn is that the four linear combinations are large because of the main effects of variables 2, 3, 6, and 4 embedded within them. However, we do not have much confidence in such a conclusion based solely on the results of these two fractions and may therefore more conservatively conclude that variable 1 and interactions 12 and 15 are important with respect to the overrun. But in this case, we previously obtained more reliable information on the importance of main effects 2, 4, and 6 (principal fraction + mirror image). That information, when combined with what we have learned above leads us now to believe that 1, 2, 4, 6, 12, and 15 are important effects.

Summary

We have just witnessed an example of a more general strategy that might be employed in running successive experimental designs by choosing judiciously among the members of a given family of fractional factorials. In this case we have seen that by folding a single column in the design (recipe) matrix we create an alternate fraction, which when combined with the principal fraction provides clear estimates of the main effect of the variable that column represents

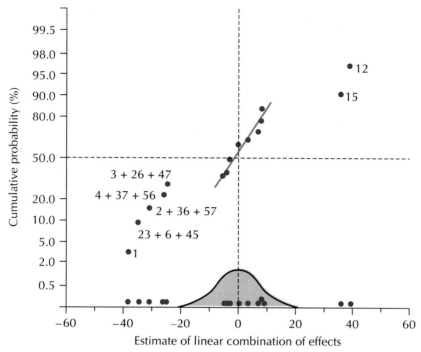

Figure 20.4 Normal Probability Plot of Effect Estimates Based on Tests 1 to 8 and 17 to 24

and all of its two-factor interactions, when third- and higher-order interactions are neglected.

Clearly, given the principal fraction we have several alternatives among the remaining alternative fractions, each providing a different set of information/effect estimates from the combined design. Table 20.11 provides some relevant information concerning the family of fractional factorials we have been examining. We have seen that running the principal fraction followed by the mirror image design (A_7) produced unconfounded estimates of all main effects and seven sets of two-factor interactions confounded in groups of three. If the estimate associated with one of these sets of two-factor interactions showed up to be large, we were forced to use our judgment to select the important two-factor interaction from a group of three.

When we combined the principal fraction with a fraction (A_{11}) to unconfound the main effect of 1 as well as all of its interactions (12, 13, 14, 15, 16, and 17), we saw that the remaining main effects were each confounded with a pair of two-factor interactions. If the estimate associated with one of these main effects plus a pair of two-factor interactions showed up to be large, it was difficult to conclude much about what term(s) were driving the estimate.

If instead of following up the principal fraction with a mirror image design or a design to clear the main and interaction effects of a single variable, we were to run one of the designs A_3, A_5, A_6, A_9, A_{10}, A_{12}, or A_{15}, the structure

TABLE 20.11	2_{III}^{7-4} Family of Fractional Factorials				

Fraction	Generators				Combined with Principal Fraction Gives Estimates of:*
Principal	$I = +124$	$I = +135$	$I = +236$	$I = +1237$	—
A_1	$I = -124$	$I = +135$	$I = +236$	$I = +1237$	4, 14, 24, 34, 45, 46, 47
A_2	$I = +124$	$I = -135$	$I = +236$	$I = +1237$	5, 15, 25, 35, 45, 56, 57
A_3	$I = -124$	$I = -135$	$I = +236$	$I = +1237$	
A_4	$I = +124$	$I = +135$	$I = -236$	$I = +1237$	6, 16, 26, 36, 46, 56, 67
A_5	$I = -124$	$I = +135$	$I = -236$	$I = +1237$	
A_6	$I = +124$	$I = -135$	$I = -236$	$I = +1237$	
A_7	$I = -124$	$I = -135$	$I = -236$	$I = +1237$	All main effects
A_8	$I = +124$	$I = +135$	$I = +236$	$I = -1237$	7, 17, 27, 37, 47, 57, 67
A_9	$I = -124$	$I = +135$	$I = +236$	$I = -1237$	
A_{10}	$I = +124$	$I = -135$	$I = +236$	$I = -1237$	
A_{11}	$I = -124$	$I = -135$	$I = +236$	$I = -1237$	1, 12, 13, 14, 15, 16, 17
A_{12}	$I = +124$	$I = +135$	$I = -236$	$I = -1237$	
A_{13}	$I = -124$	$I = +135$	$I = -236$	$I = -1237$	2, 12, 23, 24, 25, 26, 27
A_{14}	$I = +124$	$I = -135$	$I = -236$	$I = -1237$	3, 13, 23, 34, 35, 36, 37
A_{15}	$I = -124$	$I = -135$	$I = -236$	$I = -1237$	

* Assuming that third- and higher-order interactions are negligible.

shown in Table 20.12 would be obtained (assuming three-factor and higher-order interactions are negligible). The structure of the combined designs illustrated in Table 20.12 clearly is not appealing in that there are no unconfounded main effects or two-factor interactions. However, the distribution of the main and interaction effects is much more uniform across the structure. Thus if an effect estimate is large, the investigator is required only to ponder the significance of effects from two terms instead of three. In addition, two error

TABLE 20.12   Principal Fraction Combined with Fraction A$_3$

$$(l_0 + l_0^*)/2 \text{ estimates mean}$$
$$(l_1 + l_1^*)/2 \text{ estimates } 1 + 67$$
$$(l_2 + l_2^*)/2 \text{ estimates } 2 + 36$$
$$(l_3 + l_3^*)/2 \text{ estimates } 3 + 26$$
$$(l_{12} + l_{12}^*)/2 \text{ estimates } 12 + 37$$
$$(l_{13} + l_{13}^*)/2 \text{ estimates } 13 + 27$$
$$(l_{23} + l_{23}^*)/2 \text{ estimates } 23 + 6 + 17 + 45$$
$$(l_{123} + l_{123}^*)/2 \text{ estimates } 16 + 7$$
$$l_0 - l_0^* \text{ estimates error}$$
$$(l_1 - l_1^*)/2 \text{ estimates } 24 + 35$$
$$(l_2 - l_2^*)/2 \text{ estimates } 14 + 57$$
$$(l_3 - l_3^*)/2 \text{ estimates } 15 + 47$$
$$(l_{12} - l_{12}^*)/2 \text{ estimates } 4 + 56$$
$$(l_{13} - l_{13}^*)/2 \text{ estimates } 5 + 46$$
$$(l_{23} - l_{23}^*)/2 \text{ estimates error}$$
$$(l_{123} - l_{123}^*)/2 \text{ estimates } 34 + 25$$

estimates are available for this combined design as compared with one. Finally, suppose that based on a priori knowledge we know that variables 6 and 7 are not likely to interact with any other variables. Under these conditions, alternate fraction A$_3$ (with associated estimates l_i^*) will provide unconfounded estimates of variables 1, 2, 3, 4, 5, and 7 and interactions 12, 13, 14, and 15. Clearly, this alternate fraction does offer some advantages over the other designs described in detail previously.

The discussion above reinforces the value of running experiments sequentially and assembling them together to provide information deemed desirable based on the results of initial experiments. A key point here is that these related experiments can provide several different alternatives for a second experiment, depending on the results and inferences drawn from the first experiment.

20.3

Soybean Whipped Topping Example: Summary and Interpretation

Two separate analyses have been performed that build on the results obtained from an initial principal fraction. In the first analysis, the principal fraction was followed up by a mirror image design (design A_7). These two designs, when combined (Table 20.13), provided unconfounded estimates of all main effects (under the assumption that three-factor and higher-order interactions are negligible). In the second analysis, the principal fraction was followed up by a design that folded over variable 1 (design A_{11}). These two designs, when com-

TABLE 20.13	Results of Principal Fraction + Mirror Image Design (Design A_7)

$$(l_0 + l_0')/2 = 97.313 \text{ estimates mean}$$
$$(l_1 + l_1')/2 = -44.625 \text{ estimates } 1$$
$$(l_2 + l_2')/2 = -51.875 \text{ estimates } 2$$
$$(l_3 + l_3')/2 = 1.875 \text{ estimates } 3$$
$$(l_{12} + l_{12}')/2 = 37.875 \text{ estimates } 12 + 37 + 56$$
$$(l_{13} + l_{13}')/2 = -0.875 \text{ estimates } 13 + 27 + 46$$
$$(l_{23} + l_{23}')/2 = 3.375 \text{ estimates } 23 + 45 + 17$$
$$(l_{123} + l_{123}')/2 = 6.625 \text{ estimates } 7$$
$$l_0 - l_0' = 3.125 \text{ estimates error}$$
$$(l_1 - l_1')/2 = 2.875 \text{ estimates } 24 + 35 + 67$$
$$(l_2 - l_2')/2 = 15.125 \text{ estimates } 14 + 36 + 57$$
$$(l_3 - l_3')/2 = 8.375 \text{ estimates } 15 + 26 + 47$$
$$(l_{12} - l_{12}')/2 = -25.125 \text{ estimates } 4$$
$$(l_{13} - l_{13}')/2 = -3.375 \text{ estimates } 5$$
$$(l_{23} - l_{23}')/2 = -31.625 \text{ estimates } 6$$
$$(l_{123} - l_{123}')/2 = 9.625 \text{ estimates } 34 + 25 + 16$$

bined (Table 20.14), provided unconfounded estimates of the effect of variable 1 and all its interactions.

A study of Tables 20.13 and 20.14 and the normal probability plots of Figs. 20.2 and 20.3 suggests that the following effects/linear combinations are important:

Principal Fraction + Mirror Image Design (A_7)		Principal Fraction + Alternative Fraction (A_{11})	
1 estimated as	-44.625	1 estimated as	-38.625
2 estimated as	-51.875	2 + 36 + 57 estimated as	-31.375
12 + 37 + 56 estimated as	37.875	12 estimated as	39.125
4 estimated as	-25.125	4 + 37 + 56 estimated as	-26.375
6 estimated as	-31.625	23 + 6 + 45 estimated as	-35.125
		3 + 26 + 47 estimated as	-25.375
		15 estimated as	35.625

In examining the information shown above, the following observations may be made:

1. Both combined designs produce an unconfounded estimate of the main effect of variable 1. These estimates (-44.625 and -38.625) appear to be fairly close in magnitude, reinforcing the importance of this effect.
2. Both combined designs have a linear combination that involves the interaction 12. The estimates associated with these combinations (37.875 and 39.125) are quite close in magnitude. This reinforces the importance of this interaction effect.

TABLE 20.14	Results of Principal Fraction + Alternative Fraction (Design A_{11})

$$(l_0 + l_0'')/2 = \quad 97.438 \text{ estimates average}$$
$$(l_1 + l_1'')/2 = -38.625 \text{ estimates } 1$$
$$(l_2 + l_2'')/2 = -31.375 \text{ estimates } 2 + 36 + 57$$
$$(l_3 + l_3'')/2 = -25.375 \text{ estimates } 3 + 26 + 47$$
$$(l_{12} + l_{12}'')/2 = \quad 39.125 \text{ estimates } 12$$
$$(l_{13} + l_{13}'')/2 = \quad 0.125 \text{ estimates } 13$$
$$(l_{23} + l_{23}'')/2 = -35.125 \text{ estimates } 23 + \quad 6 + 45$$
$$(l_{123} + l_{123}'')/2 = \quad 8.375 \text{ estimates } 16$$
$$l_0 - l_0'' = \quad 2.875 \text{ estimates error}$$
$$(l_1 - l_1'')/2 = -3.125 \text{ estimates } 24 + 35 + 67$$
$$(l_2 - l_2'')/2 = -5.375 \text{ estimates } 14$$
$$(l_3 - l_3'')/2 = \quad 35.625 \text{ estimates } 15$$
$$(l_{12} - l_{12}'')/2 = -26.375 \text{ estimates } 4 + 37 + 56$$
$$(l_{13} - l_{13}'')/2 = -4.375 \text{ estimates } 5 + 27 + 46$$
$$(l_{23} - l_{23}'')/2 = \quad 6.875 \text{ estimates } 17$$
$$(l_{123} - l_{123}'')/2 = \quad 7.875 \text{ estimates } 34 + 25 + \quad 7$$

3. Both combined designs have a linear combination that includes the main effect of variable 4. The estimates associated with these combinations (-25.125 and -26.375) are very close in magnitude.
4. Both combined designs have a linear combination that involves the main effect of variable 6. The estimates associated with these combinations (-31.625 and -35.125) have a comparable magnitude.
5. Based on the combined design (principal fraction + A_{11}), the interaction effect 15 is important (35.625).
6. Although both combined designs have a linear combination that involves the main effect of variable 2, these estimates (-51.875 and -31.375) are not close at all in magnitude. In fact, the difference between these estimates $[-31.375 - (-51.875) = 20.5]$ is itself quite large in magnitude. This difference, 20.5, is an estimate of the linear combination $36 + 57$. Thus this linear combination also appears to be important.

As indicated in observation 6, we may integrate the information from both combined designs to further clear the confounding pattern.

$$\frac{l_{12} + l_{12}'}{2} - \frac{l_{12} + l_{12}''}{2} = -1.250 \text{ estimates } 37 + 56$$

$$\frac{l_{13} + l_{13}'}{2} - \frac{l_{13} + l_{13}''}{2} = -1.000 \text{ estimates } 27 + 46$$

$$\frac{l_{23} + l_{23}'}{2} - \frac{l_{23} + l_{23}''}{2} = -3.500 \text{ estimates } 23 + 45$$

$$\frac{l_2 - l_2'}{2} - \frac{l_2 - l_2''}{2} = \quad 20.500 \text{ estimates } 36 + 57$$

$$\frac{l_3 - l_3'}{2} - \frac{l_3 - l_3''}{2} = -27.250 \text{ estimates } 26 + 47$$

$$\frac{l_{123} - l_{123}'}{2} - \frac{l_{123} - l_{123}''}{2} = \quad 1.250 \text{ estimates } 34 + 25.$$

In examining the information shown above, it appears that the linear combinations $(36 + 57)$ and $(26 + 47)$ are important. In summary, when all three 2_{III}^{7-4} designs are examined together (combined), the following effects (and linear combinations of effects) are identified as important:

1, 2, 4, 6, 12, 15, (36 + 57), (26 + 47).

Although the linear combinations $(36 + 57)$ and $(26 + 47)$ are large, we do not know which interactions are responsible. If we consider the linear combination $(36 + 57)$, neither the main effect of variable 5 or variable 7 has been identified as significant; therefore, it is unlikely that their interaction, 57, is important either. Thus one tentative conclusion that may be reached is that this linear combination is important because of the two-factor interaction 36. If we consider the linear combination $(26 + 47)$, the main effects of both variables 2 and 6 have been identified as significant; therefore, it is reasonable to suspect that their interaction, 26, is also important. Thus one tentative conclusion that may be reached is that this linear combination is important because of the two-factor interaction 26.

In summary, based on the results of the 24 tests performed, we have evidence that the variables 1, 2, 4, and 6, the interactions 12 and 15, and the linear combinations $(26 + 47)$ and $(36 + 57)$ are important with respect to overrun. A procedure is available[2] for unconfounding the remaining ambiguities associated with the $(26 + 47)$ and $(36 + 57)$ linear combinations of effects by performing a few additional tests. Alternatively, we might use the aforementioned logic and tentatively conclude that interactions 26 and 36 are important, and therefore postulate the following model to describe the overrun response:

$$y = b_0 + b_1 x_1 + b_2 x_2 + b_4 x_4 + b_6 x_6 + b_{12} x_1 x_2 + b_{15} x_1 x_5 + b_{26} x_2 x_6$$
$$+ b_{36} x_3 x_6 + \epsilon.$$

This model may be fit to the data for all 24 tests to develop contour plots. The model and contour plots may then be used to propose some confirmatory tests that ultimately will verify the ability of the model to adequately predict the response.

Exercises

20.1. A 2^5 factorial is to be run to study the effects of five factors on the burr height in a metal stamping process. After five tests have been performed

[2] G. E. P. Box, W. G. Hunter, and J. S. Hunter, *Statistics for Experimenters*, Wiley, New York, 1978.

(in a random order), the testing program is suspended because it is realized that not enough funds are available for all 32 tests. It is then proposed to conduct a new design that requires fewer total tests (16 instead of 32), that is, conduct a 2^{5-1} fractional factorial. At worst, how many of the original five tests can be saved? Clearly explain your answer.

20.2. Consider a 2^{5-2} fractional factorial design with the generators $I = -124$ and $I = 135$.

   a. What is the defining relation for this design? Display the linear combinations of effects that can be estimated from this experiment. Assume that third- and higher-order interaction effects can be neglected. Write out the design (recipe) matrix for this design in standard order.

   b. The results of the experiment in part (a) were somewhat ambiguous; therefore, a second fractional factorial design, the mirror image design, is to be performed. Display the design matrix for the mirror image design. What are the generators for this second design? Display the linear combinations of effects that can be estimated from the mirror image design, assuming that third- and higher-order interaction effects can be neglected.

   c. When these two designs are combined, what effects, or linear combinations of effects, can be estimated? Once again, assume that third- and higher-order interactions are negligible.

20.3. Consider a 2^{5-2} fractional factorial design with the generators $I = 134$ and $I = -235$.

   a. What is the defining relation for this design? Display the linear combinations of effects that can be estimated from this experiment. Assume that third- and higher-order interaction effects can be neglected. Write out the design (recipe) matrix for this design in standard order.

   b. The results of the experiment in part (a) indicated that variable 3 and its interactions were likely to be important. A second fractional factorial design is to be performed that will unconfound the main effect of variable 3 and all of its two-factor interactions. Display the design matrix for this second design. What are the generators for this second design? Display the linear combinations of effects that can be estimated, assuming that third- and higher-order interaction effects can be neglected.

   c. When these two designs are combined, what effects, or linear combinations of effects, can be estimated? Once again, assume that third- and higher-order interactions are negligible.

20.4. It is desired to perform a 2^{6-3} fractional factorial design with the generators $4 = -12$, $5 = -23$, and $6 = 123$.

   a. What is the defining relation for this design? Display the linear combinations of effects that can be estimated from this experiment. Assume that third- and higher-order interaction effects can be neglected. Write out the design (recipe) matrix for this design in standard order.

   b. The results of the experiment in part (a) were somewhat ambiguous; therefore, a second fractional factorial design, the mirror image design, is to be performed. Display the design matrix for the mirror image design. What are the generators for this second design? Display the linear combinations of effects that can be estimated from the mirror image design, assuming that third- and higher-order interaction effects can be neglected.

     c. When these two designs are combined, what effects, or linear combinations of effects, can be estimated? Once again, assume that third- and higher-order interactions are negligible.

20.5. It is desired to perform a 2^{6-3} fractional factorial design with the generators $4 = 13$, $5 = -12$, and $6 = -23$.

     a. What is the defining relation for this design? Display the linear combinations of effects that can be estimated from this experiment. Assume that third- and higher-order interaction effects can be neglected. Write out the design (recipe) matrix for this design in standard order.

     b. The results of the experiment in part (a) indicated that variable 5 and its interactions were likely to be important. A second fractional factorial design is to be performed that will unconfound the main effect of variable 5 and all of its two-factor interactions. Display the design matrix for this second design. What are the generators for this second design? Display the linear combinations of effects that can be estimated, assuming that third- and higher-order interaction effects can be neglected.

     c. When these two designs are combined, what effects, or linear combinations of effects, can be estimated? Once again, assume that third- and higher-order interactions are negligible.

20.6. The 2^{5-2} fractional factorial design shown in the table has been performed.

Test	1	2	3	4	5	y
1	−	−	−	−	+	76
2	+	−	−	+	−	45
3	−	+	−	+	+	33
4	+	+	−	−	−	48
5	−	−	+	−	−	28
6	+	−	+	+	+	41
7	−	+	+	+	−	75
8	+	+	+	−	+	54

     a. What are the generators and defining relation for this design?

     b. Display the complete set of linear combinations of effects that can be obtained from this experiment assuming that three-factor and higher-order interactions are negligible. Calculate estimates for each of the linear combinations of effects. Prepare a normal probability plot for the estimates. Which linear combinations of effects appear to be statistically significant?

     c. It has been proposed to follow up the first design with a mirror image design. Write down the recipe matrix for this second design. What are the generators and defining relation for the mirror image design?

d. Clearly explain why it might be inadvisable to conduct the proposed mirror image design given the results of the first fractional factorial design. Propose a better design to follow up the first 2^{5-2} design.

20.7. Consider a 2^{5-2} fractional factorial design with the principal fraction defined by the generators $4 = 12$ and $5 = 13$.

a. Write down the generators for the entire family of 2^{5-2} fractional factorials associated with this principal fraction.

b. Identify the member of the family that when combined with the principal fraction unconfounds the main effects (assuming third- and higher-order interactions can be neglected).

c. Identify the member of the family that when combined with the principal fraction unconfounds the main effect of variable 1 and all its two-factor interactions. The main effect of variable 2 and its two-factor interactions. The main effect of variable 5 and its two-factor interactions. Assume that third- and higher-order interactions can be neglected.

20.8. The 2^{3-1} fractional factorial design shown in the table was run and the results of the analysis were somewhat ambiguous.

Test	1	2	3
1	−	−	−
2	+	−	+
3	−	+	+
4	+	+	−

It was decided to run the mirror image design and combine its results with the design in the table.

a. Write down the design matrix for the mirror image design.

b. What is the resulting confounding pattern when the two designs are combined (i.e., what effects or confounded effects can be estimated when the results of the two designs are combined)?

20.9. Consider a 2^{7-3} fractional factorial design with the generators $I = -1235$, $I = 1246$, and $I = -1347$.

a. Write down the defining relation for this design. What is the resolution of this design?

b. Write down all linear combinations of effects that can be estimated from this experiment. Assume that third- and higher-order interaction effects can be neglected.

c. Propose a second, follow-up design that will clear up some of the confounding associated with the first design. Assuming that three-factor interactions and higher can be neglected, which effects and linear combinations of effects can be estimated from the two designs?

20.10. An experimenter wishes to study eight variables in 16 tests using the highest resolution design possible. Physical restrictions of the experimental testing process prohibit variables 1, 3, 4, and 6 from taking on their "+" levels simultaneously. Similarly, variables 1, 2, 3, and 5 cannot take on their "−" levels simultaneously.

   a. Find generators for a 2^{8-4} resolution IV design that will satisfy the restrictions above. What is the defining relation? Write out the design matrix.

   b. Under the assumption that three-factor and higher-order interactions are negligible, display the linear combinations of confounded effects.

   c. Explain why it would be inadvisable to follow-up this 2^{8-4} design with the mirror image design. Propose a follow-up design that might be more appropriate.

20.11. A two-level fractional factorial design was run with 16 test conditions. Given that one of the linear combinations of confounded effects that can be estimated from this particular design is 123 − 245 + 6 − 13456, determine the design generators. Show all work and explain how you arrived at your answer.

20.12. [Source: Advanced experimental design course at University of Wisconsin-Madison, 1969—Professor Irwin Guttman] Ozzie Cadenza, owner and manager of Ozzie's Bar and Grill, recently decided to study the factors that influence the amount of business done at his bar. At first, he did not know which factors were important and which were not, but he drew up a list of six variables, which he decided to investigate by means of a fractional factorial experiment.

   1. The amount of lighting in the bar.
   2. The presence of free potato chips and chip dip at the bar.
   3. The volume of the jukebox.
   4. The presence of Ozzie's favorite customer, a young lady by the name of Rapunzel Freeny. Miss Freeny is a real "life of the party," continually chatting with the customers, passing around the potato chips, and so on, all of which make Ozzie feel that she has a real effect on the amount of bar business.
   5. The presence of a band of roving Gypsies, who have formed a musical group called the Roving Gypsy Band, and who have been hired by Ozzie to play a limited engagement at the bar.
   6. The effect of a particular bartender who happens to be on duty. There were originally three bartenders: Tom, Dick, and Harry. Ozzie fired Harry, however, so that each of the factors for the experiment would have only two levels.

Low and high levels were selected for each of the variables as follows:

Variable	Low Level	High Level
1	Lights dim	Lights bright
2	No chips at bar	Chips at the bar
3	Jukebox playing softly	Jukebox blaring loudly
4	Miss Freeny stays home	Miss Freeny at the bar
5	No gypsy band	Gypsy band playing
6	Tom is the bartender	Dick is the bartender

Ozzie decided to perform one "run" of the experiment every Friday night during the happy hour (4:30–6:30). He thought that he should try a fractional factorial with as few runs as possible, since he was never quite sure when the band of gypsies would pack up and leave town. He finally decided to use a member of the family of the 2^{6-3} designs with generators of $I = \pm 124$, $I = \pm 135$, and $I = \pm 236$. He had wanted to find a resolution III design, in which the jukebox would never be blaring away while the gypsies were playing, but he found this requirement impossible.

a. Why is the statement above true? Carefully explain.

Ozzie did insist, however, that no run of the experiment could simultaneously have variables 1, 3, and 5 at their high levels. This restriction was made necessary by the annoying tendency of all light fuses to blow whenever the gypsies played their electric zither at the same time the lights were turned up brightly and the jukebox was playing at full volume. Note that this restriction makes it impossible for the principal fraction generators of the specified family to be used.

b. What members of the given family of generators does the specified restriction above allow?

Ozzie finally settled on a 2^{6-3} fractional factorial design, the recipe matrix for which is given in the table. The response (income in dollars) associated with each run is also given.

1	2	3	4	5	6	y
+	+	+	+	−	+	3150
+	+	−	+	+	−	2050
+	−	+	−	−	−	2050
+	−	−	−	+	+	1550
−	−	−	+	−	+	2650
−	+	+	−	+	+	1250
−	+	−	−	−	−	1350
−	−	+	+	+	−	1950

c. What are the generators and defining relation for this design?
d. Assuming that third- and higher-order interactions are negligible, write down the estimates obtained from this experiment and tell what they estimate (e.g., "1 + 24 − 35 estimated to be 400").

Being partial to Miss Freeny, and encouraged by the results of the first fraction, Ozzie chose a second fraction that would give unaliased estimates of her and each of her two-factor interactions. The results of this second fraction in standard order are 1350, 1650, 2850, 1750, 2050, 1950, 2950, and 1450.

e. What are the estimates obtained by combining the results of both fractions?

f. Offer a brief conjecture that might explain the presence and direction of the interactions involving Miss Freeny.

20.13. A 2^{7-4} fractional factorial design was conducted on a chemical process to assess the effect of the seven factors on the chemical yield (percent reacted).

Variable	Low Level	High Level
1. Temperature (°C)	150	200
2. Pressure	Low	High
3. Chemical A concentration (%)	3	5
4. Chemical B concentration (%)	2	8
5. Catalyst type	Brand X	Brand Y
6. Reaction time (min)	2	5
7. Flow rate	Low	High

The second table summarizes the results of the 2^{7-4} design. Included in the table are the levels for each of the seven variables and the yield (percent reacted) for each test. Note that the tests are displayed in the order in which they were performed.

Run	1	2	3	4	5	6	7	y
1	−	+	+	−	−	+	−	66.1
2	−	+	−	+	−	−	+	59.6
3	+	−	+	+	−	−	−	62.3
4	+	+	−	−	+	−	−	67.1
5	−	−	+	−	+	−	+	21.1
6	+	+	+	+	+	+	+	57.8
7	−	−	−	+	+	+	−	59.7
8	+	−	−	−	−	+	+	22.5

a. What are the generators and defining relation for this 2^{7-4} design? Display the complete set of linear combinations of effects that can be obtained from this experiment assuming that three-factor and higher-order interactions are negligible. Calculate numerical values for each of the linear combinations of effects. Prepare a normal probability plot for the estimates. Which linear combinations of effects appear to be statistically significant?

b. Propose a follow-up design that you believe should be performed next. Clearly explain the logic behind your decision. When the results of the first design are combined with those of your proposed second design, what effects or linear combinations of effects can be estimated?

20.14. A chemical process has been studied through the use of a fractional factorial design to judge the effects of the following seven factors on the chemical yield (percent reacted).

Variable	Low Level	High Level
1. Temperature (°C)	150	200
2. Pressure	Low	High
3. Chemical A concentration (%)	3	5
4. Chemical B concentration (%)	2	8
5. Catalyst type	Brand X	Brand Y
6. Reaction time (min)	2	5
7. Flow rate	Low	High

Initially, a 2^{7-4} fractional factorial design was performed that utilized the following generators: $4 = 12$, $5 = 13$, $6 = 23$, and $7 = 123$. Based on the first design, the following linear combinations of effects (assuming that three-factor and higher-order interactions are negligible) and associated estimates were calculated:

$$l_0 = 51.9625 \text{ estimates mean}$$
$$l_1 = 1.2250 \text{ estimates } 1 + 24 + 35 + 67$$
$$l_2 = 21.3750 \text{ estimates } 2 + 14 + 36 + 57$$
$$l_3 = -0.4250 \text{ estimates } 3 + 15 + 26 + 47$$
$$l_{12} = -1.4250 \text{ estimates } 12 + 4 + 37 + 56$$
$$l_{13} = 15.4750 \text{ estimates } 13 + 5 + 27 + 46$$
$$l_{23} = -1.1750 \text{ estimates } 23 + 6 + 17 + 45$$
$$l_{123} = -24.3750 \text{ estimates } 34 + 25 + 16 + 7.$$

To unconfound the main effects from the two-factor interactions, the mirror image design was performed, the results of which are shown in the table (tests are displayed in the order in which they were performed).

Run	1	2	3	4	5	6	7	y
1	+	−	+	+	−	+	−	48.6
2	−	−	+	−	+	+	+	39.6
3	+	−	−	+	+	−	+	35.3
4	−	+	+	+	+	−	−	82.0
5	−	−	−	−	−	−	−	48.2
6	+	+	−	−	+	+	−	80.5
7	+	+	+	−	−	−	+	41.9
8	−	+	−	+	−	+	+	42.9

a. Display the linear combinations of effects (assuming that three-factor and higher-order interactions are negligible) and associated estimates for the mirror image design.

    b. Combine the results of the two designs and prepare a normal probability plot to identify those effects or linear combinations of effects that appear to be statistically significant.

    c. Comment on the findings of the two experiments. Specifically, which variables are important, and how may the response be maximized?

20.15. Consider a 2^{6-3} fractional factorial design with the principal fraction defined by the generators $4 = 12$, $5 = 23$, and $6 = 123$.

    a. Write down the entire family of 2^{6-3} fractional factorials associated with this principal fraction.

    b. Identify the member of the family that when combined with the principal fraction unconfounds the main effects. Assume that third- and higher-order interactions can be neglected.

    c. Identify the member of the family that when combined with the principal fraction unconfounds the main effect of variable 1 and all its two-factor interactions. The main effect of variable 3 and its two-factor interactions. The main effect of variable 6 and its two-factor interactions. Assume that third- and higher-order interactions can be neglected.

20.16. After two 2^{5-2} fractional factorials are conducted and analyzed, the following model is developed:

$$\hat{y} = 17.1 - 9.4x_2 + 7.7x_4 - 12.3x_5 + 8.5(x_2x_4 \text{ and/or } x_3x_5)$$
$$- 5.9 \ (x_3x_4 \text{ and/or } x_2x_5).$$

Why have the interaction terms in the model been expressed in the manner indicated? Clearly explain (and demonstrate) your answer.

20.17. Consider a 2^{9-5} fractional factorial design with the generators $I = 1235$, $I = 1246$, $I = 1347$, $I = 2348$, and $I = 12349$.

    a. What is the defining relation for this design? Display the linear combinations of effects that can be estimated from this experiment. Assume that third- and higher-order interaction effects can be neglected. Write out the design (recipe) matrix for this design in standard order.

    b. The results of the experiment in part (a) were somewhat ambiguous; therefore, it is proposed to perform a second fractional factorial design, the mirror image design. Display the design matrix for the mirror image design. What are the generators for this second design? Display the linear combinations of effects that can be estimated from the mirror image design, assuming that third- and higher-order interaction effects can be neglected. When the principal fraction and mirror image designs are combined, what effects, or linear combinations of effects, can be estimated? Once again, assume that third- and higher-order interactions are negligible.

    c. Instead of performing the second design described above (the mirror image design), it is proposed to follow up the principal fraction with a 2^{9-5} design having the generators $I = -1235$, $I = -1246$, $I = -1347$, $I = -2348$, and $I = 12349$. This alternative fraction was serendipitously identified by T. S. Babin and J. W. Sutherland (while they were graduate students at the University of Illinois at Urbana–Champaign). Display the linear combinations of effects that can be estimated from the Babin–Sutherland design, assuming that third- and higher-order interaction effects can be neglected. When the principal fraction and Babin–Sutherland designs are combined, what effects, or linear combinations of effects, can be estimated? Once again, assume that third- and higher-order interactions are negligible.

d. Considering the results of parts (b) and (c), which design should be performed to follow up the principal fraction? Clearly explain your answer.

20.18. A 2^{5-2} fractional factorial design is performed utilizing the generators $I = 1234$ and $I = 135$. After analyzing the results, the investigator decided to run another 2^{5-2} design with the generators $I = -1234$ and $I = -135$.

a. Write down the defining relations and linear combinations of confounded effects (assuming three-factor and higher-order interactions are negligible) for each of the designs above.

b. Combine the two designs and indicate what effects or combinations of effects can be estimated from the two designs.

c. What might have led the investigator to choose the specific second fraction that she did?

C H A P T E R

21

Robust Design Case Studies

21.1

Taguchi's Approach to Robust Design

Taguchi views the design process as evolving in three distinct phases or steps: system design, parameter design, and tolerance design. It is perhaps this broad umbrella that he places over his concepts and methods for quality design and improvement that makes his approach so readily embraceable by the engineering community. As we said earlier, Taguchi considers engineering design as the central issue and statistical methods as just one of several tools to accomplish his objectives in engineering.

System design is the initial phase of any engineering design problem wherein knowledge from specialized fields (chemistry, electronics, etc.) is applied to develop the basic design alternative. This design alternative is then refined in the phases of parameter design and tolerance design. *Parameter design* is that phase of the design process in which the best nominal values for the important design parameters are selected. Taguchi defines the "best" values for the nominals as those values that *minimize the transmitted variability resulting from the noise factors.* This is why we sometimes refer to parameter design as *robust design.* *Tolerance design* is the final phase of Taguchi's design process model, in which the tolerances on the nominal values on some of the critical design parameters are selected. In tolerance design the setting of specifications is considered to be an economic issue and the loss function model is used as a basis to determine the allowable tolerance on the nominal values.

| 21.2 |

Overview of Approaches to Robust Design

In recent years a number of different approaches have been proposed as possible ways to implement the robust design concept of Taguchi. These vary from purely analytical approaches, to computer simulation using product/process mathematical models and/or Monte Carlo methods, to the use of physical experimentation. In most of these approaches the use of experimental design strategies including two-level and multilevel factorial and fractional factorial designs has been extensive. Taguchi refers to these experimental design structures as *orthogonal arrays* and makes use of these structures for both inner and outer arrays. In Chapter 19 we discussed Taguchi's use of orthogonal arrays and their relationship to the class of two-level fractional factorial designs.

In addition to the different approaches to "generating" data on product/process performance, the issue of the specific measures of performance to use in the facilitation of parameter design needs to be considered. Taguchi and his colleagues make extensive use of the signal-to-noise ratio as a measure of performance. Others evaluate the mean and the variance or standard deviation of performance separately.

In this chapter we examine four case studies that exemplify the different ways in which the concept of parameter design can be implemented to develop robust product and process designs. These are presented under three basic approaches:

1. Experimental approaches to parameter design.
2. Analytical approaches to parameter design.
3. Computer simulation approaches to parameter design.

Experimental Approaches to Parameter Design

Two of the case studies illustrate how the parameter design method for the determination of a robust product/process can be implemented through physical experimentation. These are:

1. The explicit use of an inner array/outer array experimental design structure for robust design in the calibration and assembly of a fuel gage.
2. The exploitation of system nonlinearities in the form of variable interactions to effect a robust design in the study of glove box closing effort in an automobile assembly plant.

Analytical Approaches to Parameter Design

As the methods of Taguchi for engineering design are studied, applied, and refined it is important to keep a broad perspective on the contributions of

Taguchi and the wide variety of ways in which they can be utilized. Perhaps the single overarching concept that drives the philosophies of Taguchi is the importance placed on the consistency of performance of products and processes, that is, the importance of variation reduction in achieving quality products and processes. What this essentially means in a broad sense is that as we formulate designs, either on paper or in hard materials, and we evaluate their performance, either analytically or experimentally, we should place as much or more emphasis on variation in performance as we do on performance on average. There are many ways to invoke this overarching concept.

Analytical methods that have been proposed to date for the implementation of the parameter design concept find their roots in the foundations of classical optimization theory and statistical modeling. While much of the knowledge that we use in any given design situation is derived from experimentation, there exists an ever-expanding baseline of knowledge that is being derived from first principles/physical laws and the utilization of sophisticated modeling methods such as finite element modeling. As closed-form solutions become available, either directly or through approximations based on data, it becomes possible to evaluate them analytically or numerically using classical optimization methods. The performance function should be examined in terms of both its expected value and its variance function. Searching the variance function to find those values for the design/control variables that minimize the variance in performance then amounts to invoking the parameter design method. One of the case studies that follow uses this approach to examine automobile door-closing effort.

Computer Simulation Approaches to Robust Design

With the strong efforts of recent years in the area of mathematical modeling of processes and systems, the opportunity for in-depth study of performance through computer simulation methods has grown considerably. In addition, the drive to put product and process design analysis farther and farther upstream has been considerable in the last several years. Physical experimentation downstream is being replaced by model-based experimentation earlier in the product development life cycle. In this chapter we examine one such case study on the simultaneous engineering of automobile engine components and their associated manufacturing processes.

21.3

Case Study 1: Fuel Gage Calibration

The Problem. The following case study was developed from the results of a larger experiment and is presented at this time to demonstrate how relatively simple experiments such as two-level factorial designs can be employed to study and improve the performance of an assembly process using the concept of robust design. Customer complaints concerning the difference between the fuel gage reading and the actual amount of fuel in the tank led to an investigation of the process of gage calibration. Seven variables were tentatively identified as being of possible importance to the position (percent deflection or fallback) of the indicator in the fuel gage.

Initial Design of the Experiment.   A 2^{7-3} fractional factorial design experiment was constructed and carried out. We examine a portion of the results, which involve four of the seven variables that were determined to be the most important.

Variable	Low Level	High Level
1. Spring tension	Loose	Tight
2. Method of pointer location	Visual	Other (standard)
3. Bimetal hook twist	Vendor A	Vendor B
4. Bimetal bracket bend	90 degrees	93 degrees

Variables 1 and 2 describe adjustments that can be made during the calibration process on the assembly line. Variables 3 and 4 describe the condition (considered undesirable) of certain purchased parts which are assembled into the gage and may affect the calibration process. The 2^{7-3} fractional factorial design was collapsed into a 2^4 full factorial design to study the effects of these variables, as they are purposely changed, on the response in question.

The design matrix (Table 21.1) shows the settings for the 16 tests conducted in terms of the four variables defined above. The tests are listed in standard order, although they were actually run in a randomized test sequence. The main effects of each of the four factors, as well as the two-, three-, and four-factor interactions, were estimated from the data and are as follows:

$$
\begin{array}{llll}
& E_1 = -2.50 & * & E_{24} = 0.50 \\
* & E_2 = 0.00 & & E_{34} = -1.25 \\
& E_3 = -1.75 & & \\
& E_4 = 3.00 & * & E_{123} = 0.75 \\
& & & E_{124} = -2.50 \\
* & E_{12} = 0.50 & * & E_{134} = 0.75 \\
& E_{13} = 1.75 & * & E_{234} = 0.25 \\
& E_{14} = -1.00 & & \\
& E_{23} = 2.25 & & E_{1234} = 2.25.
\end{array}
$$

The results of the larger experiment, in fact, showed that these four variables tended to exhibit the stronger effect estimates, both main effects and interactions. A normal probability plot of the estimated effects could be interpreted as suggesting that only the "starred" effect estimates above might be considered insignificant. This interpretation leaves the investigator somewhat perplexed as

TABLE 21.1 Design Matrix					
Test	x_1	x_2	x_3	x_4	y (%)
1	Loose	Visual	Vendor A	90°	14
2	Tight	Visual	Vendor A	90°	7
3	Loose	Other	Vendor A	90°	7
4	Tight	Other	Vendor A	90°	9
5	Loose	Visual	Vendor B	90°	9
6	Tight	Visual	Vendor B	90°	7
7	Loose	Other	Vendor B	90°	9
8	Tight	Other	Vendor B	90°	10
9	Loose	Visual	Vendor A	93°	15
10	Tight	Visual	Vendor A	93°	14
11	Loose	Other	Vendor A	93°	18
12	Tight	Other	Vendor A	93°	7
13	Loose	Visual	Vendor B	93°	10
14	Tight	Visual	Vendor B	93°	8
15	Loose	Other	Vendor B	93°	12
16	Tight	Other	Vendor B	93°	12

to what exactly to do next. Many interactions seem important; it is not at all clear how best to adjust/alter the process.

It would appear that the main effects of variables 1 and 4 might be two of the more important effects.

$E_1 = -2.50$ implies that on the average going to tighter spring tension reduces fallback (deflection)

$E_4 = +3.00$ implies that larger bimetal bracket bend (93°) increases deflection, which is bad.

The latter inference drawn suggests that we might have to go to the bimetal bracket supplier and request closer adherence to the 90° bend, which is the desired nominal value. This essentially means that we would require the supplier to tighten the tolerance on this angle. Rather than taking this more costly approach, it might be useful to consider the variation in this part dimension as a noise variable and try to determine how the transmission of the variation in the bimetal bracket bend could somehow be reduced through the manipulation of control factors during the calibration process, that is, develop a more robust calibration process.

Alternate Experimental Design and Analysis.   A more thoughtful examination of the four variables under study seems to indicate that they fall into two basic categories:

Variable 1:   spring tension
Variable 2:   method of pointer location.

These two variables may be easily adjusted at the discretion of the operator/ setup person.

> *Variable 3:*   bimetal hook twist
> *Variable 4:*   bimetal bracket bend.

These two variables describe certain aspects of the condition of two purchased parts. Although we have specified certain requirements to the vendors, we do not have very strong control over these factors—the condition of the parts coming in—without tightening the specification requirements, which is a costly action. We might therefore consider these two variables to be *noise variables* and perhaps use the robust design/parameter design concept to see how adjustments in variables 1 and 2 could be made to reduce the transmitted variability due to variables 3 and 4.

Now, instead of thinking in terms of a 16-run experimental design that considers four variables at two levels each, we will think in terms of two 2^2 factorial experiments, an "inner" design and an "outer" design.

- The *inner design* is a 2^2 factorial in the two controllable variables, spring tension and method of pointer location.
- The *outer design* is a 2^2 factorial in the two noise variables, bimetal hook twist and bimetal bracket bend.

The outer design is conducted at each of the four unique variable settings of the inner design. A graphical representation of this experimental design is shown in Fig. 21.1. The numbers in the corners of each of the four outer design "squares" represent the fallback results for each of the 16 unique trials.

Referring to the data in Fig. 21.1, we now determine $\bar{y}$ and s_y at each of the four test combinations of the *inner design*.

Test 1: $x_1(-)$, $x_2(-)$

$$\bar{y} = 12.0, \qquad s_y = 2.94$$

Test 2: $x_1(+)$, $x_2(-)$

$$\bar{y} = 9.0, \qquad s_y = 3.37$$

Test 3: $x_1(-)$, $x_2(+)$

$$\bar{y} = 11.5, \qquad s_y = 4.8$$

Test 4: $x_1(+)$, $x_2(+)$

$$\bar{y} = 9.5, \qquad s_y = 2.08$$

Analyzing the responses above (average fallback and standard deviation of fallback) separately, we can determine the settings/values for the control variables that (1) minimize average fallback and (2) give rise to minimum variation in fallback (i.e., the most robust process design).

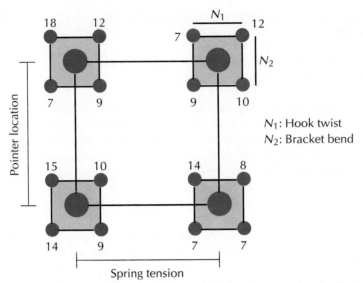

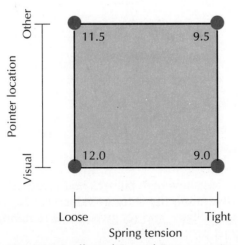

Figure 21.1 Inner/Outer Experimental Design Structure for the Fuel Gage Study

Figures 21.2 and 21.3 provide the results of the analyses. The interpretation of the results is as follows:

- In terms of average response percent deflection is reduced by increasing spring tension. This reduction is about the same for both methods of pointer location. On the average, pointer location method has little effect on percent deflection.
- *In terms of standard deviation,* it is best to use the other (standard) method of pointer location along with tighter spring tension. In the case of the

Figure 21.2 Effect of Control Factors on Average Performance

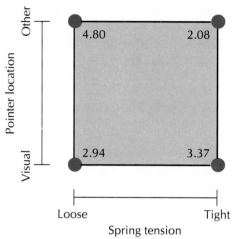

Figure 21.3 Effect of Control Factors on Variation in Performance

variation response, the interaction of the two control variables seems important.

In this case study, it was decided to adjust the spring tension to a tighter level. Such did lead to a reduction in calibration error. A more in-depth study of the method of pointer location was also initiated.

If we were to look at the results of Figs. 21.2 and 21.3 from a loss function point of view, we could calculate the loss resulting from both deviation from the target on average and variation about the average.

$$\text{loss} = k[s_y^2 + (\bar{y} - m)^2],$$

where here $m = 0$ is the target for zero deflection/fallback. For the four cases characterized by the inner array/outer array experiment, the loss is smallest for the combination, tight spring tension and visual method of pointer location. Although there is less variation resulting from the noise factors with the tight spring tension used with the standard method of pointer location, the average for this setting is farther from the target. It should be pointed out, however, that the difference in loss for these two conditions is not great.

21.4
Case Study 2: Glove Box Closing Effort Analysis

Quality Concern.   This case study deals with a classic automobile assembly problem, that of glove box fits.[1] As a result of individual component design

[1] P. M. Belaire and R. J. Deacon, "The Strategic Approach to Quality Improvement Using Design of Experiments Concepts," *Symposium on Quality: Design, Planning, and Control, ASME Winter Annual Meeting*, December 13–18, 1987, PED Vol. 27, pp. 157–167.

and the subassembly design configuration as well as manufacturing variations, quality concerns often arise. These problems manifest themselves as glove box assemblies which fit too loosely, so that a rattle condition is present, or too tightly, so that the closing effort required is excessive or the door simply cannot be closed. The issue in question here arose at the upper management level, as high warranty costs brought this quality concern to the forefront.

Figure 21.4 illustrates the situation under study. In Fig. 21.4, curve (a) represents the statistical distribution for closing effort in units of inch-pounds centered at the target specified by the design process. The spread of this distribution reflects the state of affairs at the outset of this study. Because of excessive variation about the target, the rattle condition develops too frequently. In an effort to solve this problem, an action was taken at the assembly process level that led to an overall increase in closing effort caused by shifting the mean of the distribution as shown by curve (b) in Fig. 21.4. Unfortunately, due to the excessive variation in closing effort, the problem now became one of experiencing too many assemblies with intolerably high closing effort. Further adjustment of the mean in an effort to strike a balance between the two contrasting quality problems produced only the simultaneous occurrence of both.

It is clear from the discussion that the basic issue here is not simply the proper centering of a process but rather one of being "on target with smallest variation." It is unlikely that a satisfactory solution will be found without effecting a considerable reduction in process variation, as shown by curve (c) in Fig. 21.4. This can be accomplished in several fundamental ways. Study of the assembly process itself may reveal faults that when rectified could lead to improved consistency. Improving the process consistency for vendor-supplied parts may also yield improvement. Design changes in the form of requiring the manufacturing of components to tighter specifications may also lead to improved glove box fits. It is even possible (perhaps likely) that changes in component and/or subassembly designs may affect significant quality improvement.

Design of Experiment.   In an effort to better understand the glove box closing effort problem, it was decided to plan and conduct a statistically designed

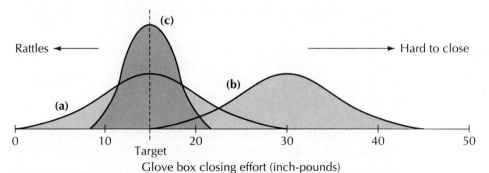

Figure 21.4 Statistical Representation of Closing Effort Problem

TABLE 21.2	Design and Process Factors and Associated Levels for Experiment		
Variable Name	Lower Level	Higher Level	Measure/Unit
1. Panel foam	Low	High	Thickness
2. Bumper durometer	Soft	Hard	Hardness
3. Ramp angle	Present	Design change	Degrees
4. Striker position	Fore	Aft	Location
5. Striker angle	Present	Design change	Degrees

experiment to reveal the effects that several thought-to-be important factors might have on closing effort. Table 21.2 lists the five factors under study together with the low and high levels for each that were used in a two-level fractional factorial experiment.

Figure 21.5 shows a sketch of the glove box assembly and illustrates the physical meaning of the five variables under study. Panel foam thickness refers to the foam-cushioned vinyl covering on the dash panel substrate. Bumper durometer is the hardness of rubber bumpers, which cushion impact and hold the door assembly firmly in place. Striker angle and ramp angle refer to two design configurations relating to the striker and the bracket it is mounted on. Striker position (fore and aft) is set by the operator during the assembly process. Three of the five factors relate to characteristics of purchased parts and can be

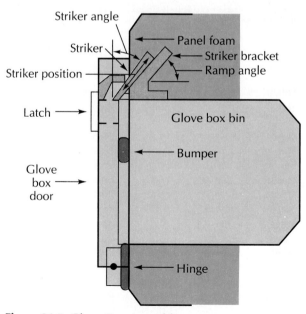

Figure 21.5 Glove Box Assembly

altered through design changes (bumper hardness, ramp angle, and striker angle). Foam thickness is a function of the foam molding process. Striker position is a function of the actual assembly process and is operator controlled.

Table 21.3 provides the design matrix and test results for the 2^{5-1} fractional factorial design employed in this study. Under the assumption (which was used in this study) that third- and higher-order interactions may be neglected, the design used provides clear estimates of all the main effects and two-factor interactions among the five variables.

A normal probability plot of the effect estimates that are given in Table 21.4 led to the conjecture that variables 3, 4, and 5 have large main effects, while interactions E_{12}, E_{34}, E_{35}, and E_{45} may also be important. It is, however, the physical interpretation of these apparently important effects that is critical to determining a permanent and cost-effective quality improvement.

Discussion of Results.   Figure 21.6 is a graphical representation of the variable main effects that appear to be most dominant. The suggestion is that the proposed design change for the ramp angle may reduce closing effort, while that for the striker angle may increase effort. The aft position for striker appears to produce lower closing effort values. There are two problems associated with the interpretation of the main effects in Fig. 21.6. First, they are average effects; that is, they are averaged over the low and high levels of the other four factors in each case. If two-factor interactions are important, as appears to be the case here, these main effects must be interpreted with great caution. Second, these

TABLE 21.3   Design Matrix and Test Results

Test	(1) Foam	(2) Bumper	(3) Ramp Angle	(4) Striker Position	(5) Striker Angle	Response (in.-lb)	Run Number
1	Low	Soft	Present	Fore	Change	150.0	8
2	High	Soft	Present	Fore	Present	150.0	2
3	Low	Hard	Present	Fore	Present	150.0	10
4	High	Hard	Present	Fore	Change	150.0	13
5	Low	Soft	Change	Fore	Present	17.3	15
6	High	Soft	Change	Fore	Change	24.7	11
7	Low	Hard	Change	Fore	Change	25.7	9
8	High	Hard	Change	Fore	Present	21.7	3
9	Low	Soft	Present	Aft	Present	22.7	5
10	High	Soft	Present	Aft	Change	150.0	7
11	Low	Hard	Present	Aft	Change	150.0	14
12	High	Hard	Present	Aft	Present	18.7	4
13	Low	Soft	Change	Aft	Change	19.7	16
14	High	Soft	Change	Aft	Present	8.0	12
15	Low	Hard	Change	Aft	Present	21.0	6
16	High	Hard	Change	Aft	Change	45.0	1

TABLE 21.4	Effect Estimates for the Glove Box Door Closing Effort Study	

Average =	70.2813	E_{23} =	5.9625
E_1 =	1.4625	E_{24} =	3.6125
E_2 =	4.9625	E_{34} =	32.8625
E_3 =	−94.7875	E_{45} =	36.3625
E_4 =	−31.7875	E_{35} =	−26.4375
E_{12} =	−29.2875	E_{25} =	1.6125
E_{13} =	2.4625	E_{15} =	4.6125
E_{14} =	0.6145	E_5 =	38.2125

effects are measures of how changes in factor levels influence the mean response. As such they do not have a direct impact on the amount of variability about the mean. However, as we will see shortly, a thoughtful interpretation of two of the important two-factor interactions will clearly reveal how improved consistency (reduced variation) about the target mean response can be achieved.

Figure 21.7 provides a graphical representation of two of the larger two-factor interaction effects. In both cases the variable striker position is involved. It was clear from the definition of the five factors under study that all are not controllable in either the same basic way (design versus process) or to the same degree of control. It may be useful to invoke Taguchi's concept of robust design/parameter design here and think of striker position as a noise variable and the ramp angle and striker angle as design/control variables.

It is clear from Fig. 21.7(a) that as the noise variable striker position varies fore to aft, the transmitted variation in closing effort is greatly reduced with a design modification to the ramp angle. One might propose that no action need be taken relative to the striker position if the ramp angle design change is made. Further, the mean response is reduced to levels very near the desired target. A similar situation is present in Fig. 21.7(b), the striker angle/striker position inter-

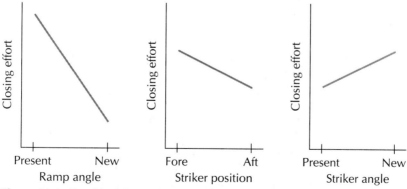

Figure 21.6 Graphical Representation of Important Main Effects

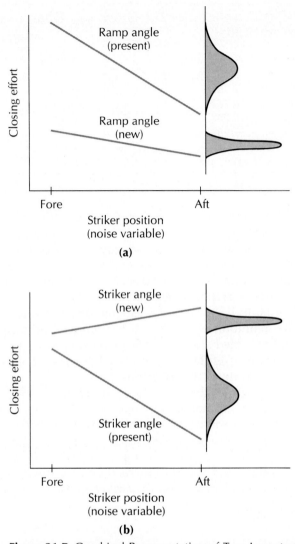

Figure 21.7 Graphical Representation of Two Important Two-Factor Interaction Effects

action. In this case, however, the mean closing effort is greatly increased, which is undesirable. Figure 21.8 shows that design changes in both ramp angle and striker angle may reduce closing effort on average.

The foregoing interpretations of some large two-factor interactions are important because they show how the discovery of such system nonlinearities can be used to develop more robust, less variable product designs in an economic fashion. Understanding Taguchi's concept of parameter design is important in using statistical design of experiments for quality improvement.

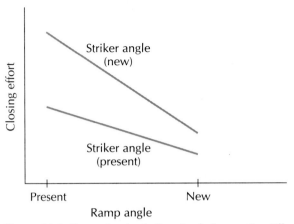

Figure 21.8 Ramp Angle–Striker Angle Interaction Effect

Improvement Action.   The results of the experimental design that relate to design changes for the striker angle and ramp angle were verified by building a number of glove box assemblies using the prescribed conditions and testing them for mean effort and variation in effort. Based on these verification tests, action was taken to implement the ramp angle change.

21.5

Case Study 3: Automobile Door Closing Effort

H. L. Oh[2] uses an explicit model for the mean square error of design performance in an automobile door closing effort problem to identify the values of the design parameters that provide the most consistent design configuration (minimum variance) while maintaining an acceptable level of average performance. In this case study the following logic/methodology is employed.

Problem Formulation.   The first step in any design optimization problem is to formulate the design performance measure of interest. In this case study, car door fit is the issue—in particular, the sensitivity of the closing effort to variations in the door positioning as it is hung in place. Figure 21.9 shows a sketch of a car door and the locations of the three positioning points used to hang the door. The closing effort is a function of the compression of the weather strip around the door periphery. The closing effort can be expressed analytically as a function of the weather strip diametral stiffness, the weather strip diameter and length, and the nominal values of the door position points that define the

[2] H. L. Oh, "Variation Tolerant Design," *Symposium on Quality: Design, Planning, and Control, ASME Winter Annual Meeting,* December 13–18, 1987, PED Vol. 27, pp. 137–147.

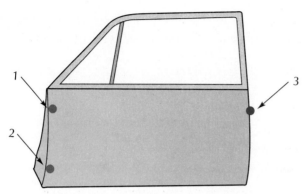

Figure 21.9 Location Points for Positioning of Door

door/frame gap. The basic design problem is then to find the nominal values for the door position points that minimize the variation in closing effort as a result of the variation in the door position points about their nominal values.

This problem formulation is very similar to the glove box closing effort problem mentioned previously in the sense that we are seeking to exploit the underlying nonlinearities in the system to improve performance consistency. This approach is an alternative to simply reducing the allowable tolerance on a more arbitrarily determined nominal value to reduce transmitted variation. Reducing the tolerance has been an altogether too common engineering design approach to improving consistency of performance, an approach that is far too costly and does not contribute substantially to the enhancement of competitive position.

Math Model Derivation.   Under the assumption that the door gap is linearly proportional to the door position point locations, it is possible to derive the door closing effort (in stored energy) as a quadratic function of the position point locations. Remember that the position point locations are the control factors in this problem. The quadratic function for closing effort is a function of the three hanging positions and has several unknown parameter values. Although the closing effort will clearly vary also as a function of variation from car to car in weather strip thickness, car body size, and so on, only the variation in the door hanging position from car to car was examined in this case study, in the interest of simplicity.

Math Model Fitting.   The parameters of the quadratic function for closing effort were estimated by first calculating the door gap at 15 points around the door as a function of variations in the door hanging positions and then using the calculated gap values to determine the closing effort values. The variations in the hanging position control factors were introduced using a three-level factorial design, which required a total of 27 (3^3) evaluations of the gap and the stored energy at each of the 15 locations around the door. These data were then used to estimate the coefficients of the quadratic closing effort function.

Design Optimization.   The variance of the closing effort quadratic function may now be derived analytically, and this variance function may be subjected to classical optimization techniques. Such an optimization algorithm will seek the values of the three door position locations that minimize the variance of the closing effort. This optimization is subject to the constraint that the mean closing effort is less than or equal to a prespecified target value. The result is the determination of nominal values for the design/control factors, the door positioning locations, which minimize the closing effort variation due to variation in these locations. In this design optimization problem the control factors are the nominal values for the door position locations, and the noise factors are really the variations in the door position locations.

Figure 21.10 shows a plot of the derived functional relationship between closing effort and the location of point 1 for positioning the door, for fixed values of location points 2 and 3. The graph shows clearly how the parameter design method exploits a basic system nonlinearity by seeking a new nominal value for the location of point 1, which reduces the transmitted variation to the closing effort that results from the variation in location of point 1 about its nominal value.

21.6

Case Study 4: Simultaneous Engineering of a Product and Its Manufacturing Process

Simultaneous Engineering Model

In this case study, the general class of products being studied is typified by castings with irregular shapes, thin-walled sections, and highly irregular and interrupted surfaces machined by processes such as face milling and boring.[3] Such components may, for example, be found in automobile engines and power trains. Such parts are generally machined in high volumes on special-purpose machining systems such as transfer lines.

The purpose of the simultaneous engineering concept is to give increased consideration to manufacturing system design and the impact of manufacturing on the product design configuration at the earliest stages of product design. It is well recognized that lead times can be significantly reduced and a more globally optimal product and process combination obtained through interactive simultaneous design rather than a more sequentially based design strategy.

Simultaneous engineering is not simply a matter of applying current prod-

---

[3] R. E. DeVor, S. G. Kapoor, R. Hayashida, G. Subramani, and T. Lindem, "A Methodology for the Simultaneous Engineering of Products and Manufacturing Processes," *Proc. 14th North American Manufacturing Research Conference*, 1986, pp. 535–542.

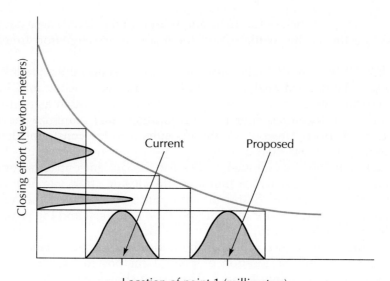

Figure 21.10 Closing Efforts Versus Upper Hinge Location

uct and process design strategies at the same time. New models and methods must be forthcoming. Figure 21.11 provides a simultaneous engineering conceptual framework. The complete framework is composed of four models, three of which describe the product and the process, while the fourth represents the design process itself.

Product Design Model.   The product design model may take several forms. Its purpose is to reduce the design to a set of parameters that are related to relevant performance measures. Simple performance measures might include size, weight, and/or shape. A finite element model may be used to define the values of the relevant structural dynamics of the product and/or characterize the stresses present in the product during field use.

Macro-Level Process Model.   The macro-level process model is a large-scale system representation of the processing system under design. The elements of the system (e.g., machines, conveyors, robotic manipulators, and gaging devices) could be modeled in a structure that allows for the simulation of the operation of the system. Simulation languages and models may be usefully employed to build up such a macro-level representation of a system consisting of a number of unit processes.

Micro-Level Process Model.   The micro-level process model provides a representation of a unit process (e.g., face milling, boring, reaming). Such a model is capable of simulating a cut or series of cuts with full consideration of the

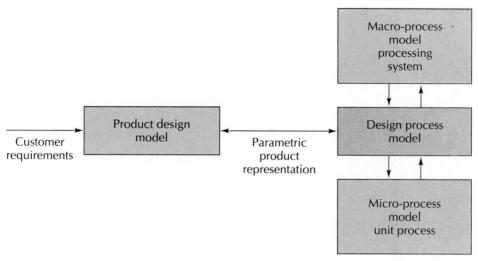

Figure 21.11 Conceptual Framework for Simultaneous Engineering

process geometry, tool geometry, part geometry, and the cutting mechanism itself. Tooling and fixturing are also part of the micro-level process model.

Design Process Model.   The design process model provides a framework and methodology to facilitate rapid convergence to a quality product and manufacturing process. Several design process models have been proposed. When merged with the classical methods of experimental design, the Taguchi design concepts provide a particularly useful model for simulation work.

Product and Process Models

To apply the Taguchi methods of parameter design through computer simulation requires performance models that can predict measures such as cutting forces and vibration/displacement magnitudes as a function of cutting conditions, cut geometry, and the cutting mechanism itself. A dynamic force model developed by Fu, DeVor, and Kapoor is used in this case study.[4] This model has three basic elements:

1. A static mechanistic model, which can predict instantaneous cutting forces based on the design parameters related to part, process, and cut geometries.
2. Finite element models of the part–fixture and cutter–spindle combinations.

[4] H. J. Fu, R. E. DeVor, and S. G. Kapoor, "A Mechanistic Model for the Prediction of the Force System in Face Milling Operations," *J. of Engr. for Ind., Trans., ASME,* Vol. 106, Feb. 1984, pp. 81–88.

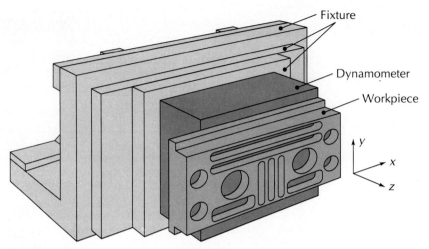

Figure 21.12 Fixture/Dynamometer/Workpiece Assembly

3. A model that represents the feedback mechanism between the cutting
   process and the structural elements.

In the dynamic cutting force model, the cutting forces act as the source of
excitation, and the structural elements of item 2 respond to this excitation
through displacement in various directions. The cutting process is affected by
the excited structural elements via a feedback loop.

Static Force Model.   The static force model may be used to predict the instantaneous X force, Y force, and Z force as a function of part geometry, tool
geometry, cut geometry, the relative positions of cutter and workpiece (critical
to certain highly irregular part designs), cutter offset, and runout (see Fig.
21.12).

Product/Part Model.   The product/part model consists of the finite element
models of the part–fixture combination, cutter–spindle combination, and a
methodology for determining the dominant modes of vibration and estimating
the corresponding stiffness and damping of these modes. The part–fixture
combination is shown in Fig. 21.12. The part–fixture was modeled with 1477
nodes and 821 elements, of which 128 were of the thin shell type and 693 were
of the solid type. This modeling was done with the assumption that the machine
table to which the fixture is bolted is perfectly rigid.
    The cutter–spindle finite element model is based on the spindle design of
a Kearney and Trecker horizontal milling machine on which the cutting tests
for model calibration were performed. This model consists of 943 nodes and
706 solid elements. The first 12 dynamic modes of the part–fixture and the first

TABLE 21.5	Dominant Frequencies and Stiffnesses for Part–Fixture and Cutter–Spindle Combinations	

Displacement Direction/Force Direction	Natural Frequency (hertz)	Stiffness (lb/in.)
XX	961.00	20,759,320.0
XY	984.00	8,964,032.0
XZ	1452.00	−5,662,946.0
YX	984.00	8,995,269.0
YY	285.00	24,641.3
YY	650.00	7,158.7
YY	961.00	24,802.7
YZ	1452.00	25,066.4
ZX	1452.00	−5,248,543.0
ZY	285.00	−2,506,015.0
ZZ	285.00	679,657.0
ZZ	417.00	800,714.7
ZZ	650.00	841,598.1
ZZ	961.00	813,324.8
ZZ	1452.00	429,528.0

three modes of the cutter–spindle were determined to be the most relevant. The information associated with these modes is shown in Table 21.5.

Experiment Design

The experimental design structure used in this simulation study was an inner/outer array. This structure is exemplified in Fig. 21.13, which illustrates the concept with four variables. In the figure, the inner design is a 2^2 factorial in two design variables, depth of cut, and workpiece–fixture stiffness. The outer design, performed for each unique test combination in the inner design, is 2^2 factorial in two noise variables, insert runout and workpiece surface error. The idea of parameter design is to find the conditions of the design variables that minimize the functional variation in performance resulting from the presence of the noise variables.

Table 21.6 provides the variable partitioning for the combined inner and outer design simulation experiments conducted in this study. The high (+) and low (−) levels for each of these variables are also given in this table. Runout was considered as a noise variable and was induced by uniformly and progressively offsetting the inserts from no runout to a maximum single insert runout of 0.0015 inch. Surface error, when present, took the form of a cylindrically based "crown" on the workpiece, with a maximum height of 0.0035 inch at the center of the workpiece and of constant cross section along the direction of

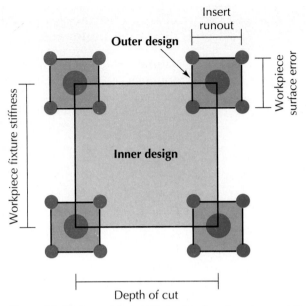

Figure 21.13 Inner Design/Outer Design Experimental Design Structure

cutter feed motion. Table 21.6 also provides relevant fixed process conditions for the simulations.

For this study the inner design was a 2^{5-1} fractional factorial (generator: $I = 12345$) and the outer design was a 2^2 factorial. As a result, $16 \times 4 = 64$ separate tests—simulations—were required (i.e., the four simulations of the outer design were conducted for each of the 16 unique variable combinations of the inner design). Table 21.7 provides the combinations of the five variables for each of the 16 tests in the inner design.

The results of the dynamic force model cutting process simulations produce a great deal of information related to cutting forces, workpiece/cutter deflections, and dynamic process behavior. In this case study we examine as measures of product/process quality and productivity performance only three responses: maximum Y-direction cutting force (MAXFY), maximum Z-direction displacement (MAXDZ), and metal removal rate (MRR). These responses have been studied in the form of a single measure (weighted objective function F).

$$F_{ij} = W_1 \left(\frac{\mathrm{MRR}_{ij} - \overline{\mathrm{MRR}}}{\hat{\sigma}_{\mathrm{MRR}}} \right) - W_2 \left(\frac{\mathrm{MAXFY}_{ij} - \overline{\mathrm{MAXFY}}}{\hat{\sigma}_{\mathrm{MAXFY}}} \right)$$

$$- W_3 \left(\frac{\mathrm{MAXDZ}_{ij} - \overline{\mathrm{MAXDZ}}}{\hat{\sigma}_{\mathrm{MAXDZ}}} \right)$$

for $i = 1$, 16 tests in the inner design and for $j = 1$, 4 tests in the outer design, and W_1, W_2, W_3 are the response weights. Several sets of weights were exam-

TABLE 21.6	Conditions of Simulation Experiments

Inner Design Control/Design Variables			Outer Design Noise Variables		
Variable	Low Level	High Level	Variable	Low Level	High Level
1. Feedrate (in./tooth)	0.00225	0.00450	1. Runout (in.)	0	0.0015
2. Depth of cut (in.)	0.020	0.040	2. Surface error (in.)	0	0.0035
3. Number of inserts	12	18			
4. Cutter offset (in.)	0	0.900			
5. Workpiece fixture stiffness*	1	1.5			

Other Process Conditions

Speed = 342 rev/min
Cutter diameter = 8 in.
Nose radius = 0.03 in.
Lead angle = 20°
Radial rake angle = 12°
Axial rake angle = 17°

Material: 390 cast aluminum
Material density: 0.0968 lb/in^3
Workpiece dimension: 13 in. ×
6 in. × 2.5 in.

* Low level of stiffness is the set of values obtained from the part model. High level equals 1.5 times the values for low level.

TABLE 21.7	Design Matrix for Inner Design Simulation Experiments

Test	Feedrate (in./tooth)	Depth of Cut (in.)	Number of Inserts	Cutter Offset (in.)	Stiffness (lb/in.)
1	0.00225	0.020	12	0.000	1.5
2	0.00450	0.020	12	0.000	1
3	0.00225	0.040	12	0.000	1
4	0.00450	0.040	12	0.000	1.5
5	0.00225	0.020	18	0.000	1
6	0.00450	0.020	18	0.000	1.5
7	0.00225	0.040	18	0.000	1.5
8	0.00450	0.040	18	0.000	1
9	0.00225	0.020	12	0.900	1
10	0.00450	0.020	12	0.900	1.5
11	0.00225	0.040	12	0.900	1.5
12	0.00450	0.040	12	0.900	1
13	0.00225	0.020	18	0.900	1.5
14	0.00450	0.020	18	0.900	1
15	0.00225	0.040	18	0.900	1
16	0.00450	0.040	18	0.900	1.5

ined. In this study the results are shown for $W_1 = 0.4$, $W_2 = 0.2$, and $W_3 = 0.4$. Each of the 3 responses was normalized using the average and standard deviation calculated over all 64 tests, in each case.

The results of the simulation experiments were examined in three basic ways in an effort to understand the manner in which the design and noise factors influence performance:

1. The objective function F was averaged over the $j = 1, 4$ conditions of the outer design for each of the $i = 1, 16$ conditions of the inner design, and these average objective function values were treated as responses to evaluate the inner design results.
2. The sample standard deviation of objective function values was determined over $j = 1, 4$ conditions of the outer design for each of the $i = 1, 16$ conditions of the inner design, and these standard deviations were treated as responses to evaluate the inner design results.
3. The inner design results in terms of variable effects on objective function values F were evaluated at each of the $j = 1, 4$ conditions of the outer design.

Discussion and Interpretation of Results

Effects of Inner Design Variables on the Average Objective Function Values. Under the assumption that third- and higher-order interactions can be neglected, the main and two-factor interaction effects among the five variables were estimated. The results indicated that all variables except feedrate have quite large (significant) main effects and that the interactions between depth of cut and number of inserts, and depth of cut and cutter offset, are also important. The smaller feedrate effect may be explained by the narrow range of feeds studied. In general, increasing stiffness and employing cutter offset tend to improve performance—increase F—while increasing depth of cut and the number of cutter inserts tend to reduce performance.

Of particular interest is the examination of the two important two-factor interaction effects. Figure 21.14 shows these interactions graphically. In the case of the depth of cut–number of inserts interaction [Fig. 21.14(a)], it appears that the deterioration of the objective function with an increase in depth of cut is greatly reduced when the number of inserts is at its lower level. However, for lower depths of cut, the use of 12 or 18 inserts causes very little change in the objective function.

Figure 21.14(b) shows that the use of cutter offset can greatly improve process performance and, in particular, can help counter the detrimental influence of increasing depth of cut. In this figure we see that the inclusion of cutter offset in the process causes a considerable improvement in the objective function when depth of cut is at its highest level. Additional experimentation (simulation) with the cutter offset variable showed two other positive results of using offset: (1) the positive offset effect was consistently observed for several other positions of the cutter along the feed direction on the workpiece, and (2) in several cases where a centered cutter gave rise to an unstable cutting condition, including

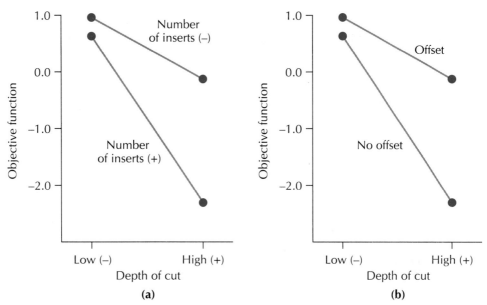

Figure 21.14 Graphical Representation of Depth of Cut/Number of Inserts Interaction Effect and Depth of Cut/Cutter Offset Interaction Effect

offset in the process stabilized the cutting condition. The use of offset (Fig. 21.15) avoids the problem of having the point of maximum instantaneous chip load coincide with the most flexible point on the workpiece. The inclusion of surface error (in the form used in this case study) only makes matters worse for a centered cutter. It is, of course, true that these results are workpiece geometry dependent and may not in all cases be reproducible for other geo-

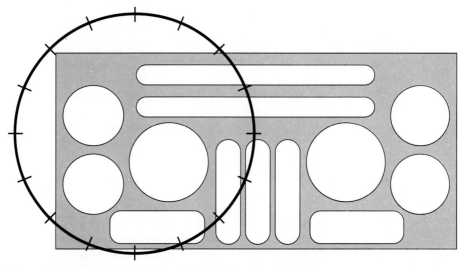

Figure 21.15 Relative Position of the Cutter on the Workpiece

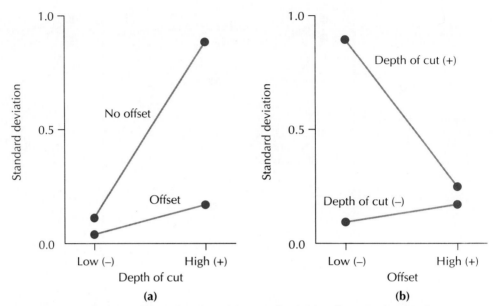

Figure 21.16 Graphical Representation of the Depth of Cut/Offset Interaction Effect

metries. In fact, for this workpiece (see Fig. 21.15), offset to the other side is actually worse than a centered cutter due to the location of points of maximum flexibility.

Effects of Inner Design Variables on the Sample Standard Deviation of Objective Function Values. An analysis of variable effects on the sample standard deviation of the objective function values for the 16 tests seems to indicate that the main effects of stiffness, offset, and, depth of cut are large. Increases in depth of cut tend to magnify the effects of the noise factors, while increased stiffness and the use of offset tend to reduce the effects of the noise factors, precipitating a more robust process and product design. Figure 21.16 shows again how the presence of cutter offset can be a positive force, this time in terms of suppressing the variation in performance due to the noise factors. Again, offset of the cutter appears to counteract the deleterious effect of increasing the depth of cut.

An analysis was also performed by examining the ratio of the average response to the sample standard deviation of the response for each condition of the inner design. This ratio may be interpreted as a signal-to-noise ratio. Analysis of the inner design variable effects on the signal-to-noise ratio shows that:

1. Increasing stiffness greatly increases the S/N ratio.
2. Increasing depth of cut greatly reduces the S/N ratio.
3. A large depth of cut-by-stiffness interaction exists, showing that increase in stiffness can offset the deleterious depth of cut effect.

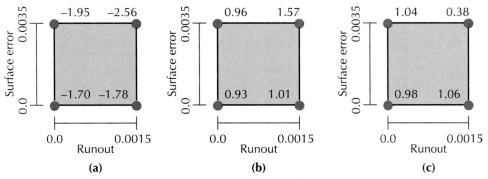

Figure 21.17 Influence of Noise Factors on (a) Depth of Cut Effect, (b) Cutter Offset Effect, and (c) Workpiece/Fixture Stiffness Effect

4. A large depth of cut-by-number of inserts interaction exists, showing that the positive influence of fewer inserts cannot be realized if the depth of cut is large.

Effect of the Inner Design Variables as a Function of the Levels of the Outer Design Variables. Figure 21.17 was constructed to show how the effects that the important design variables have on performance are influenced by the noise variables. In these figures the numbers at the corners of each "square" represent the main effect of a specified inner design variable at that particular combination of the noise variables. For example, Fig. 21.17(a) shows that the adverse effect of increasing depth of cut is worsened by the presence of higher levels of the noise variables. Conversely, Fig. 21.17(b) and (c) show that the positive effects of cutter offset and workpiece–fixture stiffness, respectively, are actually largest in the presence of higher levels of the noise variables. Such reinforces the suggestions discussed previously that these design factors may be important to the development of a robust product/process design.

Conclusions

This case study explores the use of computer simulation as a central tool in the simultaneous engineering of products and the processes used to manufacture them. Computer-based mathematical models for the part and for the machining process were used to investigate how changes in product and/or process design parameters could improve the quality and productivity of manufacture.

Taguchi's engineering design process model, in particular the method of parameter design, has been explored as a technique to understand the relationships among the design (product and process) parameters and the performance measures of interest. The method of robust design has been demonstrated by showing how manipulation of design parameters can mitigate the deleterious effects of the noise factors. In the specific context of this work it was seen that part stiffness and cutter offset are design parameters that can be usefully manipulated to reduce the effect of external/noise variables surrounding the process.

References and Further Readings

Aaron, H. B., "Quality for Today and Tomorrow," seminar presented to Lafayette, Ind., ASQC Section, Mar. 17, 1988.

Barnard, G. A., "Control Charts and Stochastic Processes," *Journal of the Royal Statistical Society, Ser. B*, Vol. 21, 1959, p. 239.

Bartlett, M. S., "The Use of Transformations," *Biometrics*, Vol. 3, 1947, pp. 39–52.

Belaire, P. M., and Deacon, R. J., "The Strategic Approach to Quality Improvement Using Design of Experiments Concepts," *Symposium on Quality: Design, Planning, and Control, ASME Winter Annual Meeting*, December 13–18, 1987, PED Vol. 27, pp. 147–157.

Bissell, A. F., *An Introduction to CuSum Charts*, The Institute of Statisticians, Bury St. Edmonds, Suffolk, England, 1984.

Box, G. E. P., and N. R. Draper, *Evolutionary Operation: A Statistical Method for Process Improvement*, Wiley, New York, 1969.

Box, G. E. P., and J. S. Hunter, "Signal-to-Noise Ratios, Performance Criteria, and Transformations," *Technometrics*, Vol. 30, No. 1, 1988, pp. 1–40.

Box, G. E. P., and J. S. Hunter, "The 2^{k-p} Fractional Factorial Designs," Parts I and II, *Technometrics*, Vol. 3, 1961, pp. 331–351 and 449–458.

Box, G. E. P., W. G. Hunter, and J. S. Hunter, *Statistics for Experimenters*, Wiley, New York, 1978.

Box, G. E. P., and R. D. Meyer, "An Analysis for Unreplicated Fractional Factorials," *Technometrics*, Vol. 28, 1986, pp. 11–18.

Box, G. E. P., and R. D. Meyer, "Dispersion Effects from Fractional Designs," *Technometrics*, Vol. 28, 1986, pp. 19–27.

Burr, I. W., *Statistical Quality Control Methods*, Marcel Dekker, New York, 1976.

Chan, L. K., S. W. Cheng, and F. A. Spiring, "A New Measure of Process Capability: C_{pm}," *Journal of Quality Technology*, Vol. 20, July 1988, pp. 162–175.

Charbonneau, H. C., and G. L. Webster, *Industrial Quality Control,* Prentice-Hall, Englewood Cliffs, N.J., 1978.

Chow, E. T. S., L. S. Wei, R. E. DeVor, and M. P. Steinberg, "Application of Two-Level Fractional Factorial Designs in Development of a Soybean Whipped Topping," *Journal of Food Science,* Vol. 48, 1983, p. 230.

Clausing, D. P., "Taguchi Methods Integrated into the Improved Total Development," *Proc. IEEE International Conference on Communications,* Philadelphia, June 1988, pp. 826–832.

Cochran, W. G., and G. M. Cox, *Experimental Designs,* 2nd ed., Wiley, New York, 1957.

"Continuing Process Control and Process Improvement," Corporate Quality Office, Ford Motor Company, Dearborn, Mich., Dec. 1987.

Conway, W., Videotape of presentation to Ford Motor Company executives, *Ford Motor Company, Radio, TV, and Film Dept.,* File No. 81V44-1, Dearborn Mich., 1981.

Corbett, J., M. Dooner, J. Meleka, and C. Pym, *Design for Manufacture: Strategies, Principles and Techniques,* Addison-Wesley, Reading, Mass., 1991.

Cox, D. R., "The Use of Range in Sequential Analysis," *Journal of the Royal Statistical Society, Ser. B,* Vol. 11, 1949.

Daniel, C., *Applications of Statistics to Industrial Experimentation,* Wiley, New York, 1976.

Davies, O. L., ed., *The Design and Analysis of Industrial Experiments,* Hafner, Macmillan, 1954.

Davis, R. B., and W. H. Woodall, "Performance of the Control Chart Trend Rule Under Linear Shift," *Journal of Quality Technology,* Vol. 20, Oct. 1988, pp. 260–262.

Deacon, R. J., "Quality Improvement in an Accounts Payable System," Independent Study Project Report, Eastern Michigan University, College of Business, Ypsilanti, Mich., 1986.

Deacon, R. J., "Reduction in Visual Defects in an Injection Molding Process," *Society of Plastics Engineers, 1983 National Technical Conference Proceedings,* p. 97.

Deacon, R. J., and P. M. Belaire, "Quality Improvement: Design of Experiments Methodology," *IMPRO 88 Conference Proceedings,* Juran Institute, Chicago, 1988.

Dehnad, K., *Quality Control, Robust Design, and the Taguchi Method,* Wadsworth & Brooks/ Cole, Advanced Book and Software, Pacific Grove, Calif., 1989.

Deming, W. E., "On the Use of Judgement-Samples," *Reports of Statistical Application Research, JUSE,* Vol. 23, No. 1, March 1976, pp. 25–31.

Deming, W. E., *Out of the Crisis,* Massachusetts Institute of Technology, Center for Advanced Engineering Study, Cambridge, Mass., 1982.

Deming, W. E., *Quality, Productivity and Competitive Position,* Massachusetts Institute of Technology, Center for Advanced Engineering Study, Cambridge, Mass., 1982.

Deming, W. E., "The Logic of Evaluation," *Handbook of Evaluation Research,* Vol. 1, E. L. Struening and M. Guttentag, eds., Sage Publications, Beverly Hills, Calif., 1975, pp. 53–68.

Dessouky M. I., S. G. Kapoor, and R. E. DeVor, "A Methodology for Integrated Quality Systems," *J. of Eng. for Ind., Trans. ASME,* Vol. 109, Feb. 1987, pp. 241–247.

DeVor, R. E., S. G. Kapoor, R. Hayashida, G. Subramani, and T. Lindem, "A Methodology for the Simultaneous Engineering of Products and Manufacturing Processes," *Proc. 14th North American Manufacturing Research Conference,* 1986, pp. 535–542.

Dodge, H. F., and H. G. Romig, *Sampling Inspection Tables: Single and Double Sampling*, 2nd ed., Wiley, New York, 1959.

Dooley, K. J., S. G. Kapoor, M. I. Dessouky, and R. E. DeVor, "An Integrated Quality Systems Approach to Quality and Productivity Improvement in Continuous Manufacturing Processes," *J. of Eng. for Ind., Trans. ASME*, Vol. 108, Nov. 1986, pp. 322–327.

Draper, N. R., and H. Smith, *Applied Regression Analysis*, Wiley, New York, 1966.

Duncan, A. J., *Quality Control and Industrial Statistics*, 4th ed., Richard D. Irwin, Homewood, Ill., 1974.

Endres, W. J., J. W. Sutherland, and R. E. DeVor, "Quality Design Using a Computer-Based Dynamic Force Model for the Turning Process," *Proc. 6th International Conference on Computer-Aided Production Engineering*, Nov. 1990, pp. 29–42.

Feigenbaum, A. V., *Total Quality Control*, McGraw-Hill, New York, 1983.

Finney, D. J., "The Fractional Replication of Factorial Arrangements," *Annals of Eugenics*, Vol. 12, No. 4, 1945, pp. 291–301.

Fisher, R. A., *The Design of Experiments*, 8th ed., Hafner, New York, 1966.

"Ford Worldwide Quality System Standard Q-101," Corporate Quality Office, Ford Motor Company, Dearborn, Mich., 1990.

Freund, R. A., "Definitions and Basic Quality Concepts," *Journal of Quality Technology*, Vol. 17, Jan. 1985, pp. 50–56.

Fu, H. J., R. E. DeVor, and S. G. Kapoor, "A Mechanistic Model for the Prediction of the Force System in Face Milling Operations," *J. Engr. for Ind., Trans. ASME.*, Vol. 106, Feb. 1984, pp. 81–88.

Gitlow, H., and S. Gitlow, *The Deming Guide to Quality and Competitive Position*, Prentice Hall, Englewoods Cliffs, N.J., 1987.

Gitlow, H., S. Gitlow, A. Oppenheim, and R. Oppenheim, *Tools and Methods for the Improvement of Quality*, Richard D. Irwin, Homewood, Ill., 1989.

Grant, E. L., and R. S. Leavenworth, *Statistical Quality Control*, 6th ed., McGraw-Hill, New York, 1988.

Guide to Data Analysis and Quality Control Using CuSum Techniques, BS5703 (4 parts), British Standards Institution, 1980–82.

Haaland, P. D., *Experimental Design in Biotechnology*, Marcel Dekker, New York, 1989.

Hahn, G. J., "Statistical Assessment of a Process Change," *Journal of Quality Technology*, Vol. 14, Jan. 1982, p. 1–9.

Hicks, C. R., *Fundamental Concepts in the Design of Experiments*, 3rd ed., Holt, Rinehart, & Winston, New York, 1982.

Hunter, J. S., "Statistical Design Applied to Product Design," *Journal of Quality Technology*, Vol. 17, No. 4, 1985, pp. 210–221.

Hunter, J. S., "The Exponentially Weighted Moving Average," *Journal of Quality Technology*, Vol. 18, Oct. 1986, pp. 203–209.

Hunter, R. G., J. W. Sutherland, and R. E. DeVor, "A Methodology for Robust Design Using Models for the Mean, Variance, and Loss," *ASME Symposium on Quality and Performance: Design, Evaluation, and Improvements*, ASME Winter Annual Meeting, December 1989, PED Vol. 42, pp. 25–42.

Ishikawa, K., *Guide to Quality Control,* Asian Productivity Organization, 2nd ed., Tokyo, 1982.

Jessup, P. T., "Process Capability, The Value of Improved Performance," *Proc. IEEE International Communications Conference,* ICC85, 1985.

Jiang, B. C., J. T. Black, D. W. H. Chen, and J. N. Hool, "Taguchi-Based Methodology for Determining/Optimizing Robust Process Capability," *IIE Transactions,* Vol. 23, 1991, pp. 169–184.

Johnson, N. L., and F. C. Leone, "Cumulative Sum Control Charts: Mathematical Principles Applied to Their Construction and Use," Parts I, II, and III, *Industrial Quality Control,* Vol. 18, No. 12, Vol. 19, No. 1, and Vol. 19, No. 2, 1962.

Juran, J. M., and F. M. Gryna, Jr., *Quality Planning and Analysis,* 2nd ed., McGraw-Hill, New York, 1980.

Juran, J. M., F. M. Gryna, Jr., and R. S. Bingham, Jr., *Quality Control Handbook,* 3rd ed., McGraw-Hill, New York, 1974.

Kackar, R. N., "Off-Line Quality Control, Parameter Design and the Taguchi Method," *Journal of Quality Technology,* Vol. 17, 1985, pp. 176–209.

Kackar, R. N., "Taguchi's Quality Philosophy: Analysis and Commentary," *Quality Progress,* Dec. 1986, pp. 21–29.

Kane, V. E., *Defect Prevention: Use of Simple Statistical Tools,* Marcel Dekker, ASQC Quality Press, Milwaukee, Wis., 1989.

Kane, V. E., "Process Capability Indices," *Journal of Quality Technology,* Vol. 18, Jan. 1986, pp. 41–52.

Keats, J. B., and N. F. Hubele, eds., *Statistical Process Control in Automated Manufacturing,* Marcel Dekker, New York, 1989.

Krismann, C., *Quality Control: An Annotated Bibliography Through 1988,* Quality Resources, White Plains, N.Y., 1990.

Lucas, J. M., "Combined Shewhart—CUSUM Quality Control Schemes," *Journal of Quality Technology,* Vol. 14, Apr. 1982, pp. 51–59.

Lucas, J. M., "The Design and Use of V-Mask Control Schemes," *Journal of Quality Technology,* Vol. 8, No. 1, 1976, pp. 1–12.

Lucas, J. M., and M. S. Saccucci, "Exponentially Weighted Moving Average Control Schemes: Properties and Enhancements," *Technometrics,* Vol. 32, 1990, pp. 1–12.

Military Standard 105D, *Sampling Procedures and Tables for Inspection by Attributes,* U.S. Government Printing Office, Washington, D.C., 1963.

Miller, I., and J. E. Freund, *Probability and Statistics for Engineers,* 3rd ed., Prentice Hall, Englewood Cliffs, N.J., 1985.

Moen, R. D., T. W. Nolan, and L. P. Provost, *Improving Quality Through Planned Experimentation,* Wiley, New York, 1989.

Montgomery, D. C., *Design and Analysis of Experiments,* 3rd ed., Wiley, New York, 1990.

Montgomery, D. C., *Introduction to Statistical Quality Control,* 2nd ed., Wiley, New York, 1990.

Murdoch, J., *Control Charts,* Macmillan, New York, 1979.

Nelson, L. S., "An Early-Warning Test for Use with the Shewhart p Control Chart," *Journal of Quality Technology,* Vol. 15, Apr. 1983, pp. 68–71.

Nelson, L. S., "Calculation of New Limits for Xbar, R Charts When Subgroup Size Is Changed," *Journal of Quality Technology*, Vol. 2, Apr. 1988, pp. 149–150.

Nelson, L. S., "Control Charts for Individual Measurements," *Journal of Quality Technology*, Vol. 14, July 1982, pp. 172–173.

Nelson, L. S., "Control Charts: Rational Subgroups and Effective Applications," *Journal of Quality Technology*, Vol. 20, Jan. 1988, p. 73–75.

Nelson, L. S., "Interpreting Shewhart Xbar Control Charts," *Journal of Quality Technology*, Vol. 17, Apr. 1985, pp. 114–116.

Nelson, L. S., "The Shewhart Control Chart: Tests for Special Causes," *Journal of Quality Technology*, Vol. 16, No. 4, 1984, pp. 237–?39.

Ng, C. H., and K. E. Case, "Development and Evaluation of Control Charts Using Exponentially Weighted Moving Averages," *Journal of Quality Technology*, Vol. 21, No. 4, 1989, pp. 242–250.

Oakland, J. S., *Statistical Process Control*, William Heinemann, London, England, 1986.

Oh, H. L., "Variation Tolerant Design," *Symposium on Quality: Design, Planning, and Control, ASME Winter Annual Meeting*, Dec. 13–18, 1987, PED, Vol. 27, pp. 137–146.

Ott, E. R., *Process Quality Control: Trouble Shooting and Interpretation of Data*, McGraw-Hill, New York, 1975.

Ott, E. R., and E. G. Schilling, *Process Quality Control*, 2nd ed., McGraw-Hill, New York, 1990.

Page, E. S., "Continuous Inspection Schemes," *Biometrika*, Vol. 41, 1954, pp. 100–114.

Phadke, M. S., *Quality Engineering Using Robust Design*, Prentice Hall, Englewood Cliffs, N.J., 1989.

Plackett, R. L., and J. P. Burman, "Design of Optimal Multifactorial Experiments," *Biometrika*, Vol. 23, 1946, pp. 305–325.

"Planning for Quality," Corporate Quality Office, Ford Motor Company, Dearborn, Mich., Apr. 1990.

Proceedings of Supplier Symposia on Taguchi Methods, Apr. 1984, Nov. 1984, Oct. 1985–1989, Nov. 1990; American Supplier Institute, Inc., Dearborn, Mich.

Pugh, S., *Tool Design: Integrated Methods for Successful Product Engineering*, Addison-Wesley, Reading, Mass., 1991.

Quesenberry, C. P., "SPC Q Charts for a Binomial Parameter p: Short or Long Runs," *Journal of Quality Technology*, Vol. 23, 1991, pp. 239–246.

Quesenberry, C. P., "SPC Q Charts for Start-Up and Short or Long Runs," *Journal of Quality Technology*, Vol. 23, 1991, pp. 217–224.

Rahn, G. S., S. G. Kapoor, and R. E. DeVor, "Development of Operational Characteristics and Diagnostics for X-Control Charts," *J. Engr. for Ind., Trans. ASME*, 1992.

Rao, C. R., "Factorial Experiments Derivable from Combinatorial Arrangements of Arrays," *Journal of the Royal Statistical Society, Ser. B*, Vol. 9, 1947, pp. 128–140.

Roberts, S. W., "Control Charts Based on Geometric Moving Averages," *Technometrics*, Vol. 1, 1959, pp. 234–250.

Robinson, A., ed., *Continuous Improvement in Operations: A Systematic Approach to Waste Reduction*, Productivity Press, Cambridge, Mass., 1991.

Ryan, T. P., *Statistical Methods for Quality Improvement*, Wiley, New York, 1989.

Sadsworth, H. M., ed., *Taguchi Methods (Special Issue of Quality and Reliability Engineering)*, Wiley-Interscience, Chichester, England, 1988.

Scheffe, H., *The Analysis of Variance*, Wiley, New York, 1959.

Scherkenbach, W. W., *The Deming Route to Quality and Productivity: Road Maps and Roadblocks*, Mercury Press/Fairchild Publications, New York, 1987.

Shewhart, W. A., *Economic Control of Quality of Manufactured Products*, D. Van Nostrand, Princeton, N.J., 1931. Reprinted by the American Society for Quality Control, Milwaukee, Wis.

Shewhart, W. A., in *Statistical Method from the Viewpoint of Quality Control*, W. E. Deming, ed., The Graduate School, Department of Agriculture, Washington, D.C., 1939.

Statistical Quality Control Handbook, Western Electric Company, Newark, N.J., 1956.

Stegner, D. A. J., S. M. Wu, and N. R. Braton, "Prediction of Heat Input for Welding," *Welding J. Res. Suppl.*, 1967.

Sullivan, L. P., "Quality Function Deployment," *Quality Progress*, June 1986, pp. 39–50.

Sullivan, L. P., "Reducing Variability: A New Approach to Quality," *Quality Progress*, Vol. 17, No. 7, 1985, pp. 15–21.

Sutherland, J. W., R. G. Hunter, and R. E. DeVor, "A Quality Engineering-Based Approach to the Simultaneous Engineering of Products and Their Manufacturing Processes," *Proc. 5th International Conference on Computer-Aided Production Engineering*, Nov. 1989, pp. 1–12.

Sweet, A. L., "Control Charts Using Coupled Exponentially Weighted Moving Averages," *Transactions of the IIE*, Vol. 18, No. 1, 1986, pp. 26–33.

Taguchi, G., *Introduction to Quality Engineering*, Asian Productivity Organization, Tokyo, 1986. Distributed by the American Supplier Institute, Inc., Dearborn, Mich.

Taguchi, G., *On-Line Quality Control During Production*, Japanese Standards Association, Tokyo, 1981.

Taguchi, G., A. Elsayed, and T. Hsiang, *Quality Engineering in Production Systems*, McGraw-Hill, New York, 1989.

Taguchi, G., and Y. Wu, *Introduction to Off-Line Quality Control*, Central Japan Quality Control Association, Meieki Nakamura-Ku Magaya, Japan, 1979.

Taylor, F. W., *Principles of Scientific Management*, Harper & Brothers, New York, 1911.

Tippett, L. H. C., *Applications of Statistical Methods to the Control of Quality in Industrial Production*, Manchester Statistical Society, Manchester, England, 1934.

The Asahi, Japanese Language Newspaper, Apr. 17, 1979. Reported by G. Taguchi during lectures at AT&T Bell Laboratories in 1980.

Vance, L. C., "A Bibliography of Statistical Quality Control Chart Techniques, 1970–1980," *Journal of Quality Technology*, Vol. 15, Apr. 1983, pp. 59–62.

Vardeman, S., and J. A. Cornell, "A Partial Inventory of Statistical Literature on Quality and Productivity Through 1985," *Journal of Quality Technology*, Vol. 19, Apr. 1987, pp. 90–97.

Wadsworth, H. M., K. S. Stephens, and A. B. Godfrey, *Modern Methods for Quality Control and Improvement*, Wiley, New York, 1986.

Wald, A., *Sequential Analysis*, Wiley, New York, 1947.

Walpole, R. E., and R. H. Myers, *Probability and Statistics for Engineers and Scientists*, 3rd ed., Macmillan, New York, 1985.

Wheeler, D. J., *Understanding Industrial Experimentation*, Statistical Process Controls, Knoxville, Tenn., 1988.

Woodall, W. H., "Conflicts Between Deming's Philosophy and the Economic Design of Control Charts," *Frontiers in Statistical Quality Control 3*, edited by H. J. Lenz, G. B. Wetherill, and P. T. Wilrich, Physica-Verlag, Heidelberg, Germany, 1987, pp. 242–248.

Woodall, W. H., "The Design of CUSUM Quality Control Charts," *Journal of Quality Technology*, Vol. 18, Apr. 1986, pp. 99–102.

Working, H., and E. G. Olds, *Manual for an Introduction to Statistical Methods of Quality Control in Industry: Outline of a Course of Lectures and Exercises*, Office of Production Research and Development, War Production Board, Washington, D.C., 1944.

Wortham, A. W., "The Use of Exponentially Smoothed Data in Continuous Process Control," *International Journal of Production Research*, Vol. 10, No. 4, 1972, pp. 393–400.

Wortham, A. W., and G. F. Heinrich, "Control Charts Using Exponential Smoothing Techniques," *Transactions of the AQC*, Vol. 26, 1972, pp. 451–458.

Wu, S. M., "Analysis of Rail Steel Bar Welds by Two-Level Factorial Designs," *Welding Journal Research Supplement*, 1964, pp. 179s–183s.

Appendix Tables

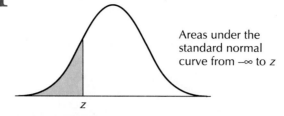

Areas under the standard normal curve from $-\infty$ to z

TABLE A.1   Areas Under the Normal Curve

z	0.09	0.08	0.07	0.06	0.05	0.04	0.03	0.02	0.01	0.00
-4.00	0.00002	0.00002	0.00002	0.00003	0.00003	0.00003	0.00003	0.00003	0.00003	0.00003
-3.90	0.00003	0.00003	0.00004	0.00004	0.00004	0.00004	0.00004	0.00004	0.00005	0.00005
-3.80	0.00005	0.00005	0.00005	0.00006	0.00006	0.00006	0.00006	0.00007	0.00007	0.00007
-3.70	0.00008	0.00008	0.00008	0.00009	0.00009	0.00009	0.00010	0.00010	0.00010	0.00011
-3.60	0.00011	0.00012	0.00012	0.00013	0.00013	0.00014	0.00014	0.00015	0.00015	0.00016
-3.50	0.00017	0.00017	0.00018	0.00019	0.00019	0.00020	0.00021	0.00022	0.00022	0.00023
-3.40	0.00024	0.00025	0.00026	0.00027	0.00028	0.00029	0.00030	0.00031	0.00033	0.00034
-3.30	0.00035	0.00036	0.00038	0.00039	0.00040	0.00042	0.00043	0.00045	0.00047	0.00048
-3.20	0.00050	0.00052	0.00054	0.00056	0.00058	0.00060	0.00062	0.00064	0.00066	0.00069
-3.10	0.00071	0.00074	0.00076	0.00079	0.00082	0.00085	0.00087	0.00090	0.00094	0.00097
-3.00	0.00100	0.00104	0.00107	0.00111	0.00114	0.00118	0.00122	0.00126	0.00131	0.00135
-2.90	0.0014	0.0014	0.0015	0.0015	0.0016	0.0016	0.0017	0.0018	0.0018	0.0019
-2.80	0.0019	0.0020	0.0021	0.0021	0.0022	0.0023	0.0023	0.0024	0.0025	0.0026
-2.70	0.0026	0.0027	0.0028	0.0029	0.0030	0.0031	0.0032	0.0033	0.0034	0.0035
-2.60	0.0036	0.0037	0.0038	0.0039	0.0040	0.0041	0.0043	0.0044	0.0045	0.0047
-2.50	0.0048	0.0049	0.0051	0.0052	0.0054	0.0055	0.0057	0.0059	0.0060	0.0062
-2.40	0.0064	0.0066	0.0068	0.0069	0.0071	0.0073	0.0075	0.0078	0.0080	0.0082
-2.30	0.0084	0.0087	0.0089	0.0091	0.0094	0.0096	0.0099	0.0102	0.0104	0.0107
-2.20	0.0110	0.0113	0.0116	0.0119	0.0122	0.0125	0.0129	0.0132	0.0136	0.0139
-2.10	0.0143	0.0146	0.0150	0.0154	0.0158	0.0162	0.0166	0.0170	0.0174	0.0179
-2.00	0.0183	0.0188	0.0192	0.0197	0.0202	0.0207	0.0212	0.0217	0.0222	0.0228
-1.90	0.0233	0.0239	0.0244	0.0250	0.0256	0.0262	0.0268	0.0274	0.0281	0.0287
-1.80	0.0294	0.0301	0.0307	0.0314	0.0322	0.0329	0.0336	0.0344	0.0351	0.0359
-1.70	0.0367	0.0375	0.0384	0.0392	0.0401	0.0409	0.0418	0.0427	0.0436	0.0446
-1.60	0.0455	0.0465	0.0475	0.0485	0.0495	0.0505	0.0516	0.0526	0.0537	0.0548
-1.50	0.0559	0.0571	0.0582	0.0594	0.0606	0.0618	0.0630	0.0643	0.0655	0.0668
-1.40	0.0681	0.0694	0.0708	0.0721	0.0735	0.0749	0.0764	0.0778	0.0793	0.0808
-1.30	0.0823	0.0838	0.0853	0.0869	0.0885	0.0901	0.0918	0.0934	0.0951	0.0968
-1.20	0.0985	0.1003	0.1020	0.1038	0.1057	0.1075	0.1093	0.1112	0.1131	0.1151
-1.10	0.1170	0.1190	0.1210	0.1230	0.1251	0.1271	0.1292	0.1314	0.1335	0.1357
-1.00	0.1379	0.1401	0.1423	0.1446	0.1469	0.1492	0.1515	0.1539	0.1562	0.1587
-0.90	0.1611	0.1635	0.1660	0.1685	0.1711	0.1736	0.1762	0.1788	0.1814	0.1841
-0.80	0.1867	0.1894	0.1922	0.1949	0.1977	0.2005	0.2033	0.2061	0.2090	0.2119
-0.70	0.2148	0.2177	0.2207	0.2236	0.2266	0.2297	0.2327	0.2358	0.2389	0.2420
-0.60	0.2451	0.2483	0.2514	0.2546	0.2578	0.2611	0.2643	0.2676	0.2709	0.2743
-0.50	0.2776	0.2810	0.2843	0.2877	0.2912	0.2946	0.2981	0.3015	0.3050	0.3085
-0.40	0.3121	0.3156	0.3192	0.3228	0.3264	0.3300	0.3336	0.3372	0.3409	0.3446
-0.30	0.3483	0.3520	0.3557	0.3594	0.3632	0.3669	0.3707	0.3745	0.3783	0.3821
-0.20	0.3859	0.3897	0.3936	0.3974	0.4013	0.4052	0.4090	0.4129	0.4168	0.4207
-0.10	0.4247	0.4286	0.4325	0.4364	0.4404	0.4443	0.4483	0.4522	0.4562	0.4602
-0.00	0.4641	0.4681	0.4721	0.4761	0.4801	0.4840	0.4880	0.4920	0.4960	0.5000

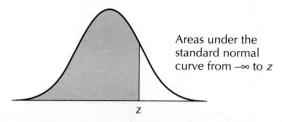

Areas under the standard normal curve from $-\infty$ to z

TABLE A.1   (continued)

z	0.00	0.01	0.02	0.03	0.04	0.05	0.06	0.07	0.08	0.09
0.00	0.5000	0.5040	0.5080	0.5120	0.5160	0.5199	0.5239	0.5279	0.5319	0.5359
0.10	0.5398	0.5438	0.5478	0.5517	0.5557	0.5596	0.5636	0.5675	0.5714	0.5753
0.20	0.5793	0.5832	0.5871	0.5910	0.5948	0.5987	0.6026	0.6064	0.6103	0.6141
0.30	0.6179	0.6217	0.6255	0.6293	0.6331	0.6368	0.6406	0.6443	0.6480	0.6517
0.40	0.6554	0.6591	0.6628	0.6664	0.6700	0.6736	0.6772	0.6808	0.6844	0.6879
0.50	0.6915	0.6950	0.6985	0.7019	0.7054	0.7088	0.7123	0.7157	0.7190	0.7224
0.60	0.7257	0.7291	0.7324	0.7357	0.7389	0.7422	0.7454	0.7486	0.7517	0.7549
0.70	0.7580	0.7611	0.7642	0.7673	0.7704	0.7734	0.7764	0.7794	0.7823	0.7852
0.80	0.7881	0.7910	0.7939	0.7967	0.7995	0.8023	0.8051	0.8079	0.8106	0.8133
0.90	0.8159	0.8186	0.8212	0.8238	0.8264	0.8289	0.8315	0.8340	0.8365	0.8389
1.00	0.8413	0.8438	0.8461	0.8485	0.8508	0.8531	0.8554	0.8577	0.8599	0.8621
1.10	0.8643	0.8665	0.8686	0.8708	0.8729	0.8749	0.8770	0.8790	0.8810	0.8830
1.20	0.8849	0.8869	0.8888	0.8907	0.8925	0.8944	0.8962	0.8980	0.8997	0.9015
1.30	0.9032	0.9049	0.9066	0.9082	0.9099	0.9115	0.9131	0.9147	0.9162	0.9177
1.40	0.9192	0.9207	0.9222	0.9236	0.9251	0.9265	0.9279	0.9292	0.9306	0.9319
1.50	0.9332	0.9345	0.9357	0.9370	0.9382	0.9394	0.9406	0.9418	0.9429	0.9441
1.60	0.9452	0.9463	0.9474	0.9484	0.9495	0.9505	0.9515	0.9525	0.9535	0.9545
1.70	0.9554	0.9564	0.9573	0.9582	0.9591	0.9599	0.9608	0.9616	0.9625	0.9633
1.80	0.9641	0.9649	0.9656	0.9664	0.9671	0.9678	0.9686	0.9693	0.9699	0.9706
1.90	0.9713	0.9719	0.9726	0.9732	0.9738	0.9744	0.9750	0.9756	0.9761	0.9767
2.00	0.9773	0.9778	0.9783	0.9788	0.9793	0.9798	0.9803	0.9808	0.9812	0.9817
2.10	0.9821	0.9826	0.9830	0.9834	0.9838	0.9842	0.9846	0.9850	0.9854	0.9857
2.20	0.9861	0.9864	0.9868	0.9871	0.9875	0.9878	0.9881	0.9884	0.9887	0.9890
2.30	0.9893	0.9896	0.9898	0.9901	0.9904	0.9906	0.9909	0.9911	0.9913	0.9916
2.40	0.9918	0.9920	0.9922	0.9925	0.9927	0.9929	0.9931	0.9932	0.9934	0.9936
2.50	0.9938	0.9940	0.9941	0.9943	0.9945	0.9946	0.9948	0.9949	0.9951	0.9952
2.60	0.9953	0.9955	0.9956	0.9957	0.9959	0.9960	0.9961	0.9962	0.9963	0.9964
2.70	0.9965	0.9966	0.9967	0.9968	0.9969	0.9970	0.9971	0.9972	0.9973	0.9974
2.80	0.9974	0.9975	0.9976	0.9977	0.9977	0.9978	0.9979	0.9979	0.9980	0.9981
2.90	0.9981	0.9982	0.9983	0.9983	0.9984	0.9984	0.9985	0.9985	0.9986	0.9986
3.00	0.99865	0.99869	0.99874	0.99878	0.99882	0.99886	0.99889	0.99893	0.99897	0.99900
3.10	0.99903	0.99907	0.99910	0.99913	0.99916	0.99918	0.99921	0.99924	0.99926	0.99929
3.20	0.99931	0.99934	0.99936	0.99938	0.99940	0.99942	0.99944	0.99946	0.99948	0.99950
3.30	0.99952	0.99953	0.99955	0.99957	0.99958	0.99960	0.99961	0.99962	0.99964	0.99965
3.40	0.99966	0.99968	0.99969	0.99970	0.99971	0.99972	0.99973	0.99974	0.99975	0.99976
3.50	0.99977	0.99978	0.99978	0.99979	0.99980	0.99981	0.99982	0.99982	0.99983	0.99984
3.60	0.99984	0.99985	0.99985	0.99986	0.99986	0.99987	0.99987	0.99988	0.99988	0.99989
3.70	0.99989	0.99990	0.99990	0.99990	0.99991	0.99991	0.99992	0.99992	0.99992	0.99993
3.80	0.99993	0.99993	0.99993	0.99994	0.99994	0.99994	0.99994	0.99995	0.99995	0.99995
3.90	0.99995	0.99995	0.99996	0.99996	0.99996	0.99996	0.99996	0.99996	0.99997	0.99997
4.00	0.99997	0.99997	0.99997	0.99997	0.99997	0.99997	0.99998	0.99998	0.99998	0.99998

TABLE A.2	Constants for Determining from $\bar{R}$ the 3 Sigma Control Limits for $\bar{X}$ and R Charts and for Estimating the Process Standard Deviation from $\bar{R}$			
Number of Observations in Subgroup/Sample	d_2	A_2	D_3	D_4
2	1.128	1.880	0	3.267
3	1.693	1.023	0	2.575
4	2.059	0.729	0	2.282
5	2.326	0.577	0	2.115
6	2.534	0.483	0	2.004
7	2.704	0.419	0.076	1.924
8	2.847	0.373	0.136	1.864
9	2.970	0.337	0.184	1.816
10	3.078	0.308	0.223	1.777
15	3.472	0.223	0.348	1.652
20	3.735	0.180	0.414	1.586

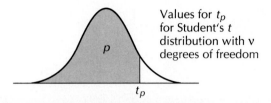

Values for t_p for Student's t distribution with ν degrees of freedom

TABLE A.3   Values of t for Various Probability Levels, p

ν	$t_{.60}$	$t_{.75}$	$t_{.85}$	$t_{.90}$	$t_{.95}$	$t_{.975}$	$t_{.99}$	$t_{.995}$	$t_{.9975}$	$t_{.999}$	$t_{.9995}$
1	0.325	1.000	1.963	3.078	6.314	12.71	31.82	63.66	127.3	318.3	636.6
2	0.289	0.816	1.386	1.886	2.920	4.303	6.965	9.925	14.09	22.33	31.60
3	0.277	0.765	1.250	1.638	2.353	3.182	4.541	5.841	7.453	10.21	12.92
4	0.271	0.741	1.190	1.533	2.132	2.776	3.747	4.604	5.598	7.173	8.610
5	0.267	0.727	1.156	1.476	2.015	2.571	3.365	4.032	4.773	5.893	6.869
6	0.265	0.718	1.134	1.440	1.943	2.447	3.143	3.707	4.317	5.208	5.959
7	0.263	0.711	1.119	1.415	1.895	2.365	2.998	3.499	4.029	4.785	5.408
8	0.262	0.706	1.108	1.397	1.860	2.306	2.896	3.355	3.833	4.501	5.041
9	0.261	0.703	1.100	1.383	1.833	2.262	2.821	3.250	3.690	4.297	4.781
10	0.260	0.700	1.093	1.372	1.812	2.228	2.764	3.169	3.581	4.144	4.587
11	0.260	0.697	1.088	1.363	1.796	2.201	2.718	3.106	3.497	4.025	4.437
12	0.259	0.695	1.083	1.356	1.782	2.179	2.681	3.055	3.428	3.930	4.318
13	0.259	0.694	1.079	1.350	1.771	2.160	2.650	3.012	3.372	3.852	4.221
14	0.258	0.692	1.076	1.345	1.761	2.145	2.624	2.977	3.326	3.787	4.140
15	0.258	0.691	1.074	1.341	1.753	2.131	2.602	2.947	3.286	3.733	4.073
16	0.258	0.690	1.071	1.337	1.746	2.120	2.583	2.921	3.252	3.686	4.015
17	0.257	0.689	1.069	1.333	1.740	2.110	2.567	2.898	3.222	3.646	3.965
18	0.257	0.688	1.067	1.330	1.734	2.101	2.552	2.878	3.197	3.610	3.922
19	0.257	0.688	1.066	1.328	1.729	2.093	2.539	2.861	3.174	3.579	3.883
20	0.257	0.687	1.064	1.325	1.725	2.086	2.528	2.845	3.153	3.552	3.850
21	0.257	0.686	1.063	1.323	1.721	2.080	2.518	2.831	3.135	3.527	3.819
22	0.256	0.686	1.061	1.321	1.717	2.074	2.508	2.819	3.119	3.505	3.792
23	0.256	0.685	1.060	1.319	1.714	2.069	2.500	2.807	3.104	3.485	3.768
24	0.256	0.685	1.059	1.318	1.711	2.064	2.492	2.797	3.091	3.467	3.745
25	0.256	0.684	1.058	1.316	1.708	2.060	2.485	2.787	3.078	3.450	3.725
26	0.256	0.684	1.058	1.315	1.706	2.056	2.479	2.779	3.067	3.435	3.707
27	0.256	0.684	1.057	1.314	1.703	2.052	2.473	2.771	3.057	3.421	3.690
28	0.256	0.683	1.056	1.313	1.701	2.048	2.467	2.763	3.047	3.408	3.674
29	0.256	0.683	1.055	1.311	1.699	2.045	2.462	2.756	3.038	3.396	3.659
30	0.256	0.683	1.055	1.310	1.697	2.042	2.457	2.750	3.030	3.385	3.646
40	0.255	0.681	1.050	1.303	1.684	2.021	2.423	2.704	2.971	3.307	3.551
50	0.255	0.679	1.047	1.299	1.676	2.009	2.403	2.678	2.937	3.261	3.496
60	0.254	0.679	1.045	1.296	1.671	2.000	2.390	2.660	2.915	3.232	3.460
70	0.254	0.678	1.044	1.294	1.667	1.994	2.381	2.648	2.899	3.211	3.435
80	0.254	0.678	1.043	1.292	1.664	1.990	2.374	2.639	2.887	3.195	3.416
90	0.254	0.677	1.042	1.291	1.662	1.987	2.368	2.632	2.878	3.183	3.402
100	0.254	0.677	1.042	1.290	1.660	1.984	2.364	2.626	2.871	3.174	3.390
110	0.254	0.677	1.041	1.289	1.659	1.982	2.361	2.621	2.865	3.166	3.381
120	0.254	0.677	1.041	1.289	1.658	1.980	2.358	2.617	2.860	3.160	3.373
∞	0.253	0.674	1.036	1.282	1.645	1.960	2.326	2.576	2.807	3.090	3.291

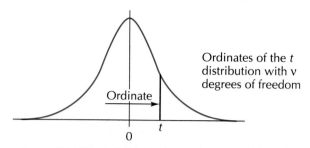

Ordinates of the t distribution with ν degrees of freedom

Ordinate

TABLE A.4	Ordinates of the t Distribution with ν Degrees of Freedom											
	t Value											
ν	0.00	0.25	0.50	0.75	1.00	1.25	1.50	1.75	2.00	2.25	2.50	2.75	3.00
1	0.318	0.300	0.255	0.204	0.159	0.124	0.098	0.078	0.064	0.053	0.044	0.037	0.032
2	0.354	0.338	0.296	0.244	0.192	0.149	0.114	0.088	0.068	0.053	0.042	0.034	0.027
3	0.368	0.353	0.313	0.261	0.207	0.159	0.120	0.090	0.068	0.051	0.039	0.030	0.023
4	0.375	0.361	0.322	0.270	0.215	0.164	0.123	0.091	0.066	0.049	0.036	0.026	0.020
5	0.380	0.366	0.328	0.276	0.220	0.168	0.125	0.091	0.065	0.047	0.033	0.024	0.017
6	0.383	0.369	0.332	0.280	0.223	0.170	0.126	0.090	0.064	0.045	0.031	0.022	0.015
7	0.385	0.372	0.335	0.283	0.226	0.172	0.126	0.090	0.063	0.044	0.030	0.021	0.014
8	0.387	0.373	0.337	0.285	0.228	0.173	0.127	0.090	0.062	0.043	0.029	0.019	0.013
9	0.388	0.375	0.338	0.287	0.229	0.174	0.127	0.090	0.062	0.042	0.028	0.018	0.012
10	0.389	0.376	0.340	0.288	0.230	0.175	0.127	0.090	0.061	0.041	0.027	0.018	0.011
11	0.390	0.377	0.341	0.289	0.231	0.176	0.128	0.089	0.061	0.040	0.026	0.017	0.011
12	0.391	0.378	0.342	0.290	0.232	0.176	0.128	0.089	0.060	0.040	0.026	0.016	0.010
13	0.391	0.378	0.342	0.291	0.233	0.177	0.128	0.089	0.060	0.039	0.025	0.016	0.010
14	0.392	0.379	0.343	0.292	0.234	0.177	0.128	0.089	0.060	0.039	0.025	0.015	0.009
15	0.392	0.380	0.344	0.292	0.234	0.178	0.128	0.089	0.059	0.038	0.024	0.015	0.009
16	0.393	0.380	0.344	0.293	0.235	0.178	0.128	0.089	0.059	0.038	0.024	0.015	0.009
17	0.393	0.380	0.345	0.293	0.235	0.178	0.128	0.089	0.059	0.038	0.023	0.014	0.009
18	0.393	0.381	0.345	0.294	0.235	0.178	0.129	0.088	0.058	0.037	0.023	0.014	0.008
19	0.394	0.381	0.345	0.294	0.236	0.179	0.129	0.088	0.058	0.037	0.023	0.014	0.008
20	0.394	0.381	0.346	0.294	0.236	0.179	0.129	0.088	0.058	0.037	0.023	0.014	0.008
22	0.394	0.382	0.346	0.295	0.237	0.179	0.129	0.088	0.058	0.036	0.022	0.013	0.008
24	0.395	0.382	0.347	0.296	0.237	0.179	0.129	0.088	0.057	0.036	0.022	0.013	0.007
26	0.395	0.383	0.347	0.296	0.237	0.180	0.129	0.088	0.057	0.036	0.022	0.013	0.007
28	0.395	0.383	0.348	0.296	0.238	0.180	0.129	0.088	0.057	0.036	0.021	0.012	0.007
30	0.396	0.383	0.348	0.297	0.238	0.180	0.129	0.088	0.057	0.035	0.021	0.012	0.007
35	0.396	0.384	0.348	0.297	0.239	0.180	0.129	0.088	0.056	0.035	0.021	0.012	0.006
40	0.396	0.384	0.349	0.298	0.239	0.181	0.129	0.087	0.056	0.034	0.020	0.011	0.006
45	0.397	0.384	0.349	0.298	0.239	0.181	0.129	0.087	0.056	0.034	0.020	0.011	0.006
50	0.397	0.385	0.350	0.298	0.240	0.181	0.129	0.087	0.056	0.034	0.020	0.011	0.006
∞	0.399	0.387	0.352	0.301	0.242	0.183	0.130	0.086	0.054	0.032	0.018	0.009	0.004

Values of χ^2_p for the chi-square distribution with ν degrees of freedom

p

χ^2_p

TABLE A.5   Values for χ^2 for Various Probability Levels

ν	$\chi^2_{.001}$	$\chi^2_{.005}$	$\chi^2_{.010}$	$\chi^2_{.025}$	$\chi^2_{.050}$	$\chi^2_{.100}$	$\chi^2_{.250}$	$\chi^2_{.500}$	$\chi^2_{.750}$	$\chi^2_{.900}$	$\chi^2_{.950}$	$\chi^2_{.975}$	$\chi^2_{.990}$	$\chi^2_{.995}$	$\chi^2_{.999}$
1	0.0000	0.0000	0.0002	0.0010	0.0039	0.0158	0.102	0.455	1.32	2.71	3.84	5.02	6.63	7.88	10.8
2	0.0020	0.0100	0.0201	0.0506	0.103	0.211	0.575	1.39	2.77	4.61	5.99	7.38	9.21	10.6	13.8
3	0.0243	0.0717	0.115	0.216	0.352	0.584	1.21	2.37	4.11	6.25	7.81	9.35	11.3	12.8	16.3
4	0.0908	0.207	0.297	0.484	0.711	1.06	1.92	3.36	5.39	7.78	9.49	11.1	13.3	14.9	18.5
5	0.210	0.412	0.554	0.831	1.15	1.61	2.67	4.35	6.63	9.24	11.1	12.8	15.1	16.7	20.5
6	0.381	0.676	0.872	1.24	1.64	2.20	3.45	5.35	7.84	10.6	12.6	14.4	16.8	18.5	22.5
7	0.598	0.989	1.24	1.69	2.17	2.83	4.25	6.35	9.04	12.0	14.1	16.0	18.5	20.3	24.3
8	0.857	1.34	1.65	2.18	2.73	3.49	5.07	7.34	10.2	13.4	15.5	17.5	20.1	22.0	26.1
9	1.15	1.73	2.09	2.70	3.33	4.17	5.90	8.34	11.4	14.7	16.9	19.0	21.7	23.6	27.9
10	1.48	2.16	2.56	3.25	3.94	4.87	6.74	9.34	12.5	16.0	18.3	20.5	23.2	25.2	29.6
11	1.83	2.60	3.05	3.82	4.57	5.58	7.58	10.3	13.7	17.3	19.7	21.9	24.7	26.8	31.3
12	2.21	3.07	3.57	4.40	5.23	6.30	8.44	11.3	14.8	18.5	21.0	23.3	26.2	28.3	32.9
13	2.62	3.57	4.11	5.01	5.89	7.04	9.30	12.3	16.0	19.8	22.4	24.7	27.7	29.8	34.5
14	3.04	4.07	4.66	5.63	6.57	7.79	10.2	13.3	17.1	21.1	23.7	26.1	29.1	31.3	36.1
15	3.48	4.60	5.23	6.26	7.26	8.55	11.0	14.3	18.2	22.3	25.0	27.5	30.6	32.8	37.7
16	3.94	5.14	5.81	6.91	7.96	9.31	11.9	15.3	19.4	23.5	26.3	28.8	32.0	34.3	39.3
17	4.42	5.70	6.41	7.56	8.67	10.1	12.8	16.3	20.5	24.8	27.6	30.2	33.4	35.7	40.8
18	4.90	6.26	7.01	8.23	9.39	10.9	13.7	17.3	21.6	26.0	28.9	31.5	34.8	37.2	42.3
19	5.41	6.84	7.63	8.91	10.1	11.7	14.6	18.3	22.7	27.2	30.1	32.9	36.2	38.6	43.8
20	5.92	7.43	8.26	9.59	10.9	12.4	15.5	19.3	23.8	28.4	31.4	34.2	37.6	40.0	45.3
21	6.45	8.03	8.90	10.3	11.6	13.2	16.3	20.3	24.9	29.6	32.7	35.5	38.9	41.4	46.8
22	6.98	8.64	9.54	11.0	12.3	14.0	17.2	21.3	26.0	30.8	33.9	36.8	40.3	42.8	48.3
23	7.53	9.26	10.2	11.7	13.1	14.8	18.1	22.3	27.1	32.0	35.2	38.1	41.6	44.2	49.7
24	8.08	9.89	10.9	12.4	13.8	15.7	19.0	23.3	28.2	33.2	36.4	39.4	43.0	45.6	51.2
25	8.65	10.5	11.5	13.1	14.6	16.5	19.9	24.3	29.3	34.4	37.7	40.6	44.3	46.9	52.6
26	9.22	11.2	12.2	13.8	15.4	17.3	20.8	25.3	30.4	35.6	38.9	41.9	45.6	48.3	54.1
27	9.80	11.8	12.9	14.6	16.2	18.1	21.7	26.3	31.5	36.7	40.1	43.2	47.0	49.6	55.5
28	10.4	12.5	13.6	15.3	16.9	18.9	22.7	27.3	32.6	37.9	41.3	44.5	48.3	51.0	56.9
29	11.0	13.1	14.3	16.0	17.7	19.8	23.6	28.3	33.7	39.1	42.6	45.7	49.6	52.3	58.3
30	11.6	13.8	15.0	16.8	18.5	20.6	24.5	29.3	34.8	40.3	43.8	47.0	50.9	53.7	59.7
40	17.9	20.7	22.2	24.4	26.5	29.1	33.7	39.3	45.6	51.8	55.8	59.3	63.7	66.8	73.4
50	24.7	28.0	29.7	32.4	34.8	37.7	42.9	49.3	56.3	63.2	67.5	71.4	76.2	79.5	86.7
60	31.7	35.5	37.5	40.5	43.2	46.5	52.3	59.3	67.0	74.4	79.1	83.3	88.4	92.0	99.6
70	39.0	43.3	45.4	48.8	51.7	55.3	61.7	69.3	77.6	85.5	90.5	95.0	100	104	112
80	46.5	51.2	53.5	57.2	60.4	64.3	71.1	79.3	88.1	96.6	102	107	112	116	125
90	54.2	59.2	61.8	65.6	69.1	73.3	80.6	89.3	98.6	108	113	118	124	128	137
100	61.9	67.3	70.1	74.2	77.9	82.4	90.1	99.3	109	118	124	130	136	140	149
110	69.8	75.6	78.5	82.9	86.8	91.5	99.7	109	120	129	135	141	147	152	162
120	77.8	83.9	86.9	91.6	95.7	101	109	119	130	140	147	152	159	164	174

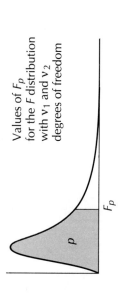

Values of F_p
for the F distribution
with ν_1 and ν_2
degrees of freedom

TABLE A.6 Critical Values of F for a Probability Level, p, of 0.75

ν_2 \ ν_1	1	2	3	4	5	6	7	8	9	10	12	15	20	24	30	40	60	120	∞
1	5.83	7.50	8.20	8.58	8.82	8.98	9.10	9.19	9.26	9.32	9.41	9.49	9.58	9.63	9.67	9.71	9.76	9.80	9.85
2	2.57	3.00	3.15	3.23	3.28	3.31	3.34	3.35	3.37	3.38	3.39	3.41	3.43	3.43	3.44	3.45	3.46	3.47	3.48
3	2.02	2.28	2.36	2.39	2.41	2.42	2.43	2.44	2.44	2.44	2.45	2.46	2.46	2.46	2.47	2.47	2.47	2.47	2.47
4	1.81	2.00	2.05	2.06	2.07	2.08	2.08	2.08	2.08	2.08	2.08	2.08	2.08	2.08	2.08	2.08	2.08	2.08	2.08
5	1.69	1.85	1.88	1.89	1.89	1.89	1.89	1.89	1.89	1.89	1.89	1.89	1.88	1.88	1.88	1.88	1.87	1.87	1.87
6	1.62	1.76	1.78	1.79	1.79	1.78	1.78	1.78	1.77	1.77	1.77	1.76	1.76	1.75	1.75	1.75	1.74	1.74	1.74
7	1.57	1.70	1.72	1.72	1.71	1.71	1.70	1.70	1.69	1.69	1.68	1.68	1.67	1.67	1.66	1.66	1.65	1.65	1.65
8	1.54	1.66	1.67	1.66	1.66	1.65	1.64	1.64	1.63	1.63	1.62	1.62	1.61	1.60	1.60	1.59	1.59	1.58	1.58
9	1.51	1.62	1.63	1.63	1.62	1.61	1.60	1.60	1.59	1.59	1.58	1.57	1.56	1.56	1.55	1.54	1.54	1.53	1.53
10	1.49	1.60	1.60	1.59	1.59	1.58	1.57	1.56	1.56	1.55	1.54	1.53	1.52	1.52	1.51	1.51	1.50	1.49	1.48
11	1.47	1.58	1.58	1.57	1.56	1.55	1.54	1.53	1.53	1.52	1.51	1.50	1.49	1.49	1.48	1.47	1.47	1.46	1.45
12	1.46	1.56	1.56	1.55	1.54	1.53	1.52	1.51	1.51	1.50	1.49	1.48	1.47	1.46	1.45	1.45	1.44	1.43	1.42
13	1.45	1.55	1.55	1.53	1.52	1.51	1.50	1.49	1.49	1.48	1.47	1.46	1.45	1.44	1.43	1.42	1.42	1.41	1.40
14	1.44	1.53	1.53	1.52	1.51	1.50	1.49	1.48	1.47	1.46	1.45	1.44	1.43	1.42	1.41	1.41	1.40	1.39	1.38
15	1.43	1.52	1.52	1.51	1.49	1.48	1.47	1.46	1.46	1.45	1.44	1.43	1.41	1.41	1.40	1.39	1.38	1.37	1.36
16	1.42	1.51	1.51	1.50	1.48	1.47	1.46	1.45	1.44	1.44	1.43	1.41	1.40	1.39	1.38	1.37	1.36	1.35	1.34
17	1.42	1.51	1.50	1.49	1.47	1.46	1.45	1.44	1.43	1.43	1.41	1.40	1.39	1.38	1.37	1.36	1.35	1.34	1.33
18	1.41	1.50	1.49	1.48	1.46	1.45	1.44	1.43	1.42	1.42	1.40	1.39	1.38	1.37	1.36	1.35	1.34	1.33	1.32
19	1.41	1.49	1.49	1.47	1.46	1.44	1.43	1.42	1.41	1.41	1.40	1.38	1.37	1.36	1.35	1.34	1.33	1.32	1.30
20	1.40	1.49	1.48	1.47	1.45	1.44	1.43	1.42	1.41	1.40	1.39	1.37	1.36	1.35	1.34	1.33	1.32	1.31	1.29
21	1.40	1.48	1.48	1.46	1.44	1.43	1.42	1.41	1.40	1.39	1.38	1.37	1.35	1.34	1.33	1.32	1.31	1.30	1.28
22	1.40	1.48	1.47	1.45	1.44	1.42	1.41	1.40	1.39	1.39	1.37	1.36	1.34	1.33	1.32	1.31	1.30	1.29	1.28
23	1.39	1.47	1.47	1.45	1.43	1.42	1.41	1.40	1.39	1.38	1.37	1.35	1.34	1.33	1.32	1.31	1.30	1.28	1.27
24	1.39	1.47	1.46	1.44	1.43	1.41	1.40	1.39	1.38	1.38	1.36	1.35	1.33	1.32	1.31	1.30	1.29	1.28	1.26
25	1.39	1.47	1.46	1.44	1.42	1.41	1.40	1.39	1.38	1.37	1.36	1.34	1.33	1.32	1.31	1.29	1.28	1.27	1.25
26	1.38	1.46	1.45	1.44	1.42	1.41	1.39	1.38	1.37	1.37	1.35	1.34	1.32	1.31	1.30	1.29	1.28	1.26	1.25
27	1.38	1.46	1.45	1.43	1.42	1.40	1.39	1.38	1.37	1.36	1.35	1.33	1.32	1.31	1.30	1.28	1.27	1.26	1.24
28	1.38	1.46	1.45	1.43	1.41	1.40	1.39	1.38	1.37	1.36	1.34	1.33	1.31	1.30	1.29	1.28	1.27	1.25	1.24
29	1.38	1.45	1.45	1.43	1.41	1.40	1.38	1.37	1.36	1.35	1.34	1.32	1.31	1.30	1.29	1.27	1.26	1.25	1.23
30	1.38	1.45	1.44	1.42	1.41	1.39	1.38	1.37	1.36	1.35	1.34	1.32	1.30	1.29	1.28	1.27	1.26	1.24	1.23
40	1.36	1.44	1.42	1.40	1.39	1.37	1.36	1.35	1.34	1.33	1.31	1.30	1.28	1.26	1.25	1.24	1.22	1.21	1.19
60	1.35	1.42	1.41	1.38	1.37	1.35	1.33	1.32	1.31	1.30	1.29	1.27	1.25	1.24	1.22	1.21	1.19	1.17	1.15
120	1.34	1.40	1.39	1.37	1.35	1.33	1.31	1.30	1.29	1.28	1.26	1.24	1.22	1.21	1.19	1.18	1.16	1.13	1.10
∞	1.32	1.39	1.37	1.35	1.33	1.31	1.29	1.28	1.27	1.25	1.24	1.22	1.19	1.18	1.16	1.14	1.12	1.08	1.00

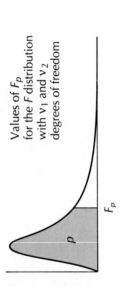

Values of F_p for the F distribution with ν_1 and ν_2 degrees of freedom

TABLE A.6 (continued) Critical Values of F for a Probability Level, p, of 0.900

ν_2 \ ν_1	1	2	3	4	5	6	7	8	9	10	12	15	20	24	30	40	60	120	∞
1	39.86	49.50	53.59	55.83	57.24	58.20	58.91	59.44	59.86	60.20	60.71	61.22	61.74	62.00	62.27	62.53	62.79	63.06	63.32
2	8.53	9.00	9.16	9.24	9.29	9.33	9.35	9.37	9.38	9.39	9.41	9.42	9.44	9.45	9.46	9.47	9.47	9.48	9.49
3	5.54	5.46	5.39	5.34	5.31	5.28	5.27	5.25	5.24	5.23	5.22	5.20	5.18	5.18	5.17	5.16	5.15	5.14	5.13
4	4.55	4.32	4.19	4.11	4.05	4.01	3.98	3.95	3.94	3.92	3.90	3.87	3.84	3.83	3.82	3.80	3.79	3.78	3.76
5	4.06	3.78	3.62	3.52	3.45	3.40	3.37	3.34	3.32	3.30	3.27	3.24	3.21	3.19	3.17	3.16	3.14	3.12	3.11
6	3.78	3.46	3.29	3.18	3.11	3.05	3.01	2.98	2.96	2.94	2.90	2.87	2.84	2.82	2.80	2.78	2.76	2.74	2.72
7	3.59	3.26	3.07	2.96	2.88	2.83	2.78	2.75	2.72	2.70	2.67	2.63	2.59	2.58	2.56	2.54	2.51	2.49	2.47
8	3.46	3.11	2.92	2.81	2.73	2.67	2.62	2.59	2.56	2.54	2.50	2.46	2.42	2.40	2.38	2.36	2.34	2.32	2.29
9	3.36	3.01	2.81	2.69	2.61	2.55	2.51	2.47	2.44	2.42	2.38	2.34	2.30	2.28	2.25	2.23	2.21	2.18	2.16
10	3.29	2.92	2.73	2.61	2.52	2.46	2.41	2.38	2.35	2.32	2.28	2.24	2.20	2.18	2.16	2.13	2.11	2.08	2.06
11	3.23	2.86	2.66	2.54	2.45	2.39	2.34	2.30	2.27	2.25	2.21	2.17	2.12	2.10	2.08	2.05	2.03	2.00	1.97
12	3.18	2.81	2.61	2.48	2.39	2.33	2.28	2.24	2.21	2.19	2.15	2.10	2.06	2.04	2.01	1.99	1.96	1.93	1.90
13	3.14	2.76	2.56	2.43	2.35	2.28	2.23	2.20	2.16	2.14	2.10	2.05	2.01	1.98	1.96	1.93	1.90	1.88	1.85
14	3.10	2.73	2.52	2.39	2.31	2.24	2.19	2.15	2.12	2.10	2.05	2.01	1.96	1.94	1.91	1.89	1.86	1.83	1.80
15	3.07	2.70	2.49	2.36	2.27	2.21	2.16	2.12	2.09	2.06	2.02	1.97	1.92	1.90	1.87	1.85	1.82	1.79	1.76
16	3.05	2.67	2.46	2.33	2.24	2.18	2.13	2.09	2.06	2.03	1.99	1.94	1.89	1.87	1.84	1.81	1.78	1.75	1.72
17	3.03	2.64	2.44	2.31	2.22	2.15	2.10	2.06	2.03	2.00	1.96	1.91	1.86	1.84	1.81	1.78	1.75	1.72	1.69
18	3.01	2.62	2.42	2.29	2.20	2.13	2.08	2.04	2.00	1.98	1.93	1.89	1.84	1.81	1.78	1.75	1.72	1.69	1.66
19	2.99	2.61	2.40	2.27	2.18	2.11	2.06	2.02	1.98	1.96	1.91	1.86	1.81	1.79	1.76	1.73	1.70	1.67	1.63
20	2.97	2.59	2.38	2.25	2.16	2.09	2.04	2.00	1.96	1.94	1.89	1.84	1.79	1.77	1.74	1.71	1.68	1.64	1.61
21	2.96	2.57	2.36	2.23	2.14	2.08	2.02	1.98	1.95	1.92	1.87	1.83	1.78	1.75	1.72	1.69	1.66	1.62	1.59
22	2.95	2.56	2.35	2.22	2.13	2.06	2.01	1.97	1.93	1.90	1.86	1.81	1.76	1.73	1.70	1.67	1.64	1.60	1.57
23	2.94	2.55	2.34	2.21	2.11	2.05	1.99	1.95	1.92	1.89	1.84	1.80	1.74	1.72	1.69	1.66	1.62	1.59	1.55
24	2.93	2.54	2.33	2.19	2.10	2.04	1.98	1.94	1.91	1.88	1.83	1.78	1.73	1.70	1.67	1.64	1.61	1.57	1.53
25	2.92	2.53	2.32	2.18	2.09	2.02	1.97	1.93	1.89	1.87	1.82	1.77	1.72	1.69	1.66	1.63	1.59	1.56	1.52
26	2.91	2.52	2.31	2.17	2.08	2.01	1.96	1.92	1.88	1.86	1.81	1.76	1.71	1.68	1.65	1.61	1.58	1.54	1.50
27	2.90	2.51	2.30	2.17	2.07	2.00	1.95	1.91	1.87	1.85	1.80	1.75	1.70	1.67	1.64	1.60	1.57	1.53	1.49
28	2.89	2.50	2.29	2.16	2.06	2.00	1.94	1.90	1.87	1.84	1.79	1.74	1.69	1.66	1.63	1.59	1.56	1.52	1.48
29	2.89	2.50	2.28	2.15	2.06	1.99	1.93	1.89	1.86	1.83	1.78	1.73	1.68	1.65	1.62	1.58	1.55	1.51	1.47
30	2.88	2.49	2.28	2.14	2.05	1.98	1.93	1.88	1.85	1.82	1.77	1.72	1.67	1.64	1.61	1.57	1.54	1.50	1.46
40	2.84	2.44	2.23	2.09	2.00	1.93	1.87	1.83	1.79	1.76	1.71	1.66	1.61	1.57	1.54	1.51	1.47	1.42	1.38
60	2.79	2.39	2.18	2.04	1.95	1.87	1.82	1.77	1.74	1.71	1.66	1.60	1.54	1.51	1.48	1.44	1.40	1.35	1.29
120	2.75	2.35	2.13	1.99	1.90	1.82	1.77	1.72	1.68	1.65	1.60	1.55	1.48	1.45	1.41	1.37	1.32	1.26	1.19
∞	2.71	2.30	2.08	1.95	1.85	1.77	1.72	1.67	1.63	1.60	1.55	1.49	1.42	1.38	1.34	1.30	1.24	1.17	1.00

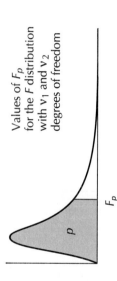

Values of F_p for the F distribution with v_1 and v_2 degrees of freedom

TABLE A.6 (continued) Critical Values of F for a Probability Level, p, of 0.950

v_2 \ v_1	1	2	3	4	5	6	7	8	9	10	12	15	20	24	30	40	60	120	∞
1	161.4	199.5	215.7	224.6	230.2	234.0	236.8	238.9	240.5	241.9	243.9	245.9	248.0	249.1	250.1	251.1	252.2	253.3	254.3
2	18.51	19.00	19.16	19.25	19.30	19.33	19.35	19.37	19.38	19.40	19.41	19.43	19.45	19.45	19.46	19.47	19.48	19.49	19.50
3	10.13	9.55	9.28	9.12	9.01	8.94	8.89	8.85	8.81	8.79	8.74	8.70	8.66	8.64	8.62	8.59	8.57	8.55	8.53
4	7.71	6.94	6.59	6.39	6.26	6.16	6.09	6.04	6.00	5.96	5.91	5.86	5.80	5.77	5.75	5.72	5.69	5.66	5.63
5	6.61	5.79	5.41	5.19	5.05	4.95	4.88	4.82	4.77	4.74	4.68	4.62	4.56	4.53	4.50	4.46	4.43	4.40	4.36
6	5.99	5.14	4.76	4.53	4.39	4.28	4.21	4.15	4.10	4.06	4.00	3.94	3.87	3.84	3.81	3.77	3.74	3.70	3.67
7	5.59	4.74	4.35	4.12	3.97	3.87	3.79	3.73	3.68	3.64	3.57	3.51	3.44	3.41	3.38	3.34	3.30	3.27	3.23
8	5.32	4.46	4.07	3.84	3.69	3.58	3.50	3.44	3.39	3.35	3.28	3.22	3.15	3.12	3.08	3.04	3.01	2.97	2.93
9	5.12	4.26	3.86	3.63	3.48	3.37	3.29	3.23	3.18	3.14	3.07	3.01	2.94	2.90	2.86	2.83	2.79	2.75	2.71
10	4.96	4.10	3.71	3.48	3.33	3.22	3.14	3.07	3.02	2.98	2.91	2.85	2.77	2.74	2.70	2.66	2.62	2.58	2.54
11	4.84	3.98	3.59	3.36	3.20	3.09	3.01	2.95	2.90	2.85	2.79	2.72	2.65	2.61	2.57	2.53	2.49	2.45	2.40
12	4.75	3.89	3.49	3.26	3.11	3.00	2.91	2.85	2.80	2.75	2.69	2.62	2.54	2.51	2.47	2.43	2.38	2.34	2.30
13	4.67	3.81	3.41	3.18	3.03	2.92	2.83	2.77	2.71	2.67	2.60	2.53	2.46	2.42	2.38	2.34	2.30	2.25	2.21
14	4.60	3.74	3.34	3.11	2.96	2.85	2.76	2.70	2.65	2.60	2.53	2.46	2.39	2.35	2.31	2.27	2.22	2.18	2.13
15	4.54	3.68	3.29	3.06	2.90	2.79	2.71	2.64	2.59	2.54	2.48	2.40	2.33	2.29	2.25	2.20	2.16	2.11	2.07
16	4.49	3.63	3.24	3.01	2.85	2.74	2.66	2.59	2.54	2.49	2.42	2.35	2.28	2.24	2.19	2.15	2.11	2.06	2.01
17	4.45	3.59	3.20	2.96	2.81	2.70	2.61	2.55	2.49	2.45	2.38	2.31	2.23	2.19	2.15	2.10	2.06	2.01	1.96
18	4.41	3.55	3.16	2.93	2.77	2.66	2.58	2.51	2.46	2.41	2.34	2.27	2.19	2.15	2.11	2.06	2.02	1.97	1.92
19	4.38	3.52	3.13	2.90	2.74	2.63	2.54	2.48	2.42	2.38	2.31	2.23	2.16	2.11	2.07	2.03	1.98	1.93	1.88
20	4.35	3.49	3.10	2.87	2.71	2.60	2.51	2.45	2.39	2.35	2.28	2.20	2.12	2.08	2.04	1.99	1.95	1.90	1.84
21	4.32	3.47	3.07	2.84	2.68	2.57	2.49	2.42	2.37	2.32	2.25	2.18	2.10	2.05	2.01	1.96	1.92	1.87	1.81
22	4.30	3.44	3.05	2.82	2.66	2.55	2.46	2.40	2.34	2.30	2.23	2.15	2.07	2.03	1.98	1.94	1.89	1.84	1.78
23	4.28	3.42	3.03	2.80	2.64	2.53	2.44	2.37	2.32	2.27	2.20	2.13	2.05	2.01	1.96	1.91	1.86	1.81	1.76
24	4.26	3.40	3.01	2.78	2.62	2.51	2.42	2.36	2.30	2.25	2.18	2.11	2.03	1.98	1.94	1.89	1.84	1.79	1.73
25	4.24	3.39	2.99	2.76	2.60	2.49	2.40	2.34	2.28	2.24	2.16	2.09	2.01	1.96	1.92	1.87	1.82	1.77	1.71
26	4.23	3.37	2.98	2.74	2.59	2.47	2.39	2.32	2.27	2.22	2.15	2.07	1.99	1.95	1.90	1.85	1.80	1.75	1.69
27	4.21	3.35	2.96	2.73	2.57	2.46	2.37	2.31	2.25	2.20	2.13	2.06	1.97	1.93	1.88	1.84	1.79	1.73	1.67
28	4.20	3.34	2.95	2.71	2.56	2.45	2.36	2.29	2.24	2.19	2.12	2.04	1.96	1.91	1.87	1.82	1.77	1.71	1.65
29	4.18	3.33	2.93	2.70	2.55	2.43	2.35	2.28	2.22	2.18	2.10	2.03	1.94	1.90	1.85	1.81	1.75	1.70	1.64
30	4.17	3.32	2.92	2.69	2.53	2.42	2.33	2.27	2.21	2.16	2.09	2.01	1.93	1.89	1.84	1.79	1.74	1.68	1.62
40	4.08	3.23	2.84	2.61	2.45	2.34	2.25	2.18	2.12	2.08	2.00	1.92	1.84	1.79	1.74	1.69	1.64	1.58	1.51
60	4.00	3.15	2.76	2.53	2.37	2.25	2.17	2.10	2.04	1.99	1.92	1.84	1.75	1.70	1.65	1.59	1.53	1.47	1.39
120	3.92	3.07	2.68	2.45	2.29	2.17	2.09	2.02	1.96	1.91	1.83	1.75	1.66	1.61	1.55	1.50	1.43	1.35	1.25
∞	3.84	3.00	2.60	2.37	2.21	2.10	2.01	1.94	1.88	1.83	1.75	1.67	1.57	1.52	1.46	1.39	1.32	1.22	1.00

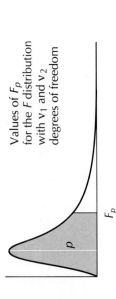

Values of F_p for the F distribution with v_1 and v_2 degrees of freedom

TABLE A.6 (continued)   Critical Values of F for a Probability Level, p, of 0.990

v_1 v_2	1	2	3	4	5	6	7	8	9	10	12	15	20	24	30	40	60	120	∞
1	4052	4999	5403	5625	5764	5859	5928	5982	6022	6056	6106	6157	6209	6235	6261	6287	6313	6339	6366
2	98.50	99.00	99.17	99.25	99.30	99.33	99.36	99.37	99.39	99.40	99.42	99.43	99.45	99.46	99.47	99.47	99.48	99.49	99.50
3	34.12	30.82	29.46	28.71	28.24	27.91	27.67	27.49	27.35	27.23	27.05	26.87	26.69	26.60	26.50	26.41	26.32	26.22	26.13
4	21.20	18.00	16.69	15.98	15.52	15.21	14.98	14.80	14.66	14.55	14.37	14.20	14.02	13.93	13.84	13.75	13.65	13.56	13.46
5	16.26	13.27	12.06	11.39	10.97	10.67	10.46	10.29	10.16	10.05	9.89	9.72	9.55	9.47	9.38	9.29	9.20	9.11	9.02
6	13.75	10.92	9.78	9.15	8.75	8.47	8.26	8.10	7.98	7.87	7.72	7.56	7.40	7.31	7.23	7.14	7.06	6.97	6.88
7	12.25	9.55	8.45	7.85	7.46	7.19	6.99	6.84	6.72	6.62	6.47	6.31	6.16	6.07	5.99	5.91	5.82	5.74	5.65
8	11.26	8.65	7.59	7.01	6.63	6.37	6.18	6.03	5.91	5.81	5.67	5.52	5.36	5.28	5.20	5.12	5.03	4.95	4.86
9	10.56	8.02	6.99	6.42	6.06	5.80	5.61	5.47	5.35	5.26	5.11	4.96	4.81	4.73	4.65	4.57	4.48	4.40	4.31
10	10.04	7.56	6.55	5.99	5.64	5.39	5.20	5.06	4.94	4.85	4.71	4.56	4.41	4.33	4.25	4.17	4.08	4.00	3.91
11	9.65	7.21	6.22	5.67	5.32	5.07	4.89	4.74	4.63	4.54	4.40	4.25	4.10	4.02	3.94	3.86	3.78	3.69	3.60
12	9.33	6.93	5.95	5.41	5.06	4.82	4.64	4.50	4.39	4.30	4.16	4.01	3.86	3.78	3.70	3.62	3.54	3.45	3.36
13	9.07	6.70	5.74	5.21	4.86	4.62	4.44	4.30	4.19	4.10	3.96	3.82	3.66	3.59	3.51	3.43	3.34	3.25	3.17
14	8.86	6.51	5.56	5.04	4.69	4.46	4.28	4.14	4.03	3.94	3.80	3.66	3.51	3.43	3.35	3.27	3.18	3.09	3.00
15	8.68	6.36	5.42	4.89	4.56	4.32	4.14	4.00	3.89	3.80	3.67	3.52	3.37	3.29	3.21	3.13	3.05	2.96	2.87
16	8.53	6.23	5.29	4.77	4.44	4.20	4.03	3.89	3.78	3.69	3.55	3.41	3.26	3.18	3.10	3.02	2.93	2.84	2.75
17	8.40	6.11	5.18	4.67	4.34	4.10	3.93	3.79	3.68	3.59	3.46	3.31	3.16	3.08	3.00	2.92	2.83	2.75	2.65
18	8.29	6.01	5.09	4.58	4.25	4.01	3.84	3.71	3.60	3.51	3.37	3.23	3.08	3.00	2.92	2.84	2.75	2.66	2.57
19	8.18	5.93	5.01	4.50	4.17	3.94	3.77	3.63	3.52	3.43	3.30	3.15	3.00	2.92	2.84	2.76	2.67	2.58	2.49
20	8.10	5.85	4.94	4.43	4.10	3.87	3.70	3.56	3.46	3.37	3.23	3.09	2.94	2.86	2.78	2.69	2.61	2.52	2.42
21	8.02	5.78	4.87	4.37	4.04	3.81	3.64	3.51	3.40	3.31	3.17	3.03	2.88	2.80	2.72	2.64	2.55	2.46	2.36
22	7.95	5.72	4.82	4.31	3.99	3.76	3.59	3.45	3.35	3.26	3.12	2.98	2.83	2.75	2.67	2.58	2.50	2.40	2.31
23	7.88	5.66	4.76	4.26	3.94	3.71	3.54	3.41	3.30	3.21	3.07	2.93	2.78	2.70	2.62	2.54	2.45	2.35	2.26
24	7.82	5.61	4.72	4.22	3.90	3.67	3.50	3.36	3.26	3.17	3.03	2.89	2.74	2.66	2.58	2.49	2.40	2.31	2.21
25	7.77	5.57	4.68	4.18	3.85	3.63	3.46	3.32	3.22	3.13	2.99	2.85	2.70	2.62	2.54	2.45	2.36	2.27	2.17
26	7.72	5.53	4.64	4.14	3.82	3.59	3.42	3.29	3.18	3.09	2.96	2.81	2.66	2.58	2.50	2.42	2.33	2.23	2.13
27	7.68	5.49	4.60	4.11	3.78	3.56	3.39	3.26	3.15	3.06	2.93	2.78	2.63	2.55	2.47	2.38	2.29	2.20	2.10
28	7.64	5.45	4.57	4.07	3.75	3.53	3.36	3.23	3.12	3.03	2.90	2.75	2.60	2.52	2.44	2.35	2.26	2.17	2.06
29	7.60	5.42	4.54	4.04	3.73	3.50	3.33	3.20	3.09	3.00	2.87	2.73	2.57	2.49	2.41	2.33	2.23	2.14	2.03
30	7.56	5.39	4.51	4.02	3.70	3.47	3.30	3.17	3.07	2.98	2.84	2.70	2.55	2.47	2.39	2.30	2.21	2.11	2.01
40	7.31	5.18	4.31	3.83	3.51	3.29	3.12	2.99	2.89	2.80	2.66	2.52	2.37	2.29	2.20	2.11	2.02	1.92	1.80
60	7.08	4.98	4.13	3.65	3.34	3.12	2.95	2.82	2.72	2.63	2.50	2.35	2.20	2.12	2.03	1.94	1.84	1.73	1.60
120	6.85	4.79	3.95	3.48	3.17	2.96	2.79	2.66	2.56	2.47	2.34	2.19	2.03	1.95	1.86	1.76	1.66	1.53	1.38
∞	6.63	4.61	3.78	3.32	3.02	2.80	2.64	2.51	2.41	2.32	2.18	2.04	1.88	1.79	1.70	1.59	1.47	1.32	1.00

Values of F_p for the F distribution with v_1 and v_2 degrees of freedom

TABLE A.6 (continued) Critical Values of F for a Probability Level, p, of 0.999

v_2 \ v_1	1	2	3	4	5	6	7	8	9	10	12	15	20	24	30	40	60	120	∞
1	4053*	5000*	5404*	5625*	5764*	5859*	5929*	5981*	6023*	6056*	6107*	6158*	6209*	6235*	6261*	6287*	6313*	6340*	6366*
2	998.5	999.0	999.2	999.2	999.3	999.3	999.4	999.4	999.4	999.4	999.4	999.4	999.4	999.5	999.5	999.5	999.5	999.5	999.5
3	167.0	148.5	141.1	137.1	134.6	132.8	131.6	130.6	129.9	129.2	128.3	127.4	126.4	125.9	125.4	125.0	124.5	124.0	123.5
4	74.14	61.25	56.18	53.44	51.71	50.53	49.66	49.00	48.47	48.05	47.41	46.76	46.10	45.77	45.43	45.09	44.75	44.40	44.05
5	47.18	37.12	33.20	31.09	29.75	28.84	28.16	27.64	27.24	26.92	26.42	25.91	25.39	25.14	24.87	24.60	24.33	24.06	23.79
6	35.51	27.00	23.70	21.92	20.81	20.03	19.46	19.03	18.69	18.41	17.99	17.56	17.12	16.89	16.67	16.44	16.21	15.99	15.75
7	29.25	21.69	18.77	17.19	16.21	15.52	15.02	14.63	14.33	14.08	13.71	13.32	12.93	12.73	12.53	12.33	12.12	11.91	11.70
8	25.42	18.49	15.83	14.39	13.49	12.86	12.40	12.04	11.77	11.54	11.19	10.84	10.48	10.30	10.11	9.92	9.73	9.53	9.33
9	22.86	16.39	13.90	12.56	11.71	11.13	10.70	10.37	10.11	9.89	9.57	9.24	8.90	8.72	8.55	8.37	8.19	8.00	7.81
10	21.04	14.91	12.55	11.28	10.48	9.92	9.52	9.20	8.96	8.75	8.45	8.13	7.80	7.64	7.47	7.30	7.12	6.94	6.76
11	19.69	13.81	11.56	10.35	9.58	9.05	8.66	8.35	8.12	7.92	7.63	7.32	7.01	6.85	6.68	6.52	6.35	6.17	6.00
12	18.64	12.97	10.80	9.63	8.89	8.38	8.00	7.71	7.48	7.29	7.00	6.71	6.40	6.25	6.09	5.93	5.76	5.59	5.42
13	17.81	12.31	10.21	9.07	8.35	7.86	7.49	7.21	6.98	6.80	6.52	6.23	5.93	5.78	5.63	5.47	5.30	5.14	4.97
14	17.14	11.78	9.73	8.62	7.92	7.43	7.08	6.80	6.58	6.40	6.13	5.85	5.56	5.41	5.25	5.10	4.94	4.77	4.60
15	16.59	11.34	9.34	8.25	7.57	7.09	6.74	6.47	6.26	6.08	5.81	5.54	5.25	5.10	4.95	4.80	4.64	4.47	4.31
16	16.12	10.97	9.00	7.94	7.27	6.81	6.46	6.19	5.98	5.81	5.55	5.27	4.99	4.85	4.70	4.54	4.39	4.23	4.06
17	15.72	10.66	8.73	7.68	7.02	6.56	6.22	5.96	5.75	5.58	5.32	5.05	4.78	4.63	4.48	4.33	4.18	4.02	3.85
18	15.38	10.39	8.49	7.46	6.81	6.35	6.02	5.76	5.56	5.39	5.13	4.87	4.59	4.45	4.30	4.15	4.00	3.84	3.67
19	15.08	10.16	8.28	7.26	6.62	6.18	5.85	5.59	5.39	5.22	4.97	4.70	4.43	4.29	4.14	3.99	3.84	3.68	3.51
20	14.82	9.95	8.10	7.10	6.46	6.02	5.69	5.44	5.24	5.08	4.82	4.56	4.29	4.15	4.00	3.86	3.70	3.54	3.38
21	14.59	9.77	7.94	6.95	6.32	5.88	5.56	5.31	5.11	4.95	4.70	4.44	4.17	4.03	3.88	3.74	3.58	3.42	3.26
22	14.38	9.61	7.80	6.81	6.19	5.76	5.44	5.19	4.99	4.83	4.58	4.33	4.06	3.92	3.78	3.63	3.48	3.32	3.15
23	14.19	9.47	7.67	6.69	6.08	5.65	5.33	5.09	4.89	4.73	4.48	4.23	3.96	3.82	3.68	3.53	3.38	3.22	3.05
24	14.03	9.34	7.55	6.59	5.98	5.55	5.23	4.99	4.80	4.64	4.39	4.14	3.87	3.74	3.59	3.45	3.29	3.14	2.97
25	13.88	9.22	7.45	6.49	5.88	5.46	5.15	4.91	4.71	4.56	4.31	4.06	3.79	3.66	3.52	3.37	3.22	3.06	2.89
26	13.74	9.12	7.36	6.41	5.80	5.38	5.07	4.83	4.64	4.48	4.24	3.99	3.72	3.59	3.44	3.30	3.15	2.99	2.82
27	13.61	9.02	7.27	6.33	5.73	5.31	5.00	4.76	4.57	4.41	4.17	3.92	3.66	3.52	3.38	3.23	3.08	2.92	2.75
28	13.50	8.93	7.19	6.25	5.66	5.24	4.93	4.69	4.50	4.35	4.11	3.86	3.60	3.46	3.32	3.18	3.02	2.86	2.69
29	13.39	8.85	7.12	6.19	5.59	5.18	4.87	4.64	4.45	4.29	4.05	3.80	3.54	3.41	3.27	3.12	2.97	2.81	2.64
30	13.29	8.77	7.05	6.12	5.53	5.12	4.82	4.58	4.39	4.24	4.00	3.75	3.49	3.36	3.22	3.07	2.92	2.76	2.59
40	12.61	8.25	6.60	5.70	5.13	4.73	4.44	4.21	4.02	3.87	3.64	3.40	3.15	3.01	2.87	2.73	2.57	2.41	2.23
60	11.97	7.76	6.17	5.31	4.76	4.37	4.09	3.87	3.69	3.54	3.32	3.08	2.83	2.69	2.55	2.41	2.25	2.08	1.89
120	11.38	7.32	5.79	4.95	4.42	4.04	3.77	3.55	3.38	3.24	3.02	2.78	2.53	2.40	2.26	2.11	1.95	1.76	1.54
∞	10.83	6.91	5.42	4.62	4.10	3.74	3.47	3.27	3.10	2.96	2.74	2.51	2.27	2.13	1.99	1.84	1.66	1.45	1.00

* Multiply entries by 100.

Index